Mathematics for electrical technicians 4/5

Second edition

Series Editor:

D. R. Browning, B.Sc., F.R.S.C., C.Chem., A.R.T.C.S.
Formerly Principal Lecturer and Head of Chemistry, Bristol
Polytechnic

Other mathematics titles in this series:

Technician mathematics 1 Third edition   *J O Bird and A J C May*
Technician mathematics 2 Third edition   *J O Bird and A J C May*
Technician mathematics 3 Second edition   *J O Bird and A J C May*
Technician mathematics 4/5 Second edition   *J O Bird and
   A J C May*

Other mathematics titles:

Mathematical formulae Second edition   *J O Bird and A J C May*
Mathematics for scientific and technical students   *H G Davies and
   G A Hicks*

Mathematics for electrical technicians 4/5

Second edition

J. O. Bird
B.Sc.(Hons), C.Math., F.I.M.A., C.Eng., M.I.E.E., F.Coll.P., F.I.E.I.E.

A. J. C. May
B.A., C.Eng., M.I.Mech.E., F.I.E.I.E., M.B.I.M.

Routledge
Taylor & Francis Group

LONDON AND NEW YORK

First published 1981 by Pearson Education Limited
Second edition 1994
Second impression 1997
Third impression 1999

Published 2013 by Routledge
2 Park Square, Milton Park, Abingdon, Oxon OX14 4RN
711 Third Avenue, New York, NY 10017, USA

Routledge is an imprint of the Taylor & Francis Group, an informa business

ISBN 13: 978-0-582-23421-5 (pbk)

British Library Cataloguing in Publication Data
A catalogue entry for this title is available from the British Library.

Set by 16SS in 10/12 Times

Contents

Preface

Mathematics for electrical technicians 4/5 is one of a series of textbooks which deal simply and carefully with the fundamental mathematics essential in the development of technicians and engineers.

The material for the second edition of this successful textbook has been updated and expanded to cover (together with *Technician mathematics 3*) the main areas of the Business and Technology Education Council's 'Mathematics for Engineering' module for Higher National Certificate and Diploma courses likely to be studied in a 'Mathematics' lesson by an Electrical or Electronic Engineering student. Four-figure tables are assumed to be redundant; the electronic calculator is now an accepted tool in technician and engineering studies. The modifications in this second edition reflect this change.

It is widely recognised that facility in the use of mathematics on entry to an Electrical or Electronic Engineering HNC, HND or degree course is a key element in determining subsequent success. First and second year undergraduates in Electrical and/or Electronic Engineering will find that this book will meet their needs.

The definition and solution of engineering problems relies on the ability to represent systems and their behaviour in mathematical terms, which in turn, depends on the use of various mathematical tools. This textbook deals with some of those mathematical skills and concepts relevant to effective performance in engineering employment at higher technician/incorporated Engineer level.

The aim of this book is to model straightforward engineering systems, to generate numerical values for system parameters, to manipulate data to determine system response in defined conditions, to evaluate the effects on systems of changes in variables and to communicate ideas mathematically.

Each topic considered in the text is presented in a way that assumes in the reader only the knowledge attained in *Technician mathematics 3*. This practical mathematics book contains over 300 detailed worked problems, followed by some 850 further problems with answers.

I would like to express my appreciation for the friendly co-operation and helpful advice given by the publishers, and to Mrs. Elaine Woolley for the excellent typing of the manuscript.

Finally, I would like to pay tribute to my colleague, friend and co-author Tony May who sadly died in 1991 after a short illness.

John O. Bird
Highbury College, Portsmouth

Chapter 1

Revision of methods of differentiation

1.1 Introduction

Calculus is a branch of mathematics involving or leading to calculations dealing with continuously varying functions. Calculus is a subject which falls into two parts, namely,

(i) **differential calculus** (which is usually abbreviated to **differentiation**), and

(ii) **integral calculus** (which is usually abbreviated to **integration**).

The central problem of the differential calculus is the investigation of the rate of change of a function with respect to changes in the variables on which it depends. Calculations involving velocity and acceleration, rates of change and maximum and minimum values of curves are typical examples that you will have met in engineering studies (see *Technician mathematics 3*)

The two main uses of integral calculus are firstly, finding such quantities as the length of a curve, the area enclosed by a curve, mean and rms values, the volume enclosed by a surface, centroids and second moments of area, (see *Technician mathematics 3*), and secondly, the problem of determining a variable quantity given its rate of change – called solving differential equations (see Chapters 21 to 25)

There is a close relationship between the processes of differentiation and integration, the latter being considered as the inverse of the former.

Calculus is a comparatively young branch of mathematics; its systematic development started in the middle of the seventeenth century. Since then there has been an enormous expansion in the scope of calculus and its is now used in every field of applied science as an instrument for the solution of problems of the most varied nature.

The ability to differentiate functions is needed for the Newton-Raphson method of solving equations (Chapter 2) and with Maclaurin's and Taylor's series (Chapter 6). Also, it is important to be sure of the methods of differentiation before proceeding with differentiating implicit functions (Chapter 11), parametric functions (Chapter 12), logarithmic functions (Chapter 13) and inverse trigonometric and hyperbolic functions (Chapter 14), together with partial differentiation (Chapters 15 and 16)

In this chapter, standard differentials, products and quotients, function of a function and successive differentiation are revised.

1.2   Differentiation of common functions

Table 1.1 shows a summary of five standard functions and their derivatives.

Table 1.1

y or $f(x)$	$\dfrac{dy}{dx}$ or $f'(x)$
ax^n	$an\,x^{n-1}$
$\sin ax$	$a \cos ax$
$\cos ax$	$-a \sin ax$
e^{ax}	$a\,e^{ax}$
$\ln ax$	$\dfrac{1}{x}$

The differential coefficient of a sum or difference is the sum or difference of the differential coefficients of the separate terms.

Thus if $f(x) = p(x) + q(x) - r(x)$

then $f'(x) = p'(x) + q'(x) - r'(x)$

Worked problems on the differentiation of common functions

Problem 1. Find the differential coefficients of (a) $y = 7x^3$, (b) $y = \dfrac{9}{x^2}$.

If $y = ax^n$ then $\dfrac{dy}{dx} = an\,x^{n-1}$

(a)   Since $y = 7x^3$, then $a = 7$ and $n = 3$,

thus $\dfrac{dy}{dx} = (7)(3)x^{3-1} = \mathbf{21x^2}$

(b)   Since $y = \dfrac{9}{x^2} = 9x^{-2}$, then $a = 9$ and $n = -2$

thus $\dfrac{dy}{dx} = (9)(-2)x^{-2-1} = -18x^{-3} = \dfrac{-18}{x^3}$

Problem 2. Differentiate (a) $y = 5$, (b) $y = 3x$.

(a) $y = 5 = 5x^\circ$, hence $\dfrac{dy}{dx} = (5)(0)x^{0-1} = 0$

Differentiation is determining the gradient of a curve. $y = 5$ is a straight horizontal line, hence the gradient is zero. In general, the differential coefficient of a constant is always zero.

(b) $y = 3x = 3x^1$, hence $\dfrac{dy}{dx} = (3)(1)x^{1-1} = 3x^0 = 3$

$y = mx + c$ is the equation of a straight line graph, where m is the gradient and c is the y-axis intercept. If $y = 3x$, the gradient is 3. Therefore, in general, if $y = kx$, where k is a constant, the differential coefficient is always k.

Problem 3. Find the derivatives of

(a) $y = 4\sqrt{x}$ (b) $y = \dfrac{2}{\sqrt[3]{x^4}}$

(a) $y = 4\sqrt{x} = 4x^{\frac{1}{2}}$

$\quad$ hence $\dfrac{dy}{dx} = (4)\left(\dfrac{1}{2}\right)x^{\frac{1}{2}-1} = 2x^{-\frac{1}{2}} = \dfrac{2}{x^{\frac{1}{2}}} = \dfrac{2}{\sqrt{x}}$

(b) $y = \dfrac{2}{\sqrt[3]{x^4}} = \dfrac{2}{x^{\frac{4}{3}}} = 2x^{-\frac{4}{3}}$,

$\quad$ hence $\dfrac{dy}{dx} = (2)\left(-\dfrac{4}{3}\right)x^{(-\frac{4}{3})-1} = \dfrac{-8}{3}x^{-\frac{7}{3}} = \dfrac{-8}{3x^{\frac{7}{3}}}$

$$= \dfrac{-8}{3\sqrt[3]{x^7}}$$

Problem 4. Differentiate $y = 2x^5 + 8x - \dfrac{1}{3x^3} + \dfrac{4}{\sqrt{x}} - 7$.

$y = 2x^5 + 8x^1 - \dfrac{1}{3}x^{-3} + 4x^{-\frac{1}{2}} - 7$

$\dfrac{dy}{dx} = (2)(5)x^{5-1} + (8)(1)x^{1-1} - \left(\dfrac{1}{3}\right)(-3)x^{-3-1}$

$\qquad + (4)\left(-\dfrac{1}{2}\right)x^{-\frac{1}{2}-1} - 0$

$\qquad = 10x^4 + 8 + x^{-4} - 2x^{-\frac{3}{2}}$

$\qquad = 10x^4 + 8 + \dfrac{1}{x^4} - \dfrac{2}{x^{\frac{3}{2}}}$

i.e. $\dfrac{dy}{dx} = 10x^4 + 8 + \dfrac{1}{x^4} - \dfrac{2}{\sqrt{x^3}}$

Problem 5. Find $f'(x)$ given (a) $f(x) = 4 \sin 3x$ (b) $f(x) = 2 \cos 5x$

(a)   When $f(x) = 4 \sin 3x$, $f'(x) = (4)(3 \cos 3x) = \mathbf{12 \cos 3x}$

(b)   When $f(x) = 2 \cos 5x$, $f'(x) = (2)(-5 \sin 5x) = \mathbf{-10 \sin 5x}$

Note that a differential coefficient of the form $f'(x)$ is called
functional notation and a differential coefficient of the form $\dfrac{dy}{dx}$ is
called **Leibniz notation**.

In effect, $f'(x) = \dfrac{dy}{dx}$

Problem 6. Determine $f'(t)$ given (a) $f(t) = 4e^{2t}$, (b) $f(t) = \dfrac{5}{2e^{3t}}$.

(a)   When $f(t) = 4e^{2t}$, $f'(t) = (4)(2e^{2t}) = \mathbf{8e^{2t}}$

(b)   When $f(t) = \dfrac{5}{2e^{3t}} = \dfrac{5}{2}e^{-3t}$, $f'(t) = \left(\dfrac{5}{2}\right)(-3\,e^{-3t})$

$$= \dfrac{-15}{2}e^{-3t} = \dfrac{-15}{2e^{3t}}$$

Problem 7. For each of the following find $\dfrac{dy}{dt}$, (a) $y = \ln 5t$,

(b) $y = 5 \ln 2t$.

(a)   When $y = \ln 5t$, $\dfrac{dy}{dt} = \dfrac{1}{t}$

(b)   When $y = 5 \ln 2t$, $\dfrac{dy}{dt} = (5)\left(\dfrac{1}{t}\right) = \dfrac{5}{t}$

Problem 8. Determine the gradient of the curve
$y = 2x^3 - 3x^2 + 4x - 1$ at the points $(0, -1)$ and $(2, 11)$.

Gradient of curve $= \dfrac{dy}{dx} = 6x^2 - 6x + 4$

At the point $(0, -1)$, $x = 0$,
$\qquad$ thus gradient $= 6(0)^2 - 6(0) + 4 = \mathbf{4}$

At the point $(2, 11)$, $x = 2$, thus gradient $= 6(2)^2 - 6(2) + 4$
$$= 24 - 12 + 4 = \mathbf{16}$$

Problem 9. Find the co-ordinates of the point on the graph $y = 3x^2 - 5x + 1$ where the gradient is 1.

Gradient of curve $= \dfrac{dy}{dx} = 6x - 5$

When the gradient is 1, then $6x - 5 = 1$

from which, $6x = 6$ and $x = 1$

When $x = 1$, $y = 3(1)^2 - 5(1) + 1 = -1$

Hence the gradient is 1 at the point $(1, -1)$

Problem 10. (a) Given $f(x) = \dfrac{3x^3 - 2\sqrt{x} + 5\sqrt[3]{x^4}}{x^2}$ find $f'(x)$,

(b) Evaluate $f'(x)$ in part (a) when $x = 1$.

(a) $f(x) = \dfrac{3x^3 - 2\sqrt{x} + 5\sqrt[3]{x^4}}{x^2} = \dfrac{3x^3 - 2x^{\frac{1}{2}} + 5x^{\frac{4}{3}}}{x^2}$

$$= \frac{3x^3}{x^2} - \frac{2x^{\frac{1}{2}}}{x^2} + \frac{5x^{\frac{4}{3}}}{x^2}$$

$$= 3x^{3-2} - 2x^{\frac{1}{2}-2} + 5x^{(\frac{4}{3})-2}$$

by the laws of indices

i.e. $f(x) = 3x - 2x^{-\frac{3}{2}} + 5x^{-\frac{2}{3}}$

Hence $f'(x) = 3 - (2)\left(-\dfrac{3}{2}\right)x^{(-\frac{3}{2})-1} + (5)\left(-\dfrac{2}{3}\right)x^{(-\frac{2}{3})-1}$

$$= 3 + 3x^{-\frac{5}{2}} - \frac{10}{3}x^{-\frac{5}{3}}$$

i.e., $f'(x) = 3 + \dfrac{3}{x^{\frac{5}{2}}} - \dfrac{10}{3x^{\frac{5}{3}}} = 3 + \dfrac{3}{\sqrt{x^5}} - \dfrac{10}{3\sqrt[3]{x^5}}$

(b)   When $x = 1$, $f'(x) = 3 + \dfrac{3}{\sqrt{1^5}} = \dfrac{10}{3\sqrt[3]{1^5}} = 3 + 3 - \dfrac{10}{3}$

$$= 2\frac{2}{3}$$

Problem 11. (a) Find $\dfrac{dx}{d\theta}$ when $x = \dfrac{2}{\theta} - 3\,(\sin 6\theta - 2\cos 3\theta) + \ln 3\theta$

$-\dfrac{11}{2e^{4\theta}}$,

(b) Evaluate $\dfrac{dx}{d\theta}$ in part (a), correct to four significant figures,

when $\theta = \dfrac{\pi}{3}$

(a) $x = \dfrac{2}{\theta} - 3(\sin 6\theta - 2 \cos 3\theta) + \ln 3\theta - \dfrac{11}{2e^{4\theta}}$

$= 2\theta^{-1} - 3 \sin 6\theta + 6 \cos 3\theta + \ln 3\theta - \dfrac{11}{2}e^{-4\theta}$

$\dfrac{dx}{d\theta} = -2\theta^{-2} - 18 \cos 6\theta - 18 \sin 3\theta + \dfrac{1}{\theta} + \dfrac{44}{2}e^{-4\theta}$

$= \dfrac{-2}{\theta^2} - 18 \cos 6\theta - 18 \sin 3\theta + \dfrac{1}{\theta} + \dfrac{22}{e^{4\theta}}$

(b)   When $\theta = \dfrac{\pi}{3}, \dfrac{dx}{d\theta} = \dfrac{-2}{\left(\dfrac{\pi}{3}\right)^2} - 18 \cos 6\left(\dfrac{\pi}{3}\right)$

$- 18 \sin 3\left(\dfrac{\pi}{3}\right) + \dfrac{1}{\left(\dfrac{\pi}{3}\right)} + \dfrac{22}{e^{4\pi/3}}$

$= -1.823\,78 - 18(1) - 18(0)$
$+ 0.954\,93 + 0.333\,62$

$= -18.54$, correct to four significant figures

Further problems on differentiation of common functions may be found in Section 1.7 (Problems 1 to 12), page 19.

1.3   Differentiation of a product

When $y = uv$, and u and v are both functions of x, then

$$\dfrac{dy}{dx} = u\dfrac{dv}{dx} + v\dfrac{du}{dx}$$

This is known as the **product rule of differentiation**. Note that the differential coefficient of a product **cannot** be obtained by merely differentiating each term and multiplying the two answers together. Whenever differentiating products the above rule **must** be used.

Worked problems on differentiating products

Problem 1. Find the differential coefficient of $y = 5x^3 \sin 2x$.

$5x^3 \sin 2x$ is a product of two terms, $5x^3$ and $\sin 2x$. Let $u = 5x^3$ and $v = \sin 2x$

Using the product rule:

$$\frac{dy}{dx} = u \quad \frac{dv}{dx} \quad + \quad v \quad \frac{du}{dx}$$

gives $\dfrac{dy}{dx} = (5x^3)(2 \cos 2x) + (\sin 2x)(15x^2)$

i.e. $\dfrac{dy}{dx} = 10x^3 \cos 2x + 15x^2 \sin 2x$

$\qquad\quad = 5x^2(2x \cos 2x + 3 \sin 2x)$

Problem 2. If $x = 2e^{3\theta} \cos 4\theta$ find $\dfrac{dx}{d\theta}$.

Let $u = 2e^{3\theta}$ and $v = \cos 4\theta$

Then $\dfrac{dx}{d\theta} = u\dfrac{dv}{d\theta} + v\dfrac{du}{d\theta}$

$\qquad = (2e^{3\theta})(-4 \sin 4\theta) + (\cos 4\theta)(6e^{3\theta})$

$\qquad = -8e^{3\theta} \sin 4\theta + 6e^{3\theta} \cos 4\theta$

i.e. $\dfrac{dx}{d\theta} = 2e^{3\theta}(3 \cos 4\theta - 4 \sin 4\theta)$

Problem 3. Determine the rate of change of y with respect to t given $y = 5\sqrt{t} \ln 3t$.

The rate of change of y with respect to t is given by $\dfrac{dy}{dt}$

$y = 5\sqrt{t} \ln 3t = 5t^{\frac{1}{2}} \ln 3t$, which is a product.

Let $u = 5t^{\frac{1}{2}}$ and $v = \ln 3t$,

then $\dfrac{dy}{dt} = u\dfrac{dv}{dt} + v\dfrac{du}{dt}$

$\qquad = (5t^{\frac{1}{2}})\left(\dfrac{1}{t}\right) + (\ln 3t)\left(\dfrac{5}{2}t^{-\frac{1}{2}}\right)$

$\qquad = 5t^{-\frac{1}{2}} + \dfrac{5}{2}t^{-\frac{1}{2}} \ln 3t$

$\qquad = 5t^{-\frac{1}{2}}\left(1 + \dfrac{1}{2} \ln 3t\right)$

i.e. $\dfrac{dy}{dt} = \dfrac{5}{\sqrt{t}}\left(1 + \dfrac{1}{2} \ln 3t\right)$

Problem 4. Differentiate $y = x^2 \cos 3x \ln 2x$.

Let $u = x^2 \cos 3x$ (i.e., a product) and $v = \ln 2x$.

$$\frac{dy}{dx} = u\frac{dv}{dx} + v\frac{du}{dx}$$

$$= (x^2 \cos 3x)\left(\frac{1}{x}\right) + (\ln 2x)\left[\frac{du}{dx}\right] \tag{1}$$

Since $u = x^2 \cos 3x$, then

$$\frac{du}{dx} = (x^2)(-3 \sin 3x) + (\cos 3x)(2x)$$ again, by the product rule

$$= -3x^2 \sin 3x + 2x \cos 3x$$

Substituting this into equation (1) above gives:

$$\frac{dy}{dx} = (x^2 \cos 3x)\left(\frac{1}{x}\right) + (\ln 2x)[-3x^2 \sin 3x + 2x \cos 3x]$$

$$= x \cos 3x - 3x^2 \sin 3x \ln 2x + 2x \cos 3x \ln 2x$$

$$= x[\cos 3x + \ln 2x(2 \cos 3x - 3x \sin 3x)]$$

Problem 5. Determine the rate of change of current, correct to three decimal places, given $i = 2t \sin 3t$ amperes, when $t = 0.10$ s

$$\text{Rate of change of current} = \frac{di}{dt} = (2t)(3 \cos 3t) + (\sin 3t)(2)$$

$$= 6t \cos 3t + 2 \sin 3t$$

When $t = 0.10$ s, $\dfrac{di}{dt} = 6(0.10) \cos 3(0.10) + 2 \sin 3(0.10)$

$$= 0.6 \cos 0.30 + 2 \sin 0.30,$$

where $\cos 0.30$ means the cosine of 0.30 radians. It is important to ensure that your calculator is on the radian mode.

Hence $\dfrac{di}{dt} = 0.57320 + 0.59104 = 1.164$ A/s,

i.e. **the rate of change of current when $t = 0.10$ s is 1.164 A/s, correct to three decimal places.**

Further problems on differentiating products may be found in Section 1.7 (Problems 13 to 20), page 20.

1.4 Differentiation of a quotient

When $y = \dfrac{u}{v}$, and u and v are both functions of x, then

$$\frac{dy}{dx} = \frac{v\,\dfrac{du}{dx} - u\,\dfrac{dv}{dx}}{v^2}$$

This is known as the **quotient rule of differentiation**. Note that the differential coefficient of a quotient **cannot** be obtained by merely differentiating each term and then dividing the numerator by the denominator. Whenever differentiating quotients the above rule **must** be used.

Worked problems on differentiating quotients

Problem 1. Find the differential coefficient of $y = \dfrac{3 \sin 2x}{2x^3}$.

$y = \dfrac{3 \sin 2x}{2x^3}$ is a quotient. Let $u = 3 \sin 2x$ and $v = 2x^3$

$$\frac{dy}{dx} = \frac{v\,\dfrac{du}{dx} - u\,\dfrac{dv}{dx}}{v^2} = \frac{(2x^3)(6 \cos 2x) - (3 \sin 2x)(6x^2)}{(2x^3)^2}$$

$$= \frac{12x^3 \cos 2x - 18x^2 \sin 2x}{4x^6}$$

$$= \frac{6x^2}{4x^6}(2x \cos 2x - 3 \sin 2x)$$

i.e. $\dfrac{dy}{dx} = \dfrac{3}{2x^4}(2x \cos 2x - 3 \sin 2x)$

Problem 2. Find the gradient of the curve $y = \dfrac{3x}{2x^2 + 5}$ at the point $\left(\sqrt{2}, \dfrac{\sqrt{2}}{3}\right)$.

$y = \dfrac{3x}{2x^2 + 5}$ is a quotient and the gradient of the curve is given by $\dfrac{dy}{dx}$. Let $u = 3x$ and $v = (2x^2 + 5)$.

$$\text{Gradient} = \frac{dy}{dx} = \frac{v\dfrac{du}{dx} - u\dfrac{dv}{dx}}{v^2} = \frac{(2x^2 + 5)(3) - (3x)(4x)}{(2x^2 + 5)^2}$$

$$= \frac{6x^2 + 15 - 12x^2}{(2x^2 + 5)^2} = \frac{15 - 6x^2}{(2x^2 + 5)^2}$$

At the point $\left(\sqrt{2}, \dfrac{\sqrt{2}}{3}\right)$, $x = \sqrt{2}$

Hence the gradient of the curve when $x = \sqrt{2}$,

$$\frac{dy}{dx} = \frac{15 - 6(\sqrt{2})^2}{[2(\sqrt{2})^2 + 5]^2} = \frac{3}{81} = \frac{1}{27}$$

Problem 3. Differentiate $y = \dfrac{2t\, e^t}{\cos 2t}$.

$y = \dfrac{2t\, e^t}{\cos 2t}$ is a quotient, whose numerator is a product.

Let $\quad u = 2t\, e^t$ and $v = \cos 2t$

$$\frac{dy}{dt} = \frac{v\dfrac{du}{dt} - u\dfrac{dv}{dt}}{v^2}$$

where $\dfrac{du}{dt} = (2t)(e^t) + (e^t)(2)$, by the product rule,

i.e. $\quad \dfrac{dy}{dt} = \dfrac{(\cos 2t)[(2t)(e^t) + (e^t)(2)] - (2t\, e^t)(-2\sin 2t)}{(\cos 2t)^2}$

$$= \frac{2t\, e^t \cos 2t + 2e^t \cos 2t + 4t\, e^t \sin 2t}{\cos^2 2t}$$

$$= \frac{2\, e^t(t \cos 2t + \cos 2t + 2t \sin 2t)}{\cos^2 2t}$$

i.e. $\quad \dfrac{dy}{dt} = \dfrac{2\, e^t}{\cos^2 2t}(t \cos 2t + \cos 2t + 2t \sin 2t)$

Problem 4. Determine the differential coefficient of $y = \tan ax$

$y = \tan ax = \dfrac{\sin ax}{\cos ax}$. Differentiation of $\tan ax$ is therefore treated
as a quotient with $u = \sin ax$ and $v = \cos ax$.

$$\frac{dy}{dx} = \frac{v\dfrac{du}{dx} - u\dfrac{dv}{dx}}{v^2}$$

$$= \frac{(\cos ax)(a\cos ax) - (\sin ax)(-a\sin ax)}{(\cos ax)^2}$$

$$= \frac{a\cos^2 ax + a\sin^2 ax}{\cos^2 ax} = \frac{a(\cos^2 ax + \sin^2 ax)}{\cos^2 ax}$$

$$= \frac{a}{\cos^2 ax} \text{ since } \cos^2 ax + \sin^2 ax = 1$$

Hence $\dfrac{dy}{dx} = a\sec^2 ax.$

Problem 5. Differentiate $y = \cot 2t$.

$$y = \cot 2t = \frac{1}{\tan 2t} = \frac{\cos 2t}{\sin 2t}$$

Let $u = \cos 2t$ and $v = \sin 2t$

$$\frac{dy}{dt} = \frac{v\dfrac{du}{dt} - u\dfrac{dv}{dt}}{v^2}$$

$$= \frac{(\sin 2t)(-2\sin 2t) - (\cos 2t)(2\cos 2t)}{(\sin 2t)^2}$$

$$= \frac{-2\sin^2 2t - 2\cos^2 2t}{\sin^2 2t} = \frac{-2(\sin^2 2t + \cos^2 2t)}{\sin^2 2t}$$

$$= \frac{-2}{\sin^2 2t}$$

Hence $\dfrac{dy}{dt} = -2\,\mathbf{cosec}^2\,2t.$

Problem 6. Determine the derivative of $y = \sec ax$

$$y = \sec ax = \frac{1}{\cos ax}, \text{ i.e., a quotient.}$$

Let $u = 1$ and $v = \cos ax$

$$\frac{dy}{dx} = \frac{v\dfrac{du}{dx} - u\dfrac{dv}{dx}}{v^2} = \frac{(\cos ax)(0) - (1)(-a\sin ax)}{(\cos ax)^2}$$

$$= \frac{a \sin ax}{\cos^2 ax} = a\left(\frac{1}{\cos ax}\right)\left(\frac{\sin ax}{\cos ax}\right)$$

Hence $\dfrac{dy}{dx} = a \sec ax \tan ax$

Problem 7. Differentiate $y = \operatorname{cosec} a\theta$

$y = \operatorname{cosec} a\theta = \dfrac{1}{\sin a\theta}$, i.e., a quotient.

Let $\quad u = 1$ and $v = \sin a\theta$

$$\frac{dy}{d\theta} = \frac{v\dfrac{du}{d\theta} - u\dfrac{dv}{d\theta}}{v^2} = \frac{(\sin a\theta)(0) - (1)(a \cos a\theta)}{(\sin a\theta)^2}$$

$$= \frac{-a \cos a\theta}{\sin^2 a\theta} = -a\left(\frac{1}{\sin a\theta}\right)\left(\frac{\cos a\theta}{\sin a\theta}\right)$$

Hence $\dfrac{dy}{d\theta} = -a \operatorname{cosec} a\theta \cot a\theta$

The differential coefficients of the six trigonometric ratios may be summarised as shown in Table 1.2.

Table 1.2

y or $f(x)$	$\dfrac{dy}{dx}$ or $f'(x)$
$\sin ax$	$a \cos ax$
$\cos ax$	$-a \sin ax$
$\tan ax$	$a \sec^2 ax$
$\sec ax$	$a \sec ax \tan ax$
$\operatorname{cosec} ax$	$-a \operatorname{cosec} ax \cot ax$
$\cot ax$	$-a \operatorname{cosec}^2 ax$

Further problems on differentiating quotients may be found in Section 1.7 (Problems 21 to 30), page 21.

1.5   Function of a function

The function $y = (3x - 1)^6$ can be differentiated by first multiplying $(3x - 1)$ by itself six times, and then differentiating each term produced in turn. This would be a long process!

In this type of function a substitution is made. Let $u = (3x - 1)$, then instead of $y = (3x - 1)^6$ we have $y = u^6$. An important rule that is used when differentiating by substitution is

$$\frac{dy}{dx} = \frac{dy}{du} \cdot \frac{du}{dx}$$

This is known as the **'function of a function'** rule (or sometimes the **chain rule**).

From above $y = (3x - 1)^6$. If $u = (3x - 1)$ then $y = u^6$

Then $\dfrac{dy}{du} = 6u^5$ and $\dfrac{du}{dx} = 3$

Hence, since $\dfrac{dy}{dx} = \dfrac{dy}{du} \cdot \dfrac{du}{dx}$

$$\frac{dy}{dx} = (6u^5)(3) = 18u^5$$

Rewriting u as $(3x - 1)$ gives $\dfrac{dy}{dx} = \mathbf{18(3x - 1)^5}$

Since y is a function of u, and u is a function of x, then y is a 'function of a function' of x – hence the name of the rule.

Worked problems on differentiating by the function of a function rule

Problem 1. Differentiate $y = 2 \sin(3x^2 + 1)$.

If $u = 3x^2 + 1$, then $y = 2 \sin u$

Hence $\dfrac{du}{dx} = 6x$ and $\dfrac{dy}{du} = 2 \cos u$

Using the function of a function rule,

$$\frac{dy}{dx} = \frac{dy}{du} \cdot \frac{du}{dx} = (2 \cos u)(6x) = 12x \cos u$$

Rewriting u as $(3x^2 + 1)$ gives: $\dfrac{dy}{dx} = \mathbf{12x \cos (3x^2 + 1)}$

(Note that this result could have been obtained 'on sight' by first differentiating the trigonometric function, i.e. differentiating $2 \sin f(x)$ giving $2 \cos f(x)$ and then multiplying by the differential coefficient of $f(x)$, i.e. $6x$.)

Problem 2. Determine the differential coefficient of $(3x^4 - 2x)^5$.

Let $y = (3x^4 - 2x)^5$

If $u = (3x^4 - 2x)$ then $y = u^5$

Hence $\dfrac{du}{dx} = (12x^3 - 2)$ and $\dfrac{dy}{du} = 5u^4$

Using the function of a function rule, $\dfrac{dy}{dx} = \dfrac{dy}{du} \cdot \dfrac{du}{dx}$

$$= (5u^4)(12x^3 - 2)$$

Rewriting u as $(3x^4 - 2x)$ gives $\dfrac{dy}{dx} = 5(3x^4 - 2x)^4(12x^3 - 2)$

(Note that this result could have been obtained 'on sight' by first differentiating the bracket giving $5[f(x)]^4$ and then multiplying this result by the differential coefficient of $f(x)$, i.e. $(12x^3 - 2)$.)

Problem 3. Find the derivative of $y = \sqrt{(5t^2 + 3t - 2)}$.

$y = \sqrt{(5t^2 + 3t - 2)} = (5t^2 + 3t - 2)^{\frac{1}{2}}$

If $u = (5t^2 + 3t - 2)$ then $y = u^{\frac{1}{2}}$

Hence $\dfrac{du}{dt} = (10t + 3)$ and $\dfrac{dy}{du} = \dfrac{1}{2} u^{-\frac{1}{2}}$

Using the function of a function rule:

$$\dfrac{dy}{dt} = \dfrac{dy}{du} \cdot \dfrac{du}{dt}$$

$$= \left(\dfrac{1}{2} u^{-\frac{1}{2}}\right)(10t + 3)$$

$$= \dfrac{(10t + 3)}{2u^{\frac{1}{2}}} = \dfrac{(10t + 3)}{2\sqrt{u}}$$

Rewriting u as $(5t^2 + 3t - 2)$ gives $\dfrac{dy}{dt} = \dfrac{(10t + 3)}{2\sqrt{(5t^2 + 3t - 2)}}$

Problem 4. Differentiate $y = 5 \tan^3 2x$.

$y = 5 \tan^3 2x = 5(\tan 2x)^3$

If $u = \tan 2x$ then $y = 5u^3$

Hence $\dfrac{du}{dx} = 2 \sec^2 2x$ and $\dfrac{dy}{du} = 15u^2$

Using the function of a function rule,

$\dfrac{dy}{dx} = \dfrac{dy}{du} \cdot \dfrac{du}{dx}$

$\quad = (15u^2)(2 \sec^2 2x)$

$\quad = 30u^2 \sec^2 2x$

Rewriting u as $\tan 2x$ gives $\dfrac{dy}{dx} = 30 (\tan 2x)^2 \sec^2 2x$

i.e. $\dfrac{dy}{dx} = \mathbf{30 \tan^2 2x \sec^2 2x}$

Problem 5. Determine the derivative of $y = \dfrac{3}{(2x^4 - 5)^6}$.

$y = \dfrac{3}{(2x^4 - 5)^6} = 3(2x^4 - 5)^{-6}$

If $u = (2x^4 - 5)$ then $y = 3u^{-6}$

Hence $\dfrac{du}{dx} = 8x^3$ and $\dfrac{dy}{du} = -18u^{-7}$

Using the function of a function rule,

$\dfrac{dy}{dx} = \dfrac{dy}{du} \cdot \dfrac{du}{dx}$

$\quad = (-18u^{-7})(8x^3)$

$\quad = \dfrac{-144x^3}{u^7}$

Rewriting u as $(2x^4 - 5)$ gives $\dfrac{dy}{dx} = \dfrac{-144x^3}{(2x^4 - 5)^7}$

Problem 6. Differentiate $y = 3e^{(2\theta^2 - 1)}$.

If $u = (2\theta^2 - 1)$ then $y = 3e^u$

Hence $\dfrac{du}{d\theta} = 4\theta$ and $\dfrac{dy}{du} = 3e^u$

Using the function of a function rule,

$$\frac{dy}{d\theta} = \frac{dy}{du} \cdot \frac{du}{d\theta}$$

$$= (3e^u)(4\theta) = 12\theta \, e^u$$

Rewriting u as $(2\theta^2 - 1)$ gives $\dfrac{dy}{d\theta} = \mathbf{12\theta \, e^{(2\theta^2 - 1)}}$

Further problems on differentiating by the function of a function rule may be found in Section 1.7 (Problems 31 to 41), page 21.

1.6   Successive differentiation

When a function, say, $y = f(x)$, is differentiated, the differential coefficient is written as $f'(x)$ or $\dfrac{dy}{dx}$. If the expression is differentiated again, the second differential coefficient or the second derivative is obtained. This is written as $f''(x)$ (pronounced 'f double-dash x') or $\dfrac{d^2y}{dx^2}$ (pronounced 'dee two y by dee x squared'). Similarly, if differentiated again the third differential coefficient or third derivative is obtained, and is written as $f'''(x)$ or $\dfrac{d^3y}{dx^3}$, and so on. Thus, if $y = 3x^4$, $\dfrac{dy}{dx} = 12x^3$, $\dfrac{d^2y}{dx^2} = 36x^2$, $\dfrac{d^3y}{dx^3} = 72x$,

$$\frac{d^4y}{dx^4} = 72 \text{ and } \frac{d^5y}{dx^5} = 0$$

Worked problems on successive differentiation

Problem 1. Given $f(x) = 2x^5 - 3x^4 + 5x^2 - x$ find $f'(x)$ and $f''(x)$

$f(x) = 2x^5 - 3x^4 + 5x^2 - x$

$f'(x) = 10x^4 - 12x^3 + 10x - 1$

$f''(x) = 40x^3 - 36x^2 + 10$

Problem 2. If $y = \dfrac{2}{3}t^3 - \dfrac{5}{t^2} + \dfrac{1}{3t} - \sqrt{t}$ determine $\dfrac{d^2y}{dt^2}$.

$$y = \frac{2}{3}t^3 - \frac{5}{t^2} + \frac{1}{3t} - \sqrt{t} = \frac{2}{3}t^3 - 5t^{-2} + \frac{1}{3}t^{-1} - t^{\frac{1}{2}}$$

$$\frac{dy}{dt} = \left(\frac{2}{3}\right)(3t^2) - (5)(-2t^{-3}) + \left(\frac{1}{3}\right)(-t^{-2}) - \frac{1}{2}t^{-\frac{1}{2}}$$

$$= 2t^2 + 10t^{-3} - \frac{1}{3}t^{-2} - \frac{1}{2}t^{-\frac{1}{2}}$$

$$\frac{d^2y}{dx^2} = 4t - 30t^{-4} + \frac{2}{3}t^{-3} + \frac{1}{4}t^{-\frac{3}{2}}$$

$$= 4t - \frac{30}{t^4} + \frac{2}{3t^3} + \frac{1}{4\sqrt{t^3}}$$

Problem 3. If $f(t) = 2\ln(\sin 2t)$ evaluate $f'(t)$ and $f''(t)$ correct to three decimal places, when $t = \dfrac{1}{2}$.

If $\qquad y = \ln[f(t)]$ then $\dfrac{dy}{dt} = \dfrac{1}{f(t)} \cdot f'(t)$

Hence if $f(t) = 2\ln(\sin 2t)$

then $\qquad f'(t) = 2\left(\dfrac{1}{\sin 2t}\right)(2\cos 2t) = \dfrac{4\cos 2t}{\sin 2t} = 4\cot 2t$

When $\qquad t = \dfrac{1}{2}$, $f'(t) = 4\cot 1 = \mathbf{2.568}$

$$f''(t) = -8\operatorname{cosec}^2 2t$$

When $\qquad t = \dfrac{1}{2}$, $f''(t) = -8\operatorname{cosec}^2 1 = -11.298$

Problem 4. If $y = \cos\theta - \sin\theta$, evaluate θ, in the range $0 \leqslant \theta \leqslant \dfrac{\pi}{2}$, when $\dfrac{d^2y}{d\theta^2}$ is zero.

Since $y = \cos\theta - \sin\theta$, $\dfrac{dy}{d\theta} = -\sin\theta - \cos\theta$

and $\dfrac{d^2y}{d\theta^2} = -\cos\theta + \sin\theta$

When $\dfrac{d^2y}{d\theta^2}$ is zero, $-\cos\theta + \sin\theta = 0$

i.e. $\sin \theta = \cos \theta$ or $\dfrac{\sin \theta}{\cos \theta} = 1$

Hence $\tan \theta = 1$ and $\theta = \arctan 1 = \mathbf{45°}$ **or** $\dfrac{\pi}{4}$ rad in the range

$0 \leqslant \theta \leqslant \dfrac{\pi}{2}$

Problem 5. If $y = (x + 5)e^{\frac{2}{5}x}$ show that $25\dfrac{d^2y}{dx^2} - 20\dfrac{dy}{dx} + 4y = 0.$

$y = (x + 5)e^{\frac{2}{5}x}$, which is a product

$\dfrac{dy}{dx} = (x + 5)\left(\dfrac{2}{5}e^{\frac{2}{5}x}\right) + (e^{\frac{2}{5}x})(1)$, by the product rule,

$\quad = \dfrac{2}{5}(x + 5)\,e^{\frac{2}{5}x} + e^{\frac{2}{5}x}$

$\dfrac{d^2y}{dx^2} = \left\{\dfrac{2}{5}(x + 5)\left(\dfrac{2}{5}e^{\frac{2}{5}x}\right) + (e^{\frac{2}{5}x})\left(\dfrac{2}{5}\right)\right\} + \dfrac{2}{5}e^{\frac{2}{5}x}$

$\quad = \dfrac{4}{25}(x + 5)\,e^{\frac{2}{5}x} + \dfrac{2}{5}e^{\frac{2}{5}x} + \dfrac{2}{5}e^{\frac{2}{5}x}$

$\quad = \dfrac{4}{25}(x + 5)\,e^{\frac{2}{5}x} + \dfrac{4}{5}e^{\frac{2}{5}x}$

Hence $25\dfrac{d^2y}{dx^2} - 20\dfrac{dy}{dx} + 4y$

$\quad = 25\left\{\dfrac{4}{25}(x + 5)\,e^{\frac{2}{5}x} + \dfrac{4}{5}e^{\frac{2}{5}x}\right\} - 20\left\{\dfrac{2}{5}(x + 5)\,e^{\frac{2}{5}x} + e^{\frac{2}{5}x}\right\}$

$\quad\quad + 4\{(x + 5)\,e^{\frac{2}{5}x}\}$

$\quad = 4(x + 5)\,e^{\frac{2}{5}x} + 20e^{\frac{2}{5}x} - 8(x + 5)\,e^{\frac{2}{5}x} - 20e^{\frac{2}{5}x} + 4(x + 5)\,e^{\frac{2}{5}x} = 0$

Problem 6. If $y = 3 \sec 4t$, evaluate $\dfrac{d^2y}{dt^2}$ when $t = 0$.

$y = 3 \sec 4t$ hence $\dfrac{dy}{dt} = 12 \sec 4t \tan 4t$, which is a product

$\dfrac{d^2y}{dt^2} = (12 \sec 4t)(4 \sec^2 4t) + (\tan 4t)(48 \sec 4t \tan 4t)$

$\quad = 48 \sec^3 4t + 48 \sec 4t \tan^2 4t$

When $t = 0$, $\dfrac{d^2y}{dt^2} = 48 \sec^3 0 + 48 \sec 0 \tan 0$

$= 48(1) + 48(1)(0)$

$= \mathbf{48}$

Further problems on successive differentiation may be found in Section 1.7 following (Problems 42 to 48), page 22.

1.7   Further problems

Differentiation of common functions

In Problems 1 to 4, find the differential coefficients, with respect to x, of the functions:

1. (a) $y = 3x^2$ (b) $y = 5x^7$ (c) $y = \dfrac{3}{x^3}$

    (a) $[6x]$ (b) $[35x^6]$ (c) $\left[\dfrac{-9}{x^4}\right]$

2. (a) $y = \dfrac{2}{5x^2}$ (b) $y = 6$ (c) $y = 4x$

    (a) $\left[\dfrac{-4}{5x^3}\right]$ (b) $[0]$ (c) $[4]$

3. (a) $\dfrac{\sqrt{x}}{3}$ (b) $y = \sqrt[3]{x^4}$ (c) $y = \dfrac{2}{\sqrt{x}}$

    (a) $\left[\dfrac{1}{6\sqrt{x}}\right]$ (b) $\left[\dfrac{4}{3}\sqrt[3]{x}\right]$ (c) $\left[\dfrac{-1}{\sqrt{x^3}}\right]$

4. (a) $y = \dfrac{5}{\sqrt[4]{x^3}}$ (b) $5 \sin 2x$ (c) $-3 \cos 4x$

    (a) $\left[\dfrac{-15}{4\sqrt[4]{x^7}}\right]$ (b) $[10 \cos 2x]$ (c) $[12 \sin 4x]$

5. (a) $y = 3e^{5x}$ (b) $y = \dfrac{2}{e^{4x}}$ (c) $y = \dfrac{5}{2} \ln 3x$

    (a) $[15e^{5x}]$ (b) $\left[\dfrac{-8}{e^{4x}}\right]$ (c) $\left[\dfrac{5}{2x}\right]$

6. If $f(t) = 3t^7 + 2t - \dfrac{2}{t^4} + \dfrac{3}{2\sqrt{t}} - 5$ find $f'(t)$.

    $\left[21t^6 + 2 + \dfrac{8}{t^5} - \dfrac{3}{4\sqrt{t^3}}\right]$

7. Given $f(\theta) = \dfrac{2\theta^2 - 3\sqrt{\theta^3} + \sqrt[4]{\theta^5}}{\theta}$, evaluate $f'(\theta)$ when

    $\theta = 1$. $\left[\dfrac{3}{4}\right]$

8. If $y = \dfrac{3}{x} - 5(3\cos 2x - 2\sin 3x) + 2\ln 2x - \dfrac{3}{2e^{3x}}$, evaluate

    $\dfrac{dy}{dx}$ when $x = \dfrac{\pi}{2}$, correct to four decimal places.   [0.0978]

9. Determine the gradient of the curve
    $y = 4x^3 - 2x^2 + 3x + 1$ at the points $(0, 1)$ and $(1, 6)$.
    [3, 11]

10. Find the co-ordinates of the points on the graph
    $y = 5x^2 - 3x + 2$ when the gradient is 7.   [(1, 4)]

11. If $f(t) = 5 + 3\ln 2t$ evaluate $f'(t)$ when $t = 0.5$.   [6]

12. Evaluate $\dfrac{ds}{dt}$, correct to four significant figures, when $t = \dfrac{\pi}{3}$

    given $s = 5\sin t - 2 + \sqrt{t}$.   [2.989]

Differentiation of a product

In Problems 13 to 18, differentiate the given functions with
respect to the variable.

13. $y = 3x^2 \cos 4x$ $[6x(\cos 4x - 2x \sin 4x)]$
14. $y = 3e^{2t} \sin 3t$ $[3e^{2t}(3\cos 3t + 2\sin 3t)]$

15. $y = 4\sqrt{x}\ln 2x$ $\left[\dfrac{2}{\sqrt{x}}(2 + \ln 2x)\right]$

16. $y = \sqrt{\theta^5}\sin 2\theta$ $\left[\sqrt{\theta^3}\left(2\theta \cos 2\theta + \dfrac{5}{2}\sin 2\theta\right)\right]$

17. $y = e^{2\theta}\ln 2\theta$ $\left[e^{2\theta}\left(\dfrac{1}{\theta} + 2\ln 2\theta\right)\right]$

18. $y = e^x \ln x \sin x$ $\left[e^x\left(\ln x \cos x + \dfrac{\sin x}{x} + \ln x \sin x\right)\right]$

19. Determine the rate of change of voltage, correct to four
    significant figures, given $v = 50t \sin 2t$ volts, when
    $t = 0.20$ s.   [37.89 V/s]

20. Evaluate $\dfrac{dz}{dt}$, correct to four significant figures, when $t = 0.25$,

    given that $z = 5e^{2t}\sin 2t$.   [22.37]

Differentiation of a quotient

In Problems 21 to 26, differentiate the given functions with respect to the variable.

21. $y = \dfrac{2 \cos 3x}{x^2}$ $\left[\dfrac{-2}{x^3}(3x \sin 3x + 2 \cos 3x)\right]$

22. $y = \dfrac{3x}{x^2 + 2}$ $\left[\dfrac{3(2 - x^2)}{(x^2 + 2)^2}\right]$

23. $y = \dfrac{2\sqrt{t^3}}{3 \sin 2t}$ $\left[\dfrac{3\sqrt{t}(3 \sin 2t - 4t \cos 2t)}{9 \sin^2 2t}\right]$

24. $y = \dfrac{\ln 2x}{\sqrt{x}}$ $\left[\dfrac{1}{\sqrt{x^3}}\left(1 - \dfrac{\ln 2x}{2}\right)\right]$

25. $y = \dfrac{2 \tan 2\theta}{e^{2\theta}}$ $[4e^{-2\theta}(\sec^2 2\theta - \tan 2\theta)]$

26. $y = \dfrac{x\, e^{2x}}{\sin 2x}$ $\left[\dfrac{e^{2x}(2x \sin 2x + \sin 2x - 2x \cos 2x)}{\sin^2 2x}\right]$

27. Determine the gradient of the curve $y = \dfrac{2t}{t^2 - 3}$ at the point $(2, 4)$. $[-14]$

28. Evaluate $\dfrac{dy}{dx}$ at $x = 1.5$, correct to three significant figures,

given $y = \dfrac{x^2 + 1}{\ln 2x}$. $[0.936]$

29. Evaluate $f'\left(\dfrac{\pi}{3}\right)$ when $f(t) = 3 \tan 3t - \cot 2t$. $\left[11\dfrac{2}{3}\right]$

30. Show that $\dfrac{\dfrac{d}{dx}(3 \sec 2x)}{\dfrac{d}{dx}(-\csc 2x)} = 3 \tan^3 2x$

Function of a function

In Problems 31 to 40, find the differential coefficients with respect to the variable.

31. $y = (3x^3 - 4x)^5$ $[5(3x^3 - 4x)^4(9x^2 - 4)]$

32. $y = 3 \cos (2x - 1)$ $[-6 \sin(2x - 1)]$

33. $y = \sqrt{(3\theta^3 - 5)}$ $\left[\dfrac{9\theta^2}{2\sqrt{(3\theta^3 - 5)}}\right]$

34. $y = 3\cos^4 t$ $[-12\cos^3 t \sin t]$

35. $y = \dfrac{1}{(2x^3 - 3x + 1)^4}$ $\left[\dfrac{12(1 - 2x^2)}{(2x^3 - 3x + 1)^5}\right]$

36. $y = 7e^{3t-2}$ $[21e^{(3t-2)}]$

37. $y = 5\sec^3 2x$ $[30\sec^3 2x \tan 2x]$

38. $y = 3\cot(3x^2 + 1)$ $[-18x \operatorname{cosec}^2(3x^2 + 1)]$

39. $y = 5\tan(2\theta - 1)$ $[10\sec^2(2\theta - 1)]$

40. $y = 3e^{\tan x}$ $[3\sec^2 x \, e^{\tan x}]$

41. Differentiate $x \sin\left(x - \dfrac{\pi}{6}\right)$ with respect to x and evaluate,

correct to three decimal places, when $x = \dfrac{\pi}{3}$. $[1.407]$

Successive differentiation

42. If $y = 2x^4 + 3x^3 - 2x^2 + 5$ find

(a) $\dfrac{d^2 y}{dx^2}$ (b) $\dfrac{d^3 y}{dx^3}$ (a) $[24x^2 + 18x - 4]$ (b) $[48x + 18]$

43. (a) Given $f(\theta) = \dfrac{2}{3}\theta^2 - \dfrac{1}{\theta^3} + \dfrac{3}{\theta} - \dfrac{\sqrt{\theta}}{2} + 2$, determine $f''(\theta)$.

(b) Evaluate $f''(\theta)$ in part (a), correct to three significant figures, when $\theta = 1$.

(a) $\left[\dfrac{4}{3} - \dfrac{12}{\theta^5} + \dfrac{6}{\theta^3} + \dfrac{1}{8\sqrt{\theta^3}}\right]$ (b) $[-4.54]$

In Problems 44 and 45, find the second differential coefficient with respect to the variable.

44. (a) $2\sin 3x + \cos x$ (b) $3\ln 5\theta$

(a) $[-(18\sin 3x + \cos x)]$ (b) $\left[-\dfrac{3}{\theta^2}\right]$

45. (a) $3\cot t$]   (b) $(3x - 1)^4$

(a) $[6\operatorname{cosec}^2 t \cot t]$ (b) $[108(3x - 1)^2]$

46. If $y = (2t + 3)e^{\frac{2}{3}t}$ prove that $9\dfrac{d^2 y}{dt^2} - 12\dfrac{dy}{dt} + 4y = 0$.

47. Evaluate $f''(t)$ when $t = \dfrac{\pi}{4}$ given $f(t) = 2 \cot 3t.$ $[-72]$

48. Show that the differential equation $\dfrac{d^2 i}{dt^2} + 500 \dfrac{di}{dt} + 62\,500i = 0$ is satisfied when $i = 100t\, e^{-250t}.$

Chapter 2

Solution of equations by iterative methods

2.1 Introduction

The solution of equations of the form $f(x) = 0$, where the function is linear, can be achieved by applying the basic rules of algebra or trigonometry. Other equations can be solved by applying a formula, as in the case of solving quadratic equations, where the equation $ax^2 + bx + c = 0$ can be solved by applying the formula $x = \dfrac{-b \pm \sqrt{(b^2 - 4ac)}}{2a}$. More difficult functions may be solved graphically, although this can be a long, tedious process if an accurate result is required. A more analytical approach is to use methods of successive approximation, in which a sequence of calculations is repeated as many times as necessary in order to give a result to the required degree of accuracy. For example, the equation $3x = 7$ can be solved as follows:

Try various integer values until the equation is approximately true; in this case, if $x = 2$, $3x = 6$ and if $x = 3$, $3x = 9$. This shows that the root lies between 2 and 3, and since $3x = 6$ is nearer to 7 than $3x = 9$, the root is nearer $x = 2$ than $x = 3$. Now try $x = 2.3$, giving $3x = 6.9$. When $x = 2.4$, $3x$ is equal to 7.2, thus the root is nearer 2.3 than 2.4, and so on. This process can be repeated as many times as it is necessary to give the required degree of accuracy and is called a **method of successive approximation**. Clearly this technique would not be used in the example given but only where straightforward algebraic methods fail. It is particularly suitable for using in cases where the solution is to be determined by a computer.

Several methods of successive approximation may be used to determine the value of the roots of an equation to a specified degree of accuracy, and two of these are introduced in this chapter.

2.2 An algebraic method of successive approximation

Methods of successive approximation may be used to solve equations of the form $a + bx + cx^2 + dx^3 + \cdots = 0$. Thus, they are suitable for solving not only polynomials of the form shown above, but also for the solution of other equations which can be expressed in polynomial form by means of, say, Maclaurin's series, introduced in Chapter 6. The method relies initially on an approximate value of a root being estimated, called the first approximation, and then a more accurate result being determined by the repetition of some procedure. The more accurately the first approximation is obtained, the less the number of repetitive cycles, called **iterations**, needed to obtain a given degree of accuracy. The first approximation can either be estimated graphically or by a method involving functional notation as shown below.

First approximation

The approximate value of x at the point where the curve of $f(x) = 0$ crosses the x-axis is used as the first approximation. This occurs when the value of $f(x)$ changes from positive to negative or from negative to positive.

Consider the equation $x^3 - 2x - 5 = 0$

When $x = 0$, $f(0) = (0)^3 - 2(0) - 5 = -5$,

when $x = 1$, $f(1) = (1)^3 - 2(1) - 5 = -6$,

when $x = 2$, $f(2) = (2)^3 - 2(2) - 5 = -1$, and

when $x = 3$, $f(3) = (3)^3 - 2(3) - 5 = 16$

Since the sign of $f(x)$ changes from a negative value at $f(2)$ to a positive value at $f(3)$, then the first approximation is between $x = 2$ and $x = 3$. If a straight line is drawn between co-ordinates $(2, -1)$ and $(3, 16)$, it will cut the x-axis very near to 2. So a **first approximation** of x is taken as, say, 2.1.

The repetitive procedure used to determine the value of the root more accurately is shown below.

Second approximation

Let the true value of the root be $x_1 = 2.1 + \delta_1$

Substituting $2.1 + \delta_1$ for x in the original equation, $x^3 - 2x - 5 = 0$, gives:

$$(2.1 + \delta_1)^3 - 2(2.1 + \delta_1) - 5 = 0$$

If the first approximation is reasonably accurate, then δ_1 is small and δ_1^2 and δ_1^3 are very small. Using the binomial expansion (see *Technician mathematics 3*, Chapter 1) to obtain the terms of $(2.1 + \delta_1)^3$ and neglecting terms containing δ_1^2 and δ_1^3 gives:

$$(2.1^3 + 3 \times 2.1^2 \times \delta_1 + \cdots) - 2(2.1 + \delta_1) - 5 \approx 0$$

i.e. $$9.261 + 13.23\,\delta_1 - 4.2 - 2\,\delta_1 - 5 \approx 0$$

$$\delta_1 \approx -\frac{0.061}{11.23} \approx -0.005$$

Thus a second (i.e. better) approximation to the root is $2.1 - 0.005$, that is, 2.095.

Third approximation

Let the true value of the root be $x_2 = 2.095 + \delta_2$
Substituting $2.095 + \delta_2$ for x in the original equation, gives

$$(2.095 + \delta_2)^3 - 2(2.095 + \delta_2) - 5 = 0$$

Using the binomial series and neglecting terms containing δ_2^2 and δ_2^3 gives:

$$(2.095^3 + 3 \times 2.095^2 \delta_2 + \cdots) - 2(2.095 + \delta_2) - 5 \approx 0$$

i.e.
$$9.195 + 13.167\delta_2 - 4.19 - 2\delta_2 - 5 \approx 0$$

$$\delta_2 \approx \frac{-0.005}{11.167} \approx -0.000\,4$$

Thus, a third approximation to the root is $2.095 - 0.000\,4$, that is, $2.094\,6$.

This procedure can be repeated until the value of the required root on two consecutive iterations does not change when expressed to the stipulated degree of accuracy. The three approximations obtained of the root of the equations $x^3 - 2x - 5 = 0$ are: first approximation 2.1, second approximation 2.095 and third approximation 2.094 6. Since the value of the root, when expressed correct to four significant figures, does not change for the second and third approximations, then the value of the root is 2.095 correct to four significant figures.

A cubic equation can have one, two or three roots.

In this equation when x is larger than 3, the x^3 term predominates and $f(x)$ becomes large and positive. Hence there are no other roots when x is positive. When x is negative

$$f(-1) = (-1)^3 - 2(-1) - 5 = -4$$
$$f(-2) = (-2)^3 - 2(-2) - 5 = -9$$

As x becomes large and negative, the x^3 term again predominates and $f(x)$ becomes large and negative. Hence $x = 2.095$ correct to four significant figures is the only root of the equation $x^3 - 2x - 5 = 0$.

When solving a polynomial equation by an algebraic successive approximation method, the procedure can be summarised as follows:

(i) Determine x, the approximate value of the root required, either graphically or by using a functional notation method.

(ii) Let the true value of the root, x_1, be $(x + \delta_1)$.

(iii) Estimate the approximate value of δ_1 by using $f(x + \delta_1) = 0$ and neglecting terms containing δ_1^2 and higher powers of δ_1.

(iv) A better approximation of the root is $(x_1 + \delta_1)$. Repeat for x_2, $x_3, \ldots$, until the value of the root does not change on two consecutive iterations when expressed to the stipulated degree of accuracy.

Worked problems on an algebraic method of successive approximation

---

Problem 1. The equation $x^3 - 3x^2 + 4x - 7 = 0$ has only one positive root. Determine the value of this root, correct to four significant figures.

$$f(x) = x^3 - 3x^2 + 4x - 7$$

When $x = 0$, $f(0) = (0)^3 - 3(0)^2 + 4(0) - 7 = -7$,

when $x = 1$, $f(1) = (1)^3 - 3(1)^2 + 4(1) - 7 = -5$,

when $x = 2$, $f(2) = (2)^3 - 3(2)^2 + 4(2) - 7 = -3$,

when $x = 3$, $f(3) = (3)^3 - 3(3)^2 + 4(3) - 7 = 5$.

Thus, the value of the positive root is between $x = 2$ and $x = 3$ and dividing this interval in the ratio of 3 to 5 gives a value of the first approximation as $x = 2.4$.

Let the true value of the root x_1 be $(2.4 + \delta_1)$.
To estimate the approximate value of δ_1,

$$(2.4 + \delta_1)^3 - 3(2.4 + \delta_1)^2 + 4(2.4 + \delta_1) - 7 = 0$$

$$2.4^3 + 3 \times 2.4^2 \delta_1 - 3 \times 2.4^2 - 3 \times 2 \times 2.4 \times \delta_1$$
$$+ 4 \times 2.4 + 4 \times \delta_1 - 7 \approx 0$$

$$13.824 + 17.28\delta_1 - 17.28 - 14.4\delta_1 + 9.6 + 4\delta_1 - 7 \approx 0$$

$$\delta_1 \approx \frac{0.856}{6.88} \approx 0.124$$

A second approximation to the root is $2.4 + 0.124 = 2.524$.
Let the true value of the root x_2 be $(2.524 + \delta_2)$

$$(2.524 + \delta_2)^3 - 3(2.524 + \delta_2)^2 + 4(2.524 + \delta_2) - 7 = 0$$

$$2.524^3 + 3 \times 2.524^2 \times \delta_2 - 3 \times 2.524^2$$
$$- 3 \times 2 \times 2.524\delta_2 + 4 \times 2.524 + 4 \times \delta_2 - 7 \approx 0$$

$$16.079 + 19.112\delta_2 - 19.112 - 15.114\delta_2$$
$$+ 10.096 + 4\delta_2 - 7 \approx 0$$

$$\delta_2 \approx \frac{-0.063}{7.998} \approx -0.008$$

A third approximation to the root is $2.524 - 0.008$, that is, 2.516, correct to four significant figures.
Let the true value of the root x_3 be $(2.516 + \delta_3)$

$$2.516^3 + 3 \times 2.516^2\delta_3 - 3 \times 2.516^2 - 6 \times 2.516\delta_3$$
$$+ 4 \times 2.516 + 4\delta_3 - 7 \approx 0$$

$$15.926\,9 + 18.990\,8\,\delta_3 - 18.990\,8 - 15.096\delta_3$$
$$+ 10.064 + 4\delta_3 - 7 \approx 0$$

$$\delta_3 \approx \frac{0.000\,1}{7.894\,8} \approx -0.000\,01$$

A fourth approximation to the root is $2.516 - 0.000\,01$, that is, 2.516 corrrect to four significant figures.

Since the third approximation is equal to the fourth approximation, correct to the stipulated degree of accuracy, then the required root is **2.516, correct to four significant figures**.

Degree of accuracy of calculations

When determining the solution of equations iteratively, the accuracy of the calculations differs. In general, the accuracy to which the various calculations are done is arrived at intuitively. For example, there is little point in calculating δ_1 and δ_2 in the above problem to an accuracy of more than two or three decimal places. However, as the true value of the root is approached more closely, greater accuracy is needed and it can be seen that δ_3 is calculated correct to five decimal places. If small errors are introduced, the worst result is that it may take, say, one extra iteration to achieve the stipulated degree of accuracy. Also, since terms containing δ^2 and higher powers of δ are neglected, the values obtained in calculations are only approximate values in any case.

Problem 2. Determine the value of the root of the equation

$$x^4 - 3x^2 - 2x + 3 = 0$$

in the interval $x = 0$ to $x = 1$, correct to three decimal places.

$$f(0) = (0)^4 - 3(0)^2 - 2(0) + 3 = 3$$
$$f(1) = (1)^4 - 3(1)^2 - 2(1) + 3 = -1$$

Let the first approximation be, say $x = 0.8$.
Let the true value of the root be $x_1 = (0.8 + \delta_1)$
Substituting $x_1 = (0.8 + \delta_1)$ in the original equation gives

$$(0.8 + \delta_1)^4 - 3(0.8 + \delta_1)^2 - 2(0.8 + \delta_1) + 3 = 0$$

Using the binomial series and neglecting terms in δ_1^2, δ_1^3 and δ_1^4, gives

$$0.8^4 + 4 \times 0.8^3\delta_1 - 3 \times 0.8^2 - 6 \times 0.8\delta_1$$
$$- 2 \times 0.8 - 2\delta_1 + 3 \approx 0$$
$$0.409\,6 + 2.048\delta_1 - 1.92 - 4.8\delta_1 - 1.6 - 2\delta_1 + 3 \approx 0$$

i.e.
$$\delta_1 \approx -\frac{0.110\,4}{4.752} \approx -0.023$$

Hence, the second approximation is $0.8 - 0.023$, i.e. 0.777.

Let the true value of the root be $x_2 = (0.777 + \delta_2)$.
Substituting $x_2 = (0.777 + \delta_2)$ in the original equation, gives

$$(0.777 + \delta_2)^4 - 3(0.777 + \delta_2)^2 - 2(0.777 + \delta_2) + 3 = 0$$

Using the binomial series and neglecting terms in δ_2^2, δ_2^3 and δ_2^4, gives

$$0.777^4 + 4 \times 0.777^3 \delta_2 - 3 \times 0.777^2 - 6 \times 0.777 \delta_2$$
$$- 2 \times 0.777 - 2\delta_2 + 3 \approx 0$$
$$0.364\,5 + 1.876\,4\delta_2 - 1.811\,2 - 4.662\delta_2$$
$$- 1.554 - 2\delta_2 + 3 \approx 0$$
$$\delta_2 \approx -\frac{0.000\,7}{4.785\,6} \approx -0.000\,1$$

Hence the third approximation is $0.777 - 0.000\,1$, i.e. 0.777 correct to three decimal places. Since the second and third approximations are the same to the degree of accuracy required, the root is **0.777, correct to three decimal places.**

Further problems on an algebraic method of successive approximation may be found in Section 2.4 (Problems 1 to 7), page 34.

2.3 The Newton–Raphson method

Another iterative method of determining the values of the roots of an equation is the Newton–Raphson method, often referred to as **'Newton's method'**. This states:

if r_1 is the approximate value for a real root of the equation $f(x) = 0$, then a closer approximation to the root, r_2, is given by:

$$r_2 = r_1 - \frac{f(r_1)}{f'(r_1)}$$

An explanation of this statement is given below.

Figure 2.1 shows part of a curve $y = f(x)$ in the region of the root, r, of the equation, shown as point A. Let the first approximation to the root be $(r_1, 0)$ at point B. A vertical line BC is drawn through $(r_1, 0)$ to meet the curve at C, hence the co-ordinates of C are $(r_1, f(r_1))$. A tangent is drawn to the curve at C, cutting the x-axis at D. Let angle CDB be θ.

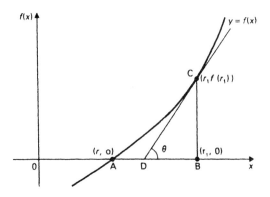

Fig. 2.1

The slope of $CD = \dfrac{BC}{BD} = \tan\theta,$

thus $BD = \dfrac{BC}{\tan\theta}$

But $BC = f(r_1)$ and the slope of the tangent drawn to the curve at C is the differential coefficient of $f(x)$ at point r_1, that is $f'(r_1)$.

Therefore, $BD = \dfrac{f(r_1)}{f'(r_1)}$

Now $OD = OB - BD$

$$= r_1 - \frac{f(r_1)}{f'(r_1)}$$

It can be seen from Fig. 2.1 that D is nearer to the root at A than B and hence is a better approximation.

It will be found in practice that Newton's method sometimes does not give a closer approximation to the root. The reason for this is illustrated in Fig. 2.2. Occasions may occur where distance AB is less than AD, depending on the curvature, and the closeness of the first approximation to the true value of the root. In cases where successive approximations obtained by Newton's method are diverging from the root, a new value of r_1 should be chosen so that $f(r_1)$ has the same sign as $f''(r_1)$. With reference to Fig. 2.2(a), as x increases the slope of the curve, i.e. $f'(x)$ is decreasing, hence $f''(x)$ (the change of $f'(x)$ with respect to x) is negative. If $f(r_1)$ is negative then a first approximation is to the left of point A, say at B'; the tangent is then C'D' and the reasoning becomes similar to that applied to Fig. 2.1. In Fig. 2.2(b), as x increases, $f'(x)$ is increasing and $f''(x)$ is positive. If $f(r_1)$ is

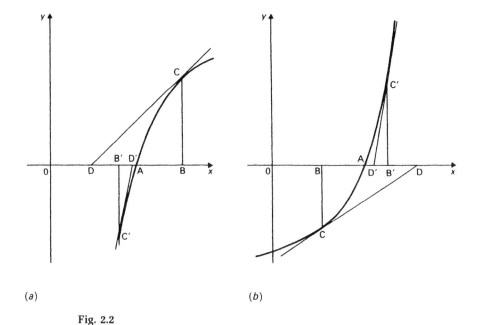

(a)   (b)

Fig. 2.2

positive, the first approximation is to the right of point A at say B'; the
tangent is then C'D' and again the reasoning becomes similar to that applied
to Fig. 2.1.

Worked problems on the Newton–Raphson method

Problem 1. Use Newton's method to determine the root of the
equation $x^4 - 11x - 5 = 0$ in the range $x = 2$ to $x = 3$, correct to
four significant figures.

$f(x) = x^4 - 11x - 5$

$f(2) = (2)^4 - 11(2) - 5 = -11$

$f(3) = (3)^4 - 11(3) - 5 = 43$

Let the first approximation be, say, $r_1 = 2.2$, obtained by
dividing the interval $x = 2$ to $x = 3$ roughly in the ratio of 11 to
43. Newton's method states that r_2 is a closer approximation to
the value of the root,

where $r_2 = r_1 - \dfrac{f(r_1)}{f'(r_1)}$

$f(x) = x^4 - 11x - 5, f(r_1) = (2.2)^4 - 11(2.2) - 5 \approx -5.774$

$f'(x) = 4x^3 - 11, f'(r_1) = 4(2.2)^3 - 11 \approx 31.59$

Hence, a second approximation.

$$r_2 = 2.2 - \frac{(-5.774)}{31.59} \approx 2.38$$

$f(r_2) = (2.38)^4 - 11(2.38) - 5 \approx 0.905\,4$

$f'(r_2) = 4(2.38)^3 - 11 \approx 42.93$

The third approximation,

$$r_3 = 2.38 - \frac{0.905\,4}{42.93} = 2.359, \text{ correct to four significant figures.}$$

$f(r_3) = (2.359)^4 - 11(2.359) - 5 \approx 0.018\,9$

$f'(r_3) = 4(2.359)^3 - 11 \approx 41.51$

The fourth approximation,

$$r_4 = 2.359 - \frac{0.018\,9}{41.51}$$

$$\approx 2.358\,5,$$

i.e. 2.359 correct to four significant figures.

Since the third and fourth approximations are equal when expressed to the stipulated degree of accuracy, then the required root is **2.359, correct to four significant figures.**

Problem 2. Use Newton's method to determine the root of the equation $3 \sin x + 4x - 5 = 0$, in the range $x = 0$ to $x = 1$, correct to four decimal places.

$f(x) = 3 \sin x + 4x - 5$

$f(0) = 3 \sin (0) + 4(0) - 5 = -5$

$f(1) = 3 \sin 1 + 4(1) - 5 \approx 1.52$

(Note, sin 1 means the sin of 1 radian)

Dividing the interval 0 to 1 roughly in the ratio of 5 to 1.5 gives a first approximation of $r_1 \approx 0.75$.

$f(x) = 3 \sin x + 4x - 5, \; f(r_1) = 3 \sin 0.75 + 4(0.75) - 5 \approx 0.044\,9$

$f'(x) = 3 \cos x + 4, \; f'(r_1) = 3 \cos 0.75 + 4 \approx 6.195$

Hence, a second approximation,

$$r_2 = 0.75 - \frac{0.044\,9}{6.195} = 0.742\,8, \text{ correct to four decimal places.}$$

$f(r_2) = 3 \sin 0.742\,8 + 4(0.742\,8) - 5 \approx 0.000\,259$

$f'(r_2) = 3 \cos 0.742\,8 + 4 \approx 6.210$

Hence, a third approximation,

$$r_3 = 0.7428 - \frac{0.000\,259}{6.210}$$

$= 0.7428$, correct to four decimal places.

Since the second and third approximations are equal when expressed to the stipulated degree of accuracy, then the required root is **0.742 8, correct to four decimal places**.
   The solution can be checked by using a calculator. When $x = 0.7428$, $\sin x = \sin (0.7428 \text{ radians}) = 0.6764$

Hence $3 \sin x + 4x - 5 = 2.0291 + 2.9712 - 5$
$$= 0.0003$$

i.e. it is very near to zero, hence the answer stated is probably correct to the required accuracy.

Problem 3. The approximate value of the root of the equation $0.95\,e^x - 35 \ln x - 3 = 0$ is $x = 4.0$. Determine the value of the root, correct to four significant figures.

Using functional notation to determine how close the root is to $x = 4.0$:

$f(4.0) = 0.95e^{4.0} - 35 \ln 4.0 - 3 \approx 0.348$
$f(4.1) = 0.95e^{4.1} - 35 \ln 4.1 - 3 \approx 4.94$

The results indicate that the root is slightly less than $x = 4.0$ and lies very near to it. Let the first approximation be, say, $r_1 = 3.99$

$f(x) = 0.95e^x - 35 \ln x - 3,$
$f(3.99) = 0.95e^{3.99} - 35 \ln 3.99 - 3 \approx -0.08$

$f'(x) = 0.95e^x - \frac{35}{x}, f'(3.99) = 0.95e^{3.99} - \frac{35}{3.99}$

$= 42.58$, correct to four significant figures.

Hence, a second approximation is

$$r_2 = 3.99 - \left(\frac{-0.08}{42.58}\right)$$

$= 3.992$, correct to four significant figures.

$f(3.992) = 0.95e^{3.992} - 35 \ln 3.992 - 3 \approx 0.00472$

$f'(3.992) = 0.95e^{3.992} - \frac{35}{3.992} \approx 42.69$

Hence, a third approximation,

$$r_3 = 3.992 - \frac{0.004\,72}{42.69}$$

= 3.992, corrrect to four significant figures.

Since the second and third approximations are the same when expressed to the stipulated degree of accuracy, the required root is **3.992, correct to four significant figures**.

Further problems on the Newton–Raphson method may be found in the following Section 2.4 (Problems 8 to 18), page 35.

2.4   Further problems

Algebraic method of successive approximation

Use an algebraic method of successive approximation to determine the value of the roots for the equations given in Problems 1 to 7.

1. $x^3 + x = 1$ in the range $x = 0$ to $x = 1$, correct to four decimal places.   [0.682 3]
2. $x^3 + 5x = 11$ in the range $x = 1$ to $x = 2$, correct to four significant figures.   [1.511]
3. The two positive roots of the equation

    $x^3 - 9x + 1 = 0$

    correct to four significant figures.   [0.111 3, 2.943]
4. The negative root of the equation

    $x^3 - 6x^2 + 12 = 0$

    correct to three decimal places.   [−1.284]
5. The root of the equation

    $x^4 - 3x^2 - 3x + 1 = 0$

    in the interval $x = 2$ to $x = 3$, correct to two decimal places. [2.06]
6. The two roots of the equation

    $x^4 - 12x = 5$

    correct to three significant figures.   [−0.414, 2.41]
7. The four roots of the equation

    $x^4 - 3x^3 - 4x^2 + 2x + 1 = 0$

    correct to three significant figures.
    [−1.19, −0.337, 0.644, 3.88]

The Newton–Raphson method

8. Determine the value of the root of the equation

   $2 \ln x + x = 2$

   correct to two decimal places.   [1.37]

9. Determine the two roots of the equation

   $x^2 = 4 \cos x$

   correct to three significant figures.   [± 1.20]

10. Determine the roots of the equation

    $5 - \theta = 5e^{-\theta}$

    correct to three significant figures.   [0, 4.97]

11. The value of the root of the equation

    $2 \tan x - 3 \ln 2x + 4e^{x/3} = 6.5$

    is roughly 1. Determine the value of this root, correct to three decimal places.   [0.978]

12. The motion of a particle in an electrostatic field is described by the equation

    $y = x^3 + 3x^2 + 5x - 28$

    When $x = 2$, y is approximately equal to 0. Determine the value of x when $y = 0$, correct to two decimal places, using Newton's method.   [1.93]

13. The velocity v mm/s of a point on an eccentric cam at a certain instant is given by $v = \frac{1}{3}x - \ln x$, where x is the displacement in mm. The velocity becomes zero when the displacement is between 4 mm and 5 mm. Determine this value of displacement for the velocity to be zero, correct to four significant figures.   [4.536 mm]

14. The amplitude of the ripple, y, on part of a waveform displayed on a C.R.O. at time t µs is of the form

    $y = (t - \pi/3) + \sin t$

    Determine the time when the amplitude of the ripple is zero, correct to four decimal places.   [0.536 3 µs]

15. The solution to a differential equation associated with the path taken by a projectile for which the resistance to motion is proportional to the velocity, is

    $y = 2.5(e^x - e^{-x}) + x - 25$

    Determine the value of x, correct to two decimal places, for which the value of y is zero.   [2.22]

16. An equation found in torsional oscillation problems is

$$\cot C = \frac{C}{2} - \frac{1}{2C}$$

Use Newton's method to determine the first positive root of the equation correct to four significant figures. [1.307]

17. The critical speeds of oscillation (λ) of a loaded beam are given by the equation $\lambda^3 - 3.250\,\lambda^2 + \lambda - 0.063 = 0$. Determine the value of λ which is approximately equal to 3.0, by Newton's method, correct to four decimal places. [2.914 3]

18. Use the Newton–Raphson method to determine the roots for the equations given in Problems 1 to 7.

Chapter 3

Partial fractions

3.1 Introduction

Consider the following addition of algebraic fractions:

$$\frac{1}{x+1} + \frac{1}{x+2} = \frac{(x+2)+(x+1)}{(x+1)(x+2)} = \frac{2x+3}{x^2+3x+2}$$

If we start with the expression $\dfrac{2x+3}{x^2+3x+2}$, and find the fractions whose sum gives this result, the two fractions obtained $\left(\text{i.e. } \dfrac{1}{x+1} \text{ and } \dfrac{1}{x+2}\right)$ are called the **partial fractions** of $\dfrac{2x+3}{x^2+3x+2}$

This process of expressing a fraction in terms of simpler fractions, called resolving into partial fractions, is used as a preliminary to integrating certain functions (see Chapter 19) and the techniques used are explained by example later in this chapter. However, before attempting to resolve an algebraic expression into partial fractions, the following points must be considered and appreciated:

(a) **The denominator of the algebraic expression must factorise.** In the above example the denominator $x^2 + 3x + 2$ factorises as $(x+1)(x+2)$.

(b) In the above example, the numerator, $2x + 3$, is said to be of degree one since the highest powered x term is x^1. The denominator, $x^2 + 3x + 2$, is said to be of degree two since the highest powered x term is x^2. **In order to resolve an algebraic expression into partial fractions, the numerator must be at least one degree less than the denominator.** When the degree of the numerator is equal to or higher than the degree of the denominator, the denominator must be divided into the numerator until the remainder is of lower degree than the denominator. For example, $\dfrac{x^2 + x - 5}{x^2 - 2x - 3}$ cannot be resolved into partial fractions as it stands, since the numerator and denominator are of the same degree. Dividing $x^2 + x - 5$ by $x^2 - 2x - 3$ gives:

$$x^2 - 2x - 3 \overline{\smash{)}\begin{array}{c} 1 \\ x^2 + x - 5 \\ \underline{x^2 - 2x - 3} \\ 3x - 2 \end{array}}$$

Thus: $\dfrac{x^2 + x - 5}{x^2 - 2x - 3} = 1 + \dfrac{3x - 2}{x^2 - 2x - 3}$

Since $x^2 - 2x - 3$ factorises as $(x + 1)(x - 3)$ then $\dfrac{3x - 2}{x^2 - 2x - 3}$ may be resolved into partial fractions (see Section 3.2).

(c) Given an identity such as:

$$5x^2 - 3x + 2 \equiv Ax^2 + Bx + C$$

(note: $\equiv$ means 'identically equal to'), then $A = 5$, $B = -3$ and $C = 2$, since the identity is true for all values of x and the coefficients of x^n (where $n = 0, 1, 2, \ldots$) on the left-hand side of the identity are equal to the coefficients of x^n on the right-hand side of the identity. Similarly, if $ax^3 + bx^2 - cx + d \equiv 2x^3 + 5x - 7$ then $a = 2$, $b = 0$, $c = -5$ and $d = -7$

3.2   Denominator containing linear factors

The corresponding partial fractions of an algebraic expression $\dfrac{f(x)}{(x - a)(x - b)}$ are of the form $\dfrac{A}{(x - a)} + \dfrac{B}{(x - b)}$, where $f(x)$ is a polynomial of degree less than 2. Similarly, the corresponding partial fractions of $\dfrac{f(x)}{(x + a)(x - b)(x - c)}$ are of the form $\dfrac{A}{(x + a)} + \dfrac{B}{(x - b)} + \dfrac{C}{(x - c)}$, where $f(x)$ is a polynomial of degree less than 3, and so on.

Problem 1. Resolve $\dfrac{x - 8}{x^2 - x - 2}$ into partial fractions.

The denominator factorises as $(x + 1)\,(x - 2)$ and the numerator is of less degree than the denominator. Thus $\dfrac{x - 8}{x^2 - x - 2}$ may be resolved into partial fractions.

Let $\dfrac{x - 8}{x^2 - x - 2} = \dfrac{x - 8}{(x + 1)(x - 2)} \equiv \dfrac{A}{x + 1} + \dfrac{B}{x - 2}$ where A and B are constants to be determined.

Adding the two fractions on the right-hand side gives:

$$\frac{x - 8}{(x + 1)(x - 2)} \equiv \frac{A(x - 2) + B(x + 1)}{(x + 1)(x - 2)}$$

Since the denominators are the same on each side of the identity then the numerators must be equal to each other.

Hence $x - 8 \equiv A(x - 2) + B(x + 1)$

There are two methods whereby A and B may be determined using the properties of identities introduced in Section 3.1.

Method 1

Since an identity is true for all real values of x, substitute into the identity a value of x to reduce one of the unknown constants to zero.

Let $x = 2$.   Then $2 - 8 = A(0) + B(3)$

               i.e. $-6 = 3B$

                       $B = -2$

Let $x = -1$.   Then $-1 - 8 = A(-3) + B(0)$

                       $-9 = -3A$

                       $A = 3$

Method 2

Since the coefficients of x^n ($n = 0, 1, 2, \ldots$) on the left-hand side of an identity equal the coefficients of x^n on the right-hand side, equate the respective coefficients on each side of the identity.

Since $x - 8 \equiv A(x - 2) + B(x + 1)$

then $x - 8 \equiv Ax - 2A + Bx + B = (A + B)x + (-2A + B)$

Thus $1 = A + B$ (by equating the coefficients of x)   (1)

and $-8 = -2A + B$ (by equating the constants)   (2)

Solving the two simultaneous equations gives $A = 3$ and $B = -2$ as before.

$$\text{Thus } \frac{x - 8}{x^2 - x - 2} \equiv \frac{3}{x + 1} + \frac{-2}{x - 2} \equiv \frac{3}{x + 1} - \frac{2}{x - 2}$$

It is usually quicker and easier to adopt the first method as far as possible although with other types of partial fractions a combination of the two methods is necessary.

Problem 2. Express $\dfrac{6x^2 + 7x - 25}{(x - 1)(x + 2)(x - 3)}$ in partial fractions.

Let $\dfrac{6x^2 + 7x - 25}{(x - 1)(x + 2)(x - 3)} \equiv \dfrac{A}{x - 1} + \dfrac{B}{x + 2} + \dfrac{C}{x - 3}$

$\equiv \dfrac{A(x + 2)(x - 3) + B(x - 1)(x - 3) + C(x - 1)(x + 2)}{(x - 1)(x + 2)(x - 3)}$

Equating the numerators gives:

$6x^2 + 7x - 25 \equiv A(x + 2)(x - 3) + B(x - 1)(x - 3)$
$+ C(x - 1)(x + 2)$

Let $x = 1$.

Then $6 + 7 - 25 = A(3)(-2) + B(0)(-2) + C(0)(3)$

i.e. $-12 = -6A$

$A = 2$

Let $x = -2$.

Then $6(-2)^2 + 7(-2) - 25 = A(0)(-5) + B(-3)(-5)$
$+ C(-3)(0)$

i.e. $-15 = 15B$

$B = -1$

Let $x = 3$.

Then $6(3)^2 + 7(3) - 25 = A(5)(0) + B(2)(0) + C(2)(5)$

i.e. $50 = 10C$

$C = 5$

Thus: $\dfrac{6x^2 + 7x - 25}{(x - 1)(x + 2)(x - 3)} \equiv \dfrac{2}{(x - 1)} - \dfrac{1}{(x + 2)} + \dfrac{5}{(x - 3)}$

Problem 3. Convert $\dfrac{x^3 - x^2 - 5x}{x^2 - 3x + 2}$ into partial fractions.

The numerator is of higher degree than the denominator, thus dividing gives:

$$
\begin{array}{r}
x + 2 \\
x^2 - 3x + 2 \overline{)x^3 - x^2 - 5x} \\
\underline{x^3 - 3x^2 + 2x} \\
2x^2 - 7x \\
\underline{2x^2 - 6x + 4} \\
-x - 4
\end{array}
$$

Thus

$$\frac{x^3 - x^2 - 5x}{x^2 - 3x + 2} = x + 2 + \frac{-x - 4}{x^2 - 3x + 2} = x + 2 - \frac{x + 4}{(x - 1)(x - 2)}$$

Let

$$\frac{x + 4}{(x - 1)(x - 2)} \equiv \frac{A}{x - 1} + \frac{B}{x - 2} \equiv \frac{A(x - 2) + B(x - 1)}{(x - 1)(x - 2)}$$

Equating the numerators gives:

$$x + 4 \equiv A(x - 2) + B(x - 1)$$

Let $x = 1$. Then $5 = -A$

$$A = -5$$

Let $x = 2$. Then $6 = B$

Thus $\dfrac{x + 4}{(x - 1)(x - 2)} \equiv \dfrac{-5}{(x - 1)} + \dfrac{6}{(x - 2)}$

Thus $\dfrac{x^3 - x^2 - 5x}{x^2 - 3x + 2} \equiv x + 2 - \dfrac{5}{(x - 1)} + \dfrac{6}{(x - 2)}$

$$\equiv x + 2 + \frac{5}{(x - 1)} - \frac{6}{(x - 2)}$$

3.3 Denominator containing repeated linear factors

When the denominator of an algebraic expression has a factor $(x - a)^n$ then the corresponding partial fractions are $\dfrac{A}{(x - a)} + \dfrac{B}{(x - a)^2} + \cdots + \dfrac{C}{(x - a)^n}$, since $(x - a)^n$ is assumed to hide the factors $(x - a)^{n-1}, (x - a)^{n-2} \ldots (x - a)$

Problem 1. Express $\dfrac{x + 5}{(x + 3)^2}$ in partial fractions.

Let $\dfrac{x + 5}{(x + 3)^2} \equiv \dfrac{A}{(x + 3)} + \dfrac{B}{(x + 3)^2} \equiv \dfrac{A(x + 3) + B}{(x + 3)^2}$

Equating the numerators gives:

$$x + 5 \equiv A(x + 3) + B$$

Let $x = -3$. Then $-3 + 5 = A(0) + B$

i.e. $B = 2$

Equating the coefficient of x gives $A = 1$.

(Check by equating constant terms gives: $5 = 3A + B$ which is true when $A = 1$ and $B = 2$.)

Thus $\dfrac{x + 5}{(x + 3)^2} \equiv \dfrac{1}{(x + 3)} + \dfrac{2}{(x + 3)^2}$

Problem 2. Resolve $\dfrac{5x^2 - 19x + 3}{(x - 2)^2(x + 1)}$ into partial fractions.

The given denominator is a combination of the linear factor type and repeated linear factor type.

Let $\dfrac{5x^2 - 19x + 3}{(x - 2)^2(x + 1)} \equiv \dfrac{A}{(x - 2)} + \dfrac{B}{(x - 2)^2} + \dfrac{C}{(x + 1)}$

$\equiv \dfrac{A(x - 2)(x + 1) + B(x + 1) + C(x - 2)^2}{(x - 2)^2(x + 1)}$

Equating the numerators gives:

$5x^2 - 19x + 3 \equiv A(x - 2)(x + 1) + B(x + 1) + C(x - 2)^2$

Let $x = 2$.   Then $5(2)^2 - 19(2) + 3 = A(0)(3) + B(3) + C(0)^2$

i.e. $-15 = 3B$

$$B = -5$$

Let $x = -1$. Then $5(-1)^2 - 19(-1) + 3 = A(-3)(0) + B(0)$
$+ C(-3)^2$

i.e. $27 = 9C$

$$C = 3$$

$5x^2 - 19x + 3 \equiv A(x - 2)(x + 1) + B(x + 1) + C(x - 2)^2$
$\equiv A(x^2 - x - 2) + B(x + 1) + C(x^2 - 4x + 4)$
$\equiv (A + C)x^2 + (-A + B - 4C)x$
$+ (-2A + B + 4C)$

Equating the coefficients of x^2 gives:

$5 = A + C$

Since $C = 3$ then $A = 2$

[Check: Equating the coefficients of x gives: $-19 = -A + B - 4C$
If $A = 2, B = -5$ and $C = 3$ then $-A + B - 4C = -19 = $ L.H.S.
Equating the constant terms gives: $3 = -2A + B + 4C$
If $A = 2, B = -5$ and $C = 3$ then
$-2A + B + 4C = 3 = $ L.H.S.]

Thus $\dfrac{5x^2 - 19x + 3}{(x-2)^2(x+1)} \equiv \dfrac{2}{(x-2)} - \dfrac{5}{(x-2)^2} + \dfrac{3}{(x+1)}$

Problem 3. Convert $\dfrac{2x^2 - 13x + 13}{(x-4)^3}$ into partial fractions.

Let $\dfrac{2x^2 - 13x + 13}{(x-4)^3} \equiv \dfrac{A}{(x-4)} + \dfrac{B}{(x-4)^2} + \dfrac{C}{(x-4)^3}$

$$\equiv \dfrac{A(x-4)^2 + B(x-4) + C}{(x-4)^3}$$

Equating the numerators gives:

$2x^2 - 13x + 13 \equiv A(x-4)^2 + B(x-4) + C$

Let $x = 4$. Then $2(4)^2 - 13(4) + 13 = A(0)^2 + B(0) + C$

i.e. $-7 = C$

Also
$2x^2 - 13x + 13 \equiv A(x^2 - 8x + 16) + B(x-4) + C$
$$\equiv Ax^2 + (-8A + B)x + (16A - 4B + C)$$

Equating the coefficients of x^2 gives: $\mathbf{2 = A}$

Equating the coefficients of x gives: $-13 = -8A + B$, from which $\mathbf{B = 3}$

(Check: Equating the constant terms gives: $13 = 16A - 4B + C$.
If $A = 2$, $B = 3$ and $C = -7$
then R.H.S. $= 16(2) - 4(3) - 7 = 13 =$ L.H.S.)

Thus: $\dfrac{2x^2 - 13x + 13}{(x-4)^3} \equiv \dfrac{2}{(x-4)} + \dfrac{3}{(x-4)^2} - \dfrac{7}{(x-4)^3}$

3.4   Denominator containing a quadratic factor

When the denominator contains a quadratic factor of the form $px^2 + qx + r$ (where p, q and r are constants), which does not factorise without containing surds or imaginary values, then the corresponding partial fraction is of the form $\dfrac{Ax + B}{px^2 + qx + r}$, i.e. the numerator is assumed to be a polynomial of degree one less than the denominator. Hence the corresponding partial fractions of an algebraic expression $\dfrac{f(x)}{(px^2 + qx + r)(x - a)}$ are of the form $\dfrac{Ax + B}{px^2 + qx + r} + \dfrac{C}{x - a}$

Problem 1. Resolve $\dfrac{8x^2 - 3x + 19}{(x^2 + 3)(x - 1)}$ into partial fractions.

Let $\dfrac{8x^2 - 3x + 19}{(x^2 + 3)(x - 1)} \equiv \dfrac{Ax + B}{(x^2 + 3)} + \dfrac{C}{(x - 1)}$

$$\equiv \dfrac{(Ax + B)(x - 1) + C(x^2 + 3)}{(x^2 + 3)(x - 1)}$$

Equating the numerators gives:

$8x^2 - 3x + 19 \equiv (Ax + B)(x - 1) + C(x^2 + 3)$

Let $x = 1$.   Then $8(1)^2 - 3(1) + 19 = (A + B)(0) + C(4)$

i.e. $24 = 4C$

$$C = 6$$

$8x^2 - 3x + 19 \equiv (Ax + B)(x - 1) + C(x^2 + 3)$
$$\equiv Ax^2 - Ax + Bx - B + Cx^2 + 3C$$
$$\equiv (A + C)x^2 + (-A + B)x + (-B + 3C)$$

Equating the coefficients of the x^2 terms gives:
$$8 = A + C$$
Since $C = 6$, $A = 2$

Equating the coefficients of the x terms gives:
$$-3 = -A + B$$
Since $A = 2$, $B = -1$

[Check: Equating the constant terms gives: $19 = -B + 3C$
R.H.S. $= -B + 3C = -(-1) + 3(6) = 19 =$ L.H.S.]

Hence: $\dfrac{8x^2 - 3x + 19}{(x^2 + 3)(x - 1)} \equiv \dfrac{2x - 1}{(x^2 + 3)} + \dfrac{6}{(x - 1)}$

Problem 2. Resolve $\dfrac{2 + x + 6x^2 - 2x^3}{x^2(x^2 + 1)}$ into partial fractions.

Terms such as x^2 may be treated as $(x + 0)^2$, i.e. it is a repeated linear factor type.

Let $\dfrac{2 + x + 6x^2 - 2x^3}{x^2(x^2 + 1)} \equiv \dfrac{A}{x} + \dfrac{B}{x^2} + \dfrac{Cx + D}{(x^2 + 1)}$

$$\equiv \dfrac{Ax(x^2 + 1) + B(x^2 + 1) + (Cx + D)x^2}{x^2(x^2 + 1)}$$

Equating the numerators gives:

$$2 + x + 6x^2 - 2x^3 \equiv Ax(x^2 + 1) + B(x^2 + 1) + (Cx + D)x^2$$
$$\equiv (A + C)x^3 + (B + D)x^2 + Ax + B$$

Let $x = 0$. Then $2 = B$

Equating the coefficients of x^3 gives:

$$-2 = A + C \qquad\qquad\qquad (1)$$

Equating the coefficients of x^2 gives:

$$6 = B + D$$

Since $B = 2, D = 4$

Equating the coefficients of x gives:

$$1 = A$$

From equation (1), $C = -3$

[Check: Equating the constant terms: $2 = B$ as before]

Hence $\dfrac{2 + x + 6x^2 - 2x^3}{x^2(x^2 + 1)} \equiv \dfrac{1}{x} + \dfrac{2}{x^2} + \dfrac{4 - 3x}{x^2 + 1}$

3.5   Summary

Denominator containing	Expression	Form of partial fractions
Linear factors	$\dfrac{f(x)}{(x + a)(x + b)(x + c)}$	$\dfrac{A}{(x + a)} + \dfrac{B}{(x + b)} + \dfrac{C}{(x + c)}$
Repeated linear factors	$\dfrac{f(x)}{(x - a)^3}$	$\dfrac{A}{(x - a)} + \dfrac{B}{(x - a)^2} + \dfrac{C}{(x - a)^3}$
Quadratic factors	$\dfrac{f(x)}{(ax^2 + bx + c)(x - d)}$	$\dfrac{Ax + B}{(ax^2 + bx + c)} + \dfrac{C}{(x - d)}$
General Example:	$\dfrac{f(x)}{(x^2 + a)(x + b)^2(x + c)}$	$\dfrac{Ax + B}{(x^2 + a)} + \dfrac{C}{(x + b)} + \dfrac{D}{(x + b)^2} + \dfrac{E}{(x + c)}$

In each of the above cases $f(x)$ must be of less degree than the relevant denominator. If it is not, then the denominator must be divided into the numerator. For every possible factor of the denominator there is a corresponding partial fraction.

3.6   Further problems

Resolve the following into partial fractions:

1. $\dfrac{8}{x^2 - 4}$ $\left[\dfrac{2}{(x-2)} - \dfrac{2}{(x+2)}\right]$

2. $\dfrac{3x + 5}{x^2 + 2x - 3}$ $\left[\dfrac{2}{(x-1)} + \dfrac{1}{(x+3)}\right]$

3. $\dfrac{y - 13}{y^2 - y - 6}$ $\left[\dfrac{3}{(y+2)} - \dfrac{2}{(y-3)}\right]$

4. $\dfrac{17x^2 - 21x - 6}{x(x+1)(x-3)}$ $\left[\dfrac{2}{x} + \dfrac{8}{(x+1)} + \dfrac{7}{(x-3)}\right]$

5. $\dfrac{6x^2 + 7x - 49}{(x-4)(x+1)(2x-3)}$ $\left[\dfrac{3}{(x-4)} - \dfrac{2}{(x+1)} + \dfrac{4}{(2x-3)}\right]$

6. $\dfrac{x^2 + 2}{(x+4)(x-2)}$ $\left[1 - \dfrac{3}{(x+4)} + \dfrac{1}{(x-2)}\right]$

7. $\dfrac{2x^2 + 4x + 19}{2(x-3)(x+4)}$ $\left[1 + \dfrac{7}{2(x-3)} - \dfrac{5}{2(x+4)}\right]$

8. $\dfrac{2x^3 + 7x^2 - 2x - 27}{(x-1)(x+4)}$ $\left[2x + 1 - \dfrac{4}{(x-1)} + \dfrac{7}{(x+4)}\right]$

9. $\dfrac{2t - 1}{(t+1)^2}$ $\left[\dfrac{2}{(t+1)} - \dfrac{3}{(t+1)^2}\right]$

10. $\dfrac{8x^2 + 12x - 3}{(x+2)^3}$ $\left[\dfrac{8}{(x+2)} - \dfrac{20}{(x+2)^2} + \dfrac{5}{(x+2)^3}\right]$

11. $\dfrac{6x + 1}{(2x+1)^2}$ $\left[\dfrac{3}{(2x+1)} - \dfrac{2}{(2x+1)^2}\right]$

12. $\dfrac{1}{x^2(x+2)}$ $\left[\dfrac{1}{2x^2} - \dfrac{1}{4x} + \dfrac{1}{4(x+2)}\right]$

13. $\dfrac{9x^2 - 73x + 150}{(x-7)(x-3)^2}$ $\left[\dfrac{5}{(x-7)} + \dfrac{4}{(x-3)} - \dfrac{3}{(x-3)^2}\right]$

14. $\dfrac{-(9x^2 + 4x + 4)}{x^2(x^2 - 4)}$ $\left[\dfrac{1}{x} + \dfrac{1}{x^2} + \dfrac{2}{(x+2)} - \dfrac{3}{(x-2)}\right]$

15. $\dfrac{-(a^2 + 5a + 13)}{(a^2 + 5)(a - 2)}$ $\left[\dfrac{2a - 1}{(a^2 + 5)} - \dfrac{3}{(a - 2)}\right]$

16. $\dfrac{3 - x}{(x^2 + 3)(x + 3)}$ $\left[\dfrac{1 - x}{2(x^2 + 3)} + \dfrac{1}{2(x + 3)}\right]$

17. $\dfrac{12 - 2x - 5x^2}{(x^2 + x + 1)(3 - x)}$ $\left[\dfrac{2x + 5}{(x^2 + x + 1)} - \dfrac{3}{(3 - x)}\right]$

18. $\dfrac{x^3 + 7x^2 + 8x + 10}{x(x^2 + 2x + 5)}$ $\left[1 + \dfrac{2}{x} + \dfrac{3x - 1}{(x^2 + 2x + 5)}\right]$

19. $\dfrac{5x^3 - 3x^2 + 41x - 64}{(x^2 + 6)(x - 1)^2}$ $\left[\dfrac{2 - 3x}{(x^2 + 6)} + \dfrac{8}{(x - 1)} - \dfrac{3}{(x - 1)^2}\right]$

20. $\dfrac{6x^3 + 5x^2 + 4x + 3}{(x^2 + x + 1)(x^2 - 1)}$ $\left[\dfrac{2x - 1}{(x^2 + x + 1)} + \dfrac{3}{(x - 1)} + \dfrac{1}{(x + 1)}\right]$

Chapter 4

Matrix arithmetic and the determinant of a matrix

4.1 Introduction

Second order matrices are introduced in *Technician mathematics 3*. In this chapter, the arithmetic of matrices is extended to third order matrices, i.e. 3 by 3 matrices, and the arithmetic operations of addition, subtraction, multiplication and inversion are introduced.

The terms used in connection with third order matrices are the same as those used for second order matrices. For the linear simultaneous equations, say,

$$2x + 3y + 4z = 7 \tag{1}$$

$$6x - 4y + 5z = 8 \tag{2}$$

$$7x + 5y - 6z = 9 \tag{3}$$

the coefficients of x, y and z are written in matrix notation as

$$\begin{pmatrix} 2 & 3 & 4 \\ 6 & -4 & 5 \\ 7 & 5 & -6 \end{pmatrix},$$

that is, occupying the same relative positions as in equations (1), (2) and (3). The grouping of the coefficients of x, y and z in this way is called an **array** and the coefficients forming the array are called the **elements** of the matrix.

If there are m rows across an array and n columns down an array, the matrix is said to be of order 'm by n'. A matrix having a single row is called a **row matrix** and one having a single column is called a **column matrix**. For example, in equation (1) above the coefficients of x, y and z form a row matrix of (2 3 4) and the coefficients of x in equations (1), (2) and (3) form a column matrix of $\begin{pmatrix} 2 \\ 6 \\ 7 \end{pmatrix}$

A matrix having the same number of rows as columns is called a **square matrix** and a square matrix having three rows and three columns is called a third order matrix. Matrices are generally denoted by capital letters and if the matrix representing the coefficients of x, y and z in equations (1), (2)

48

and (3) above is A, then

$$A = \begin{pmatrix} 2 & 3 & 4 \\ 6 & -4 & 5 \\ 7 & 5 & -6 \end{pmatrix}$$

To refer to a particular element in a matrix, the notation a_{ij} is used, where i refers to the row number and j the column number. Thus in matrix A above, the element a_{22} is -4 and element a_{31} is 7, and so on.

4.2 Addition, subtraction and multiplication of third order matrices

A matrix does not have a single numerical value and cannot be simplified to a particular numerical answer. However, by applying the laws of matrices given in this section, they can be simplified and often the number of matrices can be reduced, for example, by adding two matrices, a single matrix is produced. Also, by comparing one matrix with another similar matrix, values of unknown elements can be determined. Matrices are of very little value in themselves but an understanding of the laws of matrices can lead to a better understanding of determinants and also enables determinants to be simplified before evaluation, saving the number of arithmetic operations and reducing their size.

Addition

As with second order matrices, only matrices having the same number of rows and columns may be added. The sum of two matrices is the matrix obtained by adding the elements occupying corresponding positions in the two matrices. When two matrices are added, the result is a single matrix of the same order as those being added. Thus,

$$\begin{pmatrix} p & q & r \\ s & t & u \\ v & w & x \end{pmatrix} + \begin{pmatrix} a & b & c \\ d & e & f \\ g & h & i \end{pmatrix} = \begin{pmatrix} p+a & q+b & r+c \\ s+d & t+e & u+f \\ v+g & w+h & x+i \end{pmatrix}$$

This principle may be applied to more than two m by n matrices being added, the resulting matrix being a single m by n matrix.

Subtraction

As with second order matrices, only matrices having the same number of rows and columns may be subtracted. The difference between two matrices, say $A - B$, is the matrix obtained by subtracting the elements of matrix B

from those occupying the corresponding position in matrix A. Thus,

$$\begin{pmatrix} p & q & r \\ s & t & u \\ v & w & x \end{pmatrix} - \begin{pmatrix} a & b & c \\ d & e & f \\ g & h & i \end{pmatrix} = \begin{pmatrix} p-a & q-b & r-c \\ s-d & t-e & u-f \\ v-g & w-h & x-i \end{pmatrix}$$

This principle can be applied to the difference of two m by n matrices, the resulting matrix being a single m by n matrix.

Multiplication

(a) Scalar multiplication

When a matrix is multiplied by a number, the resulting matrix is one of the same order, having each element multiplied by the number. Thus if,

$$A = \begin{pmatrix} p & q & r \\ s & t & u \\ v & w & x \end{pmatrix}, \text{ then } kA = \begin{pmatrix} kp & kq & kr \\ ks & kt & ku \\ kv & kw & kx \end{pmatrix}$$

where k is a constant.

(b) Multiplication of matrices

When a 3 by 3 matrix, say A, is multiplied by a 3 by 3 matrix, say B, the resulting matrix is a 3 by 3 matrix, say C. The a_{11} element of C is the sum of the products of the first row of A and the first column of B, taken element by element. Thus, if $A \times B = C$

$$\begin{pmatrix} p & q & r \\ - & - & - \\ - & - & - \end{pmatrix} \times \begin{pmatrix} a & - & - \\ d & - & - \\ g & - & - \end{pmatrix} = \begin{pmatrix} pa+qd+rg & - & - \\ - & & - & - \\ - & & - & - \end{pmatrix}$$

The a_{12} element of C is the sum of the products obtained by taking the first row of A with the second column of B, element by element. Thus

$$\begin{pmatrix} p & q & r \\ - & - & - \\ - & - & - \end{pmatrix} \times \begin{pmatrix} - & b & - \\ - & e & - \\ - & h & - \end{pmatrix} = \begin{pmatrix} - & pb+qe+rh & - \\ - & & - & - \\ - & & - & - \end{pmatrix}$$

The remaining elements of C are calculated in a similar way, giving

$$\begin{pmatrix} p & q & r \\ s & t & u \\ v & w & x \end{pmatrix} \times \begin{pmatrix} a & b & c \\ d & e & f \\ g & h & i \end{pmatrix}$$

$$= \begin{pmatrix} (pa+qd+rg) & (pb+qe+rh) & (pc+qf+ri) \\ (sa+td+ug) & (sb+te+uh) & (sc+tf+ui) \\ (va+wd+xg) & (vb+we+xh) & (vc+wf+xi) \end{pmatrix}$$

Although the laws of matrices are so formulated that they follow most of the laws which govern the algebra of numbers, frequently in the multiplication of two matrices $A \times B$ is **not** equal to $B \times A$. This is shown in worked Problem 4 following.

Worked problems on addition, subtraction and multiplication of third order matrices

In Problems 1 to 4 following, matrices A, B and C refer to:

$$A = \begin{pmatrix} 2 & 3 & -1 \\ -2 & 0 & 4 \\ 3 & 1 & 5 \end{pmatrix} \quad B = \begin{pmatrix} 4 & 3 & 0 \\ 5 & -1 & 2 \\ 5 & 3 & 4 \end{pmatrix} \text{ and } C = \begin{pmatrix} 1 & -3 & 1 \\ -5 & 7 & 4 \\ 7 & 0 & 6 \end{pmatrix}$$

Problem 1. Determine the matrix D, where $D = A + B$.

$$A + B = \begin{pmatrix} 2 & 3 & -1 \\ -2 & 0 & 4 \\ 3 & 1 & 5 \end{pmatrix} + \begin{pmatrix} 4 & 3 & 0 \\ 5 & -1 & 2 \\ 5 & 3 & 4 \end{pmatrix}$$

$$= \begin{pmatrix} 2+4 & 3+3 & -1+0 \\ -2+5 & 0+(-1) & 4+2 \\ 3+5 & 1+3 & 5+4 \end{pmatrix}$$

$$D = \begin{pmatrix} 6 & 6 & -1 \\ 3 & -1 & 6 \\ 8 & 4 & 9 \end{pmatrix}$$

Problem 2. Determine the matrix E, where $E = B - A$.

$$B - A = \begin{pmatrix} 4 & 3 & 0 \\ 5 & -1 & 2 \\ 5 & 3 & 4 \end{pmatrix} - \begin{pmatrix} 2 & 3 & -1 \\ -2 & 0 & 4 \\ 3 & 1 & 5 \end{pmatrix}$$

$$= \begin{pmatrix} [4-2] & [3-3] & [0-(-1)] \\ [5-(-2)] & [-1-0] & [2-4] \\ [5-3] & [3-1] & [4-5] \end{pmatrix}$$

$$E = \begin{pmatrix} 2 & 0 & 1 \\ 7 & -1 & -2 \\ 2 & 2 & -1 \end{pmatrix}$$

Problem 3. Determine matrix F, where $F = A - B + C$.

This problem may be done in two parts, in which case the single matrix representing $A - B$ can be determined as shown in Problem 2 above, and then this single matrix can be added to matrix C, as shown in Problem 1 above. Alternatively, it may be done in one step, the elements of F being obtained from the corresponding elements of A, B and C as shown below.

$$A - B + C = \begin{pmatrix} 2 & 3 & -1 \\ -2 & 0 & 4 \\ 3 & 1 & 5 \end{pmatrix} - \begin{pmatrix} 4 & 3 & 0 \\ 5 & -1 & 2 \\ 5 & 3 & 4 \end{pmatrix} + \begin{pmatrix} 1 & -3 & 1 \\ -5 & 7 & 4 \\ 7 & 0 & 6 \end{pmatrix}$$

$$= \begin{pmatrix} [2-4+1] & [3-3+(-3)] & [-1-0+1] \\ [-2-5+(-5)] & [0-(-1)+7] & [4-2+4] \\ [3-5+7] & [1-3+0] & [5-4+6] \end{pmatrix}$$

$$F = \begin{pmatrix} -1 & -3 & 0 \\ -12 & 8 & 6 \\ 5 & -2 & 7 \end{pmatrix}$$

Problem 4. (i) Determine matrix G, where $G = A \times B$
(ii) Show that $A \times B$ is not equal to $B \times A$, in this case.

(i)   The element g_{11} is obtained from the sum of the products of the first row of A and the first column of B, taken element by element. Thus $g_{11} = a_{11} \times b_{11} + a_{12} \times b_{21} + a_{13} \times b_{31}$

Since $A = \begin{pmatrix} 2 & 3 & -1 \\ -2 & 0 & 4 \\ 3 & 1 & 5 \end{pmatrix}$ and $B = \begin{pmatrix} 4 & 3 & 0 \\ 5 & -1 & 2 \\ 5 & 3 & 4 \end{pmatrix}$

then $g_{11} = 2 \times 4 + 3 \times 5 + (-1) \times 5$
$= 8 + 15 - 5 = 18$

Similarly, $g_{12} = a_{11} \times b_{12} + a_{12} \times b_{22} + a_{13} \times b_{32}$
$= 2 \times 3 + 3 \times (-1) + (-1) \times 3$
$= 6 - 3 - 3 = 0$

The remaining seven elements are found in a similar way, giving:

$$A \times B = \begin{pmatrix} [18] & [0] & [2 \times 0 + 3 \times 2 + (-1) \times 4] \\ [-2 \times 4 + 0 \times 5 + 4 \times 5] & [-2 \times 3 + 0 \times (-1) + 4 \times 3] & [-2 \times 0 + 0 \times 2 + 4 \times 4] \\ [3 \times 4 + 1 \times 5 + 5 \times 5] & [3 \times 3 + 1 \times (-1) + 5 \times 3] & [3 \times 0 + 1 \times 2 + 5 \times 4] \end{pmatrix}$$

$$G = A \times B = \begin{pmatrix} 18 & 0 & 2 \\ 12 & 6 & 16 \\ 42 & 23 & 22 \end{pmatrix}$$

(ii) $B \times A = \begin{pmatrix} 4 & 3 & 0 \\ 5 & -1 & 2 \\ 5 & 3 & 4 \end{pmatrix} \times \begin{pmatrix} 2 & 3 & -1 \\ -2 & 0 & 4 \\ 3 & 1 & 5 \end{pmatrix}$

$$= \begin{pmatrix} [4 \times 2 + 3 \times (-2) + 0 \times 3] \\ [5 \times 2 + (-1) \times (-2) + 2 \times 3] \\ [5 \times 2 + 3 \times (-2) + 4 \times 3] \end{pmatrix}$$

$$\begin{array}{c} [4 \times 3 + 3 \times 0 + 0 \times 1] \\ [5 \times 3 + (-1) \times 0 + 2 \times 1] \\ [5 \times 3 + 3 \times 0 + 4 \times 1] \end{array}$$

$$\left. \begin{array}{c} [4 \times (-1) + 3 \times 4 + 0 \times 5] \\ [5 \times (-1) + (-1) \times 4 + 2 \times 5] \\ [5 \times (-1) + 3 \times 4 + 4 \times 5] \end{array} \right)$$

i.e. $B \times A = \begin{pmatrix} 2 & 12 & 8 \\ 18 & 17 & 1 \\ 16 & 19 & 27 \end{pmatrix}$

Since the matrix obtained for $A \times B$ is different from that obtained for $B \times A$, this shows that in this case $A \times B \neq B \times A$.

Problem 5. If $P = \begin{pmatrix} 2 & 3 & -1 & 4 \\ 4 & 2 & 7 & 8 \\ 3 & -2 & 10 & 0 \end{pmatrix}$ and $Q = \begin{pmatrix} 7 \\ 4 \\ 3 \\ 1 \end{pmatrix}$,

find matrix R, where $R = P \times Q$.

The previous problem dealt with multiplication of a square, 3×3 matrix. However, the principles introduced can be extended to any m by n matrix, provided the number of elements in the rows of matrix P are equal to the number of elements in the columns of matrix Q. Using the same principle as that used in Problem 4, gives:

$$R = P \times Q = \begin{pmatrix} 2 & 3 & -1 & 4 \\ 4 & 2 & 7 & 8 \\ 3 & -2 & 10 & 0 \end{pmatrix} \times \begin{pmatrix} 7 \\ 4 \\ 3 \\ 1 \end{pmatrix}$$

$$= \begin{pmatrix} 2 \times 7 + 3 \times 4 + (-1) \times 3 + 4 \times 1 \\ 4 \times 7 + 2 \times 4 + 7 \times 3 + 8 \times 1 \\ 3 \times 7 + (-2) \times 4 + 10 \times 3 + 0 \times 1 \end{pmatrix}$$

i.e. $R = \begin{pmatrix} 14 + 12 - 3 + 4 \\ 28 + 8 + 21 + 8 \\ 21 - 8 + 30 + 0 \end{pmatrix} = \begin{pmatrix} 27 \\ 65 \\ 43 \end{pmatrix}$

In general, when multiplying a matrix of dimension (m by n) by a matrix of dimension (n by q), the resulting matrix has a dimension of (m by q), i.e. (**rows** of first × **columns** of second). In this problem, matrix P has 3 **rows** and matrix Q has 1 **column**, hence the matrix resulting from the multiplication of these two must be a matrix having 3 rows and 1 column.

Further problems on addition, subtraction and multiplication of third order matrices may be found in Section 4.5 (Problems 1 to 8), page 61.

4.3   The determinant of a 3 by 3 matrix

For a 2 by 2 matrix, say $\begin{pmatrix} a & b \\ c & d \end{pmatrix}$, the quantity $ad - bc$ is called the **determinant** of a matrix.

The determinant of a matrix differs from the matrix itself in that it can be evaluated to a single numerical result. To distinguish it from the matrix, it is written as $\begin{vmatrix} a & b \\ c & d \end{vmatrix}$, vertical lines being used instead of brackets.

For a 3 by 3 matrix, say $\begin{pmatrix} a_1 & b_1 & c_1 \\ a_2 & b_2 & c_2 \\ a_3 & b_3 & c_3 \end{pmatrix}$, the **minor** of element a_1 is the determinant obtained if the row and column containing element a_1 are considered to be covered up, i.e. row $(a_1\, b_1\, c_1)$ and column $\begin{pmatrix} a_1 \\ a_2 \\ a_3 \end{pmatrix}$ are covered up, i.e. the minor of element a_1 is found from matrix $\begin{pmatrix} \cancel{a_1} & \cancel{b_1} & \cancel{c_1} \\ \cancel{a_2} & b_2 & c_2 \\ \cancel{a_3} & b_3 & c_3 \end{pmatrix}$ and is $\begin{vmatrix} b_2 & c_2 \\ b_3 & c_3 \end{vmatrix}$

Similarly, to find the minor of element, say c_2, row $(a_2\, b_2\, c_2)$ and column $\begin{pmatrix} c_1 \\ c_2 \\ c_3 \end{pmatrix}$ are considered to be covered, leaving exposed the determinant $\begin{vmatrix} a_1 & b_1 \\ a_3 & b_3 \end{vmatrix}$. It follows that each of the nine elements of a 3 by 3 matrix has its own minor.

The sign of the minor of an element depends on the position of the element within the matrix. Each minor has a $+$ or a $-$ sign, the signs being such that they are alternatively $+$ then $-$ in both rows and columns, element a_{11} always being $+$. Thus for a 3 by 3 matrix, the sign pattern is:

$$\begin{pmatrix} + & - & + \\ - & + & - \\ + & - & + \end{pmatrix}$$

It follows that the sign of the minor of element, say c_1 is $+$, but the sign of the element, say b_3, is $-$.

A signed-minor is called the **cofactor** of an element. Thus

the cofactor of element b_2, $B_2 = +\begin{vmatrix} a_1 & c_1 \\ a_3 & c_3 \end{vmatrix} = a_1c_3 - a_3c_1$, and

the cofactor of element b_3, $B_3 = -\begin{vmatrix} a_1 & c_1 \\ a_2 & c_2 \end{vmatrix} = -(a_1c_2 - a_2c_1)$

$$= a_2c_1 - a_1c_2$$

The value of a 3 by 3 determinant for matrix A is written as $|A|$ and is given by the sum of the products of elements and their cofactors of **any** row or **any** column. Thus, for the matrix

$$A = \begin{pmatrix} a_1 & b_1 & c_1 \\ a_2 & b_2 & c_2 \\ a_3 & b_3 & c_3 \end{pmatrix}$$

the value of the determinant $|A|$ is given by any of the six expressions:

$$|A| = a_1A_1 + b_1B_1 + c_1C_1 \quad \text{or} \quad a_2A_2 + b_2B_2 + c_2C_2$$

or

$$a_3A_3 + b_3B_3 + c_3C_3 \quad \text{or} \quad a_1A_1 + a_2A_2 + a_3A_3$$

or

$$b_1B_1 + b_2B_2 + b_3B_3 \quad \text{or} \quad c_1C_1 + c_2C_2 + c_3C_3$$

Using, say, the first row elements,

$$|A| = a_1\begin{vmatrix} b_2 & c_2 \\ b_3 & c_3 \end{vmatrix} - b_1\begin{vmatrix} a_2 & c_2 \\ a_3 & c_3 \end{vmatrix} + c_1\begin{vmatrix} a_2 & b_2 \\ a_3 & b_3 \end{vmatrix}$$

$$= a_1(b_2c_3 - b_3c_2) - b_1(a_2c_3 - a_3c_2) + c_1(a_2b_3 - a_3b_2)$$

$$= a_1b_2c_3 - a_1b_3c_2 - a_2b_1c_3 + a_3b_1c_2 + a_2b_3c_1 - a_3b_2c_1$$

Using, say, the first column elements,

$$|A| = a_1\begin{vmatrix} b_2 & c_2 \\ b_3 & c_3 \end{vmatrix} - a_2\begin{vmatrix} b_1 & c_1 \\ b_3 & c_3 \end{vmatrix} + a_3\begin{vmatrix} b_1 & c_1 \\ b_2 & c_2 \end{vmatrix}$$

$$= a_1(b_2c_3 - b_3c_2) - a_2(b_1c_3 - b_3c_1) + a_3(b_1c_2 - b_2c_1)$$

$$= a_1b_2c_3 - a_1b_3c_2 - a_2b_1c_3 + a_2b_3c_1 + a_3b_1c_2 - a_3b_2c_1$$

This result is the same as that obtained by using the first row elements, and any of the six expressions for the value of a determinant given above will give the same result. When finding the value of a determinant containing one or more 0's as elements, it is usually easier to select the row or column containing the most 0's as the basis of the expansion, since the cofactor of a zero element is zero and need not be calculated.

Worked problems on the determinant of a 3 by 3 matrix

Problem 1. Determine the value of (a) the minors of elements -1, 0 and 5, and (b) the cofactors of elements 7, 3 and -2 for the matrix

$$\begin{pmatrix} 2 & -1 & 7 \\ 3 & 4 & -2 \\ 0 & 8 & 5 \end{pmatrix}$$

(a)   The minor of the element -1 is obtained by writing the 2 by 2 determinant left when the row and column containing the element -1 are covered up (i.e. the first row and the second column).
Thus the value of the minor of the element

$$-1 = \begin{vmatrix} 3 & -2 \\ 0 & 5 \end{vmatrix} = 3 \times 5 - 0 \times (-2)$$

$$= \mathbf{15}$$

Similarly, the value of the minor of element

$$0 = \begin{vmatrix} -1 & 7 \\ 4 & -2 \end{vmatrix} = (-1) \times (-2) - 4 \times 7$$

$$= \mathbf{-26}$$

The value of the minor of element

$$5 = \begin{vmatrix} 2 & -1 \\ 3 & 4 \end{vmatrix} = 2 \times 4 - 3 \times (-1)$$

$$= \mathbf{11}$$

(b)   Cofactors are signed-minors, the signs following the pattern
$$\begin{pmatrix} + & - & + \\ - & + & - \\ + & - & + \end{pmatrix}. \text{ For the element 7,}$$
the cofactor $= +\begin{vmatrix} 3 & 4 \\ 0 & 8 \end{vmatrix} = 3 \times 8 - 0 \times 4$

$$= \mathbf{24}$$

For the element 3,
the cofactor $= -\begin{vmatrix} -1 & 7 \\ 8 & 5 \end{vmatrix} = -((-1) \times 5 - 8 \times 7)$

$$= \mathbf{61}$$

For the element -2,

$$\text{the cofactor} = -\begin{vmatrix} 2 & -1 \\ 0 & 8 \end{vmatrix} = -(2 \times 8 - 0 \times (-1))$$

$$= -16$$

Problem 2. Find the value of the determinant $|A|$ of the matrix

$$A = \begin{pmatrix} 3 & 2 & -5 \\ 0 & 1 & 7 \\ 4 & 5 & 2 \end{pmatrix}$$

The value of the determinant is the sum of the products of the elements and their cofactors of any row or any column. Since element a_{21} is 0, the arithmetic is kept to a minimum by selecting either the second row or the first column as the basis for expansion. Selecting, say, the first column, i.e. $|A| = a_1 A_1 + a_2 A_2 + a_3 A_3$, gives

$$|A| = \begin{vmatrix} 3 & 2 & -5 \\ 0 & 1 & 7 \\ 4 & 5 & 2 \end{vmatrix} = 3\begin{vmatrix} 1 & 7 \\ 5 & 2 \end{vmatrix} - 0\begin{vmatrix} 2 & -5 \\ 5 & 2 \end{vmatrix} + 4\begin{vmatrix} 2 & -5 \\ 1 & 7 \end{vmatrix}$$

$$= 3(1 \times 2 - 5 \times 7) - (0) + 4(2 \times 7 - 1 \times (-5))$$

$$= 3(-33) + 4(19)$$

$$= -23$$

Further problems on the determinant of a 3 by 3 matrix may be found in Section 4.5 (Problems 9 to 16), page 62.

4.4   The inverse or reciprocal of a matrix

To determine the inverse or reciprocal of a matrix, two new operations associated with matrices are introduced below. These refer to finding the transpose of a matrix and to finding the adjoint of a matrix.

The **transpose** of a matrix A, usually denoted by A^T, is obtained by writing the rows of matrix A as the columns of matrix A^T. Thus if

$$A = \begin{pmatrix} a_1 & b_1 & c_1 \\ a_2 & b_2 & c_2 \\ a_3 & b_3 & c_3 \end{pmatrix}, \text{ then } A^T = \begin{pmatrix} a_1 & a_2 & a_3 \\ b_1 & b_2 & b_3 \\ c_1 & c_2 & c_3 \end{pmatrix}$$

It can be seen that the elements on the leading diagonal, $(\backslash)$, of A and A^T are the same.

The algebraic expression for the determinant of A, using the first row as the basis for the expansion, is:

$$|A| = a_1 \begin{vmatrix} b_2 & c_2 \\ b_3 & c_3 \end{vmatrix} - b_1 \begin{vmatrix} a_2 & c_2 \\ a_3 & c_3 \end{vmatrix} + c_1 \begin{vmatrix} a_2 & b_2 \\ a_3 & b_3 \end{vmatrix}$$

$$= a_1 b_2 c_3 - a_1 b_3 c_2 - a_2 b_1 c_3 + a_3 b_1 c_2 + a_2 b_3 c_1 - a_3 b_2 c_1$$

and for A^T, using the first column as the basis for the expansion, is:

$$|A^T| = a_1 \begin{vmatrix} b_2 & b_3 \\ c_2 & c_3 \end{vmatrix} - b_1 \begin{vmatrix} a_2 & a_3 \\ c_2 & c_3 \end{vmatrix} + c_1 \begin{vmatrix} a_2 & a_3 \\ b_2 & b_3 \end{vmatrix}$$

$$= a_1 b_2 c_3 - a_1 b_3 c_2 - a_2 b_1 c_3 + a_3 b_1 c_2 + a_2 b_3 c_1 - a_3 b_2 c_1$$

Since the algebraic equations for $|A|$ and $|A^T|$ are the same, it follows that:

$$|A| = |A^T|$$

The **adjoint** matrix of matrix A, usually abbreviated to 'adj A', is obtained by:

(i)   determining the nine cofactors of matrix A,
(ii) forming a matrix of these cofactors, and
(iii) transposing the matrix formed in (ii) above

Thus, to determine the adjoint matrix of matrix A, where $A = \begin{pmatrix} 1 & 2 & 1 \\ 2 & 0 & 2 \\ 1 & 1 & 0 \end{pmatrix}$, and following the steps outlined above, gives:

(i)   the cofactors of the first row are:

$$+\begin{vmatrix} 0 & 2 \\ 1 & 0 \end{vmatrix} = -2, \quad -\begin{vmatrix} 2 & 2 \\ 1 & 0 \end{vmatrix} = 2 \text{ and } +\begin{vmatrix} 2 & 0 \\ 1 & 1 \end{vmatrix} = 2$$

the cofactors of the second row are:

$$-\begin{vmatrix} 2 & 1 \\ 1 & 0 \end{vmatrix} = 1, \quad +\begin{vmatrix} 1 & 1 \\ 1 & 0 \end{vmatrix} = -1 \text{ and } -\begin{vmatrix} 1 & 2 \\ 1 & 1 \end{vmatrix} = 1$$

and the cofactors of the third row are:

$$+\begin{vmatrix} 2 & 1 \\ 0 & 2 \end{vmatrix} = 4, \quad -\begin{vmatrix} 1 & 1 \\ 2 & 2 \end{vmatrix} = 0 \text{ and } +\begin{vmatrix} 1 & 2 \\ 2 & 0 \end{vmatrix} = -4$$

(ii) the matrix of these cofactors is $\begin{pmatrix} -2 & 2 & 2 \\ 1 & -1 & 1 \\ 4 & 0 & -4 \end{pmatrix}$

(iii) transposing the matrix of cofactors, gives

$$\text{adj } A = \begin{pmatrix} -2 & 1 & 4 \\ 2 & -1 & 0 \\ 2 & 1 & -4 \end{pmatrix}$$

The **unit matrix** I for a 2 by 2 matrix has been previously defined in *Technician mathematics 3* and is the matrix having elements of value 1 on its leading diagonal, all other elements being 0. It is analogous to the number 1 in ordinary algebra. For a 3 by 3 matrix, the unit matrix is

$$I = \begin{pmatrix} 1 & 0 & 0 \\ 0 & 1 & 0 \\ 0 & 0 & 1 \end{pmatrix}$$

It has also been shown previously that the inverse or reciprocal of matrix A is the matrix A^{-1}, such that

$$A \cdot A^{-1} = I$$

i.e. $$A^{-1} = \frac{I}{A} \tag{4}$$

The inverse of matrix A can also be expressed in terms of the adjoint of matrix A and the determinant of matrix A, the relationship being:

$$A^{-1} = \frac{\text{adj } A}{|A|} \tag{5}$$

Equation (5) may be verified for matrix A, where A is, say,

$$\begin{pmatrix} 1 & 2 & 1 \\ 2 & 0 & 2 \\ 1 & 1 & 0 \end{pmatrix}, \text{ as follows.}$$

It is shown above that $\text{adj } A = \begin{pmatrix} -2 & 1 & 4 \\ 2 & -1 & 0 \\ 2 & 1 & -4 \end{pmatrix}$

The product of A and adj A is

$$A \times \text{adj } A = \begin{pmatrix} 1 & 2 & 1 \\ 2 & 0 & 2 \\ 1 & 1 & 0 \end{pmatrix} \times \begin{pmatrix} -2 & 1 & 4 \\ 2 & -1 & 0 \\ 2 & 1 & -4 \end{pmatrix}$$

$$= \begin{pmatrix} 4 & 0 & 0 \\ 0 & 4 & 0 \\ 0 & 0 & 4 \end{pmatrix} = 4 \begin{pmatrix} 1 & 0 & 0 \\ 0 & 1 & 0 \\ 0 & 0 & 1 \end{pmatrix}$$

(Extracting a common factor is dealt with in Chapter 5 following.)

i.e. $A \times \text{adj } A = 4I$

Also, the determinant of A, $|A| = 1\begin{vmatrix} 0 & 2 \\ 1 & 0 \end{vmatrix} - 2\begin{vmatrix} 2 & 2 \\ 1 & 0 \end{vmatrix} + 1\begin{vmatrix} 2 & 0 \\ 1 & 1 \end{vmatrix} = 4$

i.e. $A \times \text{adj } A = |A| \times I$

However, from equation (4), $A \times A^{-1} = I$

Hence, $A \times \text{adj } A = |A| \times A \times A^{-1}$

i.e. $A^{-1} = \dfrac{\text{adj } A}{A}$

Although this relationship has been verified for a specific matrix, the proof can be done, following the same steps, for the general matrix

$$\begin{pmatrix} a_1 & b_1 & c_1 \\ a_2 & b_2 & c_2 \\ a_3 & b_3 & c_3 \end{pmatrix}$$

In this case, the algebraic expressions produced become numerous and contain a large number of factors and hence become difficult to follow.

It is now possible, using equation (5), to find the reciprocal of a matrix and since dividing by, say a, is the same as multiplying by $\dfrac{1}{a}$, the inverse matrix may be used for the operation of matrix division,

i.e. $\dfrac{A}{B} = A \times \dfrac{1}{B} = AB^{-1}$

Worked problems on the inverse or reciprocal of a matrix

Problem 1. Find the adjoint matrix of matrix A, where

$$A = \begin{pmatrix} 4 & 2 & 0 \\ 0 & 1 & 1 \\ 3 & 2 & 0 \end{pmatrix}$$

Using the procedure given in the text for determining the adjoint of a matrix:

(i)   determine the nine cofactors of A.

First row: $+\begin{vmatrix} 1 & 1 \\ 2 & 0 \end{vmatrix} = -2, \ -\begin{vmatrix} 0 & 1 \\ 3 & 0 \end{vmatrix} = 3, \ +\begin{vmatrix} 0 & 1 \\ 3 & 2 \end{vmatrix} = -3$

Second row: $-\begin{vmatrix} 2 & 0 \\ 2 & 0 \end{vmatrix} = 0, \ +\begin{vmatrix} 4 & 0 \\ 3 & 0 \end{vmatrix} = 0, \ -\begin{vmatrix} 4 & 2 \\ 3 & 2 \end{vmatrix} = -2$

Third row: $+\begin{vmatrix} 2 & 0 \\ 1 & 1 \end{vmatrix} = 2, \ -\begin{vmatrix} 4 & 0 \\ 0 & 1 \end{vmatrix} = -4, \ +\begin{vmatrix} 4 & 2 \\ 0 & 1 \end{vmatrix} = 4$

(ii) form the matrix of these cofactors, i.e. $\begin{pmatrix} -2 & 3 & -3 \\ 0 & 0 & -2 \\ 2 & -4 & 4 \end{pmatrix}$

(iii) transpose the matrix of cofactors, giving

$$\text{adj } A = \begin{pmatrix} -2 & 0 & 2 \\ 3 & 0 & -4 \\ -3 & -2 & 4 \end{pmatrix}$$

Problem 2. Determine the inverse or reciprocal matrix, A^{-1}, for the matrix A given in Problem 1 above.

From equation (5), $A^{-1} = \dfrac{\text{adj } A}{|A|}$

It is shown in Problem 1 above that the adjoint of matrix A is:

$$\text{adj } A = \begin{pmatrix} -2 & 0 & 2 \\ 3 & 0 & -4 \\ -3 & -2 & 4 \end{pmatrix}$$

The determinant of matrix A, $|A|$, using the third column as the basis of the expansion, since it contains two 0's, is: $0 - 1\begin{vmatrix} 4 & 2 \\ 3 & 2 \end{vmatrix} + 0$

i.e. $\qquad |A| = -2$

Hence, $A^{-1} = -\dfrac{1}{2}\begin{pmatrix} -2 & 0 & 2 \\ 3 & 0 & -4 \\ -3 & -2 & 4 \end{pmatrix}$

Further problems on the inverse or reciprocal of a matrix may be found in the following Section 4.5 (Problems 17 to 25), page 63.

4.5 Further problems

Addition, subtraction and multiplication of third order matrices

In Problems 1 to 4,

$$\text{matrix } M = \begin{pmatrix} 1 & 2 & 1 \\ 3 & 2 & 1 \\ 2 & 3 & 2 \end{pmatrix} \text{ and}$$

$$\text{matrix } N = \begin{pmatrix} 3 & 2 & 1 \\ 1 & 1 & 2 \\ 1 & 2 & 1 \end{pmatrix}$$

1. Determine matrix L, where $L = M + N$.
$$\left[\begin{pmatrix} 4 & 4 & 2 \\ 4 & 3 & 3 \\ 3 & 5 & 3 \end{pmatrix}\right]$$

2. Prove that $M + N = N + M$.

3. Determine the matrix $M \times N$.
$$\left[\begin{pmatrix} 6 & 6 & 6 \\ 12 & 10 & 8 \\ 11 & 11 & 10 \end{pmatrix}\right]$$

4. Prove that $M \times N \neq N \times M$.
$$\left[N \times M = \begin{pmatrix} 11 & 13 & 7 \\ 8 & 10 & 6 \\ 9 & 9 & 5 \end{pmatrix}\right]$$

In Problems 5 to 8, $A = \begin{pmatrix} 2 & 5 & 4 & 3 \\ 3 & 1 & 4 & 5 \\ 1 & 8 & 3 & 5 \end{pmatrix}$, $B = \begin{pmatrix} 1 & 2 & 3 & 1 \\ 0 & 1 & 2 & -2 \\ 1 & -3 & 1 & 0 \end{pmatrix}$

$$C = \begin{pmatrix} 2 & 2 & 1 \\ 4 & 5 & 0 \\ 3 & 5 & 7 \\ 5 & 3 & 5 \end{pmatrix} \text{ and } Q = \begin{pmatrix} 2 \\ 4 \\ 3 \\ 5 \end{pmatrix}$$

5. Determine $A + B$.
$$\left[\begin{pmatrix} 3 & 7 & 7 & 4 \\ 3 & 2 & 6 & 3 \\ 2 & 5 & 4 & 5 \end{pmatrix}\right]$$

6. Determine $A \times Q$.
$$\left[\begin{pmatrix} 51 \\ 47 \\ 68 \end{pmatrix}\right]$$

7. Determine $A \times C$.
$$\left[\begin{pmatrix} 51 & 58 & 45 \\ 47 & 46 & 56 \\ 68 & 72 & 47 \end{pmatrix}\right]$$

8. Determine $C \times A$.
$$\left[\begin{pmatrix} 11 & 20 & 19 & 21 \\ 23 & 25 & 36 & 37 \\ 28 & 76 & 53 & 69 \\ 24 & 68 & 47 & 55 \end{pmatrix}\right]$$

The determinant of a 3 by 3 matrix

Problems 9 to 12 refer to the determinant $\begin{vmatrix} a & b & c \\ m & n & p \\ x & y & z \end{vmatrix}$

9. Write down the minors for elements a, p and y.
   $[nz - yp,\ ay - xb,\ ap - mc]$

10. Write down the cofactors for elements c, m and x.
    $[my - xn,\ yc - bz,\ bp - nc]$

11. Find the algebraic expression for the determinant based on the first row elements.
    $[a(nz - yp) - b(mz - xp) + c(my - xn)]$

12. Find the algebraic expression for the determinant based on the second column elements.
    $[b(mz - xp) - n(az - xc) + y(ap - mc)]$

In Problems 13 to 16, determine the values of the determinants for the matrices given.

13. $\begin{pmatrix} 2 & -1 & 3 \\ -2 & 3 & -2 \\ 2 & 1 & 5 \end{pmatrix}$ [4]

14. $\begin{pmatrix} 3 & 5 & 1 \\ -2 & 3 & 1 \\ 4 & -2 & 1 \end{pmatrix}$ [37]

15. $\begin{pmatrix} 1 & 2 & -3 \\ 2 & -3 & 4 \\ 3 & 4 & 5 \end{pmatrix}$ [-78]

16. $\begin{pmatrix} -13 & 3 & 2 \\ 32 & -6 & 3 \\ 12 & -4 & 1 \end{pmatrix}$ [-178]

The inverse or reciprocal of a matrix

17. Write down the transpose matrix of the matrix $\begin{pmatrix} 2 & -7 & 4 \\ 3 & 2 & 5 \\ -4 & 1 & 0 \end{pmatrix}$.

    $\left[\begin{pmatrix} 2 & 3 & -4 \\ -7 & 2 & 1 \\ 4 & 5 & 0 \end{pmatrix} \right]$

18. Write down the transpose matrix of the matrix $\begin{pmatrix} 4 & 8 & 3 \\ 1 & 4 & 7 \\ -3 & -2 & 6 \end{pmatrix}$.

    $\left[\begin{pmatrix} 4 & 1 & -3 \\ 8 & 4 & -2 \\ 3 & 7 & 6 \end{pmatrix} \right]$

In Problems 19 to 21, determine the adjoint matrices for the matrices given.

19. $\begin{pmatrix} 3 & 2 & 1 \\ 4 & 2 & 2 \\ 1 & 3 & 1 \end{pmatrix}$ $\left[\begin{pmatrix} -4 & 1 & 2 \\ -2 & 2 & -2 \\ 10 & -7 & -2 \end{pmatrix} \right]$

20. $\begin{pmatrix} 1 & -3 & 0 \\ 2 & 0 & 1 \\ 4 & 1 & 3 \end{pmatrix}$ $\left[\begin{pmatrix} -1 & 9 & -3 \\ -2 & 3 & -1 \\ 2 & -13 & 6 \end{pmatrix} \right]$

21. $\begin{pmatrix} 1 & 2 & 3 \\ 2 & -1 & 4 \\ 0 & -1 & 1 \end{pmatrix}$ $\left[\begin{pmatrix} 3 & -5 & 11 \\ -2 & 1 & 2 \\ -2 & 1 & -5 \end{pmatrix} \right]$

In Problems 22 to 25, determine the reciprocals of the matrices given. Show, in each case, that the product of the matrix and its reciprocal is the unit matrix.

22. $\begin{pmatrix} -1 & 2 & -3 \\ 2 & -1 & 4 \\ 3 & 4 & 1 \end{pmatrix}$ $\left[\frac{1}{4} \begin{pmatrix} -17 & -14 & 5 \\ 10 & 8 & -2 \\ 11 & 10 & -3 \end{pmatrix} \right]$

23. $\begin{pmatrix} 0 & 0 & 1 \\ 0 & 1 & 0 \\ 1 & 0 & 0 \end{pmatrix}$ $\left[-1 \begin{pmatrix} 0 & 0 & -1 \\ 0 & -1 & 0 \\ -1 & 0 & 0 \end{pmatrix} \right]$

24. $\begin{pmatrix} 2 & 1 & 1 \\ 1 & 2 & 2 \\ 1 & 3 & 2 \end{pmatrix}$ $\left[-\frac{1}{3} \begin{pmatrix} -2 & 1 & 0 \\ 0 & 3 & -3 \\ 1 & -5 & 3 \end{pmatrix} \right]$

25. $\begin{pmatrix} 1 & 2 & 1 \\ 2 & 4 & 6 \\ 3 & 1 & 2 \end{pmatrix}$ $\left[\frac{1}{20} \begin{pmatrix} 2 & -3 & 8 \\ 14 & -1 & -4 \\ -10 & 5 & 0 \end{pmatrix} \right]$

Chapter 5

The general properties of 3 by 3 determinants and the solution of simultaneous equations

5.1 The properties of third order determinants

The amount of arithmetic used to determine the value of a determinant can often be reduced by simplifying the determinant before evaluation. The principal properties of a determinant used in the simplifying process are given below.

(a) Changing rows or columns

If all the elements in a row or a column of a determinant are interchanged with the corresponding elements in another row or column, the value of the determinant obtained by doing this is -1 multiplied by the value of the original determinant. It is shown in Chapter 4 that, using the first-row elements:

$$\begin{vmatrix} a_1 & b_1 & c_1 \\ a_2 & b_2 & c_2 \\ a_3 & b_3 & c_3 \end{vmatrix} = a_2(b_2c_3 - b_3c_2) - b_1(a_2c_3 - a_3c_2) + c_1(a_2b_3 - a_3b_2)$$

$$= a_1b_2c_3 - a_1b_3c_2 - a_2b_1c_3 + a_3b_1c_2 + a_2b_3c_1 - a_3b_2c_1$$

Suppose now that the first and third columns are interchanged. The determinant becomes:

$$\begin{vmatrix} c_1 & b_1 & a_1 \\ c_2 & b_2 & a_2 \\ c_3 & b_3 & a_3 \end{vmatrix}$$

Evaluating, using the first row elements, gives:

$$c_1(b_2a_3 - a_2b_3) - b_1(c_2a_3 - c_3a_2) + a_1(c_2b_3 - c_3b_2)$$

$$= a_3b_2c_1 - a_2b_3c_1 - a_3b_1c_2 + a_2b_1c_3 + a_1b_3c_2 - a_1b_2c_3$$

Comparing these terms with those obtained for the determinant $\begin{vmatrix} a_1 & b_1 & c_1 \\ a_2 & b_2 & c_2 \\ a_3 & b_3 & c_3 \end{vmatrix}$

shows that each term in the second expression is -1 multiplied by its corresponding term in the first expression. Thus

$$\begin{vmatrix} a_1 & b_1 & c_1 \\ a_2 & b_2 & c_2 \\ a_3 & b_3 & c_3 \end{vmatrix} = (-1) \times \begin{vmatrix} c_1 & b_1 & a_1 \\ c_2 & b_2 & a_2 \\ c_3 & b_3 & a_3 \end{vmatrix}$$

Although this has been proved by interchanging the first and third columns, it may be shown in the same way that it applies when interchanging any row with any other row and when interchanging any column with any other column.

(b) The value of a determinant if two rows or two columns are equal

Let $|A|$ be a determinant in which, say, the first and second columns are equal,

i.e. $$|A| = \begin{vmatrix} a_1 & a_1 & c_1 \\ a_2 & a_2 & c_2 \\ a_3 & a_3 & c_3 \end{vmatrix}$$

Let $|B|$ be the determinant $|A|$ in which the first and second columns have been interchanged, then $|B| = \begin{vmatrix} a_1 & a_1 & c_1 \\ a_2 & a_2 & c_2 \\ a_3 & a_3 & c_3 \end{vmatrix}$

Since the first and second columns of $|A|$ are interchanged in $|B|$, then from (a) above

$$|A| = -|B|$$

But it can be seen that $|A| = |B|$ from the elements obtained. These two statements can only be true if $|A| = 0$. It follows that the value of a determinant in which two rows or two columns are equal is 0.

This statement can also be verified by obtaining the algebraic expression for a determinant in which, say, the first and second columns are equal. For the determinant $|A|$ given above, using the first-row elements as the basis for the expansion:

$$|A| = a_1 \begin{vmatrix} a_2 & c_2 \\ a_3 & c_3 \end{vmatrix} - a_1 \begin{vmatrix} a_2 & c_2 \\ a_3 & c_3 \end{vmatrix} + c_1 \begin{vmatrix} a_2 & a_2 \\ a_3 & a_3 \end{vmatrix}$$

i.e. $|A| = 0$

(c) Common factors

All elements in a row or a column having a common factor can be divided by this factor. This factor then becomes a factor of the determinant.

Thus

$$\begin{vmatrix} a_1 & b_1 & c_1 \\ \lambda a_2 & \lambda b_2 & \lambda c_2 \\ a_3 & b_3 & c_3 \end{vmatrix} = \lambda \begin{vmatrix} a_1 & b_1 & c_1 \\ a_2 & b_2 & c_2 \\ a_3 & b_3 & c_3 \end{vmatrix}$$

This property can be verified by obtaining the algebraic expression for the determinant. Using the first-row elements to obtain this, gives:

$$\begin{vmatrix} a_1 & b_1 & c_1 \\ \lambda a_2 & \lambda b_2 & \lambda c_2 \\ a_3 & b_3 & c_3 \end{vmatrix} = a_1 \begin{vmatrix} \lambda b_2 & \lambda c_2 \\ b_3 & c_3 \end{vmatrix} - b_1 \begin{vmatrix} \lambda a_2 & \lambda c_2 \\ a_3 & c_3 \end{vmatrix} + c_1 \begin{vmatrix} \lambda a_2 & \lambda b_2 \\ a_3 & b_3 \end{vmatrix}$$

$$= a_1(\lambda b_2 c_3 - \lambda b_3 c_2) - b_1(\lambda a_2 c_3 - \lambda a_3 c_2)$$

$$+ c_1(\lambda a_2 b_3 - \lambda a_3 b_2)$$

$$= \lambda[a_1(b_2 c_3 - b_3 c_2) - b_1(a_2 c_3 - a_3 c_2)$$

$$+ c_1(a_2 b_3 - a_3 b_2)]$$

$$= \lambda \begin{vmatrix} a_1 & b_1 & c_1 \\ a_2 & b_2 & c_2 \\ a_3 & b_3 & c_3 \end{vmatrix}$$

(d)   Addition or subtraction of rows or columns

If a multiple of the elements of any row or column is added to the corresponding elements of any other row or column, then the value of the determinant so obtained is equal to the value of the original determinant. Thus multiplying the third-row elements by λ and adding them to the corresponding elements in the first row, gives:

$$\begin{vmatrix} a_1 & b_1 & c_1 \\ a_2 & b_2 & c_2 \\ a_3 & b_3 & c_3 \end{vmatrix} = \begin{vmatrix} a_1 + \lambda c_1 & b_1 & c_1 \\ a_2 + \lambda c_2 & b_2 & c_2 \\ a_3 + \lambda c_3 & b_3 & c_3 \end{vmatrix}$$

This property may be verified as follows. Expanding the determinant $\begin{vmatrix} a_1 + \lambda c_1 & b_1 & c_1 \\ a_2 + \lambda c_2 & b_2 & c_2 \\ a_3 + \lambda c_3 & b_3 & c_3 \end{vmatrix}$, using the first-row elements, gives:

$$(a_1 + \lambda c_1)(b_2 c_3 - b_3 c_2) - b_1[(a_2 + \lambda c_2)c_3 - (a_3 + \lambda c_3)c_2]$$

$$+ c_1[(a_2 + \lambda c_2)b_3 - (a_3 + \lambda c_3)b_2]$$

i.e. $a_1(b_2 c_3 - b_3 c_2) - b_1(a_2 c_3 - a_3 c_2) + c_1(a_2 b_3 - a_3 b_2)$

$$+ \lambda(c_1 b_2 c_3 - c_1 b_3 c_2 - b_1 c_2 c_3 + b_1 c_3 c_2 + c_1 c_2 b_3 - c_1 c_3 b_2)$$

All the terms containing λ cancel out and the remaining terms are equal to

$$\begin{vmatrix} a_1 & b_1 & c_1 \\ a_2 & b_2 & c_2 \\ a_3 & b_3 & c_3 \end{vmatrix}. \text{ Thus } \begin{vmatrix} a_1 + \lambda c_1 & b_1 & c_1 \\ a_2 + \lambda c_2 & b_2 & c_2 \\ a_3 + \lambda c_3 & b_3 & c_3 \end{vmatrix} = \begin{vmatrix} a_1 & b_1 & c_1 \\ a_2 & b_2 & c_2 \\ a_3 & b_3 & c_3 \end{vmatrix}.$$

When λ is equal to 1, then the elements of a row or column are being added to the corresponding elements of another row or column. When $\lambda = -1$, then the elements of a row or column are being subtracted from the corresponding elements of another row or column.

The four properties introduced above are used mainly in two ways. These are (i) to reduce the magnitude of the elements in a row or column, making subsequent arithmetic easier, and (ii) to introduce as many 0's as practical into a determinant before evaluating it. The way in which this is done is shown in the worked problems following.

Worked problems on the properties of third order determinants

Problem 1. Use the properties of determinants to simplify the determinant

$$\begin{vmatrix} 3 & 4 & 5 \\ 6 & 7 & 8 \\ 1 & 11 & 6 \end{vmatrix}, \text{ and hence find its value.}$$

There are many ways of simplifying this determinant and one method is shown below.

Taking the first row elements from the corresponding second-row elements (property (d) above) and then taking out a common factor of 3 from the second row elements (property (c) above) gives:

$$\begin{vmatrix} 3 & 4 & 5 \\ 6 & 7 & 8 \\ 1 & 11 & 6 \end{vmatrix} = \begin{vmatrix} 3 & 4 & 5 \\ 3 & 3 & 3 \\ 1 & 11 & 6 \end{vmatrix} = 3\begin{vmatrix} 3 & 4 & 5 \\ 1 & 1 & 1 \\ 1 & 11 & 5 \end{vmatrix}$$

To introduce some elements whose values are 0, property (d) is applied again. The second-row elements are taken from the corresponding third-row elements, then the first-column elements are taken from the third-column elements and finally the first-column elements from the corresponding second-column elements. These three steps are shown below.

$$3\begin{vmatrix} 3 & 4 & 5 \\ 1 & 1 & 1 \\ 1 & 11 & 6 \end{vmatrix} = 3\begin{vmatrix} 3 & 4 & 5 \\ 1 & 1 & 1 \\ 0 & 10 & 5 \end{vmatrix} = 3\begin{vmatrix} 3 & 4 & 2 \\ 1 & 1 & 0 \\ 0 & 10 & 5 \end{vmatrix} = 3\begin{vmatrix} 3 & 1 & 2 \\ 1 & 0 & 0 \\ 0 & 10 & 5 \end{vmatrix}$$

It is now fairly easy to evaluate this determinant by using the second row elements, since this row contains two 0's

$$\text{Hence } 3\begin{vmatrix} 3 & 1 & 2 \\ 1 & 0 & 0 \\ 0 & 10 & 5 \end{vmatrix} = 3\left\{ -1\begin{vmatrix} 1 & 2 \\ 10 & 5 \end{vmatrix} + 0 + 0 \right\} = -3(1 \times 5 - 10 \times 2)$$

$$\text{i.e. } \begin{vmatrix} 3 & 4 & 5 \\ 6 & 7 & 8 \\ 1 & 11 & 6 \end{vmatrix} = 45$$

Problem 2. Evaluate the determinant $\begin{vmatrix} 3 & 5 & 7 \\ 11 & 9 & 13 \\ 15 & 17 & 19 \end{vmatrix}$.

To reduce the magnitude of the elements in the second and third rows, the second-row elements are taken from the corresponding third-row elements, then the first-row elements are taken from the corresponding second-row elements, giving:

$$\begin{vmatrix} 3 & 5 & 7 \\ 11 & 9 & 13 \\ 15 & 17 & 19 \end{vmatrix} = \begin{vmatrix} 3 & 5 & 7 \\ 11 & 9 & 13 \\ 4 & 8 & 6 \end{vmatrix} = \begin{vmatrix} 3 & 5 & 7 \\ 8 & 4 & 6 \\ 4 & 8 & 6 \end{vmatrix}$$

To introduce 0's, the second-row elements are taken from the corresponding third-row elements and then the first-column elements are added to the corresponding second-column elements. This gives:

$$\begin{vmatrix} 3 & 5 & 7 \\ 8 & 4 & 6 \\ 4 & 8 & 6 \end{vmatrix} = \begin{vmatrix} 3 & 5 & 7 \\ 8 & 4 & 6 \\ -4 & 4 & 0 \end{vmatrix} = \begin{vmatrix} 3 & 8 & 7 \\ 8 & 12 & 6 \\ -4 & 0 & 0 \end{vmatrix}$$

Evaluating, using the third-row elements, gives

$$\begin{vmatrix} 3 & 8 & 7 \\ 8 & 12 & 6 \\ -4 & 0 & 0 \end{vmatrix} = -4\begin{vmatrix} 8 & 7 \\ 12 & 6 \end{vmatrix} = -4(48 - 84)$$

$$\text{i.e. } \begin{vmatrix} 3 & 5 & 7 \\ 11 & 9 & 13 \\ 15 & 17 & 19 \end{vmatrix} = 144$$

Since the use of calculators is now commonplace, the application of these properties of determinants has somewhat decreased.

Further problems on the properties of third order determinants may be found in Section 5.3 (Problems 1 to 7), page 75.

5.2   The solution of simultaneous equations having three unknowns

In engineering and science, the principal use of determinants is for the solution of simultaneous equations. Solving simultaneous equations having two unknowns by determinants is discussed in *Technician mathematics 3*. This section deals with the solution of simultaneous equations having three unknowns by determinants.

Let the simultaneous equations to be solved be

$$a_1x + b_1y + c_1z + d_1 = 0 \tag{1}$$

$$a_2x + b_2y + c_2z + d_2 = 0 \tag{2}$$

$$a_3x + b_3y + c_3z + d_3 = 0 \tag{3}$$

where a, b, c and d are constants.

The variables y and z may be eliminated from equations (1), (2) and (3) by the following procedure:

(i)   multiplying equation (1) by $(b_2c_3 - b_3c_2)$,
(ii)   multiplying equation (2) by $-(b_1c_3 - b_3c_1)$,
(iii)   multiplying equation (3) by $(b_1c_2 - b_2c_1)$ and
(iv)   adding the equations obtained in (i), (ii) and (iii) above.

(i)   Equation (1) multiplied by $(b_2c_3 - b_3c_2)$ gives:

$$a_1(b_2c_3 - b_3c_2)x + b_1(b_2c_3 - b_3c_2)y + c_1(b_2c_3 - b_3c_2)z$$
$$+ d_1(b_2c_3 - b_3c_2) = 0$$

(ii)   Equation (2) multiplied by $-(b_1c_3 - b_3c_1)$ gives:

$$-a_2(b_1c_3 - b_3c_1)x - b_2(b_1c_3 - b_3c_1)y - c_2(b_1c_3 - b_3c_1)z$$
$$- d_2(b_1c_3 - b_3c_1) = 0$$

(iii)   Equation (3) multiplied by $(b_1c_2 - b_2c_1)$ gives:

$$a_3(b_1c_2 - b_2c_1)x + b_3(b_1c_2 - b_2c_1)y + c_3(b_1c_2 - b_2c_1)z$$
$$+ d_3(b_1c_2 - b_2c_1) = 0$$

(iv)   Adding (i), (ii) and (iii) above gives:

$$x\{a_1(b_2c_3 - b_3c_2) - a_2(b_1c_3 - b_3c_1) + a_3(b_1c_2 - b_2c_1)\}$$
$$+ d_1(b_2c_3 - b_3c_2) - d_2(b_1c_3 - b_3c_1) + d_3(b_1c_2 - b_2c_1) = 0$$

Writing these algebraic expressions in determinant form gives:

$$x\begin{vmatrix} a_1 & b_1 & c_1 \\ a_2 & b_2 & c_2 \\ a_3 & b_3 & c_3 \end{vmatrix} + \begin{vmatrix} b_1 & c_1 & d_1 \\ b_2 & c_2 & d_2 \\ b_3 & c_3 & d_3 \end{vmatrix} = 0$$

Thus $x = -\dfrac{\begin{vmatrix} b_1 & c_1 & d_1 \\ b_2 & c_2 & d_2 \\ b_3 & c_3 & d_3 \end{vmatrix}}{\begin{vmatrix} a_1 & b_1 & c_1 \\ a_2 & b_2 & c_2 \\ a_3 & b_3 & c_3 \end{vmatrix}}$

Equations (1), (2) and (3) may be written as:

$$b_1 y + a_1 x + c_1 z + d_1 = 0$$
$$b_2 y + a_2 x + c_2 z + d_2 = 0$$
$$b_3 y + a_3 x + c_3 z + d_3 = 0$$

By comparison with the determinant obtained for x:

$$y = -\dfrac{\begin{vmatrix} a_1 & c_1 & d_1 \\ a_2 & c_2 & d_2 \\ a_3 & c_3 & d_3 \end{vmatrix}}{\begin{vmatrix} b_1 & a_1 & c_1 \\ b_2 & a_2 & c_2 \\ b_3 & a_3 & c_3 \end{vmatrix}}$$

Using property (a) of determinants introduced in Section 5.1, the first and second columns of the denominator can be interchanged, giving

$$y = \dfrac{+\begin{vmatrix} a_1 & c_1 & d_1 \\ a_2 & c_2 & d_2 \\ a_3 & c_3 & d_3 \end{vmatrix}}{\begin{vmatrix} a_1 & b_1 & c_1 \\ a_2 & b_2 & c_2 \\ a_3 & b_3 & c_3 \end{vmatrix}}$$

Similarly, to find the determinant for z, equations (1), (2) and (3) may be written as

$$c_1 z + b_1 y + a_1 x + d_1 = 0$$
$$c_2 z + b_2 y + a_2 x + d_2 = 0$$
$$c_3 z + b_3 y + a_3 x + d_3 = 0$$

By comparison with the determinant obtained for x:

$$z = -\dfrac{\begin{vmatrix} b_1 & a_1 & d_1 \\ b_2 & a_2 & d_2 \\ b_3 & a_3 & d_3 \end{vmatrix}}{\begin{vmatrix} c_1 & b_1 & a_1 \\ c_2 & b_2 & a_2 \\ c_3 & b_3 & a_3 \end{vmatrix}}$$

Using property (a) of determinants introduced in Section 5.1, the first and second columns of the numerator and the first and third columns of the denominator may be interchanged, giving

$$z = \dfrac{+\begin{vmatrix} a_1 & b_1 & d_1 \\ a_2 & b_2 & d_2 \\ a_3 & b_3 & d_3 \end{vmatrix}}{-\begin{vmatrix} a_1 & b_1 & c_1 \\ a_2 & b_2 & c_2 \\ a_3 & b_3 & c_3 \end{vmatrix}} = \dfrac{-\begin{vmatrix} a_1 & b_1 & d_1 \\ a_2 & b_2 & d_2 \\ a_3 & b_3 & d_3 \end{vmatrix}}{\begin{vmatrix} a_1 & b_1 & c_1 \\ a_2 & b_2 & c_2 \\ a_3 & b_3 & c_3 \end{vmatrix}}$$

These results can be combined, showing that

$$\boxed{\dfrac{x}{|D_1|} = \dfrac{-y}{|D_2|} = \dfrac{z}{|D_3|} = \dfrac{-1}{|D|}}$$

where $|D_1|$ is the determinant remaining when the x-column is 'covered up',
$|D_2|$ is the determinant remaining when the y-column is 'covered up',
$|D_3|$ is the determinant remaining when the z-column is 'covered up',
and $|D|$ is the determinant remaining when the constants-column is 'covered up'.

Thus, for the simultaneous equations:

$$a_1 x + b_1 y + c_1 z + d_1 = 0$$
$$a_2 x + b_2 y + c_2 z + d_2 = 0$$
$$a_3 x + b_3 y + c_3 z + d_3 = 0$$

the solution is given by:

$$\dfrac{x}{\begin{vmatrix} b_1 & c_1 & d_1 \\ b_2 & c_2 & d_2 \\ b_3 & c_3 & d_3 \end{vmatrix}} = \dfrac{-y}{\begin{vmatrix} a_1 & c_1 & d_1 \\ a_2 & c_2 & d_2 \\ a_3 & c_3 & d_3 \end{vmatrix}} = \dfrac{z}{\begin{vmatrix} a_1 & b_1 & d_1 \\ a_2 & b_2 & d_2 \\ a_3 & b_3 & d_3 \end{vmatrix}} = \dfrac{-1}{\begin{vmatrix} a_1 & b_1 & c_1 \\ a_2 & b_2 & c_2 \\ a_3 & b_3 & c_3 \end{vmatrix}}$$

This relationship is used to solve simultaneous equations having three unknown values, as shown in the worked problems following.

Worked problems on the solution of simultaneous equations having three unknowns

Problem 1. Use determinants to solve the simultaneous equations:

$$2x + y = 2 \tag{1}$$
$$-4y + z = 0 \tag{2}$$
$$4x + z = 6 \tag{3}$$

Writing the equations in the form $a_1 x + b_1 y + c_1 z + d_1 = 0$, gives

$$2x + \ y + 0z - 2 = 0$$

$$0x - 4y + z + 0 = 0$$

$$4x + 0y + z - 6 = 0$$

The solution can be obtained from

$$\frac{x}{|D_1|} = \frac{-y}{|D_2|} = \frac{z}{|D_3|} = \frac{-1}{|D|}$$

$|D_1|$ is the third order determinant obtained by covering up the x-column and writing down the remaining coefficients.

i.e. $|D_1| = \begin{vmatrix} 1 & 0 & -2 \\ -4 & 1 & 0 \\ 0 & 1 & -6 \end{vmatrix} = 1\begin{vmatrix} 1 & 0 \\ 1 & -6 \end{vmatrix} - 0\begin{vmatrix} -4 & 0 \\ 0 & -6 \end{vmatrix} + (-2)\begin{vmatrix} -4 & 1 \\ 0 & 1 \end{vmatrix}$

$$= 2$$

$|D_2|$ is the third order determinant remaining when the y-column is covered up, i.e.,

$$|D_2| = \begin{vmatrix} 2 & 0 & -2 \\ 0 & 1 & 0 \\ 4 & 1 & -6 \end{vmatrix} = 0 + 1\begin{vmatrix} 2 & -2 \\ 4 & -6 \end{vmatrix} + 0 = -4$$

$|D_3|$ is the third order determinant remaining when the z-column is covered up. i.e.

$$|D_3| = \begin{vmatrix} 2 & 1 & -2 \\ 0 & -4 & 0 \\ 4 & 0 & -6 \end{vmatrix} = -4\begin{vmatrix} 2 & -2 \\ 4 & -6 \end{vmatrix} = 16$$

$|D|$ is the third order determinant remaining when the constants-column is covered up, i.e.

$$|D| = \begin{vmatrix} 2 & 1 & 0 \\ 0 & -4 & 1 \\ 4 & 0 & 1 \end{vmatrix} = 2\begin{vmatrix} -4 & 1 \\ 0 & 1 \end{vmatrix} - 1\begin{vmatrix} 0 & 1 \\ 4 & 1 \end{vmatrix} + 0 = -4$$

Thus, $\dfrac{x}{2} = \dfrac{-y}{-4} = \dfrac{z}{16} = \dfrac{-1}{-4}$

i.e. $x = \frac{1}{2}, y = 1, z = 4$

Problem 2. Solve the simultaneous equations:

$$\frac{3}{x} - \frac{4}{y} - \frac{2}{z} = 1 \tag{1}$$

$$\frac{2}{x} + \frac{5}{y} - \frac{2}{z} = 3 \tag{2}$$

$$\frac{1}{x} + \frac{2}{y} + \frac{1}{z} = 2 \tag{3}$$

Letting $p = \frac{1}{x}$, $q = \frac{1}{y}$ and $r = \frac{1}{z}$ and writing the equations in the form $a_1 x + b_1 y + c_1 z + d_1 = 0$, gives

$$3p - 4q - 2r - 1 = 0$$

$$2p + 5q - 2r - 3 = 0$$

$$p + 2q + r - 2 = 0$$

The solution can be obtained from

$$\frac{p}{|D_1|} = \frac{-q}{|D_2|} = \frac{r}{|D_3|} = \frac{-1}{|D|}$$

$$|D_1| = \begin{vmatrix} -4 & -2 & -1 \\ 5 & -2 & -3 \\ 2 & 1 & -2 \end{vmatrix}$$

The arithmetic of evaluating this determinant can be simplified by taking twice the second-column elements from the first.

Thus: $\begin{vmatrix} -4 & -2 & -1 \\ 5 & -2 & -3 \\ 2 & 1 & -2 \end{vmatrix} = \begin{vmatrix} 0 & -2 & -1 \\ 9 & -2 & -3 \\ 0 & 1 & -2 \end{vmatrix}$

$$= -9 \begin{vmatrix} -2 & -1 \\ 1 & -2 \end{vmatrix} = -45$$

$$|D_2| = \begin{vmatrix} 3 & -2 & -1 \\ 2 & -2 & -3 \\ 1 & 1 & -2 \end{vmatrix}$$

Taking the second-row elements from the first and then twice the first-column elements from the third, gives:

$$|D_2| = \begin{vmatrix} 1 & 0 & 2 \\ 2 & -2 & -3 \\ 1 & 1 & -2 \end{vmatrix} = \begin{vmatrix} 1 & 0 & 0 \\ 2 & -2 & -7 \\ 1 & 1 & -4 \end{vmatrix} = 1\begin{vmatrix} -2 & -7 \\ 1 & -4 \end{vmatrix} = 15$$

$$|D_3| = \begin{vmatrix} 3 & -4 & -1 \\ 2 & 5 & -3 \\ 1 & 2 & -2 \end{vmatrix}$$

Adding the third-column elements to the second, and twice the first-column elements to the third gives:

$$|D_3| = \begin{vmatrix} 3 & -5 & -1 \\ 2 & 2 & -3 \\ 1 & 0 & -2 \end{vmatrix} = \begin{vmatrix} 3 & -5 & 5 \\ 2 & 2 & 1 \\ 1 & 0 & 0 \end{vmatrix} = 1\begin{vmatrix} -5 & 5 \\ 2 & 1 \end{vmatrix} = -15$$

$$|D| = \begin{vmatrix} 3 & -4 & -2 \\ 2 & 5 & -2 \\ 1 & 2 & 1 \end{vmatrix}$$

Adding twice the third-row elements to the first, gives:

$$|D| = \begin{vmatrix} 5 & 0 & 0 \\ 2 & 5 & -2 \\ 1 & 2 & 1 \end{vmatrix} = 5\begin{vmatrix} 5 & -2 \\ 2 & 1 \end{vmatrix} = 45$$

Thus, $\dfrac{p}{-45} = \dfrac{-q}{15} = \dfrac{r}{-15} = \dfrac{-1}{45}$

i.e. $p = 1, q = \dfrac{1}{3}, r = \dfrac{1}{3}$

Thus $x = 1$, $y = 3$ and $z = 3$

Further problems on the solution of simultaneous equations having three unknowns may be found in Section 5.3 (Problems 8 to 18), page 76.

5.3   Further problems

The properties of third order determinants

In problems 1 to 7, use the properties of determinants to simplify the determinants given and then evaluate them.

1. $\begin{vmatrix} 1 & 3 & 5 \\ 2 & 4 & 6 \\ 3 & 5 & 7 \end{vmatrix}$ [0]

2. $\begin{vmatrix} 13 & 2 & 23 \\ 30 & 7 & 53 \\ 39 & 9 & 70 \end{vmatrix}$ [34]

3. $\begin{vmatrix} 14 & 9 & 33 \\ 13 & 11 & 36 \\ 17 & 2 & 22 \end{vmatrix}$ [1]

4. $\begin{vmatrix} 1 & 11 & 16 \\ 23 & 37 & 58 \\ 16 & 9 & 1 \end{vmatrix}$ [3 310]

5. $\begin{vmatrix} 6 & 10 & 14 \\ -2 & 1 & 1 \\ -9 & -15 & -12 \end{vmatrix}$ [234]

6. $\begin{vmatrix} 22 & 25 & 28 \\ 26 & 28 & 31 \\ 24 & 27 & 29 \end{vmatrix}$ [40]

7. $\begin{vmatrix} 5.2 & 7.5 & 8.6 \\ 5.4 & 7.2 & 8.7 \\ 5.6 & 6.9 & 8.3 \end{vmatrix}$ [1.53]

The solutions of simultaneous equations

In Problems 8 to 11, solve the simultaneous equations given, using determinants.

8. $11p + 7q + 2r = 31$
  $p + q + r = 4$
  $31p + 15q + 13r = 90$ $[p = 2, q = 1, r = 1]$

9. $4l + 9m + 2n = 21$
  $13l + 5m + 7n = 1$
  $17l + 19m + 8n = 26$ $[l = -7, m = 3, n = 11]$

10. $\dfrac{1}{x} + \dfrac{2}{y} + \dfrac{2}{z} = 4$

  $\dfrac{3}{x} - \dfrac{1}{y} + \dfrac{4}{z} = 25$

  $\dfrac{3}{x} + \dfrac{2}{y} - \dfrac{1}{z} = -4$ $\left[x = \dfrac{1}{2}, y = -\dfrac{1}{3}, z = \dfrac{1}{4} \right]$

11. $\dfrac{2x}{3} - y + \dfrac{2z}{3} = 2$

$x + 8y + 3z = -31$

$\dfrac{6x}{5} - \dfrac{4y}{5} + \dfrac{2z}{5} = -2$ $[x = -5, y = -4, z = 2]$

12. Kirchhoff's laws are applied to an electrical network and the following simultaneous equations for the current flowing in amperes in various closed circuits are obtained.

$$2i_1 + \quad i_2 + \quad i_3 = 1.67$$
$$3i_1 + 4.5i_2 - \quad 1.5i_3 = 0$$
$$2.25i_1 - 1.5i_2 + 5.25i_3 = 0$$

Use determinants to find the values of i_1, i_2 and i_3, correct to three significant figures.
$[i_1 = 3.97\text{ A}, i_2 = -3.55\text{ A}, i_3 = -2.71\text{ A}]$

13. The rate of working, $\dot{W}$, is given by:

$$\dot{W} = \Sigma\, F \cdot \dot{r}$$

where F is the force in newtons, r is the distance in metres from some reference point, and $\dot{r}$ is the velocity in metres per second. The following simultaneous equations arise from experiments carried out on a system.

$$\dot{r}_1 + 3\dot{r}_2 + 2\dot{r}_3 = -13$$
$$2\dot{r}_1 - 6\dot{r}_2 + 3\dot{r}_3 = 32$$
$$3\dot{r}_1 - 4\dot{r}_2 - \dot{r}_3 = 12$$

Use determinants to find the values of $\dot{r}_1$, $\dot{r}_2$ and $\dot{r}_3$.
$[\dot{r}_1 = -2\text{ m/s}, \dot{r}_2 = -5\text{ m/s}, \dot{r}_3 = 2\text{ m/s}]$

14. The simultaneous equations representing the currents flowing in an unbalanced, three-phase, star-connected, electrical network are as follows:

$$3.6i_1 + 2.4i_2 + 4.8i_3 = 1.2$$
$$1.3i_1 - 3.9i_2 - 6.5i_3 = 2.6$$
$$11.9i_1 + 1.7i_2 + 8.5i_3 = 0$$

Use determinants to find the values of i_1, i_2 and i_3.
$[i_1 = 1, i_2 = 3, i_3 = -2]$

15. In a mass-spring-damper system, the acceleration $\ddot{x}$ m/s^2, velocity $\dot{x}$ m/s and displacement x m are related by the following simultaneous equations:

$$5.8\ddot{x} + 8.7\dot{x} + 14.5x = 23.2$$
$$8.1\ddot{x} + 5.4\dot{x} + 5.4x = 8.1$$
$$14.8\ddot{x} + 3.7\dot{x} - 14.8x = -22.2$$

By using determinants, determine the acceleration, velocity and displacement for the system, correct to two decimal places. [$\ddot{x} = -0.33$ m/s^2, $\dot{x} = 0.67$ m/s, $x = 1.33$ m]

16. The currents in a network are represented by I_1, I_2 and I_3. Application of Kirchhoff's laws to the circuit leads to the following three equations:

$$14I_1 - 5I_2 - 6I_3 = 10$$
$$-5I_1 + 14I_2 - 2I_3 = 3$$
$$-6I_1 - 2I_2 + 18I_3 = 5$$

Determine I_1, I_2 and I_3 correct to four significant figures. [1.358, 0.816 5, 0.821 1]

17. The tensions in a simple framework, T_1, T_2 and T_3, are given by the equations:

$$6T_1 + 6T_2 + 6T_3 = 8.4$$
$$T_1 + 2T_2 + 4T_3 = 2.4$$
$$4T_1 + 2T_2 \qquad = 4.0$$

Determine T_1, T_2 and T_3.   [0.8, 0.4, 0.2]

18. When a number of mass/spring systems are connected together and have a mode of oscillation, and all masses oscillate with frequency $\dfrac{n}{2\pi}$ but have different amplitudes, n can be given in terms of eigenvalues λ (where $\lambda = n^2$) by the determinant:

$$\begin{vmatrix} 1 - \lambda & -\dfrac{1}{2} & 0 \\ -\dfrac{3}{4} & \dfrac{6}{4} - \lambda & -\dfrac{3}{4} \\ 0 & -\dfrac{3}{4} & 1 - \lambda \end{vmatrix} = 0$$

Show that $16\lambda^3 - 56\lambda^2 + 49\lambda - 9 = 0$ and verify that $\lambda = 1, \dfrac{9}{4}$ and $\dfrac{1}{4}$.

Chapter 6

Maclaurin's and Taylor's series

6.1 Maclaurin's series

Conditions

Certain mathematical functions may be represented as a power series, containing terms in ascending powers of the variable. This enables mixed functions containing, say, algebraic, trigonometric and exponential functions to be expressed solely as algebraic functions. In this form, subsequent operations such as differentiation or integration often can be more readily performed. Maclaurin's theorem may be used to express any function, say $f(x)$, as a power series, provided that, at $x = 0$, the three conditions given below are met:

(i) The function is finite, i.e. $f(0) \neq \infty$. For the two functions $f(x) = \sin x$ and $g(x) = \ln x$, $f(0) = \sin 0 = 0$ and $g(0) = \ln 0 = -\infty$, hence $f(x) = \sin x$ meets this condition but $g(x) = \ln x$ does not.

(ii) The derivatives of the function are finite, i.e. $f'(0), f''(0), f'''(0), \ldots \neq \infty$. For the function $f(x) = \sin x$, $f'(0) = \cos 0 = 1$, $f''(0) = -\sin 0 = 0$, $f'''(0) = -\cos (0) = -1$ and so on. For the function $g(x) = \ln x$, $g'(x) = \dfrac{1}{x}$ and $g'(0) = \dfrac{1}{0} = \infty$. Hence the function $f(x) = \sin x$ meets this condition, but again $g(x) = \ln x$ does not meet this condition, and therefore cannot be expressed as a power series by using Maclaurin's theorem.

(iii) The power series arising from the use of Maclaurin's theorem must be convergent, that is, in general, the values of the terms (or groups of terms) must get progessively smaller and the sum of the term reach a limiting value. For example, the series $1 + \dfrac{1}{2} + \dfrac{1}{4} + \dfrac{1}{8} + \cdots$ is convergent, since the values of the terms are getting smaller and the sum of the terms is approaching a limiting value of 2. However, for the series $1 + \dfrac{1}{2} + \dfrac{1}{3} + \dfrac{1}{4} + \dfrac{1}{5} + \cdots$, groups of terms are getting successively larger, i.e. $\left(\dfrac{1}{3} + \dfrac{1}{4}\right) > \dfrac{1}{2}$, $\left(\dfrac{1}{5} + \dfrac{1}{6} + \dfrac{1}{7} + \dfrac{1}{8}\right) > \dfrac{1}{2}$ and so on, so although the values of the terms are getting smaller, groups of terms are not and this series diverges to an infinite number. A full study of the conditions for a series to be convergent is a topic in its own right but one test which applies to most series is as follows. If two consecutive terms in a series are u_n and u_{n+1}, then for the series to be convergent, the limiting value of $\dfrac{u_{n+1}}{u_n}$ as $n \to \infty$ must be less than 1, i.e.

$$\lim_{n \to \infty} \left| \frac{u_{n+1}}{u_n} \right| < 1, \text{ for the series to converge,}$$

where the vertical lines indicate 'the modulus of' or 'the positive value' of the quantity. For example, in the series $1 + \dfrac{1}{2!} + \dfrac{1}{3!} + \dfrac{1}{4!} + \cdots$, the third term is $\dfrac{1}{3!}$, the fourth term is $\dfrac{1}{4!}$ and the nth term is $\dfrac{1}{n!}$, i.e. $u_n = \dfrac{1}{n!}$. The term after the nth term is the $(n + 1)$th term, hence $u_{n+1} = \dfrac{1}{(n + 1)!}$. Thus $\dfrac{u_{n+1}}{u_n} = \dfrac{\dfrac{1}{(n + 1)!}}{\dfrac{1}{n!}} = \dfrac{n!}{(n + 1)!} = \dfrac{1}{n + 1}$

As $n \to \infty$, $\dfrac{1}{n + 1} \to 0$, hence this series is convergent.

Derivation of Maclaurin's theorem

Let the power series for $f(x)$ be:

$$f(x) = a_0 + a_1 x + a_2 x^2 + a_3 x^3 + a_4 x^4 + a_5 x^5 + \cdots, \tag{1}$$

where $a_0, a_1, a_2, \ldots$ are constants.
When $x = 0$, $f(0) = a_0$, since all terms having a factor of x^n in them become zero. Thus, $a_0 = f(0)$.

Differentiating equation (1) with respect to x, gives:

$$f'(x) = a_1 + 2a_2 x + 3a_3 x^2 + 4a_4 x^3 + 5a_5 x^4 + \cdots \tag{2}$$

When $x = 0$, $f'(0) = a_1$, i.e. $a_1 = \boldsymbol{f'(0)}$

Differentiating equation (2) with respect to x, gives:

$$f''(x) = 2a_2 + 3 \cdot 2a_3 x + 4 \cdot 3a_4 x^2 + 5 \cdot 4a_5 x^3 + \cdots \tag{3}$$

When $x = 0$, $f''(0) = 2a_2 = 2!a_2$, i.e. $a_2 = \dfrac{f''(0)}{2!}$

Differentiating equation (3) with respect to x, gives:

$$f'''(x) = 3 \cdot 2a_3 + 4 \cdot 3 \cdot 2a_4 x + 5 \cdot 4 \cdot 3a_5 x^2 + \cdots \qquad (4)$$

When $x = 0$, $f'''(0) = 3 \cdot 2a_3 = 3!a_3$, i.e. $a_3 = \dfrac{f'''(0)}{3!}$

The same procedure may be repeated to find $f^{iv}(x)$, $f^{v}(x), \ldots$ giving $f^{iv}(0) = 4!a_4$, $f^{v}(0) = 5!a_5$ and so on. Thus

$$a_4 = \frac{f^{iv}(0)}{4!}, \; a_5 = \frac{f^{v}(0)}{5!} \cdots$$

Substituting for a_0, a_1, a_2, $a_3, \ldots$ in equation (1), gives

$$f(x) = f(0) + f'(0)x + \frac{f''(0)}{2!}x^2 + \frac{f'''(0)}{3!}x^3 + \cdots$$

i.e. $$f(x) = f(0) + xf'(0) + \frac{x^2}{2!}f''(0) + \frac{x^3}{3!}f'''(0) + \cdots \qquad (5)$$

This mathematical statement is called **Maclaurin's theorem** or **Maclaurin's series.** The uses of Maclaurin's series include expressing mixed functions, containing some or all of trigonometric, exponential, algebraic and logarithmic terms, in a form where such techniques as Newton's method (see Chapter 2), or approximate integration (see Section 6.3), can be employed. At a more advanced level, such theories as Webb's buckling of struts are based on an understanding of Maclaurin's series.

Worked problems on Maclaurin's series

Problem 1. Determine the first four terms of the power series for $\sin x$. Hence determine the value of $\sin 0.5$ radians correct to three decimal places. What will be the power series of $\sin 2x$?

The values of $f(0)$, $f'(0)$, $f''(0) \ldots$ in Maclaurin's series are determined as follows:

$$f(x) = \sin x \qquad\qquad f(0) = \sin 0 = 0$$
$$f'(x) = \cos x \qquad\qquad f'(0) = \cos 0 = 1$$
$$f''(x) = -\sin x \qquad\quad f''(0) = -\sin 0 = 0$$
$$f'''(x) = -\cos x \qquad\quad f'''(0) = -\cos 0 = -1$$
$$f^{iv}(x) = \sin x \qquad\qquad f^{iv}(0) = \sin 0 = 0$$

Since $f^{iv}(x) = f(x)$, then $f^{v}(x) = f'(x)$, $f^{vi}(x) = f''(x)$ and so on.

Thus the coefficients are 0, 1, 0, -1, 0, 1, 0, -1, ... Substituting the values of $f(0)$, $f'(0)$, $f''(0)$, ... in equation (5) gives:

$$\sin x = 0 + (1)x + 0 + (-1)\frac{x^3}{3!} + 0$$

$$+ (1)\frac{x^5}{5!} + 0 + (-1)\frac{x^7}{7!} + \cdots$$

i.e. $\sin x = x - \dfrac{x^3}{3!} + \dfrac{x^5}{5!} - \dfrac{x^7}{7!} + \cdots$

When $x = 0.5$ radians

$$\sin 0.5 = 0.5 - \frac{0.5^3}{6} + \frac{0.5^5}{120} - \frac{0.5^7}{5040} + \cdots$$

$$= 0.500\,00 - 0.020\,83 + 0.000\,26 - 0.000\,00$$

$$= \textbf{0.479 correct to three decimal places.}$$

Since $\sin x = x - \dfrac{x^3}{3!} + \dfrac{x^5}{5!} - \dfrac{x^7}{7!} + \cdots$

then $\sin u = u - \dfrac{u^3}{3!} + \dfrac{u^5}{5!} - \dfrac{u^7}{7!} + \cdots$

Replacing u by $2x$, gives

$$\sin 2x = 2x - \frac{(2x)^3}{3!} + \frac{(2x)^5}{5!} - \frac{(2x)^7}{7!} + \cdots$$

i.e. $\sin 2x = 2x - \dfrac{4x^3}{3} + \dfrac{4x^5}{15} - \dfrac{8x^7}{315} + \cdots$

Problem 2. Determine the first five terms of the series for $\ln(1 + x)$

The values of $f(0)$, $f'(0)$, $f''(0)$, ... in Maclaurin's series are defined as follows:

$f(x) = \ln(1 + x)$ $f(0) = \ln(1 + 0) = 0$

$f'(x) = \dfrac{1}{1 + x}$ $f'(0) = \dfrac{1}{1 + 0} = 1$

$f''(x) = -\dfrac{1}{(1 + x)^2}$ $f''(0) = -\dfrac{1}{(1 + 0)^2} = -1$

$f'''(x) = \dfrac{2}{(1 + x)^3}$ $f'''(0) = \dfrac{2}{(1 + 0)^3} = 2!$

$$f^{iv}(x) = -\frac{3 \cdot 2}{(1+x)^4} \qquad f^{iv}(0) = -\frac{3!}{(1+0)^4} = (-1)(3!)$$

$$f^{v}(x) = \frac{4 \cdot 3 \cdot 2}{(1+x)^5} \qquad f^{v}(0) = \frac{4!}{(1+0)^5} = 4!$$

Substituting the values of $f(0)$, $f'(0)$, $f''(0)$, ... in equation (5), gives:

$$\ln(1+x) = 0 + (1)x + (-1)\frac{x^2}{2!} + \frac{(2!)x^3}{3!}$$

$$+ \frac{(-1)(3!)x^4}{4!} + \frac{4!x^5}{5!} + \cdots$$

i.e. $\ln(1+x) = x - \dfrac{x^2}{2} + \dfrac{x^3}{3} - \dfrac{x^4}{4} + \dfrac{x^5}{5} - \cdots$

In this series, $|U_n| = \dfrac{x^n}{n}$, i.e. when n is, say, 4 the term in x^4 is $\dfrac{x^4}{4}$, and so on.

Also $|U_{n+1}| = \dfrac{x^{n+1}}{n+1}$

Testing for convergence (see page 80),

$$\lim_{n \to \infty} \left| \frac{U_{n+1}}{U_n} \right| = \lim_{n \to \infty} \left| \frac{x^{n+1}}{n+1} \times \frac{n}{x^n} \right|$$

$$= \lim_{n \to \infty} \left| \frac{nx}{n+1} \right|$$

As $n \to \infty$, $n \to n+1$ and $\dfrac{n}{n+1} \to 1$

Hence, $\displaystyle\lim_{n \to \infty} \left| \frac{nx}{n+1} \right| = |x|$

The modulus of x must be less than 1 for the series to converge, hence this series only converges for values of x such that $-1 < x < 1$ and hence is only valid for these values of x.

Problem 3. Use the power series for $\ln\left(\dfrac{1+x}{1-x}\right)$ to determine the value of ln 3 correct to three decimal places.

The power series for $\ln(1 + x)$ is derived in Problem 2 above, and is

$$\ln(1 + x) = x - \frac{x^2}{2} + \frac{x^3}{3} - \frac{x^4}{4} + \frac{x^5}{5} - \cdots$$

Replacing x by $(-x)$ in this power series, gives:

$$\ln(1 + (-x)) = (-x) - \frac{(-x)^2}{2} + \frac{(-x)^3}{3} - \frac{(-x)^4}{4} + \frac{(-x)^5}{5} - \cdots$$

i.e. $\ln(1 - x) = -x - \frac{x^2}{2} - \frac{x^3}{3} - \frac{x^4}{4} - \frac{x^5}{5} - \cdots$

A law of logarithms states that $\ln\left(\dfrac{A}{B}\right) = \ln A - \ln B$

Applying this law to $\ln\left(\dfrac{1 + x}{1 - x}\right)$ gives:

$$\ln\left(\frac{1 + x}{1 - x}\right) = \ln(1 + x) - \ln(1 - x)$$

$$= \left(x - \frac{x^2}{2} + \frac{x^3}{3} - \frac{x^4}{4} + \frac{x^5}{5} - \cdots\right)$$

$$- \left(-x - \frac{x^2}{2} - \frac{x^3}{3} - \frac{x^4}{4} - \frac{x^5}{5} - \cdots\right)$$

$$= 2x + \frac{2x^3}{3} + \frac{2x^5}{5} + \frac{2x^7}{7} + \cdots$$

i.e. $\ln\left(\dfrac{1 + x}{1 - x}\right) = 2\left(x + \dfrac{x^3}{3} + \dfrac{x^5}{5} + \dfrac{x^7}{7} + \cdots\right)$

The series for $\ln\left(\dfrac{1 + x}{1 - x}\right)$ is convergent provided $-1 < x < 1$, and is far more suitable than the power series for $\ln(1 + x)$ or $\ln(1 - x)$ when determining the values of Naperian logarithms of numbers. This is because it converges far more rapidly since its terms contain only odd powers of x and it reaches the term containing, say, x^7 in four terms instead of the seven terms in the expansion of $\ln(1 \pm x)$. Hence less terms have to be calculated to give the value of a Naperian logarithm to a required degree of accuracy; also it may be used to determine the values of Naperian logarithms for numbers equal to or greater than 2.

Since the series for $\ln\left(\dfrac{1+x}{1-x}\right)$ is being used to find the value of $\ln 3$, then

$$\frac{1+x}{1-x} = 3, \text{ and } 1+x = 3-3x$$

i.e. $x = \dfrac{1}{2}$

Thus $\ln\left(\dfrac{1+\frac{1}{2}}{1-\frac{1}{2}}\right) = \ln 3 = 2\left(\dfrac{1}{2} + \dfrac{(\frac{1}{2})^3}{3} + \dfrac{(\frac{1}{2})^5}{5} + \dfrac{(\frac{1}{2})^7}{7} + \dfrac{(\frac{1}{2})^9}{9} + \cdots\right)$

$$= 2(0.500\,0 + 0.041\,7 + 0.006\,3 + 0.001\,1$$
$$\qquad + 0.000\,2 + \cdots)$$

$$= 2(0.549\,3) = 1.098\,6$$

i.e.   **$\ln 3 = 1.099$ correct to three decimal places.**

Problem 4. Use Maclaurin's series to find the expansion of $(3+x)^4$

The values of $f(0), f'(0), f''(0), \ldots$ in Maclaurin's series are determined as follows:

$f(x) = (3+x)^4$ $f(0) = 3^4$

$f'(x) = 4(3+x)^3$ $f'(0) = 4\cdot 3^3$

$f''(x) = 4\cdot 3(3+x)^2$ $f''(0) = 4\cdot 3\cdot 3^2$

$f'''(x) = 4\cdot 3\cdot 2(3+x)$ $f'''(0) = 4\cdot 3\cdot 2\cdot 3^1$

$f^{\text{iv}}(x) = 4\cdot 3\cdot 2\cdot 1(3+x)^0$ $f^{\text{iv}}(0) = 4\cdot 3\cdot 2\cdot 1\cdot 3^0$

Substituting the values of $f(0), f'(0), f''(0), \ldots$ in equation (5) gives:

$$(3+x)^4 = 3^4 + 4\cdot 3^3 x + \frac{4\cdot 3\cdot 3^2\cdot x^2}{2!} + \frac{4\cdot 3\cdot 2\cdot 3^1\cdot x^3}{3!} + \frac{4\cdot 3\cdot 2\cdot 1\cdot x^4}{4!}$$

i.e. the expansion which would have been obtained by applying the binomial series. Thus

$(3+x)^4 = 81 + 108x + 54x^2 + 12x^3 + x^4$

Problem 5. Use Maclaurin's series to determine the terms of the power series for $\ln(1+e^x)$ as far as the term in x^3.

$$f(x) = \ln(1 + e^x) \qquad\qquad f(0) = \ln(1 + 1) = \ln 2$$

$$f'(x) = \frac{e^x}{1 + e^x} \qquad\qquad f'(0) = \frac{1}{1 + 1} = \frac{1}{2}$$

$$f''(x) = \frac{(1 + e^x)e^x - e^x(e^x)}{(1 + e^x)^2}$$

$$= \frac{e^x(1 + e^x - e^x)}{(1 + e^x)^2} = \frac{e^x}{(1 + e^x)^2} \qquad f''(0) = \frac{1}{2^2} = \frac{1}{4}$$

$$f'''(x) = \frac{(1 + e^x)^2 e^x - e^x \cdot 2(1 + e^x)e^x}{(1 + e^x)^4} \qquad f'''(0) = \frac{2^2 - 2(2)}{2^4} = 0$$

Substituting the values of $f(0)$, $f'(0)$, $f''(0)$, ... in equation (5) gives:

$$\ln(1 + e^x) = \ln 2 + \frac{x}{2} + \frac{x^2}{4(2!)} + 0 \cdot x^3$$

i.e. $\ln(1 + e^x) = \ln 2 + \dfrac{x}{2} + \dfrac{x^2}{8}$, as far as the term in x^3.

Power series

Many mathematical functions may be determined along similar lines to those shown in the worked problems above. Some of the results which may be obtained by applying Maclaurin's theorem to various functions are listed below.

$$\sin x = x - \frac{x^3}{3!} + \frac{x^5}{5!} - \frac{x^7}{7!} + \cdots$$

$$\cos x = 1 - \frac{x^2}{2!} + \frac{x^4}{4!} - \frac{x^6}{6!} + \cdots$$

$$e^x = 1 + x + \frac{x^2}{2!} + \frac{x^3}{3!} + \cdots$$

$$\sinh x = x + \frac{x^3}{3!} + \frac{x^5}{5!} + \frac{x^7}{7!} + \cdots$$

$$\cosh x = 1 + \frac{x^2}{2!} + \frac{x^4}{4!} + \frac{x^6}{6!} + \cdots$$

$$\ln(1 + x) = x - \frac{x^2}{2} + \frac{x^3}{3} - \frac{x^4}{4} + \cdots \quad \text{(Only valid for } -1 < x < 1)$$

$$(a + x)^n = a^n + na^{n-1}x + \frac{n(n - 1)}{2!} a^{n-2}x^2$$

$$+ \frac{n(n - 1)(n - 2)}{3!} a^{n-3}x^3 + \cdots$$

(This last power series is called the **binomial series** which is discussed in Chapter 1 of *Technician mathematics 3*.)

These results can be used to determine the power series for more complicated functions. For example, since the series for e^x is $1 + x + \dfrac{x^2}{2!} + \dfrac{x^3}{3!} + \cdots$ the series for e^{2x} is $1 + (2x) + \dfrac{(2x)^2}{2!} + \dfrac{(2x)^3}{3!} + \cdots$

A power series for the product of two functions can be determined by multiplying the corresponding power series. For example, the power series for $e^x \cos x$ is $\left(1 + x + \dfrac{x^2}{2!}\right)\left(1 - \dfrac{x^2}{2!}\right)$, that is, $1 - \dfrac{x^2}{2!} + x - \dfrac{x^3}{2!} + \dfrac{x^2}{2!}$, i.e. $1 + x - \dfrac{x^3}{2!}$, as far as the term in x^3. Functions involving quotients may be determined in a similar way.

Further problems on Maclaurin's series may be found in Section 6.4 (Problems 1 to 10), page 94.

6.2   Taylor's series

Maclaurin's series, introduced in Section 6.1, expresses the height of a curve at any point x in terms of its height at the origin. With reference to Fig. 6.1(a), the height of the curve at B is AB, that is, $f(x) = $ AB and the height at the origin is EC, that is, $f(0) = $ EC. Maclaurin's series, introduced in equation (5), expresses $f(x)$ in terms of $f(0)$ and is

$$f(x) = f(0) + xf'(0) + \frac{x^2}{2!} f''(0) + \cdots$$

Taylor's series expresses the height of a curve at distance $(a + h)$ from the origin in terms of its height at distance a from the origin. With reference to Fig. 6.1(b), the origin of the vertical axis is moved until it lies along GD, that is, it is moved distance a to the left. Since the shape of ECBA in Fig. 6.1(b) is the same as in Fig. 6.1(a), then Maclaurin's series must still relate the height of AB to the height of EC. However, since the origin has been moved, EC is now of height $f(a)$ and distance x in Fig. 6.1(a) now becomes $(x - a)$ with reference to the new axis. Writing $f(a)$ for $f(0)$ and $(x - a)$ for x in Maclaurin's series gives the height of the curve at AB as:

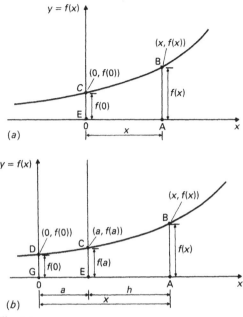

Fig. 6.1

$$f(x) = f(a) + (x - a)f'(a) + \frac{(x - a)^2}{2!} f''(a)$$

$$+ \frac{(x - a)^3}{3!} f'''(a) + \cdots \qquad (6)$$

This equation is called **Taylor's theorem** or **Taylor's series**.

If a is made zero in Taylor's series given in equation (6), then:

$$f(x) = f(0) + xf'(0) + \frac{x^2}{2!} f''(0) + \cdots$$

i.e. Maclaurin's series is a special case of Taylor's series when a is equal to 0. Some of the applications of Taylor's series are to such techniques as numerical differentiation, limits, small errors and the numerical solution of certain differential equations.

An alternative way of writing Taylor's series is obtained by writing the series in terms of h rather than x. Since, from Fig. 6.1(b), $x = (a + h)$, and $x - a = h$, then substituting for x in equation (6) gives:

$$f(a + h) = f(a) + hf'(a) + \frac{h^2}{2!} f''(a) + \frac{h^3}{3!} f'''(a) + \cdots \qquad (7)$$

Taylor's series may be used for the following two applications:

(i) To determine the approximate values of functions which are close to known values of that function. For example, the value of $\cos 60°$ is known (i.e. $\cos 60° = 0.5$). Taylor's series may be used to determine the value of, say, $\cos 61°$. In the power series for $f(a + h)$, given in equation (7), a is taken as $60°$, i.e. $\pi/3$ radians and h is taken as $1°$, i.e. $\dfrac{1 \times \pi}{180}$ radians. These values are then substituted in the power series for $\cos(a + h)$. Naturally, with electronic calculators readily available, this application of Taylor's theorem has decreased in use.

(ii) Newton's method of finding the approximate value of the roots of an equation is introduced in Chapter 2, and may be derived as follows. Let the equation $f(x) = 0$ have a root which is approximately equal to a. Also, let the true value of the root be (a + a small quantity), say, $(a + h)$. Then $f(a + h) = 0$. By Taylor's series:

$$f(a + h) = f(a) + hf'(a) + \frac{h^2}{2!}f''(a) + \frac{h^3}{3!}f'''(a) + \cdots$$

Since $f(a + h) = 0$, then

$$f(a) + hf'(a) + \frac{h^2}{2!}f''(a) + \cdots = 0$$

Because h is small, terms containing h^2 and higher powers of h are very small and may be neglected, hence

$$f(a) + hf'(a) \approx 0$$

i.e. $\qquad h \approx \dfrac{-f(a)}{f'(a)}$

Since the true value of the root is at $(a + h)$ and the value obtained for h is only an approximate value, then

$$a + h \approx a - \frac{f(a)}{f'(a)}$$

Let this latter approximation to the value of the root be a_1 then

$$a_1 = a - \frac{f(a)}{f'(a)} \tag{8}$$

This is the basis of Newton's method of determining the approximate value of roots of equations.

Worked problems on Taylor's series

Problem 1. Determine the power series for $\sin(a + h)$ and hence determine the value of $\sin 30° \ 30'$ correct to five decimal places.

Taylor's series for $f(a + h)$ is given in equation (7) and is:

$$f(a + h) = f(a) + hf'(a) + \frac{h^2}{2!}f''(a) + \frac{h^3}{3!}f'''(a) + \cdots$$

The function f is the sine function in this problem, hence

$$f(a + h) = \sin(a + h) \text{ and } f(a) = \sin a$$

By repeated differentiation, $f'(a) = \cos a$, $f''(a) = -\sin a$, $f'''(a) = -\cos a$, $f^{iv}(a) = \sin a$, and so on.

Substituting for $f(a)$, $f'(a)$, $f''(a)$, ... in Taylor's series, gives:

$$\sin(a + h) = \sin a + h \cos a - \frac{h^2}{2!} \sin a - \frac{h^3}{3!} \cos a + \cdots$$

To evaluate $\sin 30° \, 30'$, radian measure must be used and a is taken as $30°$, i.e. $\pi/6$ radians, and h as $30'$ or $0.5°$, i.e. $\dfrac{0.5 \times \pi}{180}$ radians.

Thus, $\sin 30° \, 30' = \sin \dfrac{\pi}{6} + \dfrac{0.5\pi}{180} \cos \dfrac{\pi}{6} - \dfrac{1}{2!} \left(\dfrac{0.5\pi}{180}\right)^2 \sin \dfrac{\pi}{6}$

$$- \frac{1}{3!}\left(\frac{0.5\pi}{180}\right)^3 \cos \pi/6 + \cdots$$

$$= 0.500\,000 + 0.008\,727 \cdot \frac{\sqrt{3}}{2} - \frac{1}{2} \cdot 0.000\,076 \cdot \frac{1}{2}$$

$$- \frac{1}{6} \cdot 0.000\,000\,7 \frac{\sqrt{3}}{2}$$

$$= 0.500\,000 + 0.007\,558 - 0.000\,019 - 0.000\,000\,1,$$

i.e. $\sin 30° \, 30' = 0.507\,54$, correct to five decimal places.

Problem 2. Given that $\cos 60° = 0.5$, determine the value of $\cos 70°$ by using Taylor's series, correct to four decimal places.

Taylor's series for $f(x)$ is given in equation (6) and is:

$$f(x) = f(a) + (x - a)f'(a) + \frac{(x - a)^2}{2!}f''(a) + \frac{(x - a)^3}{3!}f'''(a) + \cdots$$

With reference to Fig. 6.1(b), x corresponds to $70°$, i.e. $\dfrac{7\pi}{18}$ radians, a corresponds to $60°$, i.e. $\dfrac{\pi}{3}$ radians, and $(x - a)$ corresponds to $10°$, i.e. $\dfrac{\pi}{18}$ radians. Also, $f(a) = \cos \dfrac{\pi}{3}$, and by repeated differentiation,

$$f'(a) = -\sin\frac{\pi}{3}, \quad f''(a) = -\cos\frac{\pi}{3}, \quad f'''(a) = \sin\frac{\pi}{3} \text{ and so on.}$$

Thus, substituting these values in Taylor's series, gives:

$$f(x) = \cos 70° = \cos\frac{\pi}{3} - \left(\frac{\pi}{18}\right)\sin\frac{\pi}{3} - \frac{1}{2!}\left(\frac{\pi}{18}\right)^2\cos\frac{\pi}{3}$$

$$+ \frac{1}{3!}\left(\frac{\pi}{18}\right)^3\sin\frac{\pi}{3} + \frac{1}{4!}\left(\frac{\pi}{18}\right)^4\cos\frac{\pi}{3} - \cdots$$

$$= 0.500\,00 - 0.151\,15 - 0.007\,62$$
$$+ 0.000\,77 + 0.000\,02$$

$$= 0.342\,02$$

i.e.   **cos 70° = 0.342 0, correct to four decimal places.**

Problem 3. Determine the value of the positive root of the equation $x^3 - 2x - 5 = 0$, correct to three decimal places, using Newton's method.

The first approximation to the value of the root is found by determining the approximate point where the graph cuts the x-axis. Using functional notation:

$$f(x) = x^3 - 2x - 5$$
$$f(0) = (0)^3 - 2(0) - 5 = -5$$
$$f(1) = (1)^3 - 2(1) - 5 = -6$$
$$f(2) = (2)^3 - 2(2) - 5 = -1$$
$$f(3) = (3)^3 - 2(3) - 5 = +16$$

This shows that $x^3 - 2x - 5$ is very nearly equal to 0 at $x = 2$, i.e. the first approximation is $a = 2$.

Since $f(x) = x^3 - 2x - 5$, $f'(x) = 3x^2 - 2$.

From equation (8), a second approximation is given by

$$a_1 = a - \frac{f(a)}{f'(a)}$$

When $a = 2$, $a_1 = 2 - \dfrac{(-1)}{3(2)^2 - 2} = 2.1$

The third approximation is given by:

$$a_2 = a_1 - \frac{f(a_1)}{f'(a_1)}$$

When $a_1 = 2.1$, $a_2 = 2.1 - \left(\dfrac{(2.1)^3 - 2(2.1) - 5}{3(2.1)^2 - 2}\right)$

$$= 2.1 - \frac{0.06}{11.23} = 2.095$$

The fourth approximation is given by:

$$a_3 = a_2 - \frac{f(a_2)}{f'(a_2)}$$

When $a_2 = 2.095$, $a_3 = 2.095 - \left(\frac{(2.095)^3 - 2(2.095) - 5}{3(2.095)^2 - 2}\right)$

$$= 2.095 - \frac{0.005}{11.17}$$

$$= 2.095 - 0.000\,4 = 2.094\,6$$

Since the values of the third and fourth approximations are the same, when expressed to the required degree of accuracy (i.e. three decimal places), then

the positive root of $x^3 - 2x - 5 = 0$ is 2.095, correct to three decimal places.

Further problems on Taylor's series may be found in Section 6.4 (Problems 11 to 17), page 95.

6.3   Approximate values of definite integrals by using series expansions

The value of many definite integrals cannot be determined by the various analytical methods introduced in *Technician mathematics 3* and in Chapters 17 to 21. Several methods exist of finding the approximate value of definite integrals. One of these is to express the function as a power series and then to integrate the terms of the power series. This method is particularly suitable for use with functions whose power series converge fairly rapidly, where evaluation of a few terms gives the required degree of accuracy. This method is shown in the worked problems following.

Worked problems on approximate values of definite integrals by using series expansions

---

Problem 1. Determine the value of $\int_{0.1}^{0.5} e^{\sin x}\, dx$, correct to three significant figures.

The power series for $e^{\sin x}$ is obtained by applying Maclaurin's theorem given in equation (5), i.e.

$$f(x) = f(0) + f'(0)x + f''(0)\frac{x^2}{2!} + f'''(0)\frac{x^3}{3!} + \cdots$$

$f(x) = e^{\sin x}$. Hence, $f(0) = e^{\sin 0} = 1$

$f'(x) = \cos x \, e^{\sin x}$. Hence, $f'(0) = 1 \cdot e^0 = 1$

$f''(x) = \cos^2 x \, e^{\sin x} + (-\sin x) \, e^{\sin x}$
$\qquad = e^{\sin x} (\cos^2 x - \sin x)$. Hence, $f''(0) = 1(1 - 0) = 1$

$f'''(x) = e^{\sin x}(2 \cos x \sin x - \cos x) + \cos x \, e^{\sin x} (\cos^2 x - \sin x)$
$\qquad = \cos x \, e^{\sin x} (\sin x - 1 + \cos^2 x)$. Hence,
$f'''(0) = 1(0 - 1 + 1) = 0$

Thus, the power series for $e^{\sin x}$ is $1 + x + \dfrac{x^2}{2} + 0 + \cdots$

It follows that

$$\int_{0.1}^{0.5} e^{\sin x} \, dx \approx \int_{0.1}^{0.5} \left(1 + x + \frac{x^2}{2}\right) dx$$

$$\approx \left[x + \frac{x^2}{2} + \frac{x^3}{6}\right]_{0.1}^{0.5}$$

$$\approx 0.645\,8 - 0.105\,2$$

$$\approx 0.540\,6$$

i.e. $\displaystyle\int_{0.1}^{0.5} e^{\sin x} \, dx = \mathbf{0.541, \ correct \ to \ three \ significant \ figures.}$

Problem 2. Determine the value of $\displaystyle\int_0^1 \frac{\cos 2x}{x^{\frac{1}{3}}} \, dx$, correct to two decimal places.

Maclaurin's theorem is used to express $\cos 2x$ as a power series.

$f(x) = \cos 2x \qquad\qquad f(0) = \cos 0 = 1$

$f'(x) = -2 \sin 2x \qquad\quad f'(0) = -2(0) = 0$

$f''(x) = -4 \cos 2x \qquad\quad f''(0) = -4(1) = -4$

$f'''(x) = 8 \sin 2x \qquad\qquad f'''(0) = 8(0) = 0$

$f^{iv}(x) = 16 \cos 2x \qquad\quad f^{iv}(0) = 16(1) = 16$

$f^v(x) = -32 \sin 2x \qquad\quad f^v(0) = -32(0) = 0$

$f^{vi}(x) = -64 \cos 2x \qquad\, f^{vi}(0) = -64(1) = -64$

Hence, $\cos 2x = 1 - \dfrac{4x^2}{2!} + \dfrac{16x^4}{4!} - \dfrac{64x^6}{6!} + \cdots$

$$= 1 - 2x^2 + \frac{2x^4}{3} - \frac{4x^6}{45} + \cdots$$

Thus, $\displaystyle \int_0^1 \frac{\cos 2x}{x^{\frac{1}{3}}}\, dx = \int_0^1 x^{-\frac{1}{3}}(\cos 2x)\, dx$

$$\approx \int_0^1 \left\{ x^{-\frac{1}{3}}\left(1 - 2x^2 + \frac{2x^4}{3} - \frac{4x^6}{45}\right)\right\} dx$$

$$\approx \int_0^1 \left(x^{-\frac{1}{3}} - 2x^{\frac{5}{3}} + \frac{2x^{\frac{11}{3}}}{3} - \frac{4x^{\frac{17}{3}}}{45}\right) dx$$

$$\approx \left[\frac{3x^{\frac{2}{3}}}{2} - \frac{(2)3x^{\frac{8}{3}}}{8} + \frac{(2)3x^{\frac{14}{3}}}{(3)(14)} - \frac{(4)3x^{\frac{20}{3}}}{(45)(20)}\right]_0^1$$

$$\approx \left[\frac{3x^{\frac{2}{3}}}{2} - \frac{3x^{\frac{8}{3}}}{4} + \frac{x^{\frac{14}{3}}}{7} - \frac{x^{\frac{20}{3}}}{75}\right]_0^1$$

$$\approx \left(\frac{3}{2} - \frac{3}{4} + \frac{1}{7} - \frac{1}{75}\right)$$

$$\approx 0.879\,5$$

i.e. $\displaystyle \int_0^1 x^{-\frac{1}{3}}(\cos 2x)\, dx = \textbf{0.88, correct to two decimal places.}$

Further problems on approximate values of definite integrals by series expansions may be found in the following Section 6.4 (Problems 18 to 27), page 96.

6.4   Further problems

Maclaurin's series

In Problems 1 to 9, use Maclaurin's theorem or the series given on page 86, to determine the power series stated.

1. cos x, as far as the term in x^6. Hence determine the value of cos 0.38 radians correct to six decimal places.

$$\left[1 - \frac{x^2}{2!} + \frac{x^4}{4!} - \frac{x^6}{6!};\ 0.928\,665 \right]$$

2. $e^{x/2}$, as far as the term in x^5.

$$\left[1 + \frac{x}{2} + \frac{x^2}{8} + \frac{x^3}{48} + \frac{x^4}{384} + \frac{x^5}{3\,840} \right]$$

3. tan ax, as far as the term in x^5.

$$\left[ax + \frac{a^3x^3}{3} + \frac{2a^5x^5}{15} \right]$$

4. $\sec \dfrac{x}{2}$, as far as the term in x^4. $\left[1 + \dfrac{x^2}{8} + \dfrac{5x^4}{384}\right]$

5. $\cos^2 2x$, as far as the term in x^6.
$$\left[1 - 4x^2 + \dfrac{16x^4}{3} - \dfrac{128x^6}{45}\right].$$

6. $(1 - x)^{-3}$, as far as the term in x^4.
$$[1 + 3x + 6x^2 + 10x^3 + 15x^4]$$

7. $e^{2x} \cos 3x$, as far as the term in x^2. $\left[1 + 2x - \dfrac{5}{2}x^2\right]$

8. $\dfrac{1 + x}{(1 - x)^3}$, as far as the term in x^3. $[1 + 4x + 9x^2 + 16x^3]$

9. $\ln(1 + 3x) \sin \dfrac{3x}{2}$, as far as the term in x^4.
$$\left[\dfrac{9x^2}{2} - \dfrac{27x^3}{4} + \dfrac{189x^4}{16}\right]$$

10. By expressing $\ln\left(\dfrac{1 + x}{1 - x}\right)$ as a power series, determine the
value of $\ln 0.8$, correct to five decimal places. $[-0.223\,14]$

Taylor's series

11. Determine the power series for $\cos(a + h)$ and hence determine
the value of $\cos 31°$, correct to five decimal places.
$$\left[\cos(a + h) = \cos a - h \sin a\right.$$
$$\left. - \dfrac{h^2}{2!} \cos a + \dfrac{h^3}{3!} \sin a + \cdots; 0.857\,17\right]$$

12. Find the value of $\tan 31°$ correct to five decimal places by
determining the power series for $\tan(a + h)$.
$$[\tan(a + h) = \tan a + h \sec^2 a + h^2 \sec^2 a \tan a$$
$$+ \dfrac{h}{3} \sec^2 a(\sec^2 a + 2 \tan^2 a) + \cdots; 0.600\,86]$$

13. Given that $\ln 10 = 2.302\,585$, determine the value of $\ln 12$,
correct to five decimal places. $[2.484\,91]$

14. Use Taylor's series to determine the value of $\tan 50°$, correct
to six significant figures, given that $\tan 40° = 0.839\,100$.
$$[1.191\,75]$$

15. Determine the three roots of $x^3 = 7x + 1$, correct to four
significant figures, by Newton's method.
$$[2.714, -0.143\,3, -2.571]$$

16. Find the value of the real root of the equation

$$9 - 9\left(\frac{7}{6}x - 1\right)^{\frac{8}{7}} - 10x = 0,$$

correct to three significant figures, by using Newton's method.
[0.883]

17. Determine the value of the root of the equation

$$1 - \frac{x}{6} + \frac{x^2}{180} - \frac{x^3}{12\,960} = 0,$$

which lies near to 8, correct to four significant figures.
[7.815]

Approximate values of definite integrals by using series expansions

In Problems 18 to 22, determine the values of the definite integrals given, correct to two significant figures.

18. $\displaystyle\int_0^1 x^{\frac{1}{2}} \cos x \, dx$ [0.53]

19. $\displaystyle\int_0^{\frac{1}{2}} \sqrt{(x)} \ln(x + 1) \, dx$ [0.061]

20. $\displaystyle\int_0^{\pi/3} (\sec x)^{\frac{1}{2}} \, dx$ [1.2]

21. $\displaystyle\int_1^2 \tfrac{1}{2} \cos 3\theta \, d\theta$ [−0.070]

22. $\displaystyle\int_{\frac{1}{3}}^1 \sec^2(3x - 1) \, dx$ [−0.73]

23. Determine the value of $\displaystyle\int_0^{\frac{1}{2}} x^2 \ln(1 + \sin x) \, dx$, correct to two significant figures.   [0.013]

24. By determining the power series for $\sec \theta$ as far as the term in θ^4, determine the approximate value of $\displaystyle\int_0^{\pi/6} \sec \theta \, d\theta$.

$$\left[\int_0^{\pi/6}\left(1 + \frac{\theta^2}{2} + \frac{5}{24}\theta^4\right) d\theta \approx 0.549 \text{ correct to three decimal places.}\right]$$

25. Use Maclaurin's theorem to determine the power series for e^{-x^2} as far as the term in x^4 and hence determine the value of $\displaystyle\int_0^1 e^{-x^2} \, dx$, correct to three significant figures.   [0.767]

26. Determine the value of $\int_0^{0.3} \dfrac{(1 - y^2)^{\frac{1}{2}}}{4\sqrt{y}}\, dy$, correct to three

significant figures.   [0.271]

27. Use the power series for $\sin \phi$ to determine the value of

$\int_0^{\pi/6} \phi^{\frac{1}{2}} \sin \phi \, d\phi$, correct to three decimal places.   [0.077]

Chapter 7

Complex numbers

7.1 Introduction

The solutions of the quadratic equation $ax^2 + bx + c = 0$ may be obtained by using the quadratic formula:

$$x = \frac{-b \pm \sqrt{(b^2 - 4ac)}}{2a}$$

Hence the solutions of $2x^2 + x - 3 = 0$ are

$$x = \frac{-1 \pm \sqrt{[1^2 - 4(2)(-3)]}}{2(2)}$$

$$= \frac{-1 \pm \sqrt{25}}{4} = -\frac{1}{4} \pm \frac{5}{4}$$

i.e. $\quad x = 1$ or $x = -1\frac{1}{2}$

However, a problem exists if $b^2 - 4ac$ in the quadratic formula results in a negative number, since we cannot obtain in real terms the square root of a negative number. The only numbers we have met to date have been real numbers. These are either integers (such as $+1$, $+5$, -7, etc.) or rational numbers (such as $\frac{4}{1}$ or 4.000, $-\frac{7}{9}$ or $-0.7\dot{7}$, $\frac{1}{2}$ or 0.5000, etc.) or irrational numbers (such as $\pi = 3.141\,592\ldots$, $\sqrt{3} = 1.732,\ldots$, etc.)

If $x^2 - 2x + 5 = 0$, then

$$x = \frac{2 \pm \sqrt{[(-2)^2 - 4(1)(5)]}}{2(1)}$$

$$= \frac{2 \pm \sqrt{-16}}{2}$$

In order to deal with such a problem as determining $\sqrt{-16}$ the concept of complex numbers has been evolved.

7.2   Definition of a complex number

(a) Imaginary numbers

Let b and c be real numbers. An imaginary number is written in the form jb or jc (i.e. $j \times b$ or $j \times c$) where operator j is defined by the following two rules:

(i)   For addition, $jb + jc = j(b + c)$ (1)
(ii) For multiplication, $jb \times jc = -bc$ (2)
      i.e. $j^2 bc = -bc$
      Thus $j^2 = -1$
      or $j = \sqrt{-1}$

It is immaterial whether the operator j is placed in front of, or after, the number, i.e. $j4 = 4j$, and so on. An imaginary number $j3$ means $(\sqrt{-1}) \times 3$ and $-j2$ means $-(\sqrt{-1}) \times 2$ or $(-2) \times (\sqrt{-1})$
  Similarly, from equation (1), $j2 + j5 = j7$
  and $j5 - j2 = j3$

(b) Complex numbers

From section 7.1, $x^2 - 2x + 5 = 0$

and $x = \dfrac{2 \pm \sqrt{-16}}{2}$

$\sqrt{-16}$ can be split into $(\sqrt{-1}) \times \sqrt{16}$, i.e. $j\sqrt{16}$ or $\pm j4$
Hence $x = \dfrac{2 \pm j4}{2} = 1 \pm j2$

i.e. $x = 1 + j2$ or $x = 1 - j2$

The solutions of the quadratic equation $x^2 - 2x + 5 = 0$ are of the form $a + jb$, where 'a' is a real number and 'jb' an imaginary number. Numbers in the form $a + jb$ are called complex numbers. Hence $1 + j2$, $5 - j7$, $-2 + j3$ and $-\pi + j\sqrt{2}$ are all examples of complex numbers.
  In algebra, if a quantity x is added to a quantity $3y$ the result is written as $x + 3y$ since x and y are separate quantities. Similarly, it is important to appreciate that real and imaginary numbers are different types of numbers and must be kept separate.
  When imaginary numbers were first introduced the symbol i (i.e. the first letter of the word imaginary) was used to indicate $\sqrt{-1}$ and this symbol is still used in pure mathematics. However, in engineering, the symbol i indicates electric current, and to avoid any possible confusion the next letter in the alphabet, i.e. j, is used to represent $\sqrt{-1}$.

7.3   The Argand diagram

A complex number can be represented pictorially on rectangular or cartesian axes. The horizontal (or x) axis is used to represent the real axis and the vertical (or y) axis is used to represent the imaginary axis. Such a diagram is called an **Argand diagram** (named after an eighteenth-century French mathematician), and is shown in Fig. 7.1(a).

In Fig. 7.1(b) the point A represents the complex number $2 + j3$ and is obtained by plotting the coordinates (2, $j3$) as in graphical work. Often an arrow is drawn from the origin to the point as shown. Figure 7.1(b) also

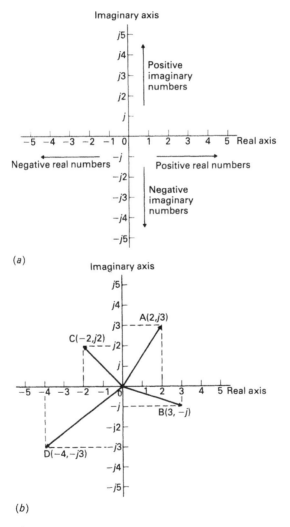

Fig. 7.1

shows the Argand points B, C and D representing the complex numbers $3-j$, $-2+j2$ and $-4-j3$ respectively.

A complex number of the form $a+jb$ is called a **cartesian complex number**.

7.4 Operations involving cartesian complex numbers

(a) Addition and subtraction

Let two complex numbers be represented by $Z_1 = a+jb$ and $Z_2 = c+jd$. Two complex numbers are added/subtracted by adding/subtracting separately the two real parts and the two imaginary parts.

Hence $Z_1 + Z_2 = (a+jb)+(c+jd)$

$$= (a+c)+j(b+d)$$

and $Z_1 - Z_2 = (a+jb)-(c+jd)$

$$= (a-c)+j(b-d)$$

Thus, if $Z_1 = 3+j2$ and $Z_2 = 2-j4$ then:

$$Z_1 + Z_2 = (3+j2)+(2-j4) = 3+j2+2-j4 = 5-j2$$

and $Z_1 - Z_2 = (3+j2)-(2-j4) = 3+j2-2+j4 = 1+j6$

The addition and subtraction of complex numbers may be achieved graphically as shown in the Argand diagram in Fig. 7.2.

In Fig. 7.2(a), by vector addition, $\mathbf{OP} + \mathbf{OQ} = \mathbf{OR_1}$. R_1 is found to be the Argand point $(5, -j2)$, i.e. $(3+j2)+(2-j4) = 5-j2$ (as above).

In Fig. 7.2(b), vector $\mathbf{OQ}$ is reversed (shown as $\mathbf{OQ'}$) since it is being subtracted.

(Note that $\mathbf{OQ} = 2-j4$
Thus $-\mathbf{OQ} = -(2-j4)$
$$= -2+j4, \text{ shown as } \mathbf{OQ'}.)$$

Thus $\mathbf{OP} - \mathbf{OQ} = \mathbf{OP} + \mathbf{OQ'} = \mathbf{OR_2}$. R_2 is found to be the Argand point $(1, j6)$, i.e. $(3+j2)-(2-j4) = 1+j6$ (as above)

(b) Multiplication

Two complex numbers are multiplied by assuming that all quantities involved are real and then, by using $j^2 = -1$, expressing the product in the form $a+jb$.

(a)

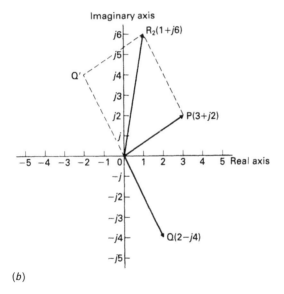

(b)

Fig. 7.2   (a) **OP + OQ = OR₁** (b) **OP − OQ = OR₂**

Hence $(a + jb)(c + jd) = ac + a(jd) + (jb)c + (jb)(jd)$

$$= ac + jad + jbc + j^2bd$$

But $j^2 = -1$, thus

$(a + jb)(c + jd) = (ac - bd) + j(ad + bc)$.
If $Z_1 = 3 + j2$ and $Z_2 = 2 - j4$ then

$$Z_1Z_2 = (3 + j2)(2 - j4) = 6 - j12 + j4 - j^28$$
$$= (6 - (-1)8) + j(-12 + 4)$$
$$= 14 + j(-8)$$
$$= 14 - j8$$

(c) Complex conjugate

The complex conjugate of a complex number is obtained by changing the sign of the imaginary part. Hence $a + jb$ is the complex conjugate of $a - jb$ and $-2 + j$ is the complex conjugate of $-2 - j$. (Note that only the sign of the imaginary part is changed.)

The product of a complex number Z and its complex conjugate $\bar{Z}$ is always a real number, and this is an important property used when dividing complex numbers.

Thus if $Z = a + jb$ then $\bar{Z} = a - jb$

and $Z\bar{Z} = (a + jb)(a - jb)$

$$= a^2 - jab + jab - j^2b^2$$

$$= a^2 - (-b^2) = a^2 + b^2 \text{ (i.e. a real number)}.$$

Similarly, if $Z = 1 + j2$ then $\bar{Z} = 1 - j2$

and $Z\bar{Z} = (1 + j2)(1 - j2) = 1^2 + 2^2 = 5$

(d) Division

Expressing one complex number divided by another, in the form $a + jb$, is accomplished by multiplying the numerator and denominator by the complex conjugate of the denominator. This has the effect of making the denominator a real number.

Hence, for example, $\dfrac{2 + j4}{3 - j4} = \dfrac{2 + j4}{3 - j4} \times \dfrac{3 + j4}{3 + j4}$

$$= \frac{6 + j8 + j12 + j^216}{3^2 + 4^2}$$

$$= \frac{-10 + j20}{25} = \frac{-10}{25} + j\frac{20}{25} \text{ or } -0.4 + j0.8$$

(e) Complex equations

If two complex numbers are equal, then their real parts are equal and their imaginary parts are equal.

Hence, if $a + jb = c + jd$

then $a = c$ and $b = d$

This is a useful property, since equations having two unknown quantities can be solved from one equation (see Worked Problem 4 following).

Worked problems on operations involving cartesian complex numbers

---

Problem 1. Solve the following quadratic equations: (a) $x^2 + 9 = 0$ (b) $4y^2 - 3y + 5 = 0$

(a) $x^2 + 9 = 0$

$$x^2 = -9$$

$$x = \sqrt{-9} = (\sqrt{-1})(\sqrt{9}) = (\sqrt{-1})(\pm 3) = (j)(\pm 3) = \pm j3$$

(b) $4y^2 - 3y + 5 = 0$

Using the quadratic formula:

$$y = \frac{-(-3) \pm \sqrt{[(-3)^2 - 4(4)(5)]}}{2(4)} = \frac{3 \pm \sqrt{(9 - 80)}}{8}$$

$$= \frac{3 \pm \sqrt{-71}}{8} = \frac{3 \pm (\sqrt{-1})(\sqrt{71})}{8} = \frac{3 \pm j\sqrt{71}}{8}$$

$$= \frac{3}{8} \pm j\frac{\sqrt{71}}{8} = \mathbf{0.375 \pm j1.053}, \text{ correct to three decimal places.}$$

Problem 2. Evaluate: (a) j^3 (b) j^4 (c) j^5 (d) j^6 (e) j^7 (f) $\dfrac{3}{j^9}$

(a) $j = \sqrt{-1}$

$j^2 = -1$

$j^3 = j \times j^2 = (j)(-1) = -j$

(b) $j^4 = j \times j^3 = (j)(-j) = -j^2 = +1$

(c) $j^5 = j \times j^4 = (j) \times (1) = j$

(d) $j^6 = j \times j^5 = (j) \times (j) = j^2 = -1$

(e) $j^7 = j \times j^6 = (j) \times (-1) = -j$

(f) $j^9 = j^2 \times j^7 = (-1) \times (-j) = j$

Hence $\dfrac{3}{j^9} = \dfrac{3}{j} = \dfrac{3}{j} \times \dfrac{(-j)}{(-j)} = \dfrac{-j3}{-j^2} = \dfrac{-j3}{+1} = -j3$

Problem 3. If $Z_1 = 1 + j2$, $Z_2 = 2 - j3$, $Z_3 = -4 + j$ and $Z_4 = -3 - j2$ evaluate in $a + jb$ form the following:
(a) $Z_1 + Z_2 - Z_3$ (b) Z_1Z_3 (c) $Z_2Z_3Z_4$
(d) $\dfrac{Z_2}{Z_4}$ (e) $\dfrac{Z_1 - Z_4}{Z_2 + Z_3}$

(a) $Z_1 + Z_2 - Z_3 = (1 + j2) + (2 - j3) - (-4 + j)$
$\qquad\qquad\qquad = 1 + j2 + 2 - j3 + 4 - j$
$\qquad\qquad\qquad = \mathbf{7 - j2}$

(b) $\qquad Z_1Z_3 = (1 + j2)(-4 + j)$
$\qquad\qquad\qquad = -4 + j - j8 + j^2 2$
$\qquad\qquad\qquad = \mathbf{-6 - j7}$

(c) $\qquad Z_2Z_3Z_4 = (2 - j3)(-4 + j)(-3 - j2)$
$\qquad\qquad\qquad = (-8 + j2 + j12 - j^2 3)(-3 - j2)$
$\qquad\qquad\qquad = (-5 + j14)(-3 - j2)$
$\qquad\qquad\qquad = 15 + j10 - j42 - j^2 28$
$\qquad\qquad\qquad = \mathbf{43 - j32}$

(d) $\qquad \dfrac{Z_2}{Z_4} = \dfrac{2 - j3}{-3 - j2}$
$\qquad\qquad\qquad = \dfrac{2 - j3}{-3 - j2} \times \dfrac{-3 + j2}{-3 + j2}$
$\qquad\qquad\qquad = \dfrac{-6 + j4 + j9 - j^2 6}{3^2 + 2^2}$
$\qquad\qquad\qquad = \dfrac{0 + j13}{13}$
$\qquad\qquad\qquad = \mathbf{0 + j1 \ or \ j}$

(e) $\qquad \dfrac{Z_1 - Z_4}{Z_2 + Z_3} = \dfrac{(1 + j2) - (-3 - j2)}{(2 - j3) + (-4 + j)}$
$\qquad\qquad\qquad = \dfrac{4 + j4}{-2 - j2}$
$\qquad\qquad\qquad = \dfrac{4(1 + j)}{-2(1 + j)}$
$\qquad\qquad\qquad = \mathbf{-2}$

Problem 4. Solve the following complex equations:
(a) $3(a + jb) = 9 - j2$ (b) $(2 + j)(-2 + j) = x + jy$
(c) $(a - j2b) + (b - j3a) = 5 + j2$

(a) $3(a + jb) = 9 - j2$

$3a + j3b = 9 - j2$

Equating real parts gives: $3a = 9$, i.e. **a = 3**

Equating imaginary parts gives: $3b = -2$, i.e. $b = -\dfrac{2}{3}$

(b) $(2 + j)(-2 + j) = x + jy$

$-4 + j2 - j2 + j^2 = x + jy$

$-5 + j0 = x + jy$

Equating real and imaginary parts gives: **x = −5, y = 0**

(c) $(a - j2b) + (b - j3a) = 5 + j2$

$(a + b) + j(-2b - 3a) = 5 + j2$

Hence $a + b = 5$ (3)

and $-2b - 3a = 2$ (4)

We have two simultaneous equations to solve.
Multiplying equation (3) by 2 gives:

$2a + 2b = 10$ (5)

Adding equations (4) and (5) gives:

$-a = 12$, i.e. **a = −12**

From equation (1) **b = 17**

Further problems on operations involving cartesian complex numbers may be found in Section 7.8 (Problems 1 to 23), page 121.

7.5   The polar form of a complex number

Let a complex number Z, in cartesian form, be $x + jy$. This is shown in the Argand diagram of Fig. 7.3(a). Let r be the distance OZ and θ the angle OZ makes with the positive real axis.

From trigonometry, $\cos \theta = \dfrac{x}{r}$, i.e. $x = r \cos \theta$

and $\sin \theta = \dfrac{y}{r}$, i.e. $y = r \sin \theta$

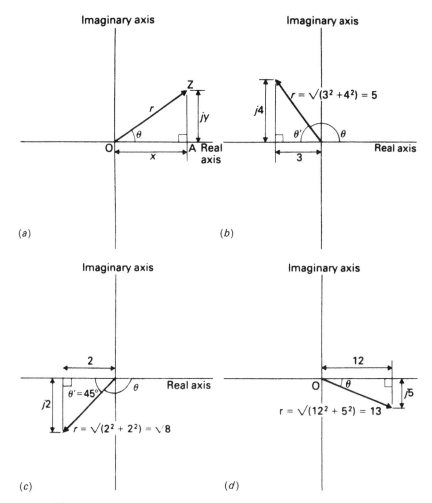

Fig. 7.3

$$Z = x + jy$$
$$= r \cos \theta + jr \sin \theta$$
$$= r (\cos \theta + j \sin \theta)$$

This latter form is usually abbreviated to $Z = r \angle \theta$, and is called the **polar form** of a complex number. The complex number is now specified in terms of r and θ instead of x and y.

r is called the **modulus** (or magnitude) of Z and is written as mod Z or $|Z|$. r is determined from Pythagoras's theorem on triangle OAZ,

i.e. $|Z| = r = \sqrt{(x^2 + y^2)}$

The modulus is represented on the Argand diagram by the distance OZ. θ is called the **argument** (or amplitude) of Z and is written as arg Z. θ is also deduced from triangle OAZ, giving **arg $Z = \theta = $ arctan $\dfrac{y}{x}$.**

In Fig. 7.3(a) the Argand point Z is shown in the first quadrant. However, the above results apply to any point in the Argand diagram.

By convention, the principal value of θ is used, i.e. the numerically least value such that $-\pi \leqslant \theta \leqslant \pi$. For example, in Fig. 7.3(b),

$\theta' = \arctan \dfrac{4}{3} = 53° \, 8'$. Hence $\theta = 180° - 53° \, 8' = 126° \, 52'$. Therefore

$-3 + j4 = 5 \angle 126° \, 52'$. Similarly, in Fig. 7.3(c), θ is $180° - 45°$, i.e. $135°$ measured in the negative direction. Hence $-2 - j2 = \sqrt{8} \angle -135°$. (This is the same as $\sqrt{8} \angle 225°$. However, the principal value of $\sqrt{8} \angle -135°$ is normally used.) In Fig. 7.3(d), $\theta = \arctan \dfrac{5}{12} = 22° \, 37'$. Hence $12 - j5 = 13 \angle -22° \, 37'$.

Whenever changing from a cartesian form of complex number to a polar form, or vice versa, a sketch is invaluable for deciding the quadrant in which the complex number occurs. There are always two possible values of $\theta = \arctan \dfrac{y}{x}$, only one of which is correct for a particular complex number.

7.6   Multiplication and division using complex numbers in polar form

An important use of the polar form of a complex number is in multiplication and division which is achieved more easily than with cartesian form.

(a) Multiplication

Let $Z_1 = r_1 \angle \theta_1$ and $Z_2 = r_2 \angle \theta_2$

Then $Z_1 Z_2 = [r_1 \angle \theta_1][r_2 \angle \theta_2]$

$\qquad\qquad = [r_1(\cos \theta_1 + j \sin \theta_1)] \times [r_2(\cos \theta_2 + j \sin \theta_2)]$

$\qquad\qquad = r_1 r_2(\cos \theta_1 \cos \theta_2 + j \sin \theta_2 \cos \theta_1 + j \cos \theta_2 \sin \theta_1$
$\qquad\qquad\qquad + j^2 \sin \theta_1 \sin \theta_2)$

$\qquad\qquad = r_1 r_2[(\cos \theta_1 \cos \theta_2 - \sin \theta_1 \sin \theta_2)$
$\qquad\qquad\qquad + j(\sin \theta_1 \cos \theta_2 + \cos \theta_1 \sin \theta_2)]$

$\qquad\qquad = r_1 r_2[\cos(\theta_1 + \theta_2) + j \sin(\theta_1 + \theta_2)]$ (from compound
$\qquad\qquad\qquad$ angle formulae – see *Technician mathematics 3*, Chapter 16)

$\qquad\qquad = r_1 r_2 \angle (\theta_1 + \theta_2)$

Hence to obtain the product of complex numbers in polar form their moduli are multiplied together and their arguments are added. This result is true for all polar complex numbers.

Thus $3 \angle 25° \times 2 \angle 32° = 6 \angle 57°$,

$\qquad 4 \angle 11° \times 5 \angle -18° = 20 \angle -7°$, and

$$2 \angle \frac{\pi}{3} \times 7 \angle \frac{\pi}{6} = 14 \angle \frac{\pi}{2}, \text{ and so on.}$$

(b) Division

Let $\qquad Z_1 = r_1 \angle \theta_1$ and $Z_2 = r_2 \angle \theta_2$

Then $\qquad \dfrac{Z_1}{Z_2} = \dfrac{r_1 \angle \theta_1}{r_2 \angle \theta_2} = \dfrac{r_1(\cos \theta_1 + j \sin \theta_1)}{r_2(\cos \theta_2 + j \sin \theta_2)}$

$\qquad = \dfrac{r_1(\cos \theta_1 + j \sin \theta_1)}{r_2(\cos \theta_2 + j \sin \theta_2)} \times \dfrac{(\cos \theta_2 - j \sin \theta_2)}{(\cos \theta_2 - j \sin \theta_2)}$

$\qquad = \dfrac{\begin{array}{c} r_1(\cos \theta_1 \cos \theta_2 - j \sin \theta_2 \cos \theta_1 \\ + j \sin \theta_1 \cos \theta_2 - j^2 \sin \theta_1 \sin \theta_2) \end{array}}{r_2(\cos^2 \theta_2 + \sin^2 \theta_2)}$

$\qquad = \dfrac{\begin{array}{c} r_1[(\cos \theta_1 \cos \theta_2 + \sin \theta_1 \sin \theta_2) \\ + j(\sin \theta_1 \cos \theta_2 - \sin \theta_2 \cos \theta_1)] \end{array}}{r_2(1)}$

$\qquad = \dfrac{r_1}{r_2}[\cos(\theta_1 - \theta_2) + j \sin(\theta_1 - \theta_2)]$ (from compound angle

$\qquad\qquad\qquad\qquad\qquad\qquad\qquad\qquad\qquad\qquad$ formulae)

$\qquad = \dfrac{r_1}{r_2} \angle (\theta_1 - \theta_2)$

Hence to obtain the ratio of two complex numbers in polar form their moduli are divided and their arguments subtracted. This result is true for all polar complex numbers.

Thus $\qquad \dfrac{8 \angle 58°}{2 \angle 11°} = 4 \angle 47°$,

$\qquad \dfrac{9 \angle 136°}{3 \angle -60°} = 3 \angle (136° - -60°) = 3 \angle 196° = 3 \angle -164°$,

and $\qquad \dfrac{10 \angle \dfrac{\pi}{2}}{5 \angle -\dfrac{\pi}{4}} = 2 \angle \dfrac{3\pi}{4}, \text{ and so on.}$

Worked problems on the polar form of complex numbers

Problem 1. Determine the modulus and argument of the complex number $Z = -5 + j8$ and express Z in polar form.

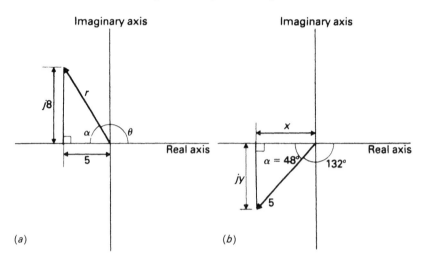

Fig. 7.4

A sketch, shown in Fig. 7.4(a), indicates that the complex number $-5 + j8$ lies in the second quadrant of the Argand diagram. Modulus, $|Z| = r = \sqrt{[(-5)^2 + (8)^2]} = \mathbf{9.434}$, correct to four significant figures.

$$\alpha = \arctan \frac{8}{5} = 58°$$

Argument, $\arg Z = \theta = 180° - 58° = 122°$

In polar form $-5 + j8$ is written as $\mathbf{9.434 \: \angle \: 122°}$

Problem 2. Convert $5 \angle -132°$ into $a + jb$ form correct to four significant figures.

A sketch, shown in Fig. 7.4(b), indicates that the polar complex number $5 \angle -132°$ lies in the third quadrant of the Argand diagram.

Using trigonometrical ratios, $x = 5 \cos 48° = 3.346$

and $y = 5 \sin 48° = 3.716$

Hence $\mathbf{5 \angle -132° = -3.346 - j3.716}$

Problem 3. Evaluate (a) $\dfrac{15\angle 30° \times 8 \angle 45°}{4 \angle 60°}$ and

(b) $4 \angle 30° + 3 \angle -60° - 5 \angle -135°$

giving answers in polar and in cartesian forms, correct to three significant figures.

(a) $\dfrac{15 \angle 30° \times 8 \angle 45°}{4 \angle 60°} = \dfrac{15 \times 8}{4} \angle (30° + 45° - 60°)$

$$= \mathbf{30.0 \angle 15°} \quad \text{in polar form}$$

$$= 30.0\,(\cos 15° + j \sin 15°)$$

$$= \mathbf{29.0 + j7.76} \quad \text{in cartesian form.}$$

(b) $4 \angle 30° + 3 \angle -60° - 5 \angle -135°$

The advantages of polar form are seen when multiplying and dividing complex numbers. Addition or subtraction in polar form is not possible on sight. Each polar complex number has to be converted into cartesian form first. Thus:

$$4 \angle 30° = 4(\cos 30° + j \sin 30°) = 4 \cos 30° + j4 \sin 30°$$

$$= 3.464 + j2.000$$

$$3 \angle -60° = 3[\cos(-60°) + j \sin(-60°)] = 3 \cos 60° - j3 \sin 60°$$

$$= 1.500 - j2.598$$

$$5 \angle -135° = 5[\cos(-135°) + j \sin(-135°)]$$

$$= 5 \cos(-135°) + j5 \sin(-135°)$$

$$= -3.536 - j3.536$$

Hence $4 \angle 30° + 3 \angle -60° - 5 \angle -135°$

$$= (3.464 + j2.000) + (1.500 - j2.598) - (-3.536 - j3.536)$$

$$= \mathbf{8.50 + j2.94} \quad \text{in cartesian form}$$

$$= \sqrt{[(8.50)^2 + (2.94)^2]} \angle \arctan \dfrac{2.94}{8.50},$$

$$= \mathbf{8.99 \angle 19° 5'} \quad \text{in polar form}$$

since the complex number is in the first quadrant

Further problems on the polar form of complex numbers may be found in Section 7.8 (Problems 24 to 38), page 122.

7.7   Applications of complex numbers

Complex numbers are widely used in the analysis of electrical circuits supplied by an alternating voltage and can also be used for solving problems involving coplanar vectors.

(a)   Phasor applications to series connected circuits

In an electrical circuit containing resistance only (R ohms), the current (I amperes) is in phase with the applied voltage (V volts) and the ratio of voltage to current is given by $\dfrac{V}{I} = R$. When an electrical circuit contains inductance only (L henrys), the ratio of voltage to current is called the inductive reactance, (X_L ohms), and

$$\frac{V}{I} = X_L = 2\pi f L \text{ ohms} \tag{6}$$

where f is the frequency in hertz of the applied voltage. The current in a purely inductive circuit lags 90° behind the applied voltage.

When both resistance and inductance are present, as shown in Fig. 7.5(a), the phasor diagram, (a phasor is a rotating vector), is as shown in Fig. 7.5(b).

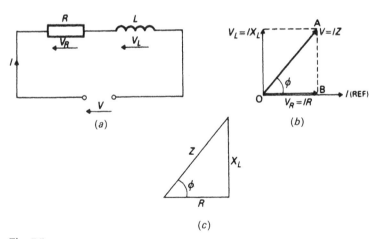

Fig. 7.5

The ratio of $\dfrac{V}{I}$ in any a.c. electrical circuit containing both resistance and reactance is called the **impedance** Z of the circuit and is measured in ohms. With reference to triangle OAB in Fig. 7.5(b), the impedance triangle shown in Fig. 7.5(c) is produced by dividing each quantity by current I. From the impedance triangle, it can be seen that:

$$\text{modulus of impedance, } |Z| = \sqrt{(R^2 + X_L{}^2)} \text{ ohms} \tag{7}$$

$$\text{and circuit phase angle, } \quad \phi = \arctan \frac{X_L}{R} \tag{8}$$

If the impedance triangle is superimposed on an Argand diagram as shown in Fig. 7.6, then:

$$\text{impedance, } Z = (R + jX_L) \text{ ohms} \tag{9}$$

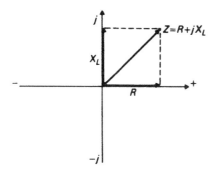

Fig. 7.6

When an electrical circuit contains capacitance (C farads) only, the ratio of voltage to current is called the capacitive reactance, X_C ohms, and

$$\frac{V}{I} = X_C = \frac{1}{2\pi f C} \text{ ohms} \tag{10}$$

The current in a purely capacitive circuit leads the applied voltage by 90°. When both resistance and capacitance are present, as shown in Fig. 7.7(a), the phasor diagram is as shown in Fig. 7.7(b) and the impedance triangle is obtained, as for the circuit containing inductance, and is shown in Fig. 7.7(c). From the impedance triangle, it can be seen that:

$$\text{modulus of impedance, } |Z| = \sqrt{(R^2 + X_C{}^2)} \text{ ohms} \tag{11}$$

$$\text{and circuit phase angle, } \quad \phi = \arctan \frac{X_C}{R} \tag{12}$$

If the impedance triangle is superimposed on an Argand diagram, as shown

Fig. 7.7

Fig. 7.8

in Fig. 7.8:

$$\text{impedance, } Z = R - jX_C \tag{13}$$

For an electrical circuit containing resistance, inductance and capacitance connected in series, the circuit diagram is as shown in Fig. 7.9(a), the phasor diagram is as shown in Fig. 7.9(b) and the impedance triangle is as shown in Fig. 7.9(c).

From the impedance triangle, it can be seen that:

$$\text{modulus of impedance, } |Z| = \sqrt{[R^2 + (X_L - X_C)^2]} \text{ ohms} \tag{14}$$

$$\text{and circuit phase angle, } \phi = \arctan \frac{X_L - X_C}{R} \tag{15}$$

When superimposing the impedance triangle on an Argand diagram, as shown in Fig. 7.10, the resultant impedance is given by

$$Z = R + j(X_L - X_C) \text{ ohms} \tag{16}$$

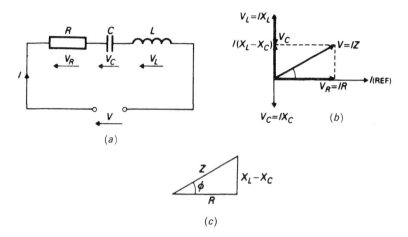

(a)

(b)

(c)

Fig. 7.9

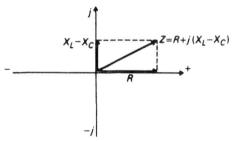

Fig. 7.10

(See Worked Problems 1 to 4)

(b) Vector applications

Problems involving coplanar vector quantities, such as those dealing with displacement, velocity, acceleration, force, moment and momentum may be resolved using complex number theory. Examples of the application of complex number theory to such problems are shown in worked Problems 5 to 7.

Worked problems on the applications of complex numbers

Problem 1. Determine the resistance and the series connected inductance or capacitance for each of the following impedances:
(a) $12 + j5$, (b) $-j40$ and (c) $30 \angle 60°$.
Assume a frequency of 50 Hz.

(a) From equation (9), $Z = R + jX_L$
Hence, $12 + j5$ shows that the **resistance is 12 ohms** and the inductive reactance is 5 ohms.

From equation (6), since $X_L = 2\pi fL$, $L = \dfrac{X_L}{2\pi f}$

$$= \frac{5}{2\pi(50)} = 0.016 \text{ H}$$

i.e. **the inductance is 16 mH.**

(b) From equation (13), $Z = R - jX_C$
Hence, $0 - j40$ shows that the **resistance is zero** and the capacitive reactance is 40 ohms.

From equation (10), $X_C = \dfrac{1}{2\pi fC}$, $C = \dfrac{1}{2\pi fX_C}$

$$= \frac{1}{2\pi(50)(40)} \text{ F} = \frac{10^6}{2\pi(50)(40)} \text{ μF}$$

$$= 79.6 \text{ μF}$$

i.e. **the capacitance is 79.6 μF.**

(c) $30 \angle 60° = 30(\cos 60° + j \sin 60°) = 15 + j25.98$
From equation (9), $15 + j25.98$ shows that the **resistance is 15 ohms** and the inductive reactance is 25.98 ohms.

From equation (6), the inductance, $L = \dfrac{X_L}{2\pi f}$

$$= \frac{25.98}{2\pi(50)} = 0.0827 \text{ H},$$

i.e. **the inductance is 82.7 mH.**

Problem 2. The impedance of an electrical circuit is $30 - j50$ ohms. Determine (a) the resistance, (b) the capacitance, (c) the modulus of the impedance and (d) the current flowing, when the circuit is connected to a 240 V, 50 Hz supply.

(a) Since $Z = R - jX_C$, the **resistance is 30 ohms**

(b) Since $Z = R - jX_C$, the capacitance reactance is 50 ohms

From equation (10), capacitance, $C = \dfrac{1}{2\pi fX_C} = \dfrac{10^6}{2\pi(50)(50)} \text{ μF}$

$$= 63.7 \text{ μF}$$

(c) The modulus of the impedance,

$$|Z| = \sqrt{[R^2 + (X_C)^2]} = \sqrt{[30^2 + (-50)^2]}$$

= **58.31 ohms**

(d) From equation (12), the circuit phase angle,

$$\phi = \arctan \frac{X_C}{R} = \arctan \frac{50}{30} = 59° \, 2'.$$

Since $Z = R - jX_C$, this angle is in the fourth quadrant, i.e., $-59° \, 2'$. Thus an alternative way of expressing the impedance is $Z = 58.31 \angle -59° \, 2'$

The current flowing, $I = \dfrac{V}{Z} = \dfrac{240 \angle 0°}{58.31 \angle -59° \, 2'}$

$$= 4.116 \angle 59° \, 2'$$

[Since the voltage is given as 240 volts, it is $240 + j0$ volts in rectangular form and $240 \angle 0°$ volts in polar form.]

Problem 3. A series connected electrical circuit has a resistance of 32 ohms and an inductance of 0.15 H. It is connected to a 200 V, 50 Hz supply. Determine (a) the inductive reactance, (b) the impedance in rectangular and polar forms, (c) the current and the circuit phase angle, (d) the voltage drop across the resistor and (e) the voltage drop across the inductor.

(a) From equation (6), inductive reactance, $X_L = 2\pi f L$

$$= 2\pi(50)(0.15) = \textbf{47.1 ohms}$$

(b) From equation (9), impedance, $Z = R + jX_L$

i.e. $Z = 32 + j47.1$

Thus, from equation (2), $|Z| = \sqrt{(32^2 + 47.1^2)} = 57$ and from equation (8), circuit phase angle, $\phi = \arctan \dfrac{47.1}{32} = 55° \, 48'$

Thus $Z = 57 \angle 55° \, 48'$ **ohms**, in polar form.

(c)   Current $I = \dfrac{V}{Z} = \dfrac{200 \angle 0°}{57 \angle 55° 48'} = 3.51 \angle -55° 48'$

i.e. **the current is 3.51 A lagging V by 55° 48'.**

(d)   The voltage drop across the 32 ohm resistor,

$V_R = IR$

$\quad = (3.51 \angle -55° 48')(32)$

$\quad =$ **112.3 $\angle -55°$ 48' volts**

(e)   The voltage drop across the 0.15 H inductor,

$V_L = IX_L$

$\quad = (3.51 \angle -55° 48')(47.1 \angle 90°)$

$\quad =$ **165.3 $\angle 34°$ 12' volts**

Problem 4. A 240 V, 50 Hz voltage is applied across a series connected circuit having a resistance of 12 Ω, an inductance of 0.10 H and a capacitance of 120 μF. Determine the current flowing in the circuit.

From equation (6), inductive reactance,

$X_L = 2\pi f L$

$\quad = 2\pi(50)\ 0.10 = 31.4\ \Omega$

From equation (10), capacitance reactance,

$X_C = \dfrac{1}{2\pi f C}$

$\quad = \dfrac{10^6}{2\pi(50)(120)} = 26.5\ \Omega$

From equation (16), the impedance,

$Z = R + j(X_L - X_C)$

$\quad = 12 + j(31.4 - 26.5)$

$\quad = 12 + j4.9$

In polar form,

$Z = \sqrt{(12^2 + 4.9^2)} \angle \arctan\dfrac{4.9}{12}$

$\quad = 13 \angle 22° 13'$ ohms

Hence, current $I = \dfrac{V}{Z} = \dfrac{240 \angle 0°}{13 \angle 22° \, 13'}$

$\qquad\qquad = 18.5 \angle -22° \, 13' \text{ amperes}$

i.e. **the current flowing is 18.5 amperes, lagging the voltage by 22° 13'.**

Problem 5. Coplanar forces of $(7 + jx)$ newtons, $(3 - j5)$ newtons and $(y + j3)$ newtons act on a body which is in equilibrium. Determine the values of the three forces, both in rectangular and polar co-ordinate forms.

Since the body is in equilibrium, the algebraic sum of the forces is zero, i.e.

$$(7 + jx) + (3 - j5) + (y + j3) = 0 \qquad\qquad (1)$$

Also, since the resultant force is zero, then the horizontal and vertical components of the force are zero. It follows that the real and imaginary parts of equation (1) are zero. Thus:

$$7 + 3 + y = 0, \quad \text{i.e. } y = -10$$

and $j(x - 5 + 3) = j(0)$, i.e. $x = 2$

Thus, the three forces are $(7 + j2)$, $(3 - j5)$ and $(-10 + j3)$

In polar form, $(7 + j2) = \sqrt{(7^2 + 2^2)} \angle \arctan \dfrac{2}{7} = 7.280 \angle 15° \, 57'$

$$(3 - j5) = \sqrt{(3^2 + 5^2)} \angle \arctan -\dfrac{5}{3}$$

$$= 5.831 \angle -59° \, 2'$$

$$(-10 + j3) = \sqrt{(10^2 + 3^2)} \angle \arctan -\dfrac{3}{10}$$

$$= 10.440 \angle 163° \, 18'$$

Problem 6. The velocity of ship A is $(4 - j5)$ knots and that of ship B is $(-6 - j6)$ knots. Find the magnitude and direction of the velocity of ship B relative to ship A.

Let v_A be the velocity of ship A, i.e. $v_A = (4 - j5)$ and let v_B be the velocity of ship B, i.e. $v_B = (-6 - j6)$
The velocity of B relative to A is $v_B - v_A$

$$v_B - v_A = (-6 - j6) - (4 - j5)$$

$$= -10 - j = \sqrt{(10^2 + 1^2)} \angle \arctan \dfrac{-1}{-10} = 10.05 \angle 185° \, 43'$$

Thus, the magnitude of the velocity of B relative to A is **10.05 knots** and the direction of B relative to A is **185° 43′**

Problem 7. A projectile is given an initial velocity of $u = (21 + j8)$ m/s. Determine, (a) the magnitude and direction of its velocity and (b) the distance and direction relative to the origin after 1.5 seconds. [Take g as 9.81 m/s^2, velocity as $v = u - jgt$ and displacement as $s = ut - j\dfrac{1}{2}gt^2$, where t is the time in seconds.]

(a) The velocity at time t seconds is $v = u - jgt$, (since g acts vertically downwards),

hence $\qquad v = (21 + j8) - jgt$

$$= 21 + j(8 - 9.81\ t)$$

When $t = 1.5$ s, $v = 21 + j[8 - (9.81)(1.5)] = 21 - j6.715$

In polar form, $v = \sqrt{(21^2 + 6.715^2)} \angle \arctan \dfrac{-6.715}{21}$

$$= 22.05 \angle -17°\ 44'$$

Thus the magnitude of the velocity after 1.5 seconds is **22.05 m/s** and the direction is **−17° 44′**

(b) The displacement, $s = ut - j\dfrac{1}{2}gt^2$

$$= (21 + j8)t - j\left(\frac{1}{2}\right)(9.81)\ t^2$$

$$= 21t + j\left(8t - \left(\frac{1}{2}\right)(9.81)t^2\right)$$

When $t = 1.5$ s, $\quad s = 21(1.5) + j\left[(8)(1.5) - \left(\frac{1}{2}\right)(9.81)(1.5)^2\right]$

$$= 31.5 + j0.963\ 8$$

In polar form, $\quad s = \sqrt{[(31.5)^2 + (0.963\ 8)^2]}$
$$\angle \arctan \dfrac{0.963\ 8}{31.5}$$

$$= 31.51 \angle 1°\ 45'$$

Thus the distance from the origin is **31.51 m** and the direction is **1° 45′**

Further problems on the applications of complex numbers may be found in the following Section 7.8 (Problems 39 to 52), page 124.

7.8   Further problems

Operations on cartesian complex numbers

In Problems 1 to 5 solve the quadratic equations.

1. $x^2 + 16 = 0$ $[x = \pm j4]$

2. $x^2 - 2x + 2 = 0$ $[x = 1 \pm j]$

3. $2x^2 + 3x + 4 = 0$ $\left[x = -\dfrac{3}{4} \pm j \dfrac{\sqrt{23}}{4} \right]$

4. $5y^2 + 2y = -3$ $\left[y = -\dfrac{1}{5} \pm j \dfrac{\sqrt{14}}{5} \right]$

5. $4t^2 = t - 1$ $\left[t = \dfrac{1}{8} \pm j \dfrac{\sqrt{15}}{8} \right]$

6. Show on an Argand diagram the following complex numbers:
   (a) $3 + j6$ (b) $2 - j3$ (c) $-3 + j4$ (d) $-1 - j5$

7. Write down the complex conjugates of the following complex numbers:
   (a) $4 + j$ (b) $3 - j2$ (c) $-5 - j$
   (a) $[4 - j]$ (b) $[3 + j2]$ (c) $[-5 + j]$

In Problems 8 to 12, evaluate in $a + jb$ form assuming that
$Z_1 = 2 + j3$, $Z_2 = 3 - j4$, $Z_3 = -1 + j2$ and $Z_4 = -2 - j5$

8. (a) $Z_1 - Z_2$ (b) $Z_2 + Z_3 - Z_4$
   (a) $[-1 + j7]$ (b) $[4 + j3]$

9. (a) $Z_1 Z_2$ (b) $Z_3 Z_4$ (a) $[18 + j]$ (b) $[12 + j]$

10. (a) $Z_1 Z_3 Z_4$ (b) $Z_2 Z_3 + Z_4$
    (a) $[21 + j38]$ (b) $[3 + j5]$

11. (a) $\dfrac{Z_1}{Z_2}$ (b) $\dfrac{Z_1 + Z_2}{Z_3 + Z_4}$ (a) $\left[-\dfrac{6}{25} + j\dfrac{17}{25} \right]$ (b) $\left[-\dfrac{2}{3} + j \right]$

12. (a) $\dfrac{Z_1 Z_2}{Z_1 + Z_2}$ (b) $Z_1 + \dfrac{Z_2}{Z_3} + Z_4$
    (a) $\left[\dfrac{89}{26} + j\dfrac{23}{26} \right]$ (b) $\left[-\dfrac{11}{5} - j\dfrac{12}{5} \right]$

13. Evaluate $\left[\dfrac{(1 + j)^2 - (1 - j)^2}{j} \right]$ $[4]$

14. If $Z_1 = 4 - j3$ and $Z_2 = 2 + j$ evaluate x and y given
    $$x + jy = \dfrac{1}{Z_1 - Z_2} + \dfrac{1}{Z_1 Z_2}$$ $[x = 0.188, \; y = 0.216]$

15. Evaluate (a) j^8 (b) j^{11} (c) $\dfrac{3}{j^3}$ (d) $\dfrac{5}{j^6}$

(a) $[1]$ (b) $[-j]$ (c) $[j3]$ (d) $[-5]$

16. Evaluate (a) $(1+j)^4$ (b) $\dfrac{2-j}{2+j}$ (c) $\dfrac{1}{2+j3}$

(a) $[-4]$ (b) $\left[\dfrac{3}{5} - j\dfrac{4}{5}\right]$ (c) $\left[\dfrac{2}{13} - j\dfrac{3}{13}\right]$

17. If $Z = \dfrac{1+j3}{1-j2}$, evaluate Z^2 in $a+jb$ form $[0-j2]$

18. Evaluate (a) j^{33} (b) $\dfrac{1}{(2-j2)^4}$ (c) $\dfrac{1+j3}{2+j4} + \dfrac{3-j2}{5-j}$

(a) $[j]$ (b) $\left[-\dfrac{1}{64}\right]$ (c) $[1.354 - j0.169]$

In Problems 19 to 23 solve the given complex equations:

19. $4(a+jb) = 7 - j3$ $\left[a = \dfrac{7}{4}, b = -\dfrac{3}{4}\right]$

20. $(3+j4)(2-j3) = x + jy$ $[x = 18, y = -1]$
21. $(1+j)(2-j) = j(p+jq)$ $[p = 1, q = -3]$
22. $(a - j3b) + (b - j2a) = 4 + j6$ $[a = 18, b = -14]$
23. $5 + j2 = \sqrt{(e+jf)}$ $[e = 21, f = 20]$

Polar form of complex numbers

In Problems 24 to 26 determine the modulus and the argument of each of the complex numbers given.

24. (a) $3 + j4$ (b) $2 - j5$
  (a) $[5, 53° 8']$ (b) $[5.385, -68° 12']$
25. (a) $-4 + j$ (b) $-5 - j3$
  (a) $[4.123, 165° 58']$ (b) $[5.831, -149° 2']$
26. (a) $(2+j)^2$ (b) $j(3-j)$
  (a) $[5, 53° 8']$ (b) $[3.162, 71° 34']$

In Problems 27 to 29 express the given cartesian complex numbers in polar form, leaving answers in surd form.

27. (a) $6 + j5$ (b) $3 - j2$ (c) -3
  (a) $[\sqrt{61} \angle 39° 48']$ (b) $[\sqrt{13} \angle -33° 41']$
  (c) $[3 \angle 180°$ or $3 \angle \pi]$

28. (a) $-5 + j$ (b) $-4 - j3$ (c) $-j2$

(a) $[\sqrt{26} \angle 168° 41']$ (b) $[5 \angle -143° 8']$

(c) $\left[2 \angle -90° \text{ or } 2 \angle -\dfrac{\pi}{2}\right]$

29. (a) $(-1 + j)^3$ (b) $-j(1 - j)$ (c) $j^3(2 - j3)$

(a) $[\sqrt{8} \angle 45°]$ (b) $[\sqrt{2} \angle -135°]$

(c) $[\sqrt{13} \angle -146° 19']$

In Problems 30 to 32 convert the given polar complex numbers into $(a + jb)$ form giving answers correct to four significant figures.

30. (a) $6 \angle 30°$ (b) $4 \angle 60°$ (c) $3 \angle 45°$

(a) $[5.196 + j3.000]$ (b) $[2.000 + j3.464]$

(c) $[2.121 + j2.121]$

31. (a) $2 \angle \dfrac{\pi}{2}$ (b) $3 \angle \pi$ (c) $5 \angle \dfrac{5\pi}{6}$

(a) $[0 + j2.000]$ (b) $[-3.000 + j0]$

(c) $[-4.330 + j2.500]$

32. (a) $8 \angle 150°$ (b) $4.2 \angle -120°$ (c) $3.6 \angle -25°$

(a) $[-6.928 + j4.000]$ (b) $[-2.100 - j3.637]$

(c) $[3.263 - j1.521]$

33. Using an Argand diagram evaluate in polar form

(a) $2 \angle 30° + 3 \angle 40°$ (b) $5.5 \angle 120° - 2.5 \angle -50°$

(a) $[4.982 \angle 36°]$ (b) $[7.974 \angle 123° 7']$

In Problems 34 to 36 evaluate in polar form.

34. (a) $2 \angle 40° \times 5 \angle 20°$ (b) $2.6 \angle 72° \times 4.3 \angle 45°$

(a) $[10 \angle 60°]$ (b) $[11.18 \angle 117°]$

35. (a) $5.8 \angle 35° \div 2 \angle -10°$ (b) $4 \angle 30° \times 3 \angle 70° \div 2 \angle -15°$

(a) $[2.9 \angle 45°]$ (b) $[6 \angle 115°]$

36. (a) $\dfrac{4.1 \angle 20° \times 3.2 \angle -62°}{1.2 \angle 150°}$ (b) $6 \angle 25° + 3 \angle -36° - 4 \angle 72°$

(a) $[10.93 \angle 168°]$ (b) $[7.289 \angle -24° 35']$

37. Solve the complex equations, giving answers correct to four significant figures.

(a) $\dfrac{12 \angle \dfrac{\pi}{2} \times 3 \angle \dfrac{3\pi}{4}}{2 \angle -\dfrac{\pi}{3}} = x + jy$

(b) $15 \angle \dfrac{\pi}{3} + 12 \angle \dfrac{\pi}{2} - 6 \angle -\dfrac{\pi}{3} = r \angle \theta$

(a) $[x = 4.659, y = -17.39]$ (b) $[r = 30.52, \theta = 81° 31']$

38. Three vectors are represented by P, $2 \angle 30°$, Q, $3 \angle 90°$ and R, $4 \angle -60°$. Determine in polar form the vectors represented by (a) $P + Q + R$ and   (b) $P - Q - R$ (a) [3.770 $\angle$ 8° 10′] (b) [1.488 $\angle$ 100° 22′]

Application of complex numbers

39. Determine the resistance R and series inductance L (or capacitance C) for each of the following impedances, assuming the frequency to be 50 Hz.

(a) $4 + j7$ (b) $3 - j2$ (c) $j10$ (d) $-j200$ (e) $15 \angle \dfrac{\pi}{3}$

(f) $6 \angle -45°$ (a) [$R = 4\,\Omega$, $L = 22.3$ mH]
(b) [$R = 3\,\Omega$, $C = 1\,592\,\mu F$]   (c) [$R = 0$, $L = 31.8$ mH]
(d) [$R = 0$, $C = 15.92\,\mu F$]   (e) [$R = 7.5\,\Omega$, $L = 41.3$ mH]
(f) [$R = 4.243\,\Omega$, $C = 750.3\,\mu F$]

40. An alternating voltage of 100 V, 50 Hz is applied across an impedance of $(20 - j30)$ ohms. Calculate: (a) the resistance, (b) the capacitance; (c) the current; and (d) the phase angle between current and voltage.
(a) [20 Ω]   (b) [106.1 μF]   (c) [2.774 A]   (d) [56° 19′]

41. Two voltages are represented by $(15 + j10)$ and $(12 - j4)$ volts. Determine the magnitude of the resultant voltage when these voltages are added.   [27.66 V]

42. Two impedances, $Z_1 = (2 + j6)$ ohms and $Z_2 = (5 - j2)$ ohms are connected in series to a supply voltage of 100 V. Determine the magnitude of the current and its phase angle relative to the voltage.   [12.40 A; 29° 45′ lagging]

43. A resistance of 45 ohms is connected in series with a capacitor of 42 μF. If the applied voltage is 250 V, 50 Hz determine: (a) the capacitive reactance; (b) the impedance; (c) the current, and its phase relative to the applied voltage; (d) the voltage across the resistance; and (e) the voltage across the capacitance.   (a) [75.79 Ω]   (b) [88.14 Ω] (c) [2.836 A at 59° 18′ leading]   (d) [127.6 V]   (e) [214.9 V]

44. Forces of $(-3 - j5)$ N, $(13 + j2)$ N, $(-8 + j4)$ N and $(x + jy)$ N are in equilibrium. Find x and y.   [$x = -2$ N, $y = -1$ N]

45. Find the resultant of forces $F_1 = (3 - j17)$ N, $F_2 = (10 - j2)$ N and $F_3 = (-8 + j2)$ N. Express the resultant force in polar form.   [17.72 N $\angle$ -73° 37′]

46. A body moves from $(0 + j0)$ to point A at $(9 - j10)$. It then moves to points B and C at $(22 + j4)$ and $(-7 - j)$ respectively. Determine its distance from the origin and its angle relative to the real positive axis. $[25, \angle -16° \; 16']$

47. An aircraft, A, flying at a constant height, has a velocity of $(380 + j270)$ km/h. Another aircraft, B, at the same height, has a velocity of $(290 - j417)$ km/h. Determine (a) the velocity of A relative to B, and (b) the velocity of B relative to A, expressing the results in polar form.
    (a) $[692.9 \text{ km/h at } 82° \; 32']$ (b) $[692.9 \text{ km/h at } -97° \; 28']$

48. A projectile is given an initial velocity, u, of $(24 + j27)$ m/s. A second projectile launched from the same place in the same vertical plane at the same time has a velocity of $(31 + j17)$ m/s. After launching, both projectiles move freely under gravity. Determine the magnitude and direction of the velocity of the second projectile relative to the first.
    $(v = u - jgt)$ $[12.21 \text{ m at } \angle -55° \; 0']$

49. In the hydrogen atom, the angular momentum, (p), of the de Broglie wave is given by:

    $$p \, \Psi = -(jh/2\pi)(\pm jm \, \Psi). \text{ Find } p. \qquad \left[p = \pm \frac{mh}{2\pi} \right]$$

50. In an electrical circuit the equivalent impedance Z is given by

    $$Z = Z_1 + \frac{Z_2 Z_3}{Z_2 + Z_3}$$

    Determine Z in rectangular form, correct to two decimal places, when
    $Z_1 = (4.98 + j3.17) \, \Omega$, $Z_2 = (5 - j12) \, \Omega$ and $Z_3 = (8 + j15) \, \Omega$ $[Z = (20.70 - j2.07) \, \Omega]$

51. A star-connected impedance Z_1 is given by:

    $$Z_1 = \frac{Z_A Z_B}{Z_A + Z_B + Z_C}$$

    Evaluate Z_1, in both cartesian and polar form, given $Z_A = (10 + j0) \, \Omega$, $Z_B = (0 - j20) \, \Omega$ and $Z_C = (10 + j10) \, \Omega$
    $[Z_1 = (4 - j8) \, \Omega$ or $8.94 \angle -63.43° \; \Omega]$

52. A delta-connected impedance Z_A is given by

    $$Z_A = \frac{Z_1 Z_2 + Z_2 Z_3 + Z_3 Z_1}{Z_2}$$

    Determine Z_A, in both cartesian and polar form, given $Z_1 = (20 + j0) \, \Omega$, $Z_2 = (0 - j20) \, \Omega$ and $Z_3 = (20 + j20) \, \Omega$
    $[Z_A = (20 + j40) \, \Omega$ or $44.72 \angle 63.43° \; \Omega]$

Chapter 8

De Moivre's theorem

8.1 The exponential form of a complex number

By using Maclaurin's theorem, the power series for e^θ, $\sin\theta$ and $\cos\theta$ may be derived, and are:

$$e^\theta = 1 + \theta + \frac{\theta^2}{2!} + \frac{\theta^3}{3!} + \frac{\theta^4}{4!} + \frac{\theta^5}{5!} + \cdots \tag{1}$$

$$\sin\theta = \theta - \frac{\theta^3}{3!} + \frac{\theta^5}{5!} - \frac{\theta^7}{7!} + \cdots \tag{2}$$

and $\qquad \cos\theta = 1 - \frac{\theta^2}{2!} + \frac{\theta^4}{4!} - \frac{\theta^6}{6!} + \cdots \tag{3}$

Replacing θ by the imaginary number $j\theta$ in equation (1) gives:

$$e^{j\theta} = 1 + j\theta + \frac{(j\theta)^2}{2!} + \frac{(j\theta)^3}{3!} + \frac{(j\theta)^4}{4!} + \frac{(j\theta)^5}{5!} + \cdots$$

$$= 1 + j\theta + j^2\frac{\theta^2}{2!} + j^3\frac{\theta^3}{3!} + j^4\frac{\theta^4}{4!} + j^5\frac{\theta^5}{5!} + \cdots$$

Since $\quad j^2 = -1, j^3 = -j, j^4 = 1$ and $j^5 = j,$

then $\quad e^{j\theta} = 1 + j\theta - \frac{\theta^2}{2!} - j\frac{\theta^3}{3!} + \frac{\theta^4}{4!} + j\frac{\theta^5}{5!} - \cdots$

Grouping real and imaginary terms gives:

$$e^{j\theta} = \left(1 - \frac{\theta^2}{2!} + \frac{\theta^4}{4!} - \cdots\right) + j\left(\theta - \frac{\theta^3}{3!} + \frac{\theta^5}{5!} - \cdots\right)$$

$$= \cos\theta + j\sin\theta, \text{ from equations (2) and (3) above.}$$

Thus $\quad e^{j\theta} = \cos\theta + j\sin\theta \tag{4}$

If each side of equation (4) is multiplied by r, then

$$r\,e^{j\theta} = r(\cos\theta + j\sin\theta), \text{ often abbreviated to } r\angle\theta$$

Thus $z = r\,e^{j\theta}$ is a complex number having a modulus, $|z| = r$ and argument, $\arg z = \theta$.

Writing $-\theta$ for θ in equation (4) gives:

$$e^{j(-\theta)} = \cos(-\theta) + j\,\sin(-\theta)$$

Since $\cos(-\theta) = \cos\theta$ and $\sin(-\theta) = -\sin\theta$, then

$$e^{-j\theta} = \cos\theta - j\,\sin\theta \tag{5}$$

By adding equations (4) and (5)

$$e^{j\theta} + e^{-j\theta} = 2\cos\theta$$

i.e.
$$\cos\theta = \frac{e^{j\theta} + e^{-j\theta}}{2} \tag{6}$$

By subtracting equation (5) from equation (4)

$$e^{j\theta} - e^{-j\theta} = 2j\,\sin\theta$$

i.e.
$$\sin\theta = \frac{e^{j\theta} - e^{-j\theta}}{j2} \tag{7}$$

The identities given in equations (6) and (7) are used when establishing the relationship between trigonometric and hyperbolic functions (see Chapter 10).

8.2   De Moivre's theorem – powers and roots of complex numbers

From multiplication in polar form, $r \angle \theta \times r \angle \theta = r^2 \angle 2\theta$
Also, $r \angle \theta \times r \angle \theta \times r \angle \theta = r^3 \angle 3\theta$, and so on.

In general, $\boxed{[r \angle \theta]^n = r^n \angle n\theta}$ $(= r^n(\cos n\theta + j\sin n\theta))$

This is a statement of De Moivre's theorem and is true for all positive, negative or fractional values of n. De Moivre's theorem is therefore useful for determining powers and roots of complex numbers.

The theorem may also be deduced from the exponential form of a complex number.

For a complex number, z, expressed in exponential form

$$z = r\,e^{j\theta}$$

Then, $z^2 = z \times z = r\,e^{j\theta} \times r\,e^{j\theta}$

$$= r^2\,e^{(j\theta + j\theta)} \text{ by the laws of indices}$$

i.e.
$$z^2 = r^2\,e^{j2\theta}$$

Similarly,
$$z^3 = r\,e^{j\theta} \times r\,e^{j\theta} \times r\,e^{j\theta}$$

$$= r^3\,e^{(j\theta + j\theta + j\theta)}$$

$$= r^3\,e^{j3\theta}$$

A general relationship is that:

$$z^n = r^n e^{jn\theta}$$

But from equation (4) in section 8.1, $e^{jn\theta} = \cos n\theta + j \sin n\theta$

Thus, $z^n = r^n(\cos n\theta + j \sin n\theta) = r^n \angle n\theta$, as above   (8)

For example, $[2 \angle 15°]^5 = 2^5 \angle 5 \times 15° = 32 \angle 75° = 8.282 + j30.910$
A square root of a complex number is determined as follows:

$$\sqrt{[r \angle \theta]} = [r \angle \theta]^{\frac{1}{2}} = r^{\frac{1}{2}} \angle \tfrac{1}{2}\theta = \sqrt{r} \angle \tfrac{\theta}{2}$$

It is important to appreciate that a real number has two square roots, equal in size but opposite in sign. On an Argand diagram the roots are 180° apart.
A cubed root of a complex number is determined as follows:

$$\sqrt[3]{[r \angle \theta]} = [r \angle \theta]^{\frac{1}{3}} = r^{\frac{1}{3}} \angle \tfrac{1}{3}\theta = \sqrt[3]{r} \angle \tfrac{\theta}{3}$$

It is important to appreciate that a real number has three cube roots, equal in size and equally displaced about the Argand diagram, i.e., three roots each displaced by 120°.
For example, $\sqrt[3]{8} = \sqrt[3]{[8 \angle 0°]} = [8 \angle 0°]^{\frac{1}{3}} = 8^{\frac{1}{3}} \angle 0°$ and the solutions are $2 \angle 0°$, $2 \angle 120°$ and $2 \angle 240°$.
The uses of De Moivre's theorem include the solution of problems associated with electrical circuit theory, simplifying certain types of integrals and verifying and determining certain types of trigonometric identities.

Worked problems on De Moivre's theorem

Problem 1. Determine the square of the complex number $3 - j4$ (a) in cartesian form and (b) in polar form, using De Moivre's theorem. Compare the results obtained and show the roots on an Argand diagram.

(a) In cartesian form, $(3 - j4)^2 = (3 - j4)(3 - j4)$

$$= 9 - j12 - j12 + j^2 16$$

$$= -7 - j24$$

(b) In polar form, $(3 - j4) = \sqrt{[(3)^2 + (4)^2]} \angle \arctan \dfrac{4}{3}$

(see Fig.8.1)

$$= 5 \angle -53° 8'$$

$$[5 \angle -53° 8']^2 = 5^2 \angle (2 \times -53° 8') = 25 \angle -106° 16'$$

The complex number $(3 - j4)$, together with its square,
i.e. $(3 - j4)^2$, is shown in Fig. 8.1.
$25 \angle -106°\ 16'$ in cartesian form is
$25 \cos(-106°\ 16') + j25 \sin(-106°\ 16')$, i.e. $-7 - j24$, as in part (a)

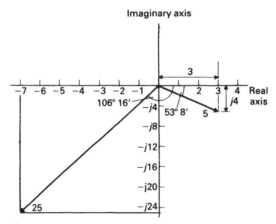

Fig. 8.1

Problem 2. Determine $(-2 + j3)^5$ in polar and in cartesian form.

$Z = -2 + j3$ is situated in the second quadrant of the Argand
diagram.

Thus $r = \sqrt{[(2)^2 + (3)^2]} = \sqrt{13}$

and $\alpha = \arctan \dfrac{3}{2} = 56°\ 19'$

Hence the argument $\theta = 180° - 56°\ 19' = 123°\ 41'$
Thus $2 - j3$ in polar form is $\sqrt{13} \angle 123°\ 41'$

$$(-2 + j3)^5 = [\sqrt{13} \angle 123°\ 41']^5$$
$$= (\sqrt{13})^5 \angle (5 \times 123°\ 41'),\ \text{from De Moivre's theorem}$$
$$= 13^{\frac{5}{2}} \angle 618°\ 25'$$
$$= 13^{\frac{5}{2}} \angle 258°\ 25'\ (\text{since } 618°\ 25' = 618°\ 25' - 360°)$$
$$= 13^{\frac{5}{2}} \angle -101°\ 35'$$
$$= \mathbf{609.3 \angle -101°\ 35'}$$

In cartesian form

$$609.3 \angle -101°\ 35' = 609.3 \cos(-101°\ 35') + j609.3 \sin(-101°\ 35')$$
$$= \mathbf{-122.3 - j596.9}$$

Problem 3. Determine the two square roots of the complex number $12 + j5$ in cartesian and polar form, correct to three significant figures. Show the roots on an Argand diagram.

In polar form $12 + j5 = \sqrt{(12^2 + 5^2)} \angle \arctan \dfrac{5}{12}$, since $12 + j5$ is in the first quadrant of the Argand diagram,

i.e. $12 + j5 = 13 \angle 22° 37'$

Since we are finding the square roots of $13 \angle 22° 37'$ there will be two solutions. To obtain the second solution it is helpful to express $13 \angle 22° 37'$ also as $13 \angle (360° + 22° 37')$, i.e. $13 \angle 382° 37'$ (we have merely rotated one revolution to obtain this result). The reason for doing this is that when we divide the angles by 2 we still obtain angles less than $360°$, as shown below.

Hence $\sqrt{(12 + j5)} = \sqrt{[13 \angle 22° 37']}$ or $\sqrt{[13 \angle 382°37']}$

$\qquad\qquad = [13 \angle 22° 37']^{\frac{1}{2}}$ or $[13 \angle 382° 37']^{\frac{1}{2}}$

$\qquad\qquad = 13^{\frac{1}{2}} \angle (\frac{1}{2} \times 22° 37')$ or $13^{\frac{1}{2}} \angle (\frac{1}{2} \times 382° 37')$,
$\qquad\qquad$ from De Moivre's theorem,

$\qquad\qquad = \sqrt{13} \angle 11° 19'$ or $\sqrt{13} \angle 191° 19'$

$\qquad\qquad = 3.61 \angle 11° 19'$ or $3.61 \angle -168° 41'$

These two solutions of $\sqrt{(12 + j5)}$ are shown in the Argand diagram of Fig. 8.2.

$3.61 \angle 11° 19'$ is in the first quadrant of the Argand diagram.

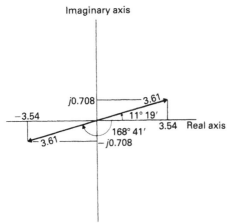

Fig. 8.2

Thus $3.61 \angle 11° 19' = 3.61(\cos 11° 19' + j \sin 11° 19')$

$$= 3.54 + j0.708$$

$3.61 \angle -168° 41'$ is in the third quadrant of the Argand diagram.

Thus $3.61 \angle -168° 41' = 3.61[\cos(-168° 41') + j \sin(-168° 41')]$

$$= -3.54 - j0.708$$

Thus in cartesian form the two roots are $\pm(\mathbf{3.54 + j0.708})$.

From the Argand diagram the roots are seen to be 180° apart, i.e. they lie on a straight line. This is always true when finding square roots of complex numbers.

Problem 4. Use De Moivre's theorem to find the value of $(2 - j\sqrt{5})^7$ in the form $r \angle \theta$ and in the form $a + jb$.

The general relationship between a complex number in $a + jb$ and polar forms is that if $z = x + jy$, then the modulus of z, i.e. $|z| = \sqrt{(x^2 + y^2)}$ and the argument of z, i.e. arg $z = \arctan \dfrac{y}{x}$. Thus, since $z = 2 - j\sqrt{5}$, $|z| = \sqrt{(2^2 + \sqrt{5}^2)} = \sqrt{9} = 3$, since a modulus cannot have a negative value.

$$\text{arg } z = \arctan -\frac{\sqrt{5}}{2} = \arctan -1.118$$

An angle having a negative value for its tangent can lie in either the second or fourth quadrants. However, visualising the Argand diagram, $+2$ is to the right ($\rightarrow$) and $-j\sqrt{5}$ is downwards, ($\downarrow$) showing that only the fourth quadrant value for the argument is possible. Thus:

$$\text{arg } z = \arctan - 1.118 = -48° 11'$$

Then, $\qquad (2 - j\sqrt{5}) = 3 \angle -48° 11'$

Applying De Moivre's theorem (equation (8)) gives:

$$(2 - j\sqrt{5})^7 = 3^7 \angle [7 \times (-48° 11')]$$

$$= 2\,187 \angle -337° 17'$$

It is usual to express angles in terms of their principal values, that is, so that $-180° \leqslant \theta \leqslant 180°$, hence

$(2 - j\sqrt{5})^7 = \mathbf{2\,187 \angle 22° 43'}$ **in polar form.**

The general relationship between a complex number in polar

form and the $a + jb$ form is that if

$$r(\cos \theta + j \sin \theta) = a + jb, \text{ then } a = r \cos \theta \text{ and } b = r \sin \theta$$

Thus, $2\,187 \angle 22° 43' = (2\,187 \cos 22° 43') + j(2\,187 \sin 22° 43')$

giving $\quad (2 - j\sqrt{5})^7 = 2\,017 + j844.6$ in $a + jb$ form,
correct to four significant figures.

Problem 5. Determine the four roots of $(-3 - j7)^{\frac{1}{4}}$ in polar form.

Let $z = -3 - j7$, then $|z| = \sqrt{[(-3)^2 + (-7)^2]}$

$$= 7.616$$

$$\arg z = \arctan \frac{-7}{-3} = \arctan 2.333$$

$$= 66° 48' \text{ or } 246° 48'$$

Visualising the Argand diagram, -3 is to the left ($\leftarrow$) and $-j7$ is downward ($\downarrow$), showing that only the third quadrant value for the argument is possible. Expressing the argument of $246° 48'$ as a principal value gives

$$-3 - j7 = 7.616 \angle -113° 12'$$

By De Moivre's thorem,

$$(-3 - j7)^{\frac{1}{4}} = 7.616^{\frac{1}{4}} \angle \left[\frac{1}{4} \times (-113° 12') \right]$$

$$= 1.661 \angle -28° 18'$$

The remaining three roots have the same modulus and are calculated by adding $360°$, $720°$ and $1\,080°$ to the argument of the complex number in turn. This is done because when dividing these arguments by 4, angles of less than $360°$ are still obtained.
    Thus the arguments of the other three roots are

$$\frac{1}{4} \times (-113° 12' + 360°) = 61° 42'$$

$$\frac{1}{4} \times (-113° 12' + 720°) = 151° 42' \text{ and}$$

$$\frac{1}{4} \times (-113° 12' + 1\,080°) = 241° 42' \text{ or } -118° 18'$$

(The argument of the root obtained by adding $4 \times 360°$, that

is, $1\,440°$, is $\dfrac{1}{4} \times (-113° \, 12' + 1\,440°) = 331° \, 42'$ or $-28° \, 18'$,

which is the same value as the first of the arguments shown above. Hence all the roots have been calculated.) The values of the arguments can be easily checked, since they are symmetrically

placed on an Argand diagram $\left(\dfrac{360}{n}\right)°$ apart where n is the

number of roots required. Thus for a fourth root problem, the

arguments are $\left(\dfrac{360}{4}\right)°$, i.e. $90°$ apart, and it can be seen that the

four arguments obtained in this problem are displaced from one another by $90°$.

Thus the four roots of $(-3 - j7)^{\frac{1}{4}}$ are:

1.661 $\angle$ $-28°$ 18′,   1.661 $\angle$ 61° 42′
1.661 $\angle$ 151° 42′ and 1.661 $\angle$ $-118°$ 18′

Problem 6. Calculate the five roots of $(4 \angle 21°)^{-\frac{2}{5}}$ in the $a + jb$ form, correct to four decimal places.

For fractional powers of a complex number, the relationship between rectangular and polar forms is:

$$(a + jb)^{p/q} = [r \angle \theta]^{p/q} = r^{p/q} \angle \left[\frac{p}{q}(\theta + 360\,n)\right]$$

where $r = \sqrt{(a^2 + b^2)}$, $\theta = \arctan \dfrac{b}{a}$ and $n = 0, 1, 2, \ldots, q - 1$,

since the three values of a cube root are given by $n = 0, 1, 2$, the four values of a fourth root by $n = 0, 1, 2$ and 3 and q values of a qth root by $n = 0, 1, 2, \ldots, q - 1$.

By De Moivre's theorem,

$$(4 \angle 21°)^{-\frac{2}{5}} = 4^{-\frac{2}{5}} \angle \left[-\frac{2}{5}(21° + 360\,n°)\right]$$

where $n = 0, 1, 2, 3$ and 4, since there are 5 roots.

When $n = 0$, $(4 \angle 21°)^{-\frac{2}{5}} = 0.5743 \angle \left[-\frac{2}{5}(21°)\right]$

$$= 0.5743 \angle 351° \, 36'$$

When $n = 1$, $(4 \angle 21°)^{-\frac{2}{5}} = 0.5743 \angle \left[-\frac{2}{5}(21 + 360)°\right]$

$$= 0.5743 \angle 207° \, 36'$$

When $n = 2$, $(4 \angle 21°)^{-\frac{2}{5}} = 0.5743 \angle \left[-\frac{2}{5}(21 + 360 \times 2)° \right]$

$= 0.5743 \angle 63° \, 36'$

When $n = 3$, $(4 \angle 21°)^{-\frac{2}{5}} = 0.5743 \angle \left[-\frac{2}{5}(21 + 360 \times 3)° \right]$

$= 0.5743 \angle 279° \, 36'$

When $n = 4$, $(4 \angle 21°)^{-\frac{2}{5}} = 0.5743 \angle \left[-\frac{2}{5}(21 + 360 \times 4)° \right]$

$= 0.5743 \angle 135° \, 36'$

Checking: the 5 roots should be $\dfrac{360}{5}$, i.e. 72° apart. The values of the arguments obtained are spaced 72° apart, so no error has been made. (The arguments have been expressed in $0 \leqslant \theta \leqslant 360°$ form to enable this check to be easily carried out).

Since $a = r \cos \theta$ and $b = r \sin \theta$, the five roots expressed in the $a + jb$ form are:

$0.5743 \cos(351° \, 36') + j0.5743 \sin(351° \, 36')$
$$= 0.5681 - j0.0839$$

$0.5743 \cos(207° \, 36') + j0.5743 \sin(207° \, 36')$
$$= -0.5089 - j0.2661$$

$0.5743 \cos(63° \, 36') + j0.5743 \sin(63° \, 36')$
$$= 0.2554 + j0.5144$$

$0.5743 \cos(279° \, 36') + j0.5743 \sin(279° \, 36')$
$$= 0.0958 - j0.5663 \text{ and}$$

$0.5743 \cos(135° \, 36') + j0.5743 \sin(135° \, 36')$
$$= -0.4103 + j0.4018$$

Further problems on De Moivre's theorem may be found in Section 8.5 (Problems 1 to 20), page 140.

8.3   Expressing cos $n\theta$ and sin $n\theta$ in terms of powers of cos θ and sin θ

The mathematical statement of De Moivre's theorem is

$$z^n = r^n(\cos n\theta + j \sin n\theta)$$

But $z^n = [r(\cos \theta + j \sin \theta)]^n$

Hence, $r^n(\cos n\theta + j \sin n\theta) = [r(\cos \theta + j \sin \theta)]^n$

When $r = 1$, this equation becomes

$$\cos n\theta + j \sin n\theta = (\cos \theta + j \sin \theta)^n$$

The right-hand side of this equation can be expanded using the binomial series. (The general binomial expansion is given by:

$$(a + b)^n = a^n + na^{n-1}b + \frac{n(n-1)}{2!} a^{n-2}b^2 + \cdots)$$

Since the terms produced when doing this are long, abbreviations are used below, where

(i) $\cos^n \theta$ is written as C^n
(ii) $\sin^n \theta$ is written as S^n, and
(iii)   binomial coefficients such as $\dfrac{n(n-1)(n-2)}{3!}$ are written as $\dbinom{n}{3}$

Using this notation:

$$(\cos n\theta + j \sin n\theta) = C^n + nC^{n-1} \cdot jS + \binom{n}{2} C^{n-2} \cdot j^2 S^2$$

$$+ \binom{n}{3} C^{n-3} \cdot j^3 S^3 + \binom{n}{4} C^{n-4} \cdot j^4 S^4 + \cdots$$

But $j^2 = -1$, $j^3 = -j$ and $j^4 = 1$, hence:

$$(\cos n\theta + j \sin n\theta) = C^n + jnC^{n-1}S - \binom{n}{2} C^{n-2}S^2 - j\binom{n}{3} C^{n-3}S^3$$

$$+ \binom{n}{4} C^{n-4}S^4 + \cdots$$

Grouping real and imaginary terms gives:

$$(\cos n\theta + j \sin n\theta) = \left[C^n - \binom{n}{2} C^{n-2}S^2 + \binom{n}{4} C^{n-4}S^4 - \cdots \right]$$

$$+ j\left[nC^{n-1}S - \binom{n}{3} C^{n-3}S^3 \right.$$

$$\left. + \binom{n}{5} C^{n-5}S^5 - \cdots \right]$$

Equating real parts gives:

$$\cos n\theta = C^n - \binom{n}{2} C^{n-2}S^2 + \binom{n}{4} C^{n-4}S^4 - \cdots$$

That is,

$$\cos n\theta = \cos^n\theta - \frac{n(n-1)}{2!}\cos^{n-2}\theta \sin^2\theta$$

$$+\frac{n(n-1)(n-2)(n-3)}{4!}\cos^{n-4}\theta \sin^4\theta - \cdots \quad (9)$$

Equating imaginary parts gives:

$$\sin n\theta = nC^{n-1}S - \binom{n}{3}C^{n-3}S^3 + \binom{n}{5}C^{n-5}S^5 - \cdots$$

That is,

$$\sin n\theta = n\cos^{n-1}\theta \sin\theta - \frac{n(n-1)(n-2)}{3!}\cos^{n-3}\theta \sin^3\theta$$

$$+\frac{n(n-1)(n-2)(n-3)(n-4)}{5!}\cos^{n-5}\theta \sin^5\theta - \cdots \quad (10)$$

Worked problems on expressing cos *nθ* and sin *nθ* in terms of powers of cos *θ* and sin *θ*

Problem 1. Express $\cos 3\theta$ in terms of powers of $\cos\theta$

Substituting $n = 3$ in equation (9) gives:

$$\cos 3\theta = \cos^3\theta - \frac{3(3-1)}{2!}\cos^{3-2}\theta \sin^2\theta$$

(when $n = 3$, $\frac{n(n-1)(n-2)(n-3)}{4!} = \frac{3 \times 2 \times 1 \times 0}{4 \times 3 \times 2 \times 1} = 0$

Hence all terms after the second in equation (9) are zero when $n = 3$.)

Thus $\cos 3\theta = \cos^3\theta - \frac{3 \times 2}{2 \times 1}\cos\theta \sin^2\theta$

$$= \cos^3\theta - 3\cos\theta \sin^2\theta$$

But $\sin^2\theta = 1 - \cos^2\theta$,

Hence, $\cos 3\theta = \cos^3\theta - 3\cos\theta(1 - \cos^2\theta)$

$$= \cos^3\theta - 3\cos\theta + 3\cos^3\theta$$

i.e. $\cos 3\theta = 4\cos^3\theta - 3\cos\theta$ in terms of powers of $\cos\theta$.

Problem 2. Express $\sin 5\theta$ in terms of powers of $\cos \theta$ and $\sin \theta$

Substituting $n = 5$ in equation (10) gives:

$$\sin 5\theta = 5 \cos^{5-1} \theta \sin \theta - \frac{5 \times 4 \times 3}{3 \times 2 \times 1} \cos^{5-3} \theta \sin^3 \theta$$

$$+ \frac{5 \times 4 \times 3 \times 2 \times 1}{5 \times 4 \times 3 \times 2 \times 1} \cos^{5-5} \theta \sin^5 \theta$$

i.e. $\sin 5\theta = 5 \cos^4 \theta \sin \theta - 10 \cos^2 \theta \sin^3 \theta + \sin^5 \theta$

Further problems on expressing $\cos n\theta$ *and* $\sin n\theta$ *in terms of powers of* $\cos \theta$ *and* $\sin \theta$ *may be found in Section 8.5 (Problems 21 to 26), page 142.*

8.4   Expressing $\cos^n \theta$ and $\sin^n \theta$ in terms of sines and cosines of multiples of θ

A complex number, z, may be written as

$$z = re^{j\theta} = r(\cos \theta + j \sin \theta) \tag{11}$$

When $r = 1$, then $z = \cos \theta + j \sin \theta = e^{j\theta}$
Also, from section 1, equation (5), $e^{-j\theta} = \cos \theta - j \sin \theta$
But, if $e^{j\theta} = z$, then $e^{-j\theta} = \dfrac{1}{z}$

Hence $\dfrac{1}{z} = \cos \theta - j \sin \theta \tag{12}$

Adding equations (11) and (12) gives:

$$z + \frac{1}{z} = 2 \cos \theta \tag{13}$$

Subtracting equation (12) from equation (11) gives:

$$z - \frac{1}{z} = j2 \sin \theta \tag{14}$$

Raising equation (13) to a power of n gives:

$$\left(z + \frac{1}{z}\right)^n = 2^n \cos^n \theta \tag{15}$$

where $z = \cos \theta + j \sin \theta$ and $\dfrac{1}{z} = \cos \theta - j \sin \theta$

Raising equation (14) to a power of n gives:

$$\left(z - \frac{1}{z}\right)^n = j^n 2^n \sin^n \theta \tag{16}$$

By expanding $\left(z + \frac{1}{z}\right)^n$ by the binomial theorem, equation (15) can be used to express $\cos^n \theta$ in terms of $\cos n\theta$ and $\sin n\theta$. Similarly, equation (16) can be used to express $\sin^n \theta$ in terms of $\cos n\theta$ and $\sin n\theta$. This technique is used to express integrals of $\cos^n x$ and $\sin^n x$ in a form which enables integration to be achieved.

Worked problems on expressing $\cos^n \theta$ and $\sin^n \theta$ in terms of sines and cosines of multiples of θ

Problem 1. Express $\cos^2 \theta$ in terms of cosines of multiples of θ.

From equation (15): $2^n \cos^n \theta = \left(z + \frac{1}{z}\right)^n$, where

$$z = \cos \theta + j \sin \theta \text{ and } \frac{1}{z} = \cos \theta - j \sin \theta$$

When $n = 2$, $2^2 \cos^2 \theta = \left(z + \frac{1}{z}\right)^2 = z^2 + 2 + \frac{1}{z^2}$

$$= \left(z^2 + \frac{1}{z^2}\right) + 2$$

However, $z^n + \frac{1}{z^n} = (\cos n\theta + j \sin n\theta) + (\cos n\theta - j \sin n\theta)$

$$= 2 \cos n\theta$$

Hence, $\left(z^2 + \frac{1}{z^2}\right) + 2 = 2 \cos 2\theta + 2$

Thus, $2^2 \cos^2 \theta = 2 \cos 2\theta + 2$

$$\cos^2 \theta = \frac{1}{2}(\cos 2\theta + 1)$$

Problem 2. Express $\sin^4 \theta$ in terms of $\cos n\theta$.

From equation (16): $j^n 2^n \sin^n \theta = \left(z - \frac{1}{z}\right)^n$

When $n = 4$: $\qquad j^4 2^4 \sin^4 \theta = \left(z - \dfrac{1}{z} \right)^4$

Using the binomial theorem to expand $\left(z - \dfrac{1}{z} \right)^4$ gives:

$$\left(z - \frac{1}{z} \right)^4 = z^4 - 4z^3 \cdot \frac{1}{z} + 6z^2 \cdot \frac{1}{z^2} - 4z \cdot \frac{1}{z^3} + \frac{1}{z^4}$$

$$= \left(z^4 + \frac{1}{z^4} \right) - 4 \left(z^2 + \frac{1}{z^2} \right) + 6$$

But it is shown in problem 1 above that $\left(z^n + \dfrac{1}{z^n} \right) = 2 \cos n\theta$

Hence, $j^4 2^4 \sin^4 \theta = 2 \cos 4\theta - 8 \cos 2\theta + 6$

But, $j^4 = 1$, hence $\sin^4 \theta = \dfrac{1}{8} \cos 4\theta - \dfrac{1}{2} \cos 2\theta + \dfrac{3}{8}$

Problem 3. Express $\sin^7 \theta$ in terms of $\sin n\theta$.

From equation (16): $j^n 2^n \sin^n \theta = \left(z - \dfrac{1}{z} \right)^n$

When $n = 7$, $\qquad j^7 2^7 \sin^7 \theta = \left(z - \dfrac{1}{z} \right)^7$

Using the binomial theorem to expand the right-hand side of this equation gives:

$$\left(z - \frac{1}{z} \right)^7 = z^7 - 7z^6 \cdot \frac{1}{z} + 21z^5 \cdot \frac{1}{z^2} - 35z^4 \cdot \frac{1}{z^3} + 35z^3 \cdot \frac{1}{z^4}$$

$$- 21z^2 \cdot \frac{1}{z^5} + 7z \cdot \frac{1}{z^6} - \frac{1}{z^7}$$

$$= \left(z^7 - \frac{1}{z^7} \right) - 7 \left(z^5 - \frac{1}{z^5} \right) + 21 \left(z^3 - \frac{1}{z^3} \right) - 35 \left(z - \frac{1}{z} \right)$$

However, $\left(z^n - \dfrac{1}{z^n} \right) = (\cos n\theta + j \sin n\theta) - (\cos n\theta - j \sin n\theta)$

$$= 2j \sin n\theta$$

Hence, $j^7 2^7 \sin^7 \theta = 2j(\sin 7\theta - 7 \sin 5\theta + 21 \sin 3\theta - 35 \sin \theta)$

But $\qquad\qquad j^7 = -j$, hence

$$-j 2^7 \sin^7 \theta = 2j(\sin 7\theta - 7 \sin 5\theta + 21 \sin 3\theta - 35 \sin \theta)$$

i.e. $\sin^7 \theta = \dfrac{1}{2^6}(-\sin 7\theta + 7 \sin 5\theta - 21 \sin 3\theta + 35 \sin \theta)$

Further problems on expressing $\cos^n \theta$ *and* $\sin^n \theta$ *in terms of sines and cosines of multiples of* θ *may be found in the following Section 8.5 (Problems 27 to 32), page 142.*

8.5   Further problems

De Moivre's theorem

In Problems 1 to 4 evaluate in cartesian and in polar form.

1. (a) $(2 + j3)^2$ (b) $(4 - j5)^2$
    (a) $[-5 + j12; 13 \angle 112° 37']$
    (b) $[-9 - j40; 41 \angle -102° 41']$
2. (a) $(-3 + j2)^5$ (b) $(-2 - j)^3$
    (a) $[597 + j122; 609.3 \angle 11° 33']$
    (b) $[-2 - j11; 11.18 \angle -100° 18']$
3. (a) $[4 \angle 32°]^4$ (b) $[2 \angle 125°]^5$
    (a) $[-157.6 + j201.7; 256 \angle 128°]$
    (b) $[-2.789 - j31.88; 32 \angle -95°]$

4. (a) $\left[3 \angle -\dfrac{\pi}{3} \right]^3$ (b) $(1.5 \angle -160°)^4$

    (a) $[-27 + j0; 27 \angle -\pi]$
    (b) $[0.879\,2 + j4.986; 5.063 \angle 80°]$
5. Determine the value of $(3 + j4)^3$, expressing the result in the $a + jb$ form. $[-117 + j44]$
6. Find the value of $(\sqrt{3} - j)^9$, expressing the result in both polar and $a + jb$ forms. $[512 \angle 90°; j512]$
7. Determine the values of (a) $(3 - j4)^5$ and (b) $(3 + j4)^5$, giving the answers in polar form.   (a) $[3\,125 \angle 94° 21']$
    (b) $[3\,125 \angle -94° 21']$

In Problems 8 to 10 determine the two square roots of the given complex numbers in cartesian form and show the results on an Argand diagram.

8. (a) $2 + j$ (b) $3 - j2$
    (a) $[\pm(1.455 + j0.344)]$ (b) $[\pm(1.817 - j0.550)]$
9. (a) $-3 + j4$ (b) $-1 - j3$
    (a) $[\pm(1 + j2)]$ (b) $[\pm(1.040 - j1.443)]$

10. (a) $5 \angle 36°$ (b) $14 \angle \dfrac{3\pi}{2}$

    (a) $[\pm(2.127 + j0.691)]$ (b) $[\pm(-2.646 + j2.646)]$

11. If $Z = 3 \angle 30°$ evaluate in polar form (a) Z^2, (b) $\sqrt{Z}$.
    (a) $[9 \angle 60°]$ (b) $[\sqrt{3} \angle 15°$ and $\sqrt{3} \angle -165°]$

12. Convert $2 - j$ into polar form and hence evaluate $(2 - j)^7$ in polar form. $[\sqrt{5} \angle -26° 34'; 279.5 \angle 174° 3']$

13. Simplify $\dfrac{\left(\cos \dfrac{\pi}{9} - j \sin \dfrac{\pi}{9}\right)^4}{\left(\cos \dfrac{\pi}{9} + j \sin \dfrac{\pi}{9}\right)^5}$ $[-1]$

14. Find the roots of $(-2 + j2)^{\frac{1}{2}}$ in polar form, correct to four significant figures. $[1.682 \angle 67° 30', 1.682 \angle -112° 30']$

15. Determine the three cube roots of $8(1 - j)$ in polar form.
    $[2.245 \angle -15°, 2.245 \angle 105°, 2.245 \angle -135°]$

16. Find the four fourth roots of $3 - j4$, expressing them in the $a + jb$ form, correct to four significant figures.
    $$\begin{bmatrix} 1.455 - j0.343\,6, -0.343\,6 - j1.455 \\ -1.455 + j0.343\,6, 0.343\,6 + j1.455 \end{bmatrix}$$

17. Express the roots of $(-2 + j3)^{-\frac{2}{3}}$ in the $a + jb$ form, correct to four decimal places.
    $$\begin{bmatrix} 0.055\,8 - j0.421\,6, 0.337\,2 + j0.259\,1 \\ -0.393\,0 + j0.1625 \end{bmatrix}$$

18. Evaluate $(-2 - j)^{-\frac{1}{5}}$, expressing the answers in both polar and $a + jb$ forms, correct to four decimal places.
    $$\begin{bmatrix} 0.851\,3 \angle 30° 41', 0.851\,3 \angle 102° 41', 0.851\,3 \angle 174° 41', \\ 0.851\,3 \angle -41° 19', 0.851\,3 \angle -113° 19'; \\ 0.732\,1 + j0.434\,4, -0.186\,9 + j0.830\,5, -0.847\,6 + j0.078\,9, \\ 0.639\,4 - j0.562\,0, -0.337\,0 - j0.781\,8 \end{bmatrix}$$

19. The characteristic impedance (Z_0) and the propagation constant (P) of a transmission line are given by the equations:

    $$Z_0 = (0.3 + j0.5)^{\frac{1}{2}}(2 + j500)^{-\frac{1}{2}} \times 10^4$$

    $$P = (0.3 + j0.5)^{\frac{1}{2}}(2 + j500)^{\frac{1}{2}} \times 10^{-4}$$

    Evaluate Z_0 and P. $\begin{bmatrix} Z_0 = (329 - j90.5) \\ P = (0.000\,46 + j0.001\,64) \end{bmatrix}$

20. When displaced electrons oscillate about an equilibrium position the displacement x is given by the equation:

    $$x = A \exp\left\{-\frac{ht}{2m} + j\frac{\sqrt{(4\,mf - h^2)}}{2m - a}\right\}$$

Determine the real part of x assuming $(4\,mf - h^2)$ is positive.

$$\left[A\,e^{-ht/2m} \cos\left(\frac{\sqrt{(4\,mf - h^2)}}{2\,m - a}\right)t \right]$$

Expressing cos $n\theta$ and sin $n\theta$ in terms of powers of cos θ and sin θ

21. Express $\cos 4\theta$ in terms of powers of $\cos\theta$ and $\sin\theta$.
    $[\cos^4\theta - 6\cos^2\theta\sin^2\theta + \sin^4\theta]$

22. Express $\sin 4\theta$ in terms of powers of $\cos\theta$ and $\sin\theta$.
    $[4\cos^3\theta\sin\theta - 4\cos\theta\sin^3\theta]$

23. Express $\cos 6\theta$ in terms of powers of $\cos\theta$.
    $[32\cos^6\theta - 48\cos^4\theta + 18\cos^2\theta - 1]$

24. Express $\dfrac{\sin 5\theta}{\sin\theta}$ in terms of powers of $\sin\theta$.
    $[16\sin^4\theta - 20\sin^2\theta + 5]$

25. Express $\dfrac{\sin 7\theta}{\sin\theta}$ in terms of powers of $\cos\theta$ and $\sin\theta$.
    $[7\cos^6\theta - 35\cos^4\theta\sin^2\theta + 21\cos^2\theta\sin^4\theta - \sin^6\theta]$

26. Express $\cos 7\theta$ in terms of powers of $\cos\theta$.
    $[64\cos^7\theta - 112\cos^5\theta + 56\cos^3\theta - 7\cos\theta]$

Expressing cos^n θ and sin^n θ in terms of sines and cosines of multiples of θ

27. Express $\cos^6\theta$ in terms of cosines of multiples of θ.
    $$\left[\frac{1}{2^5}(\cos 6\theta + 6\cos 4\theta + 15\cos 2\theta + 10) \right]$$

28. Express $\sin^6\theta$ in terms of cosines of multiples of θ.
    $$\left[\frac{1}{2^5}(-\cos 6\theta + 6\cos 4\theta - 15\cos 2\theta + 10) \right]$$

29. Express $\cos^8\theta$ in terms of cosines of multiples of θ.
    $$\left[\frac{1}{2^7}(\cos 8\theta + 8\cos 6\theta + 28\cos 4\theta + 56\cos 2\theta + 35) \right]$$

30. Show that
    $$\sin^2\theta\cos^5\theta = \frac{1}{2^6}(-\cos 7\theta - 3\cos 5\theta - \cos 3\theta + 5\cos\theta).$$

31. Show that

$$\sin^4 \theta \cos^3 \theta = \frac{1}{2^6} (\cos 7\theta - \cos 5\theta - 3 \cos 3\theta + 3 \cos \theta).$$

32. Express $\sin^5 \theta$ in terms of sines of multiples of θ.

$$\left[\frac{1}{2^4} (\sin 5\theta - 5 \sin 3\theta + 10 \sin \theta) \right]$$

Chapter 9

Hyperbolic functions

9.1 Definitions of hyperbolic functions

Trigonometric functions are called **'circular functions'** since they arise naturally in connection with the geometry of the circle. There are other functions which are associated with the geometry of the conic section called a **hyperbola** (see page 187) and are thus classified as **hyperbolic functions**. Such functions have several applications, in particular with transmission line theory and with catenary problems, a catenary being a curve formed by a chain or rope of uniform density hanging freely from two fixed points not in the same vertical line.

Six hyperbolic functions are defined below, each of them being closely connected with the six trigonometrical ratios.

By definition:

(*a*) Hyperbolic sine of *x*, **sinh *x*** $= \dfrac{e^x - e^{-x}}{2}$ (1)

'sinh *x*' is often abbreviated to 'sh *x*' and is pronounced as 'shine *x*'.

(*b*) Hyperbolic cosine of *x*, **cosh *x*** $= \dfrac{e^x + e^{-x}}{2}$ (2)

'cosh *x*' is often abbreviated to 'ch *x*' and is pronounced as 'kosh *x*'.

(*c*) Hyperbolic tangent of *x*, **tanh *x*** $= \dfrac{\sinh x}{\cosh x} = \dfrac{e^x - e^{-x}}{e^x + e^{-x}}$ (3)

'tanh *x*' is often abbreviated to 'th *x*' and is pronounced as 'than *x*'.

(*d*) Hyperbolic cosecant of *x*, **cosech *x*** $= \dfrac{1}{\sinh x} = \dfrac{2}{e^x - e^{-x}}$ (4)

'cosech *x*' is pronounced as 'coshec *x*'.

(*e*) Hyperbolic secant of *x*, **sech *x*** $= \dfrac{1}{\cosh x} = \dfrac{2}{e^x + e^{-x}}$ (5)

'sech *x*' is pronounced as 'shec *x*'.

(*f*) Hyperbolic contangent of *x*, **coth *x*** $= \dfrac{1}{\tanh x} = \dfrac{e^x + e^{-x}}{e^x - e^{-x}}$ (6)

'coth *x*' is pronounced as 'koth *x*'.

9.2   Some properties of hyperbolic functions

(a) If for a function of x, $f(-x) = f(x)$, i.e. $f(x)$ is unchanged when x is replaced by $-x$, then $f(x)$ is called an **even function** of x. $\text{Cos}(-x) = \cos x$, thus the cosine function is an even function. Replacing x by $-x$ in the formula for $\cosh x$ gives:

$$\cosh(-x) = \frac{e^{-x} + e^{-(-x)}}{2} = \frac{e^{-x} + e^{x}}{2} = \cosh x$$

Thus the hyperbolic cosine, **cosh x, is an even function**. The graph of an even function is symmetrical about the y-axis (see Section 9.4).

Since $\cosh x$ is an even function then $\text{sech } x = \dfrac{1}{\cosh x}$ is an even function also.

(b) If for a function of x, $f(-x) = -f(x)$ then $f(x)$ is called an **odd function** of x. $\text{Sin}(-x) = -\sin x$, thus the sine function is an odd function. Replacing x by $-x$ in the formula for $\sinh x$ gives:

$$\sinh(-x) = \frac{e^{-x} - e^{-(-x)}}{2} = \frac{e^{-x} - e^{x}}{2} = -\left(\frac{e^{x} - e^{-x}}{2}\right) = -\sinh x$$

Thus the hyperbolic sine, **sinh x, is an odd function**. The graph of an odd function is symmetrical about the origin (see Section 9.4).

Since $\sinh x$ is an odd function then $\text{cosech } x = \dfrac{1}{\sinh x}$ is an odd function also.

(c) Replacing x by $-x$ in the formula for $\tanh x$ gives:

$$\tanh(-x) = \frac{e^{-x} - e^{-(-x)}}{e^{-x} + e^{-(-x)}} = \frac{e^{-x} - e^{x}}{e^{-x} + e^{x}} = -\left(\frac{e^{x} - e^{-x}}{e^{x} + e^{-x}}\right) = -\tanh x$$

Thus the hyperbolic tangent, **tanh x, is an odd function**. It follows that $\coth x = \dfrac{1}{\tanh x}$ is also an odd function.

(d) Replacing x by 0 in the formula for $\sinh x$ gives:

$$\sinh 0 = \frac{e^{0} - e^{-0}}{2} = \frac{1 - 1}{2} = 0$$

(e) Replacing x by 0 in the formula for $\cosh$ gives:

$$\cosh 0 = \frac{e^{0} + e^{-0}}{2} = \frac{1 + 1}{2} = 1$$

9.3   Evaluation of hyperbolic functions

Hyperbolic functions may be evaluated readily using a calculator. Some scientific notation calculators possess sinh, cosh and tanh functions; however, if a calculator does not have these functions then the hyperbolic functions are evaluated from their definitions. For example, $\sinh 0.32 = \dfrac{e^{0.32} - e^{-0.32}}{2}$ from the definition of sinh, and this may be evaluated since all scientific calculators possess an exponential function. It may be checked that $\sinh 0.32 = 0.325\,489\,36\ldots$ or **0.3255**, correct to four decimal places. Use your calculator to check the answers to the following worked problems.

Worked problems on evaluating hyperbolic functions

Problem 1. Evaluate correct to four significant figures:
(a) $\sinh 3.162$ (b) $\cosh 0.798\,3$

(a) $\sinh 3.162 = \dfrac{e^{3.162} - e^{-3.162}}{2} = 11.787\,721\,6\ldots$

$= \mathbf{11.79}$, correct to four significant figures.

(b) $\cosh 0.798\,3 = \dfrac{e^{0.798\,3} + e^{-0.798\,3}}{2} = 1.335\,927\,09\ldots$

$= \mathbf{1.336}$, correct to four significant figures.

Problem 2. Evaluate, correct to four decimal places:
(a) $\tanh 1.469$ (b) $\operatorname{cosech} 2.748$

(a) $\tanh 1.469 = \dfrac{e^{1.469} - e^{-1.469}}{e^{1.469} + e^{-1.469}} = \dfrac{4.114\,732\,55\ldots}{4.575\,043\,60\ldots}$

$= \mathbf{0.899\,4}$, correct to four decimal places.

(b) $\operatorname{cosech} 2.748 = \dfrac{1}{\sinh 2.748} = \dfrac{2}{e^{2.748} - e^{-2.748}}$

$= \mathbf{0.128\,6}$, correct to four decimal places.

Problem 3. Evaluate, correct to five significant figures:
(a) $\operatorname{sech} 0.963\,2$ (b) $\coth 0.443\,7$

(a) $\operatorname{sech} 0.963\,2 = \dfrac{1}{\cosh 0.963\,2} = \dfrac{2}{e^{0.963\,2} + e^{-0.963\,2}}$

$= \mathbf{0.666\,28}$, correct to five significant figures.

(b) $\coth 0.443\,7 = \dfrac{1}{\tanh 0.443\,7} = \dfrac{e^{0.443\,7} + e^{-0.443\,7}}{e^{0.443\,7} - e^{-0.443\,7}}$

$= \dfrac{2.200\,120\,76\ldots}{0.916\,804\,98\ldots}$

$= \mathbf{2.399\,8}$, correct to five significant figures.

Further problems on evaluating hyperbolic functions may be found in Section 9.9 (Problems 1 to 14), page 161.

9.4   Graphs of hyperbolic functions

(a) $y = \cosh x$

A graph of $y = \cosh x$ may be plotted using values obtained by calculator, or by taking the average of the graphs of $y = e^x$ and $y = e^{-x}$. Either method will produce the curve shown in Fig. 9.1.

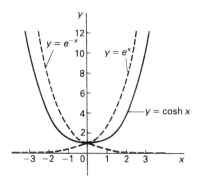

Fig. 9.1   Graphs of $y = e^x$, $y = e^{-x}$ and $y = \cosh x$

It is noted that:
  (i) $\operatorname{ch} 0 = 1$
  (ii)   The graph is symmetrical about the y-axis, thus, $\cosh x$ is an even function
  (iii)   As $x \to \infty$, $\operatorname{ch} x \to \frac{1}{2} e^x$ (since $e^{-x} \to 0$)
       and as $x \to -\infty$, $\operatorname{ch} x \to \frac{1}{2} e^{-x}$ (since $e^x \to 0$)

The shape of the curve $y = \cosh x$ is that of a heavy rope or chain hanging freely under gravity and is called a **catenary**. Examples include a telegraph wire, a fisherman's line or a transmission line.

(b) $y = \sinh x$

A graph of $y = \sinh x$ may be plotted using values obtained by calculator or by taking half the difference of the graphs $y = e^x$ and $y = e^{-x}$. Either of these methods will produce the curve shown in Fig. 9.2.

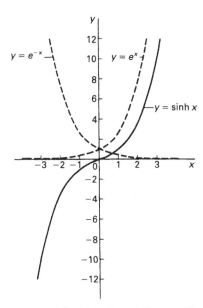

Fig. 9.2   Graphs of $y = e^x$, $y = e^{-x}$ and $y = \sinh x$

It is noted that:

(i)   sh $0 = 0$

(ii)   The graph is symmetrical about the origin, thus $\sinh x$ is an odd function.

(iii)   As $x \to \infty$, sh $x \to \frac{1}{2} e^x$ (since $e^{-x} \to 0$)
and as $x \to -\infty$, sh $x \to -\frac{1}{2} e^{-x}$ (since $e^x \to 0$)

(c) $y = \tanh x$

A calculator may be used to draw up the table of values shown below.

x	-4	-3	-2	-1	0	1	2	3	4
$y = \text{th } x = \dfrac{e^x - e^{-x}}{e^x + e^{-x}}$	-0.999	-0.995	-0.97	-0.77	0	0.77	0.97	0.995	0.999

The graph $y = \tanh x$ is shown in Fig. 9.3.

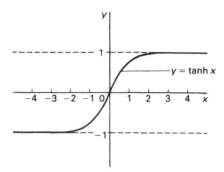

Fig. 9.3 Graph of $y = \tanh x$

It is noted that:

(i) $\operatorname{th} 0 = \dfrac{\operatorname{sh} 0}{\operatorname{ch} 0} = \dfrac{0}{1} = 0$

(ii) The graph is symmetrical about the origin, thus $\tanh x$ is an odd function.

(iii) As $x \to \infty$, $\tanh x \to \dfrac{\operatorname{sh} \infty}{\operatorname{ch} \infty} \to \dfrac{\frac{1}{2} e^x}{\frac{1}{2} e^x} \to 1$,

and as $x \to -\infty$, $\tanh x \to \dfrac{\operatorname{sh} - \infty}{\operatorname{ch} - \infty} \to \dfrac{-\frac{1}{2} e^{-x}}{\frac{1}{2} e^{-x}} \to -1$

(d) $y = \operatorname{cosech} x$

A calculator may be used to draw up the table of values shown below

x		-4	-3	-2	-1	0	1	2	3	4
$\operatorname{cosech} x = \dfrac{2}{e^x - e^{-x}}$		-0.04	-0.10	-0.28	-0.85	$\pm\infty$	0.85	0.28	0.10	0.04

The graph $y = \operatorname{cosech} x$ is shown in Fig. 9.4
It is noted that:

(i) $\operatorname{cosech} 0 = \pm\infty$

(ii) The graph is symmetrical about the origin, thus $\operatorname{cosech} x$ is an odd function.

(iii) As $x \to \infty$, $\operatorname{cosech} x \to 0$ and as $x \to -\infty$, $\operatorname{cosech} x \to 0$.

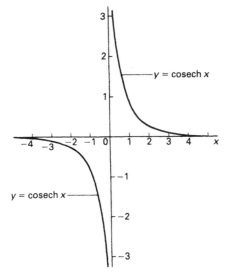

Fig. 9.4   Graph of $y = \text{cosech } x$

(e) $y = \text{sech } x$

A calculator may be used to draw up the table of values shown below.

x		-4	-3	-2	-1	0	1	2	3	4
$\text{sech } x = \dfrac{2}{e^x + e^{-x}}$		0.04	0.10	0.27	0.65	1	0.65	0.27	0.10	0.04

The graph $y = \text{sech } x$ is shown in Fig. 9.5

It is noted that:
  (i) $\text{sech } 0 = 1$
 (ii)   The graph is symmetrical about the y-axis, thus $\text{sech } x$ is an even function.
(iii)   As $x \to \infty$, $\text{sech } x \to 0$ and as $x \to -\infty$, $\text{sech } x \to 0$

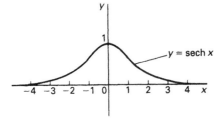

Fig. 9.5   Graph of $y = \text{sech } x$

(f) $y = \coth x$

A calculator may be used to draw up the table of values shown below.

x	-4	-3	-2	-1	0	1	2	3	4
$\coth x = \dfrac{e^x + e^{-x}}{e^x - e^{-x}}$	-1.0007	-1.005	-1.04	-1.31	$\pm\infty$	1.31	1.04	1.005	1.0007

The graph $y = \coth x$ is shown in Fig. 9.6

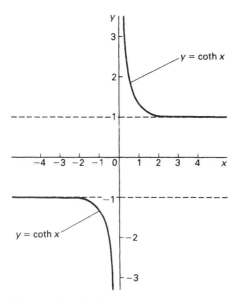

Fig. 9.6   Graph of $y = \coth x$

It is noted that:

(i) $\coth 0 = \pm\infty$

(ii)   The graph is symmetrical about the origin, thus $\coth x$ is an odd function.

(iii)   As $x \to \infty$, $\coth x \to 1$ and as $x \to -\infty$, $\coth x \to -1$

9.5   Hyperbolic identities – Osborne's rule

$$\cosh x + \sinh x = \left(\frac{e^x + e^{-x}}{2}\right) + \left(\frac{e^x - e^{-x}}{2}\right) = e^x$$

$$\cosh x - \sinh x = \left(\frac{e^x + e^{-x}}{2}\right) - \left(\frac{e^x - e^{-x}}{2}\right) = e^{-x}$$

$$(\text{ch } x + \text{sh } x)(\text{ch } x - \text{sh } x) = (e^x)(e^{-x}) = e^0 = 1$$

i.e. $\text{ch}^2 x - \text{sh}^2 x = 1$ (7)

Dividing each term of equation (7) by $\text{ch}^2 x$ gives:

$$\frac{\text{ch}^2 x}{\text{ch}^2 x} - \frac{\text{sh}^2 x}{\text{ch}^2 x} = \frac{1}{\text{ch}^2 x}$$

i.e. $1 - \text{th}^2 x = \text{sech}^2 x$ (8)

Dividing each term of equation (7) by $\text{sh}^2 x$ gives:

$$\frac{\text{ch}^2 x}{\text{sh}^2 x} - \frac{\text{sh}^2 x}{\text{sh}^2 x} = \frac{1}{\text{sh}^2 x}$$

i.e. $\text{coth}^2 x - 1 = \text{cosech}^2 x$ (9)

Osborne's rule states that *the six trigonometrical ratios used in trigonometrical identities relating general angles may be replaced by their corresponding hyperbolic functions, but the sign of any direct or implied product of two sines must be changed.*

Hence, since $\cos^2 x + \sin^2 x = 1$, by Osborne's rule $\text{ch}^2 x - \text{sh}^2 x = 1$, i.e. the trigonometrical functions have been changed to their corresponding hyperbolic functions and since $\sin^2 x$ is a direct production of two sines the sign is changed from $+$ to $-$.

Similarly, since $1 + \tan^2 x = \sec^2 x$, then $1 - \text{th}^2 x = \text{sech}^2 x$.

In this case $\tan^2 x = \left(\dfrac{\sin^2 x}{\cos^2 x}\right)$ is considered as an implied product of two sines, hence the sign change.

Also, since

$$\cot^2 x + 1 = \text{cosec}^2 x, \text{ then, from Osborne's rule,}$$

$$-\text{coth}^2 x + 1 = -\text{cosech}^2 x$$

i.e. $\text{coth}^2 x - 1 = \text{cosech}^2 x$

In this case, $\cot^2 x = \dfrac{\cos^2 x}{\sin^2 x}$ and $\text{cosec}^2 x = \dfrac{1}{\sin^2 x}$ are both implied products of two sines, hence the sign changes.

Below are listed some trigonometrical identities, with their corresponding hyperbolic identities which may be proved either by replacing sh A by $\left(\dfrac{e^A - e^{-A}}{2}\right)$ and ch A by $\left(\dfrac{e^A + e^{-A}}{2}\right)$ or by using Osborne's rule. Those identities marked with an asterisk involve sign changes due to a product of two sines.

Trigonometrical identities	Corresponding hyperbolic identities	
$\cos^2 A + \sin^2 A = 1$	$\mathrm{ch}^2 A - \mathrm{sh}^2 A = 1$	*
$1 + \tan^2 A = \sec^2 A$	$1 - \tanh^2 A = \mathrm{sech}^2 A$	*
$\cot^2 A + 1 = \mathrm{cosec}^2 A$	$\coth^2 A - 1 = \mathrm{cosech}^2 A$	*

Compound angle addition and subtraction formulae

$\sin(A + B) = \sin A \cos B + \cos A \sin B$	$\mathrm{sh}(A + B) = \mathrm{sh}\,A\,\mathrm{ch}\,B + \mathrm{ch}\,A\,\mathrm{sh}\,B$	
$\sin(A - B) = \sin A \cos B - \cos A \sin B$	$\mathrm{sh}(A - B) = \mathrm{sh}\,A\,\mathrm{ch}\,B - \mathrm{ch}\,A\,\mathrm{sh}\,B$	
$\cos(A + B) = \cos A \cos B - \sin A \sin B$	$\mathrm{ch}(A + B) = \mathrm{ch}\,A\,\mathrm{ch}\,B + \mathrm{sh}\,A\,\mathrm{sh}\,B$	*
$\cos(A - B) = \cos A \cos B + \sin A \sin B$	$\mathrm{ch}(A - B) = \mathrm{ch}\,A\,\mathrm{ch}\,B - \mathrm{sh}\,A\,\mathrm{sh}\,B$	*

Double angles

$\sin 2A = 2 \sin A \cos A$	$\mathrm{sh}\,2A = 2\,\mathrm{sh}\,A\,\mathrm{ch}\,A$	
$\cos 2A = \cos^2 A - \sin^2 A$	$\mathrm{ch}\,2A = \mathrm{ch}^2 A + \mathrm{sh}^2 A$	*
$\quad = 2\cos^2 A - 1$	$\quad = 2\,\mathrm{ch}^2 A - 1$	
$\quad = 1 - 2\sin^2 A$	$\quad = 1 + 2\,\mathrm{sh}^2 A$	*

Worked problems on hyperbolic identities

Problem 1. Prove that (a) $2\,\mathrm{ch}^2 \theta - 1 \equiv \mathrm{ch}\,2\theta$
and (b) $\dfrac{\mathrm{sech}\,x - \mathrm{cosech}\,x}{\mathrm{th}\,x - \coth x} \equiv \mathrm{ch}\,x - \mathrm{sh}\,x.$

(a) Left hand side (L.H.S.) $\equiv 2\,\mathrm{ch}^2 \theta - 1$

$$\equiv 2\left(\frac{e^\theta + e^{-\theta}}{2}\right)^2 - 1$$

$$\equiv 2\left(\frac{e^{2\theta} + 2e^\theta e^{-\theta} + e^{-2\theta}}{4}\right) - 1$$

$$\equiv \frac{e^{2\theta} + 2 + e^{-2\theta}}{2} - 1$$

$$\equiv \frac{e^{2\theta} + e^{-2\theta}}{2} + \frac{2}{2} - 1$$

$$\equiv \frac{e^{2\theta} + e^{-2\theta}}{2} \equiv \mathrm{ch}\,2\theta = \text{R.H.S.}$$

(b) L.H.S. $\equiv \dfrac{\mathrm{sech}\,x - \mathrm{cosech}\,x}{\mathrm{th}\,x - \coth x}$

$$\equiv \frac{\dfrac{1}{\mathrm{ch}\,x} - \dfrac{1}{\mathrm{sh}\,x}}{\dfrac{\mathrm{sh}\,x}{\mathrm{ch}\,x} - \dfrac{\mathrm{ch}\,x}{\mathrm{sh}\,x}} \equiv \frac{\dfrac{\mathrm{sh}\,x - \mathrm{ch}\,x}{\mathrm{ch}\,x\,\mathrm{sh}\,x}}{\dfrac{\mathrm{sh}^2 x - \mathrm{ch}^2 x}{\mathrm{ch}\,x\,\mathrm{sh}\,x}}$$

$$\equiv \frac{\text{sh } x - \text{ch } x}{\text{sh}^2 x - \text{ch}^2 x} \equiv \frac{\text{sh } x - \text{ch } x}{-1}$$

$$\left(\begin{array}{l} \text{since ch}^2 x - \text{sh}^2 x = 1 \\ \text{then sh}^2 x - \text{ch}^2 x = -1 \end{array} \right)$$

$$\equiv \text{ch } x - \text{sh } x = \text{R.H.S.}$$

Problem 2. Prove that ch A ch B − sh A sh $B \equiv$ ch $(A − B)$.

L.H.S. $\equiv$ ch A ch B − sh A sh B

$$\equiv \left(\frac{e^A + e^{-A}}{2} \right)\left(\frac{e^B + e^{-B}}{2} \right) - \left(\frac{e^A - e^{-A}}{2} \right)\left(\frac{e^B - e^{-B}}{2} \right)$$

$$\equiv \tfrac{1}{4}(e^A e^B + e^A e^{-B} + e^{-A} e^B + e^{-A} e^{-B})$$
$$\qquad - \tfrac{1}{4}(e^A e^B - e^A e^{-B} - e^{-A} e^B + e^{-A} e^{-B})$$

$$\equiv \tfrac{1}{4}(e^{A+B} + e^{A-B} + e^{-A+B} + e^{-A-B}$$
$$\qquad - e^{A+B} + e^{A-B} + e^{-A-B} - e^{-A-B})$$

$$\equiv \tfrac{1}{2}e^{A-B} + \tfrac{1}{2}e^{-A+B}$$

$$\equiv \tfrac{1}{2}(e^{A-B} + e^{-(A-B)}) \equiv \text{ch}(A - B) = \text{R.H.S.}$$

Problem 3. (a) If $P e^x + Q e^{-x} \equiv 3$ ch x − 4 sh x find the values of P and Q. (b) If $5e^x − 2e^{-x} \equiv A$ sh $x + B$ ch x find the values of A and B.

(a) $P e^x + Q e^{-x} \equiv 3$ ch x − 4 sh x

$$\equiv 3\left(\frac{e^x + e^{-x}}{2} \right) - 4\left(\frac{e^x - e^{-x}}{2} \right)$$

$$P e^x + Q e^{-x} \equiv \tfrac{3}{2} e^x + \tfrac{3}{2}e^{-x} - 2e^x + 2e^{-x}$$

$$P e^x + Q e^{-x} \equiv -\tfrac{1}{2}e^x + \tfrac{7}{2}e^{-x}$$

Equating coefficients gives $P = -\tfrac{1}{2}$ and $Q = 3\tfrac{1}{2}$

(b) $5e^x − 2e^{-x} \equiv A$ sh $x + B$ ch x

$$\equiv A\left(\frac{e^x - e^{-x}}{2} \right) + B\left(\frac{e^x + e^{-x}}{2} \right)$$

$$\equiv \frac{A}{2} e^x - \frac{A}{2} e^{-x} + \frac{B}{2} e^x + \frac{B}{2} e^{-x}$$

$$5e^x − 2e^{-x} = \left(\frac{A + B}{2} \right)e^x + \left(\frac{B - A}{2} \right)e^{-x}$$

Equating coefficients gives: $5 = \dfrac{A + B}{2}$ and $-2 = \dfrac{B - A}{2}$

i.e. $A + B = 10$ (10)

and $-A + B = -4$ (11)

Adding equations (10) and (11) gives: $2B = 6$, i.e. $B = 3$
Substituting in equation (10) gives: $A = 7$

Further problems on hyperbolic identities may be found in Section 9.9
(Problems 15 to 21), page 163.

9.6   Differentiation of hyperbolic functions

(a) $\dfrac{d}{dx}(\sinh x) = \dfrac{d}{dx}\left(\dfrac{e^x - e^{-x}}{2}\right) = \frac{1}{2}[e^x - (-e^{-x})]$

$= \frac{1}{2}(e^x + e^{-x}) = \cosh x$

More generally, if $y = \sinh ax$ then $\dfrac{dy}{dx} = a \cosh ax$.

(b) $\dfrac{d}{dx}(\cosh x) = \dfrac{d}{dx}\left(\dfrac{e^x + e^{-x}}{2}\right) = \frac{1}{2}[e^x + (-e^{-x})]$

$= \frac{1}{2}(e^x - e^{-x}) = \sinh x$

More generally, if $y = \cosh ax$ then $\dfrac{dy}{dx} = a \sinh ax$.

(c)   Since $\tanh x = \dfrac{\sinh x}{\cosh x}$ then by the quotient rule:

$\dfrac{d}{dx}(\tanh x) = \dfrac{d}{dx}\left(\dfrac{\sinh x}{\cosh x}\right) = \dfrac{(\cosh x)(\cosh x) - (\sinh x)(\sinh x)}{\cosh^2 x}$

$= \dfrac{\cosh^2 x - \sinh^2 x}{\cosh^2 x} = \dfrac{1}{\cosh^2 x} = \text{sech}^2 x$

More generally, if $y = \tanh ax$ then $\dfrac{dy}{dx} = a \, \text{sech}^2 ax$.

(d) $\dfrac{d}{dx}(\text{sech } x) = \dfrac{d}{dx}\left(\dfrac{1}{\cosh x}\right) = \dfrac{(\cosh x)(0) - (1)(\sinh x)}{\cosh^2 x} = \dfrac{-\sinh x}{\cosh^2 x}$

$= -\left(\dfrac{1}{\cosh x}\right)\left(\dfrac{\sinh x}{\cosh x}\right) = -\text{sech } x \tanh x$

More generally, if $y = \text{sech } ax$ then $\dfrac{dy}{dx} = -a \, \text{sech } ax \tanh ax$.

(e) $\dfrac{d}{dx}(\text{cosech } x) = \dfrac{d}{dx}\left(\dfrac{1}{\sinh x}\right) = \dfrac{(\sinh x)(0) - (1)(\cosh x)}{\sinh^2 x} = -\dfrac{\cosh x}{\sinh^2 x}$

$\qquad\qquad = -\left(\dfrac{1}{\sinh x}\right)\left(\dfrac{\cosh x}{\sinh x}\right) = -\text{cosech } x \coth x$

More generally, if $y = \text{cosech } ax$ then $\dfrac{dy}{dx} = -a\,\text{cosech } ax \coth ax$.

(f) $\dfrac{d}{dx}(\coth x) = \dfrac{d}{dx}\left(\dfrac{\cosh x}{\sinh x}\right) = \dfrac{(\sinh x)(\sinh x) - (\cosh x)(\cosh x)}{\sinh^2 x}$

$\qquad\qquad = \dfrac{\sinh^2 x - \cosh^2 x}{\sinh^2 x} = -\dfrac{(\cosh^2 x - \sinh^2 x)}{\sinh^2 x}$

$\qquad\qquad = \dfrac{-1}{\sinh^2 x} = -\text{cosech}^2 x$

More generally, if $y = \coth ax$ then $\dfrac{dy}{dx} = -a\,\text{cosech}^2 ax$.

Summary

y or $f(x)$	$\dfrac{dy}{dx}$ or $f'(x)$
$\sinh ax$	$a \cosh ax$
$\cosh ax$	$a \sinh ax$
$\tanh ax$	$a \,\text{sech}^2 ax$
$\text{sech } ax$	$-a \,\text{sech } ax \tanh ax$
$\text{cosech } ax$	$-a \,\text{cosech } ax \coth ax$
$\coth ax$	$-a \,\text{cosech}^2 ax$

Worked problems on differentiation of hyperbolic functions

Problem 1. Differentiate the following with respect to x:

(a) $y = 3 \,\text{sh}\, 4x + \frac{2}{5} \,\text{ch}\, 2x$ (b) $y = 5 \,\text{th}\, 3x - 4 \coth \dfrac{x}{2}$

(c) $y = \frac{1}{2}(\text{sech}\, 6x - 3 \,\text{cosech}\, 5x)$

(a) $y = 3 \,\text{sh}\, 4x + \frac{2}{5} \,\text{ch}\, 2x$

$\dfrac{dy}{dx} = (3)(4) \,\text{ch}\, 4x + (\frac{2}{5})(2) \,\text{sh}\, 2x = \mathbf{12\, ch\, 4x + \frac{4}{5}\, sh\, 2x}$

(b) $y = 5 \text{ th } 3x - 4 \coth \dfrac{x}{2}$

$\dfrac{dy}{dx} = (5)(3) \operatorname{sech}^2 3x - (4)(-\tfrac{1}{2}) \operatorname{cosech}^2 \dfrac{x}{2}$

$\qquad = 15 \operatorname{sech}^2 3x + 2 \operatorname{cosech}^2 \dfrac{x}{2}$

(c) $y = \tfrac{1}{2}(\operatorname{sech} 6x - 3 \operatorname{cosech} 5x)$

$\dfrac{dy}{dx} = \tfrac{1}{2}[-6 \operatorname{sech} 6x \text{ th } 6x - (3)(-5) \operatorname{cosech} 5x \coth 5x]$

$\qquad = \tfrac{3}{2}[5 \operatorname{cosech} 5x \coth 5x - 2 \operatorname{sech} 6x \text{ th } 6x]$

Problem 2. Find the differential coefficients of the following with respect to the variable:

(a) $\ln(\text{sh } 2x) + 3 \text{ ch}^2 4x$ (b) $3 \sin 2t \text{ ch } 5t$ (c) $\dfrac{2\theta^3}{\text{ch } 3\theta}$

(a) $f(x) = \ln(\text{sh } 2x) + 3 \text{ ch}^2 4x$ (i.e. 'a function of a function')

$f'(x) = \left(\dfrac{1}{\text{sh } 2x}\right)(2 \text{ ch } 2x) + (3)(2 \text{ ch } 4x)(4 \text{ sh } 4x)$

$\qquad = 2(\coth 2x + 12 \text{ sh } 4x \text{ ch } 4x)$

(b) $f(t) = 3 \sin 2t \text{ ch } 5t$ (i.e. a product)

$f'(t) = (3 \sin 2t)(5 \text{ sh } 5t) + (\text{ch } 5t)(6 \cos 2t)$

$\qquad = 3(5 \sin 2t \text{ sh } 5t + 2 \cos 2t \text{ ch } 5t)$

(c) $f(\theta) = \dfrac{2\theta^3}{\text{ch } 3\theta}$ (i.e. a quotient)

$f'(\theta) = \dfrac{(\text{ch } 3\theta)(6\theta^2) - (2\theta^3)(3 \text{ sh } 3\theta)}{(\text{ch } 3\theta)^2} = 6\theta^2\left[\dfrac{\text{ch } 3\theta}{\text{ch}^2 3\theta} - \dfrac{\theta \text{ sh } 3\theta}{\text{ch}^2 3\theta}\right]$

$\qquad = 6\theta^2\left[\dfrac{1}{\text{ch } 3\theta} - \dfrac{\theta \text{ sh } 3\theta}{\text{ch } 3\theta}\left(\dfrac{1}{\text{ch } 3\theta}\right)\right]$

$\qquad = 6\theta^2 \operatorname{sech} 3\theta(1 - \theta \text{th } 3\theta)$

Further problems on differentiation of hyperbolic functions may be found in Section 9.9 (Problems 22 to 27), page 163.

9.7   Solution of equations of the form $a \cosh x + b \sinh x = c$

Equations of the form $a\,\text{ch}\,x + b\,\text{sh}\,x = c$, where a, b and c are constants, may be solved either by plotting graphs of $y = a\,\text{ch}\,x + b\,\text{sh}\,x$ and $y = c$ and noting the points of intersection or, more accurately, as follows:

(i)   Change ch x to $\left(\dfrac{e^x + e^{-x}}{2}\right)$ and sh x to $\left(\dfrac{e^x - e^{-x}}{2}\right)$.

(ii) Rearrange the equation into the form $m\,e^x + n\,e^{-x} + p = 0$, where m, n and p are constants.

(iii) Multiply each term by e^x. This produces an equation of the form: $m(e^x)^2 + p\,e^x + n = 0$, since $(e^{-x})(e^x) = 1$.

(iv) Solve the quadratic equation $m(e^x)^2 + p\,e^x + n = 0$ for e^x either by factorising or by using the quadratic formula.

(v)   Given $e^x = k$, take logarithms to base e of both sides to give $x = \ln k$.

Worked problems on solving equations of the form $a\,\text{ch}\,x + b\,\text{sh}\,x = c$

Problem 1. Solve the equation sh $x = 1$ correct to four significant figures.

Following the above procedure:

(i)   sh $x = \dfrac{e^x - e^{-x}}{2} = 1$

(ii) $e^x - e^{-x} = 2$

i.e. $e^x - e^{-x} - 2 = 0$

(iii) $(e^x)^2 - (e^{-x})e^x - 2e^x = 0$

$(e^x)^2 - 2e^x - 1 = 0$

(iv) $e^x = \dfrac{-(-2) \pm \sqrt{[(-2)^2 - 4(1)(-1)]}}{2(1)} = \dfrac{2 \pm \sqrt{(4 + 4)}}{2}$

$= \dfrac{2 \pm \sqrt{8}}{2} = \dfrac{2 \pm 2.828\,4}{2} = \dfrac{4.828\,4}{2}$ or $\dfrac{-0.828\,4}{2}$

Hence $e^x = 2.414\,2$ or $-0.414\,2$

(v) $x = \ln 2.414\,2$ or $x = \ln(-0.414\,2)$ which has no real solution.

Hence $x = 0.881\,4$ correct to four significant figures.

Problem 2. Solve the equation $3\,\text{ch}\,\theta - 5 = 0$ correct to four decimal places.

(i) $3 \operatorname{ch} \theta - 5 = 0$

$$3\left(\frac{e^\theta + e^{-\theta}}{2}\right) - 5 = 0$$

(ii) $\frac{3}{2} e^\theta + \frac{3}{2} e^{-\theta} - 5 = 0$

i.e. $3e^\theta + 3e^{-\theta} - 10 = 0$

(iii) $3(e^\theta)^2 + 3 - 10e^\theta = 0$

$3(e^\theta)^2 - 10e^\theta + 3 = 0$

(iv) $(3e^\theta - 1)(e^\theta - 3) = 0$

$e^\theta = \frac{1}{3}$ or $e^\theta = 3$

(v) $\theta = \ln(\frac{1}{3})$ or $\theta = \ln 3$

Hence $\theta = -1.098\,6$ or $1.098\,6$ correct to four decimal places.

Problem 3. Solve the equation $3.0 \operatorname{ch} x + 2.0 \operatorname{sh} x = 14.31$, correct to four decimal places.

(i) $3.0 \operatorname{ch} x + 2.0 \operatorname{sh} x = 14.31$

$$3.0\left(\frac{e^x + e^{-x}}{2}\right) + 2.0\left(\frac{e^x - e^{-x}}{2}\right) = 14.31$$

(ii) $1.5e^x + 1.5e^{-x} + e^x - e^{-x} = 14.31$

$2.5e^x + 0.5e^{-x} - 14.31 = 0$

(iii) $2.5(e^x)^2 - 14.31e^x + 0.5 = 0$

(iv) $e^x = \dfrac{-(-14.31) \pm \sqrt{[(-14.31)^2 - 4(2.5)(0.5)]}}{2(2.5)}$

$$= \frac{14.31 \pm 14.134\,22}{5} = \frac{28.444\,22}{5} \text{ or } \frac{0.175\,78}{5}$$

Hence $e^x = 5.688\,844$ or $0.035\,156$

(v) $x = \ln 5.688\,844$ or $x = \ln 0.035\,156$

Hence $x = 1.738\,5$ or $-3.348\,0$ correct to four decimal places.

Further problems on solving equations of the form $a \operatorname{ch} x + b \operatorname{sh} x = c$ may be found in Section 9.9 (Problems 28 to 38), page 164.

9.8   Series expansions for cosh x and sinh x

e^x is defined by the power series:

$$e^x = 1 + x + \frac{x^2}{2!} + \frac{x^3}{3!} + \frac{x^4}{4!} + \frac{x^5}{5!} + \cdots$$

If x is replaced by $-x$ then $e^{-x} = 1 - x + \dfrac{x^2}{2!} - \dfrac{x^3}{3!} + \dfrac{x^4}{4!} - \dfrac{x^5}{5!} + \cdots$

$$\cosh x = \tfrac{1}{2}(e^x + e^{-x}) = \tfrac{1}{2}\left[\left(1 + x + \dfrac{x^2}{2!} + \dfrac{x^3}{3!} + \dfrac{x^4}{4!} + \dfrac{x^5}{5!} + \cdots\right)\right.$$

$$\left. + \left(1 - x + \dfrac{x^2}{2!} - \dfrac{x^3}{3!} + \dfrac{x^4}{4!} - \dfrac{x^5}{5!} + \cdots\right)\right]$$

$$= \tfrac{1}{2}\left[2 + \dfrac{2x^2}{2!} + \dfrac{2x^4}{4!} + \cdots\right]$$

i.e.
$$\cosh x = 1 + \dfrac{x^2}{2!} + \dfrac{x^4}{4!} + \cdots$$

$\cosh x$ is an even function and contains only even powers of x
Similarly,

$$\sinh x = \tfrac{1}{2}(e^x - e^{-x})$$

$$= \tfrac{1}{2}\left[\left(1 + x + \dfrac{x^2}{2!} + \dfrac{x^3}{3!} + \dfrac{x^4}{4!} + \dfrac{x^5}{5!} + \cdots\right)\right.$$

$$\left. - \left(1 - x + \dfrac{x^2}{2!} - \dfrac{x^3}{3!} + \dfrac{x^4}{4!} - \dfrac{x^5}{5!} + \cdots\right)\right]$$

$$= \tfrac{1}{2}\left[2x + \dfrac{2x^3}{3!} + \dfrac{2x^5}{5!} + \cdots\right]$$

i.e.
$$\sinh x = x + \dfrac{x^3}{3!} + \dfrac{x^5}{5!} + \cdots$$

$\sinh x$ is an odd function and contains only odd powers of x.
The series expansions for $\cosh x$ and $\sinh x$ are true for all values of x.

Worked problems on series expansions for cosh x and sinh x

Problem 1. Using the series expansion for ch x, evaluate ch 1 correct to four decimal places.

$$\cosh x = 1 + \dfrac{x^2}{2!} + \dfrac{x^4}{4!} + \cdots$$

Let $x = 1$ then

$$\cosh 1 = 1 + \dfrac{1}{2 \times 1} + \dfrac{1}{4 \times 3 \times 2 \times 1} + \dfrac{1}{6 \times 5 \times 4 \times 3 \times 2 \times 1}$$

$$= 1 + 0.5 + 0.041\,67 + 0.001\,389$$

i.e. **ch 1 = 1.543 1 correct to four decimal places,**

which is the same as the value of ch 1 obtained by calculator.

Problem 2. Using the series expansion for sh x, evaluate sh 2 correct to four decimal places.

$$\sinh x = x + \frac{x^3}{3!} + \frac{x^5}{5!} + \cdots$$

Let $x = 2$ then sh $2 = 2 + \dfrac{2^3}{3 \times 2 \times 1} + \dfrac{2^5}{5 \times 4 \times 3 \times 2 \times 1}$

$$+ \frac{2^7}{7 \times 6 \times 5 \times 4 \times 3 \times 2 \times 1}$$

$$+ \frac{2^9}{9 \times 8 \times 7 \times 6 \times 5 \times 4 \times 3 \times 2 \times 1}$$

$$+ \frac{2^{11}}{11 \times 10 \times 9 \times 8 \times 7 \times 6 \times 5 \times 4 \times 3 \times 2 \times 1}$$

$$= 2 + 1.333\,33 + 0.266\,67 + 0.025\,40 + 0.001\,41 + 0.000\,05$$

i.e. **sh 2 = 3.626 9 correct to four decimal places,**

which is the same as the value of sh 2 obtained by calculator.

Problem 3. Find the series expansion for ch 2θ − sh 2θ as far as the term in θ^5.

In the series expansions for sh x and ch x let $x = 2\theta$ then:

$$\text{ch } 2\theta = 1 + \frac{(2\theta)^2}{2!} + \frac{(2\theta)^4}{4!} + \cdots = 1 + 2\theta^2 + \frac{2\theta^4}{3} + \cdots$$

$$\text{sh } 2\theta = 2\theta + \frac{(2\theta)^3}{3!} + \frac{(2\theta)^5}{5!} + \cdots = 2\theta + \tfrac{4}{3}\theta^3 + \tfrac{4}{15}\theta^5 + \cdots$$

Hence ch 2θ − sh $2\theta = (1 + 2\theta^2 + \tfrac{2}{3}\theta^4 + \cdots)$

$$- (2\theta + \tfrac{4}{3}\theta^3 + \tfrac{4}{15}\theta^5 + \cdots)$$

$$= 1 - 2\theta + 2\theta^2 - \tfrac{4}{3}\theta^3 + \tfrac{2}{3}\theta^4 - \tfrac{4}{15}\theta^5 + \cdots$$

as far as the term in θ^5.

Further problems on series expansions for cosh x *and* sinh x *may be found in the following Section 9.9 (Problems 39 to 43), page 164.*

9.9 Further problems

Evaluation of hyperbolic functions

In Problems 1 to 10, evaluate correct to four significant figures.

1. (a) sh 0.52   (b) sh 2.24   (c) sh 5.36
   (a) [0.543 8]   (b) [4.643]   (c) [106.4]
2. (a) sh 0.212   (b) sh 7.7   (c) sh 8.6
   (a) [0.213 6]   (b) [1 104]   (c) [2 716]
3. (a) ch 0.67   (b) ch 1.84   (c) ch 4.78
   (a) [1.233]   (b) [3.228]   (c) [59.56]
4. (a) ch 0.346   (b) ch 6.5   (c) ch 7.4
   (a) [1.060]   (b) [332.6]   (c) [818.0]
5. (a) sh 1.527 3   (b) sh 2.689 1   (c) sh 5.326 4
   (a) [2.194]   (b) [7.325]   (c) [102.8]
6. (a) ch 0.478 8   (b) ch 1.723 1   (c) ch 4.681 5
   (a) [1.117]   (b) [2.890]   (c) [53.97]
7. (a) th 0.43   (b) th 0.76   (c) th 1.54
   (a) [0.405 3]   (b) [0.641 1]   (c) [0.912 1]
8. (a) cosech 0.26   (b) cosech 0.624   (c) cosech 2.45
   (a) [3.804]   (b) [1.503]   (c) [0.173 9]
9. (a) sech 0.15   (b) sech 0.82   (c) sech 2.324
   (a) [0.988 8]   (b) [0.737 8]   (c) [0.193 9]
10. (a) coth 0.33   (b) coth 0.746   (c) coth 1.168
    (a) [3.140]   (b) [1.580]   (c) [1.214]
11. A telegraph wire hangs so that its shape is described by
    $y = 60 \text{ ch } \dfrac{x}{60}$. Evaluate, correct to five significant figures, y
    when x is 30.   [67.658]
12. The length, l of a heavy cable hanging under gravity is given by:
    $$l = 2c \sinh \frac{L}{2c}$$
    Find l when $c = 50$ and $L = 40$.   [41.08]
13. The speed, V, of waves over the bottom of shallow water is
    given by the formula: $V^2 = 0.55 \, L \tanh \left(\dfrac{6.3 \, d}{L} \right)$ where d is
    the depth and L is the wavelength. If $d = 10$ and $L = 100$,
    calculate the value of V.   [5.540]
14. The increase in resistance of strip conductors due to eddy
    currents at power frequencies is given by:
    $$\lambda = \frac{\alpha t}{2} \left[\frac{\sinh \alpha t + \sin \alpha t}{\cosh \alpha t - \cos \alpha t} \right].$$ Calculate λ correct to five
    significant figures when $\alpha = 1.08$ and $t = 1$.   [1.007 5]

Hyperbolic identities

In Problems 15 to 20 prove the given identities.

15. (a) $\operatorname{ch}(A + B) \equiv \operatorname{ch} A \operatorname{ch} B + \operatorname{sh} A \operatorname{sh} B$
    (b) $\operatorname{ch} 2\theta \equiv \operatorname{ch}^2 \theta + \operatorname{sh}^2 \theta$
    (c) $1 + 2 \operatorname{sh}^2 y \equiv \operatorname{ch} 2y$
16. (a) $1 + \operatorname{sh}^2 A \equiv \operatorname{ch}^2 A$
    (b) $\operatorname{sech}^2 B + \operatorname{th}^2 B \equiv 1$
    (c) $\operatorname{cosech}^2 C + 1 \equiv \operatorname{coth}^2 C$
17. (a) $\operatorname{th} \alpha (1 + \operatorname{sh}^2 \alpha) \equiv \frac{1}{2} \operatorname{sh} 2\alpha$
    (b) $1 + \operatorname{ch} 2\beta \equiv \operatorname{sh} 2\beta \operatorname{coth} \beta$
    (c) $\operatorname{coth} t - 2 \operatorname{cosech} 2t \equiv \operatorname{th} t$
18. (a) $\operatorname{th} 2\theta \equiv \dfrac{2 \operatorname{th} \theta}{1 + \operatorname{th}^2 \theta}$

    (b) $\operatorname{ch} 2\theta + \operatorname{sh} 2\theta \equiv \dfrac{1 + \operatorname{th} \theta}{1 - \operatorname{th} \theta}$

    (c) $\operatorname{sh} 2\theta \equiv 2 \operatorname{sh} \theta \operatorname{ch} \theta$
19. (a) $\operatorname{sh} A \operatorname{ch} B + \operatorname{ch} A \operatorname{sh} B \equiv \operatorname{sh}(A + B)$

    (b) $\operatorname{th}(A + B) \equiv \dfrac{\operatorname{th} A + \operatorname{th} B}{1 + \operatorname{th} A \operatorname{th} B}$

    (c) $\operatorname{sh}^2 x \equiv \frac{1}{2}(\operatorname{ch} 2x - 1)$
20. (a) $\operatorname{ch}(A + B) + \operatorname{ch}(A - B) \equiv 2 \operatorname{ch} A \operatorname{ch} B$
    (b) $\operatorname{ch}(A + B) - \operatorname{ch}(A - B) \equiv 2 \operatorname{sh} A \operatorname{sh} B$
    (c) $\operatorname{sh} 3A - 3 \operatorname{sh} A \equiv 4 \operatorname{sh}^3 A$
21. (a) If $4e^x + 3e^{-x} \equiv P \operatorname{sh} x + Q \operatorname{ch} x$, find P and Q
    $[P = 1, Q = 7]$
    (b) If $A \operatorname{ch} x - B \operatorname{sh} x \equiv 3e^x - 4e^{-x}$, find A and B
    $[A = -1, B = -7]$
    (c) If $\alpha e^x - \beta e^{-x} \equiv 5 \operatorname{ch} x - 2 \operatorname{sh} x$, find α and β
    $[\alpha = 1\frac{1}{2}, \beta = -3\frac{1}{2}]$

Differentiation of hyperbolic functions

In Problems 22 to 27, differentiate the given functions with respect to the variable.

22. (a) $4 \operatorname{sh} 3x$ (b) $2 \operatorname{ch} 4t$ (c) $5 \operatorname{th} 7x$
    (a) $[12 \operatorname{ch} 3x]$ (b) $[8 \operatorname{sh} 4t]$ (c) $[35 \operatorname{sech}^2 7x]$
23. (a) $\frac{3}{4} \operatorname{cosech} \frac{3}{2}\theta$ (b) $\frac{5}{8}\operatorname{sech} 4x$ (c) $3 \operatorname{coth} 5y$

    (a) $\left[-\frac{9}{8} \operatorname{cosech} \dfrac{3\theta}{2} \operatorname{coth} \dfrac{3\theta}{2} \right]$ (b) $[-\frac{5}{2} \operatorname{sech} 4x \tanh 4x]$

    (c) $[-15 \operatorname{cosech}^2 5y]$
24. (a) $3 \ln(\operatorname{ch} x)$ (b) $2 \ln\left(\operatorname{th} \dfrac{3x}{2} \right)$ (c) $\operatorname{cosech}^3 2t$

    (a) $[3 \operatorname{th} x]$ (b) $[3 \operatorname{sech} \frac{3}{2} x \operatorname{cosech} \frac{3}{2} x]$
    (c) $[-6 \operatorname{cosech}^3 2t \operatorname{coth} 2t]$

25. (a) $7 \operatorname{sh}^2 x$ (b) $2 \operatorname{coth}^4 3p$ (c) $\operatorname{ch}^3 4x + 3 \operatorname{th}^2 2x$
    (a) $[14 \operatorname{sh} x \operatorname{ch} x \text{ or } 7 \operatorname{sh} 2x]$ (b) $[-24 \operatorname{coth}^3 3p \operatorname{cosech}^2 3p]$
    (c) $[12 \operatorname{ch}^2 4x \operatorname{sh} 4x + \operatorname{th} 2x \operatorname{sech}^2 2x]$

26. (a) $\operatorname{sh} 3x \operatorname{ch} 3x$ (b) $2 \cos 3t \operatorname{ch} 3t$ (c) $e^{2x} \operatorname{ch} 3x$
    (a) $[3(\operatorname{sh}^2 3x + \operatorname{ch}^2 3x) \text{ or } 3 \operatorname{ch} 6x]$
    (b) $[6(\cos 3t \operatorname{sh} 3t - \operatorname{ch} 3t \sin 3t)]$
    (c) $[e^{2x}(3 \operatorname{sh} 3x + 2 \operatorname{ch} 3x)]$

27. (a) $\dfrac{\operatorname{sh} 2\theta}{\sin 2\theta}$ (b) $\dfrac{2 \operatorname{ch} 5t}{3t^4}$ (c) $x^{\operatorname{ch} x}$

    (a) $\left[\dfrac{2(\sin 2\theta \operatorname{ch} 2\theta - \operatorname{sh} 2\theta \cos 2\theta)}{\sin^2 2\theta} \right]$

    (b) $\left[\dfrac{2}{3t^5}(5t \operatorname{sh} 5t - 4 \operatorname{ch} 5t) \right]$ (c) $\left[x^{\operatorname{ch} x}\left(\dfrac{\operatorname{ch} x}{x} + \operatorname{sh} x \ln x \right) \right]$

Solution of equations of the form $a \operatorname{ch} x + b \operatorname{sh} x = c$

In Problems 28 to 37 solve the equations correct to four decimal places.

28. $\operatorname{sh} x = 2$ $[1.4436]$
29. $1.52 \operatorname{ch} x = 6.70$ $[\pm 2.1634]$
30. $3.0 \operatorname{sh} x + 2.0 \operatorname{ch} x = 0$ $[-0.8047]$
31. $\operatorname{ch} x + 2 \operatorname{sh} x = 4$ $[1.0247]$
32. $6.80 \operatorname{sh} \theta - 3.72 \operatorname{ch} \theta = 1.68$ $[0.9052]$
33. $2.97 - 3.16 \operatorname{sh} x - 4 \operatorname{ch} x = 0$ $[-0.4327 \text{ or } -1.7102]$
34. $3.94 \operatorname{ch} x + 10.82 \operatorname{sh} x = 4.77$ $[0.0756]$
35. $20.2 \operatorname{sh} y - 16.4 \operatorname{ch} y = 11.7$ $[2.0083]$
36. $\operatorname{th} x = \operatorname{sh} x$ $[0]$
37. $7 \operatorname{th} x - 3 = 0$ $[0.4581]$

38. A chain hangs in the form $y = 50 \operatorname{ch} \dfrac{x}{50}$. Find, correct to four
    significant figures, (i) the value of y when $x = 30$, and (ii) the
    value of x when $y = 64.73$.   (i) $[59.28]$ (ii) $[37.49]$

Series expansions for cosh x and sinh x

39. Use the series expansion for cosh x to evaluate correct to
    four decimal places: (a) ch 2 (b) ch 0.5 (c) ch 2.5
    (a) $[3.7622]$ (b) $[1.1276]$ (c) $[6.1323]$

40. Use the series expansion for sinh x to evaluate correct to
    four decimal places: (a) sh 0.4 (b) sh 1.5 (c) sh 2.4
    (a) $[0.4108]$ (b) $[2.1293]$ (c) $[5.4662]$

41. Expand the following as a power series as far as the term in x^5:

(a) sh $2x$ (b) ch $4x$ (c) sh $3x$

(a) $\left[2x + \dfrac{4}{3}x^3 + \dfrac{4}{15}x^5 \right]$ (b) $\left[1 + 8x^2 + \dfrac{32}{3}x^4 \right]$

(c) $\left[3x + \dfrac{9}{2}x^3 + \dfrac{81}{40}x^5 \right]$

42. Prove the following identities, the series being taken as far as the term in x^5 only.

(a) ch $2x$ − ch $x \equiv \dfrac{x^2}{2}\left(3 + \dfrac{5}{4}x^2 \right)$

(b) sh x + sh $2x \equiv 3x\left(1 + \dfrac{x^2}{2} + \dfrac{11}{120}x^4 \right)$

(c) ch $3x$ + ch $2x$ + ch $x \equiv 3 + 7x^2 + \dfrac{49}{12}x^4$

43. The value of C_v for an Einstein crystal is given by

$$C_v = \frac{3No\varepsilon^2 k}{(kT)^2} \cdot \frac{e^{\varepsilon/kT}}{(e^{\varepsilon/kT} - 1)^2}$$

Express C_v in terms of sinh $\dfrac{\varepsilon}{2kT}$, and by using Taylor's series

show that when $\dfrac{\varepsilon}{2kT}$ is very small $C_v = 3Nok$.

$$\left[C_v = \frac{3Nok}{4} \cdot \frac{(\varepsilon/kT)^2}{\sinh^2(\varepsilon/2kT)} \right]$$

Chapter 10

The relationship between trigonometric and hyperbolic functions and hyperbolic identities

10.1 The relationship between trigonometric and hyperbolic functions

In Chapter 8, the relationship between trigonometric and exponential functions are developed, showing that:

$$\cos\theta + j\sin\theta = e^{j\theta} \text{ and } \cos\theta - j\sin\theta = e^{-j\theta}$$

Addition of these two equations gives:

$$\cos\theta = \frac{1}{2}(e^{j\theta} + e^{-j\theta}) \tag{1}$$

Subtraction gives:

$$\sin\theta = \frac{1}{2j}(e^{j\theta} - e^{-j\theta}) \tag{2}$$

In Chapter 9, the hyperbolic functions $\cosh\theta$ and $\sinh\theta$ are defined in terms of exponential functions and are:

$$\cosh\theta = \frac{1}{2}(e^{\theta} + e^{-\theta}) \tag{3}$$

$$\sinh\theta = \frac{1}{2}(e^{\theta} - e^{-\theta}) \tag{4}$$

Substituting $j\theta$ for θ in equation (1) gives:

$$\cos(j\theta) = \frac{1}{2}(e^{j(j\theta)} + e^{-j(j\theta)})$$

Since $j^2 = -1, \cos(j\theta) = \frac{1}{2}(e^{-\theta} + e^{\theta})$

$$= \cosh\theta \text{ from equation (3).}$$

Hence $\cos(j\theta) = \cosh\theta$ (5)

Substituting $j\theta$ for θ in equation (2) gives:

$$\sin(j\theta) = \frac{1}{2j}(e^{j(j\theta)} - e^{-j(j\theta)})$$

$$= \frac{1}{2j}(e^{-\theta} - e^{\theta})$$

$$= -\frac{1}{2j}(e^{\theta} - e^{-\theta})$$

$$= -\frac{1}{j}\sinh\theta$$

Multiplying both numerator and denominator of $\left(-\dfrac{1}{j}\right)$ by j gives:

$$-\frac{1}{j} \times \frac{j}{j} = -\frac{j}{j^2} = -\frac{j}{-1} = j,$$

hence $\sin(j\theta) = j \sinh\theta$ (6)

Substituting $j\theta$ for θ in equation (3) gives:

$$\cosh(j\theta) = \frac{1}{2}(e^{(j\theta)} + e^{-(j\theta)}) = \cos\theta \text{ from equation (1),}$$

i.e. $\cosh(j\theta) = \cos\theta$ (7)

Substituting $j\theta$ for θ in equation (4) gives:

$$\sinh(j\theta) = \frac{1}{2}(e^{(j\theta)} - e^{-(j\theta)})$$

$$= j \cdot \frac{1}{2j}(e^{(j\theta)} - e^{-(j\theta)})$$

$$= j \cdot \sin\theta \text{ from equation (2),}$$

i.e. $\sinh(j\theta) = j \sin\theta$ (8)

Dividing equation (6) by equation (5) gives:

$$\frac{\sin(j\theta)}{\cos(j\theta)} = \tan(j\theta) = \frac{j \sinh\theta}{\cosh\theta} = j \tanh\theta,$$

i.e. $\tan(j\theta) = j \tanh\theta$ (9)

Dividing equation (8) by equation (7) gives:

$$\frac{\sinh(j\theta)}{\cosh(j\theta)} = \tanh(j\theta) = \frac{j\sin\theta}{\cos\theta} = j\tan\theta,$$

i.e. $\boldsymbol{\tanh(j\theta) = j\tan\theta}$ (10)

10.2   Hyperbolic identities

The equations developed in Section 10.1 may be used to determine hyperbolic identities corresponding to standard trigonometric identities.

(i)   Hyperbolic identities corresponding to the type $\cos^2 A + \sin^2 A = 1$

Substituting $j\theta$ for A gives:

$$\cos^2(j\theta) + \sin^2(j\theta) = 1$$

From equations (5) and (6) above, $\cos(j\theta) = \cosh\theta$ and $\sin(j\theta) = j\sinh\theta$. Hence, $\cosh^2\theta + j^2\sinh^2\theta = 1$. However, $j^2 = -1$, hence

$$\boldsymbol{\cosh^2\theta - \sinh^2\theta = 1}$$

Hyperbolic identities corresponding to $1 + \tan^2 A = \sec^2 A$ and $\cot^2 A + 1 = \operatorname{cosec}^2 A$ may be developed in a similar way and are listed in Section 10.3.

(ii)   Hyperbolic identities corresponding to the compound angle addition and subtraction formulae

One of this group of formulae (developed in Chapter 16 of *Technician mathematics 3*) is:

$$\sin(A + B) = \sin A \cos B + \cos A \sin B$$

Substituting $j\theta$ for A and $j\phi$ for B gives:

$$\sin[j(\theta + \phi)] = \sin(j\theta)\cos(j\phi) + \cos(j\theta)\sin(j\phi)$$

From equations (5) and (6), $\cos(j\theta) = \cosh\theta$ and $\sin(j\theta) = j\sinh\theta$, hence $j\sinh(\theta + \phi) = j\sinh\theta\cosh\phi + \cosh(\theta j)\sinh\phi$. Dividing throughout by j gives:

$$\boldsymbol{\sinh(\theta + \phi) = \sinh\theta\cosh\phi + \cosh\theta\sinh\phi}$$

Identities for $\sinh(\theta - \phi)$, $\cosh(\theta + \phi)$ and $\cosh(\theta - \phi)$ may be developed in a similar way. Identities for $\tanh(\theta + \phi)$ and $\tanh(\theta - \phi)$ may be developed from $\dfrac{\sinh(\theta + \phi)}{\cosh(\theta + \phi)}$ and $\dfrac{\sinh(\theta - \phi)}{\cosh(\theta - \phi)}$. These identities are listed in Section 10.3.

(iii) Hyperbolic identities corresponding to double angle formulae

One of this group of formulae is:

$\sin 2A = 2 \sin A \cos A$

Substituting $j\theta$ for A gives:

$\sin[2(j\theta)] = 2 \sin(j\theta) \cos(j\theta)$

From equations (5) and (6), $\cos(j\theta) = \cosh \theta$ and $\sin(j\theta) = j \sinh \theta$, hence $j \sinh 2\theta = 2j \sinh \theta \cosh \theta$.
Dividing throughout by j gives:

$\sinh 2\theta = 2 \sinh \theta \cosh \theta$

An identity for $\cosh 2\theta$ may be developed in a similar way. An identity for $\tanh 2\theta$ may be developed from $\dfrac{\sinh 2\theta}{\cosh 2\theta}$. These identities are listed in Section 10.3.

(iv) Hyperbolic identities corresponding to changing products to sums or difference formulae

One of this group of formulae is:

$\sin A \cos B = \frac{1}{2}[\sin(A + B) + \sin(A - B)]$

Substituting $j\theta$ for A and $j\phi$ for B gives:

$\sin(j\theta) \cos(j\phi) = \frac{1}{2}[\sin j(\theta + \phi) + \sin j(\theta - \phi)]$

From equations (5) and (6), $\cos(j\theta) = \cosh \theta$ and $\sin j\theta = j \sinh \theta$, hence, $j \sinh \theta \cosh \phi = \frac{1}{2}[j \sinh(\theta + \phi) + j \sinh(\theta - \phi)]$
Dividing throughout by j gives:

$\sinh \theta \cosh \phi = \frac{1}{2}[\sinh(\theta + \phi) + \sinh(\theta - \phi)]$

Identities for $\cosh \theta \sinh \phi$, $\cosh \theta \cosh \phi$ and $\sinh \theta \sinh \phi$ may be developed in a similar way, and are listed in Section 10.3.

(v) Hyperbolic identities corresponding to changing sums or differences to product formulae

One of this group of formulae is:

$$\sin A + \sin B = 2 \sin\left(\frac{A + B}{2}\right) \cos\left(\frac{A - B}{2}\right)$$

Substituting $j\theta$ for A and $j\phi$ for B gives:

$$\sin(j\theta) + \sin(j\phi) = 2\sin j\left(\frac{\theta + \phi}{2}\right)\cos j\left(\frac{\theta - \phi}{2}\right)$$

From equations (5) and (6), $\cos(j\theta) = \cosh\theta$ and $\sin(j\theta) = j\sinh\theta$, hence,

$$j\sinh\theta + j\sinh\phi = 2j\sinh\left(\frac{\theta + \phi}{2}\right)\cosh\left(\frac{\theta - \phi}{2}\right)$$

Dividing throughout by j gives:

$$\mathbf{\sinh\theta + \sinh\phi = 2\sinh\left(\frac{\theta + \phi}{2}\right)\cosh\left(\frac{\theta - \phi}{2}\right)}$$

Identities for $\sinh\theta - \sinh\phi$, $\cosh\theta + \cosh\phi$ and $\cosh\theta - \cosh\phi$ may be developed in a similar way and are listed in Section 10.3.

There are three ways in which any hyperbolic identity may be verified. These are:

(a)   by substituting $j\theta$ and if necessary $j\phi$ in the corresponding trigonometric identity, as shown above,

(b)   by expressing the identity in terms of exponential functions, or

(c)   by applying **Osborne's rule**. This is introduced in Chapter 9, and may be stated as follows:

> '*in formulae connecting trigonometric functions, replace each circular function by the corresponding hyperbolic function and change the sign of every product (or implied product) of two sines*'.

Each of these three ways is demonstrated in the worked problems following. When the corresponding trigonometric identity is known or can be readily proved, then Osborne's rule is the easiest method of determining the corresponding hyperbolic identity.

The relationship between trigonometric and hyperbolic functions and hyperbolic identities will be met in technological subjects dealing with topics such as transmission line theory, wave guide theory and filters.

Worked problems on hyperbolic identities

Problem 1. By substituting $j\theta$ for A, develop the hyperbolic identity corresponding to the trigonometric identity
$$\frac{1 + \sin^4 A - \cos^4 A}{\sin^2 A} = 2.$$ Check the result by applying Osborne's rule.

Writing $j\theta$ for A gives:

$$\frac{1 + \sin^4(j\theta) - \cos^4(j\theta)}{\sin^2(j\theta)} = 2$$

However, $\sin(j\theta) = j \sinh \theta$ and $\cos(j\theta) = \cosh \theta$.
Thus,

$$\frac{1 + j^4 \sinh^4 \theta - \cosh^4 \theta}{j^2 \sinh^2 \theta} = 2$$

However, $j^4 = 1$ and $j^2 = -1$, hence the required identity is

$$\frac{1 + \sinh^4 \theta - \cosh^4 \theta}{-\sinh^2 \theta} = 2$$

i.e. $\dfrac{1 + \sinh^4 \theta - \cosh^4 \theta}{\sinh^2 \theta} = -2$

Rewriting the trigonometric identity in terms of the products of two sines gives:

$$\frac{1 + (\sin^2 A)(\sin^2 A) - \cos^4 A}{\sin^2 A} = 2$$

Osborne's rule states that each trigonometric function is replaced by its corresponding hyperbolic function, the sign being changed for any terms which are the products of two sines. Thus, the identity becomes:

$$\frac{1 + (-\sinh^2 \theta)(-\sinh^2 \theta) - \cosh^4 \theta}{(-\sinh^2 \theta)} = 2$$

i.e. $\dfrac{1 + \sinh^4 \theta - \cosh^4 \theta}{\sinh^2 \theta} = -2$, as shown above.

Problem 2. Use the exponential form of a hyperbolic function to prove the identity:

$$\frac{1}{\sqrt{(1 + \sinh^2 \theta)}} = \operatorname{sech} \theta$$

From the basic definitions of hyperbolic functions given in equations (3) and (4), $\cosh \theta = \frac{1}{2}(e^\theta + e^{-\theta})$ and $\sinh \theta = \frac{1}{2}(e^\theta - e^{-\theta})$. Substituting for $\sinh \theta$ in the left-hand side of the identity gives:

$$\frac{1}{\sqrt{(1 + \sinh^2 \theta)}} = \frac{1}{\sqrt{\{1 + [\frac{1}{2}(e^\theta - e^{-\theta})]^2\}}}$$

$$= \frac{1}{\sqrt{\{1 + \frac{1}{4}(e^{2\theta} - 2 + e^{-2\theta})\}}}$$

$$= \frac{1}{\sqrt{\{1 + \frac{1}{4}e^{2\theta} - \frac{1}{2} + \frac{1}{4}e^{-2\theta}\}}}$$

$$= \frac{1}{\sqrt{\{\frac{1}{4}e^{2\theta} + \frac{1}{2} + \frac{1}{4}e^{-2\theta}\}}}$$

$$= \frac{1}{\sqrt{\frac{1}{4}(e^{2\theta} + 2 + e^{-2\theta})}}$$

$$= \frac{1}{\sqrt{[\frac{1}{2}(e^{\theta} + e^{-\theta})]^2}}$$

$$= \frac{1}{\sqrt{(\cosh \theta)^2}}$$

$$= \frac{1}{\cosh \theta} = \operatorname{sech} \theta$$

$$= \text{R.H.S.}$$

Problem 3. Prove that:
$$\sin^2 \theta \cosh^2 \theta + \cos^2 \theta \sinh^2 \theta = \tfrac{1}{2}(\cosh 2\theta - \cos 2\theta)$$

Writing the left-hand side in terms of $\cos \theta$ and $\cosh \theta$ only, using the identities $\cos^2 \theta + \sin^2 \theta = 1$ and $\cosh^2 \theta - \sinh^2 \theta = 1$, gives:

$$(1 - \cos^2 \theta) \cosh^2 \theta + \cos^2 \theta (\cosh^2 \theta - 1)$$

i.e. L.H.S. $= \cosh^2 \theta - \cos^2 \theta \cosh^2 \theta + \cos^2 \theta \cosh^2 \theta - \cos^2 \theta$

$$= \cosh^2 \theta - \cos^2 \theta$$

From the double angle formulae listed in the summary in Section 10.3,

$$\cos^2 \theta = \tfrac{1}{2}(1 + \cos 2\theta) \text{ and } \cosh^2 \theta = \tfrac{1}{2}(1 + \cosh 2\theta)$$

Hence, L.H.S. $= \tfrac{1}{2}(1 + \cosh 2\theta) - \tfrac{1}{2}(1 + \cos 2\theta)$

$$= \tfrac{1}{2} + \tfrac{1}{2}\cosh 2\theta - \tfrac{1}{2} - \tfrac{1}{2}\cos 2\theta$$

$$= \tfrac{1}{2}(\cosh 2\theta - \cos 2\theta) = \text{R.H.S.}$$

Further problems on hyperbolic identities may be found in Section 10.4 (Problems 1 to 20), page 174.

10.3   Summary of trigonometric and hyperbolic identities

Trigonometric identity	Corresponding hyperbolic identity

(i)   *Standards*

$$\cos^2 A + \sin^2 A = 1 \qquad\qquad \cosh^2\theta - \sinh^2\theta = 1$$
$$1 + \tan^2 A = \sec^2 A \qquad\qquad 1 - \tanh^2\theta = \operatorname{sech}^2\theta$$
$$\cot^2 A + 1 = \operatorname{cosec}^2 A \qquad\qquad \coth^2\theta - 1 = \operatorname{cosech}^2\theta$$

(ii)   *Compound angle addition and subtraction formulae*

$$\sin(A+B) = \sin A \cos B + \cos A \sin B \qquad \sinh(\theta+\phi) = \sinh\theta\cosh\phi + \cosh\theta\sinh\phi$$
$$\sin(A-B) = \sin A \cos B - \cos A \sin B \qquad \sinh(\theta-\phi) = \sinh\theta\cosh\phi - \cosh\theta\sinh\phi$$
$$\cos(A+B) = \cos A \cos B - \sin A \sin B \qquad \cosh(\theta+\phi) = \cosh\theta\cosh\phi + \sinh\theta\sinh\phi$$
$$\cos(A-B) = \cos A \cos B + \sin A \sin B \qquad \cosh(\theta-\phi) = \cosh\theta\cosh\phi - \sinh\theta\sinh\phi$$
$$\tan(A+B) = \frac{\tan A + \tan B}{1 - \tan A \tan B} \qquad \tanh(\theta+\phi) = \frac{\tanh\theta + \tanh\phi}{1 + \tanh\theta\tanh\phi}$$
$$\tan(A-B) = \frac{\tan A - \tan B}{1 + \tan A \tan B} \qquad \tanh(\theta-\phi) = \frac{\tanh\theta - \tanh\phi}{1 - \tanh\theta\tanh\phi}$$

(iii)   *Double angle formulae*

$$\sin 2A = 2\sin A \cos A \qquad\qquad \sinh 2\theta = 2\sinh\theta\cosh\theta$$
$$\cos 2A = \cos^2 A - \sin^2 A \qquad\qquad \cosh 2\theta = \cosh^2\theta + \sinh^2\theta$$
$$= 2\cos^2 A - 1 \qquad\qquad\qquad = 2\cosh^2\theta - 1$$
$$= 1 - 2\sin^2 A \qquad\qquad\qquad = 1 + 2\sinh^2\theta$$
$$\tan 2A = \frac{2\tan A}{1 - \tan^2 A} \qquad\qquad \tanh 2\theta = \frac{2\tanh\theta}{1 + \tanh^2\theta}$$

(iv)   *Changing products into sums or differences formulae*

$$\sin A \cos B = \tfrac{1}{2}[\sin(A+B) + \sin(A-B)] \qquad \sinh\theta\cosh\phi = \tfrac{1}{2}[\sinh(\theta+\phi) + \sinh(\theta-\phi)]$$
$$\cos A \sin B = \tfrac{1}{2}[\sin(A+B) - \sin(A-B)] \qquad \cosh\theta\sinh\phi = \tfrac{1}{2}[\sinh(\theta+\phi) - \sinh(\theta-\phi)]$$
$$\cos A \cos B = \tfrac{1}{2}[\cos(A+B) + \cos(A-B)] \qquad \cosh\theta\cosh\phi = \tfrac{1}{2}[\cosh(\theta+\phi) + \cosh(\theta-\phi)]$$
$$\sin A \sin B = -\tfrac{1}{2}[\cos(A+B) - \cos(A-B)] \qquad \sinh\theta\sinh\phi = \tfrac{1}{2}[\cosh(\theta+\phi) - \cosh(\theta-\phi)]$$

(v)   *Changing sums or differences into products formulae*

$$\sin A + \sin B \qquad\qquad\qquad \sinh\theta + \sinh\phi$$
$$= 2\sin\left(\frac{A+B}{2}\right)\cos\left(\frac{A-B}{2}\right) \qquad = 2\sinh\left(\frac{\theta+\phi}{2}\right)\cosh\left(\frac{\theta-\phi}{2}\right)$$
$$\sin A - \sin B \qquad\qquad\qquad \sinh\theta - \sinh\phi$$
$$= 2\cos\left(\frac{A+B}{2}\right)\sin\left(\frac{A-B}{2}\right) \qquad = 2\cosh\left(\frac{\theta+\phi}{2}\right)\sinh\left(\frac{\theta-\phi}{2}\right)$$
$$\cos A + \cos B \qquad\qquad\qquad \cosh\theta + \cosh\phi$$
$$= 2\cos\left(\frac{A+B}{2}\right)\cos\left(\frac{A-B}{2}\right) \qquad = 2\cosh\left(\frac{\theta+\phi}{2}\right)\cosh\left(\frac{\theta-\phi}{2}\right)$$
$$\cos A - \cos B \qquad\qquad\qquad \cosh\theta - \cosh\phi$$
$$= -2\sin\left(\frac{A+B}{2}\right)\sin\left(\frac{A-B}{2}\right) \qquad = 2\sinh\left(\frac{\theta+\phi}{2}\right)\sinh\left(\frac{\theta-\phi}{2}\right)$$

10.4   Further problems

In Problems 1 to 6, state the trigonometric identities corresponding to the hyperbolic identities given in terms of A (and B) and by substituting $j\theta$ for A (and $j\phi$ for B) in these trigonometric identities, prove the hyperbolic identities.

1. (a) $1 - \tanh^2 \theta = \operatorname{sech}^2 \theta$ (b) $\coth^2 \theta - 1 = \operatorname{cosech}^2 \theta$
2. (a) $\sinh(\theta - \phi) = \sinh \theta \cosh \phi - \cosh \theta \sinh \phi$
    (b) $\cosh(\theta + \phi) = \cosh \theta \cosh \phi + \sinh \theta \sinh \phi$
    (c) $\cosh(\theta - \phi) = \cosh \theta \cosh \phi - \sinh \theta \sinh \phi$
3. (a) $\sinh \theta - \sinh \phi = 2 \cosh\left(\dfrac{\theta + \phi}{2}\right) \sinh\left(\dfrac{\theta - \phi}{2}\right)$

    (b) $\cosh \theta + \cosh \phi = 2 \cosh\left(\dfrac{\theta + \phi}{2}\right) \cosh\left(\dfrac{\theta - \phi}{2}\right)$

    (c) $\cosh \theta - \cosh \theta = 2 \sinh\left(\dfrac{\theta + \phi}{2}\right) \sinh\left(\dfrac{\theta - \phi}{2}\right)$

4. (a) $\tanh(\theta + \phi) = \dfrac{\tanh \theta + \tanh \phi}{1 + \tanh \theta \tanh \phi}$

    (b) $\tanh(\theta - \phi) = \dfrac{\tanh \theta - \tanh \phi}{1 - \tanh \theta \tanh \phi}$

5. (a) $\cosh 2\theta = \cosh^2 \theta + \sinh^2 \theta$ (b) $\cosh 2\theta = 2 \cosh^2 \theta - 1$

    (c) $\cosh 2\theta = 1 + 2 \sinh^2 \theta$ (d) $\tanh 2\theta = \dfrac{2 \tanh \theta}{1 + \tanh^2 \theta}$

6. (a) $\cosh \theta \sinh \phi = \frac{1}{2}[\sinh(\theta + \phi) - \sinh(\theta - \phi)]$
    (b) $\cosh \theta \cosh \phi = \frac{1}{2}[\cosh(\theta + \phi) + \cosh(\theta - \phi)]$
    (c) $\sinh \theta \sinh \phi = \frac{1}{2}[\cosh(\theta + \phi) - \cosh(\theta - \phi)]$

In Problems 7 to 10, prove the hyperbolic identities given by expressing them in terms of exponential functions.

7. (a) $1 + \cosh x = 2 \cosh^2 \dfrac{x}{2}$ (b) $1 - \cosh x = -2 \sinh^2 \dfrac{x}{2}$

8. $\coth \dfrac{\theta}{2} - \coth \theta = \operatorname{cosech} \theta$

9. $\cosh^3 \theta = \frac{1}{4}(\cosh 3\theta + 3 \cosh \theta)$

10. $\dfrac{1 - \tanh^2 \phi}{1 - \coth^2 \phi} = -\tanh^2 \phi$

Prove the hyperbolic identities given in Problems 11 to 18.

11. $-\left(\dfrac{1 - \cosh A}{1 + \cosh A}\right) = (\operatorname{cosech} A - \coth A)^2$

12. $\cosh \theta \sqrt{(1 - \operatorname{sech}^2 \theta)} = \sinh \theta$

13. $\dfrac{\text{cosech } x}{\text{sech } x} - \dfrac{\text{sech } x}{\text{cosech } x} = \dfrac{2}{\sinh 2x}$

14. $\dfrac{1 + \coth \phi}{1 + \tanh \phi} = \coth \phi$

15. $(\sinh A + \cosh A)^2 + (\sinh A - \cosh A)^2 = 2 \cosh 2A$

16. $\cosh^2(A + B) - \cosh^2(A - B) = \sinh 2A \sinh 2B$

17. $\cosh^2 A \sin^2 B + \sinh^2 A \cos^2 B = \frac{1}{2}(\cosh 2B - \cos 2A)$

18. $\cosh x + \cosh 2x = \dfrac{1}{2} \sinh \dfrac{5x}{2} \text{ cosech } \dfrac{x}{2} - \dfrac{1}{2}$

   (Hint: use changing sums or differences to products formulae.)

19.   If $\sin A \cosh B = \cos \theta$ and $\cos A \sinh B = \sin \theta$, show that $\sinh^2 B = \cos^2 A = \pm \sin \theta$.

   (Hint: use the identities $\cos^2 \theta + \sin^2 \theta = 1$ and $\cosh^2 B - \sinh^2 B = 1$.)

20.   If $\tan A = \tan \theta \tanh \phi$ and $\tan B = \cot \theta \tanh \phi$, show that
$$\tan(A + B) = \dfrac{\sinh 2\phi}{\sin 2\theta}.$$

   (Hint: use the double angle formulae.)

Chapter 11

Differentiation of implicit functions

11.1 Implicit functions

In all previous work involving differentiation, the starting point has been equations of the form $y = f(x)$. When an equation is expressed in this form, it is said to be an **explicit** function. Thus:

$$y = 3x^3 + 4x - 5$$

$$y = e^x \sin x$$

and $\quad y = 3(x^2 + 1) \ln \dfrac{x}{2}$

are all examples of explicit functions. In many equations it is difficult, or indeed impossible, to make y the subject of the formula. The equation is then called an **implicit** function, there being an implied relationship between x and y. Thus:

$$x^2 - 4xy + 3y^3 = 8$$

$$e^x \sin y + 4xy^2 = 0$$

and $\quad 2 \sin \dfrac{x}{2} \cos y^2 = x^2 - y$

are all examples of implicit functions.

11.2 Differentiating implicit functions

When differentiating an explicit function, the function of a variable, say x, is differentiated with respect to x. However, when differentiating an implicit function, it becomes necessary to differentiate a function of one variable, say y, with respect to another variable, say x. This is achieved by using the 'chain' rule of differentiation, i.e.

$$\frac{du}{dx} = \frac{du}{dy} \times \frac{dy}{dx}$$

Thus, to differentiate, say, y^2 with respect to x, let $u = y^2$

Then $\dfrac{du}{dy} = 2y$ and $\dfrac{d(y^2)}{dx} = 2y\dfrac{dy}{dx}$ by using the 'chain' rule.

The rule when differentiating a function of, say y, with respect to, say x, is:

(i) differentiate the function of y with respect to y, and

(ii) always multiply by $\dfrac{dy}{dx}$.

This rule can be expressed mathematically as

$$\frac{dy}{dx}[f(y)] = \frac{d}{dy}[f(y)] \times \frac{dy}{dx}$$
(1)

When a term is a product or a quotient of two functions of, say, x, the product or quotient rules of differentiation must be applied to find its differential coefficient. These same rules apply to products and quotients of terms having two variables (say $f(x, y)$).

Thus if $z = x^2y^3$, the product rule of differentiation must be used to find $\dfrac{dz}{dx}$, i.e. $\dfrac{dz}{dx} = u\dfrac{dv}{dx} + v\dfrac{du}{dx}$ where $u = x^2$ and $v = y^3$. Since $u = x^2$, $\dfrac{du}{dx} = 2x$

and since $v = y^3$, $\dfrac{dv}{dx} = 3y^2\dfrac{dy}{dx}$ from equation (1) above.

Hence $\dfrac{dz}{dx} = x^2 \cdot 3y^2\dfrac{dy}{dx} + y^3 \cdot 2x$

$$= xy^2\left(3x\frac{dy}{dx} + 2y\right)$$

For a quotient, say $z = \dfrac{3y^2}{\cos x}$, the quotient rule of differentiation is used,

i.e. $\dfrac{dz}{dx} = \dfrac{v\dfrac{du}{dx} - u\dfrac{dv}{dx}}{v^2}$, where $u = 3y^2$ and $v = \cos x$

Since $u = 3y^2$, $\dfrac{du}{dx} = 6y\dfrac{dy}{dx}$ from equation (1) above and since $v = \cos x$,

$\dfrac{dv}{dx} = -\sin x$,

thus,

$$\frac{dz}{dx} = \frac{(\cos x)\left(6y\,\dfrac{dy}{dx}\right) - (3y^2)(-\sin x)}{\cos^2 x}$$

$$= \frac{3y\left(2\cos x\,\dfrac{dy}{dx} + y\sin x\right)}{\cos^2 x}$$

When differentiating an implicit function having several terms, each term is differentiated with respect to the variable. Thus, to differentiate $x^2 + 2xy + 3y^2 = 0$ with respect to x, it becomes:

$$\frac{d(x^2)}{dx} + \frac{d(2xy)}{dx} + \frac{d(3y^2)}{dx} = \frac{d(0)}{dx}$$

i.e. $$2x + \left[\left(2x\frac{dy}{dx}\right) + y\cdot 2\right] + 6y\frac{dy}{dx} = 0$$

To express $\dfrac{dy}{dx}$ in terms of x and y, terms containing $\dfrac{dy}{dx}$ are grouped on the left-hand side of the equation, and those not containing $\dfrac{dy}{dx}$ are grouped on the right-hand side. Thus

$$2x\frac{dy}{dx} + 6y\frac{dy}{dx} = -2x - 2y$$

$$\frac{dy}{dx}(2x + 6y) = -2x - 2y$$

$$\frac{dy}{dx} = \frac{-2x - 2y}{2x + 6y} = \frac{-(x + y)}{x + 3y}$$

Worked problems on differentiating implicit functions

Problem 1. Differentiate the following functions with respect to x:

(a) y^5 (b) $\cos 2\theta$ (c) $3\ln\dfrac{p}{2}$ (d) $\dfrac{1}{4}e^{2q+3}$

(a) Let $u = y^5$, then

$$\frac{du}{dx} = \frac{du}{dy} \times \frac{dy}{dx} = \frac{d}{dy}(y^5) \times \frac{dy}{dx} = 5y^4\frac{dy}{dx}$$

(b) Let $u = \cos 2\theta$

$$\frac{du}{dx} = \frac{du}{d\theta} \times \frac{d\theta}{dx} = \frac{d}{d\theta}(\cos 2\theta) \times \frac{d\theta}{dx}$$

$$= -2 \sin 2\theta \frac{d\theta}{dx}$$

(c) Let $u = 3 \ln \dfrac{p}{2}$

$$\frac{du}{dx} = \frac{du}{dp} \times \frac{dp}{dx} = \frac{d}{dp}\left(3 \ln \frac{p}{2}\right) \times \frac{dp}{dx}$$

$$= \frac{3}{p} \frac{dp}{dx}$$

(d) Let $u = \dfrac{1}{4} e^{2q+3}$

$$\frac{du}{dx} = \frac{du}{dq} \times \frac{dq}{dx} = \frac{d}{dq}\left(\frac{1}{4} e^{2q+3}\right) \times \frac{dq}{dx}$$

$$= \frac{1}{2} e^{2q+3} \frac{dq}{dx}$$

Problem 2. If $y = f(m, p)$, differentiate with respect to p:

(a) $y = m^2 \cos 2p$ and (b) $y = \dfrac{(1 - 3p)^{\frac{1}{2}}}{2 \tan 2m}$

(a) Since y is a function of two variables, m and p, the product rule of differentiation is used.

Let $u = m^2$. Now $\dfrac{du}{dp} = \dfrac{du}{dm} \times \dfrac{dm}{dp} = \dfrac{d}{dm}(m^2) \times \dfrac{dm}{dp}$

$$= 2m \frac{dm}{dp}$$

Let $v = \cos 2p$, $\dfrac{dv}{dp} = -2 \sin 2p$

Applying the product rule, $\dfrac{dy}{dp} = u \dfrac{dv}{dp} + v \dfrac{du}{dp}$ gives

$$\frac{dy}{dp} = m^2(-2 \sin 2p) + \cos 2p \left(2m \frac{dm}{dp}\right)$$

$$= 2m \left(\cos 2p \frac{dm}{dp} - m \sin 2p\right)$$

(b) Since y is a function of both m and p, the quotient rule of differentiation is used.

$$\text{Let } u = (1 - 3p)^{\frac{1}{2}}, \frac{du}{dp} = -\frac{3}{2}(1 - 3p)^{-\frac{1}{2}}$$

$$\text{Let } v = 2 \tan 2m, \frac{dv}{dp} = \frac{dv}{dm} \times \frac{dm}{dp}$$

$$= 4 \sec^2 2m \frac{dm}{dp}$$

Applying the quotient rule, $\dfrac{dy}{dp} = \dfrac{v\dfrac{du}{dp} - u\dfrac{dv}{dp}}{v^2}$, gives

$$\frac{dy}{dp} = \frac{2 \tan 2m \left(-\dfrac{3}{2}(1 - 3p)^{-\frac{1}{2}}\right) - (1 - 3p)^{\frac{1}{2}}\left(4 \sec^2 2m \dfrac{dm}{dp}\right)}{4 \tan^2 2m}$$

$$= -\frac{3}{4(\tan 2m)(1 - 3p)^{\frac{1}{2}}} - \frac{(1 - 3p)^{\frac{1}{2}} \sec^2 2m \dfrac{dm}{dp}}{\tan^2 2m}$$

Problem 3. Given that $x^2 \sin \theta - 3x^3 = \sec \theta$, determine the value of $\dfrac{dx}{d\theta}$ when $\theta = \pi$.

$$\frac{d}{d\theta}(x^2 \sin \theta) = x^2 \cos \theta + \sin \theta \left(2x \frac{dx}{d\theta}\right)$$

$$\frac{d}{d\theta}(3x^3) = 9x^2 \frac{dx}{d\theta}$$

$$\frac{d}{d\theta}(\sec \theta) = \sec \theta \tan \theta$$

Hence, $x^2 \cos \theta + 2x \sin \theta \dfrac{dx}{d\theta} - 9x^2 \dfrac{dx}{d\theta} = \sec \theta \tan \theta$

Grouping only terms containing $\dfrac{dx}{d\theta}$ on the left-hand side gives:

$$2x \sin \theta \, \frac{dx}{d\theta} - 9x^2 \, \frac{dx}{d\theta} = \sec \theta \tan \theta - x^2 \cos \theta$$

$$\frac{dx}{d\theta}(2x \sin \theta - 9x^2) = \sec \theta \tan \theta - x^2 \cos \theta$$

$$\frac{dx}{d\theta} = \frac{\sec \theta \tan \theta - x^2 \cos \theta}{2x \sin \theta - 9x^2}$$

To determine the value of this expression, the value of x when $\theta = \pi$ is needed.

Substituting $\theta = \pi$ in the original equation, gives:

$$x^2 \sin \pi - 3x^3 = \sec \pi$$

Now, $\sin \pi = 0$ and $\sec \pi = -1$.

Hence, $0 - 3x^3 = -1$, i.e. $x^3 = \dfrac{1}{3}$ i.e. $x = \sqrt[3]{\tfrac{1}{3}}$

Thus $\dfrac{dx}{d\theta} = \dfrac{\sec \pi \tan \pi - \sqrt[3]{(\tfrac{1}{3})^2} \cos \pi}{2\sqrt[3]{(\tfrac{1}{3})} \sin \pi - 9\sqrt[3]{(\tfrac{1}{3})^2}}$

$$= \frac{(-1)(0) - \sqrt[3]{(\tfrac{1}{3})^2} \cdot (-1)}{2\sqrt[3]{(\tfrac{1}{3})}(0) - 9\sqrt[3]{(\tfrac{1}{3})^2}}$$

$$= \frac{\sqrt[3]{(\tfrac{1}{3})^2}}{-9\sqrt[3]{(\tfrac{1}{3})^2}} = -\frac{1}{9}$$

Problem 4. Determine the values of the gradients of the tangents drawn to the circle $x^2 + y^2 - 3x + 4y + 1 = 0$ at $x = 1$, correct to four significant figures.

The equation of the gradient of the tangent is given by $\dfrac{dy}{dx}$

Differentiating each term with respect to x, gives.

$$2x + 2y \frac{dy}{dx} - 3 + 4 \frac{dy}{dx} + 0 = 0$$

i.e. $\dfrac{dy}{dx}(2y + 4) = 3 - 2x$

$$\frac{dy}{dx} = \frac{3 - 2x}{2y + 4}$$

The value of y when $x = 1$ is obtained substituting $x = 1$ in the original equation. Thus:

$(1)^2 + y^2 - 3(1) + 4y + 1 = 0$

i.e. $y^2 + 4y - 1 = 0$

$$y = \frac{-4 \pm \sqrt{[4^2 - 4 \times 1 \times (-1)]}}{2}$$

$y = 0.236\,07$ or $-4.236\,07$

Substituting $x = 1$, $y = 0.236\,07$ in the equation for $\dfrac{dy}{dx}$ gives:

$$\frac{dy}{dx} = \frac{3 - 2(1)}{2 \times 0.236\,07 + 4} = 0.233\,6, \text{ correct to four significant figures.}$$

Substituting $x = 1$ and $y = -4.236\,07$ in the equation for $\dfrac{dy}{dx}$ gives:

$$\frac{dy}{dx} = \frac{3 - 2(1)}{2 \times (-4.236\,07) + 4} = -0.223\,6, \text{ correct to four significant figures.}$$

That is, the gradients to the tangents are $\pm 0.223\,6$, correct to four significant figures.

Further problems on differentiating implicit functions may be found in the following Section 11.3 (Problems 1 to 20).

11.3   Further problems

Differentiating implicit functions

1. Differentiate with respect to x:

    (a) y^6 (b) $3z^{\frac{1}{2}}$ (c) $\sin \dfrac{p}{3}$

    (a) $\left[6y^5 \dfrac{dy}{dx} \right]$ (b) $\left[\dfrac{3}{2} z^{-\frac{1}{2}} \dfrac{dz}{dx} \right]$ (c) $\left[\dfrac{1}{3} \cos \dfrac{p}{3} \dfrac{dp}{dx} \right]$

2. Differentiate with respect to y:

    (a) $\sin 2p$ (b) $\dfrac{1}{2} \cos 3x$ (c) $2e^{2m+4}$

    (a) $\left[2 \cos 2p \dfrac{dp}{dy} \right]$ (b) $\left[-\dfrac{3}{2} \sin 3x \dfrac{dx}{dy} \right]$

    (c) $\left[4e^{2m+4} \dfrac{dm}{dy} \right]$

3. Differentiate with respect to u:

   (a) $(3 - 4x)^{\frac{1}{2}}$ (b) $2 \ln(3y + 2)$ (c) $\dfrac{1}{4} \tan(4u + \pi)$

   (a) $\left[-2(3 - 4x)^{-\frac{1}{2}} \dfrac{dx}{du} \right]$ (b) $\left[\dfrac{6}{3y + 2} \dfrac{dy}{du} \right]$

   (c) $[\sec^2(4u + \pi)]$

4. Differentiate with respect to w:

   (a) $-\dfrac{1}{2} \sec 2\theta$ (b) $\dfrac{6}{3y + 2}$ (c) $\dfrac{3}{2} \cot\left(\dfrac{\pi}{2} - \alpha\right)$

   (a) $\left[-\sec 2\theta \tan 2\theta \dfrac{d\theta}{dw} \right]$ (b) $\left[\dfrac{-18}{(3y + 2)^2} \dfrac{dy}{dw} \right]$

   (c) $\left[\dfrac{3}{2} \operatorname{cosec}^2 \left(\dfrac{\pi}{2} - \alpha\right) \dfrac{d\alpha}{dw} \right]$

In Problems 5 to 8, find $\dfrac{dz}{dx}$

5. $z = x^2 y^3$ $\left[3x^2 y^2 \dfrac{dy}{dx} + 2xy^3 \right]$

6. $z = 3y^2 \cos x$ $\left[-3y^2 \sin x + 6y \cos x \dfrac{dy}{dx} \right]$

7. $z = x^2 \ln y$ $\left[\dfrac{x^2}{y} \dfrac{dy}{dx} + 2x \ln y \right]$

8. $z = \dfrac{2x^{\frac{1}{3}}}{\tan y}$ $\left[\dfrac{\dfrac{2}{3}(\tan y)x^{-\frac{2}{3}} - 2x^{\frac{1}{3}} \sec^2 y \dfrac{dy}{dx}}{\tan^2 y} \right]$

In Problems 9 to 12, differentiate with respect to y.

9. $x^2 + \sin x \cos y$ $\left[2x \dfrac{dx}{dy} - \sin x \sin y + \cos x \cos y \dfrac{dx}{dy} \right]$

10. $e^{2x} \ln y - 4x^2$ $\left[\dfrac{e^{2x}}{y} + 2e^{2x} \ln y \dfrac{dx}{dy} - 8x \dfrac{dx}{dy} \right]$

11. $x^2 + 3xy - y^2$ $\left[2x \dfrac{dx}{dy} + 3x + 3y \dfrac{dx}{dy} - 2y \right]$

12. $y^3 - 4x^2 \cos y + \dfrac{\ln x}{\sec y}$

$$\left[\,3y^2 + 4x^2 \sin y - 8x \cos y\,\dfrac{dx}{dy} + \dfrac{\dfrac{1}{x}\dfrac{dx}{dy} - \ln x \tan y}{\sec y}\,\right]$$

In problems 13 to 16, find $\dfrac{dy}{dx}$ in terms of x and y.

13. $x^2 + y^2 + 6x + 7y + 3 = 0$ $\qquad \left[\dfrac{dy}{dx} = \dfrac{-2(x+3)}{2y+7}\right]$

14. $2x^3 - 4x^2 y = \cos y$ $\qquad \left[\dfrac{dy}{dx} = \dfrac{2x(3x - 4y)}{4x^2 - \sin y}\right]$

15. $\sin 2x \cos 3y = x^{\frac{1}{2}} y^{\frac{1}{3}}$ $\qquad \left[\dfrac{dy}{dx} = \dfrac{2 \cos 2x \cos 3y - y^{\frac{1}{3}}/(2x^{\frac{1}{2}})}{3 \sin 2x \sin 3y + x^{\frac{1}{2}}/(3y^{\frac{2}{3}})}\right]$

16. $2y - \dfrac{x^3}{\sec y} = y^3$ $\qquad \left[\dfrac{dy}{dx} = \dfrac{3x^2}{x^3 \tan y + \sec y(2 - 3y^2)}\right]$

17. Determine the gradients of the tangents drawn to the circle $x^2 + y^2 = 4$ at the point $x = 1$. $\qquad \left[\pm\dfrac{1}{\sqrt{3}} = \pm 0.5774\right]$

18. Find the gradients of the tangents drawn to the ellipse $2x^2 + y^2 = 9$ at the point $x = 2$ and to the hyperbola $x^2 - y^2 = 8$ at the point $x = 3$. $\qquad [\pm 4, \pm 3]$

19. If the distance moved by a body is given by $x = 3 \tan \theta$, the angular velocity, w, is $\dfrac{d\theta}{dt}$ and the velocity v is $\dfrac{dx}{dt}$, show that

$$w = \dfrac{v}{3}\cos^2 \theta.$$

20. The pressure, p, and volume, V, of a gas are related by the law $pV^n = C$, when n and C are constants. Show that the rate of change of pressure, $\dfrac{dp}{dt} = -n\dfrac{p}{V}\dfrac{dV}{dt}$.

Chapter 12

Differentiation of functions defined parametrically

12.1 Parametric representation of points

Certain mathematical relationships can be expressed more simply by stating both x and y in terms of a third variable, say, θ. By doing this, subsequent work is frequently simplified. When both x and y are expressed in terms of the same variable, this variable is called a **parameter**. The equation of any point on a circle, centre at $x = 0$ and $y = 0$, and of radius r, is given by $x^2 + y^2 = r^2$.

Such a circle is shown in Fig. 12.1. Any point on the circumference of the circle may be expressed in terms of the radius and the angle θ and when this is done, θ is the parameter used.

With reference to Fig. 12.1, from triangle OAB, $x = r \cos \theta$ and $y = r \sin \theta$. These two equations are called the **parametric equations** for a circle. A check may be made when parametric equations have been formed by substituting for x and y in the given equation. Thus, substituting $x = r \cos \theta$ and $y = r \sin \theta$ in the left-hand side of the equation $x^2 + y^2 = r^2$ gives:

$$\text{L.H.S.} = r^2 \cos^2 \theta + r^2 \sin^2 \theta = r^2(\cos^2 \theta + \sin^2 \theta)$$

$$= r^2,$$

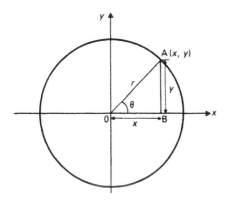

Fig. 12.1

which is equal to the right-hand side of the original equation. Hence $x = r \cos \theta$, $y = r \sin \theta$ are suitable parametric equations of circle $x^2 + y^2 = r^2$

When the parametric equations of a curve are given, the curve may be plotted as shown below. The parametric equations for a parabola are of the form $x = at^2$, $y = 2at$. Checking these parametric equations in the equation of a parabola $y^2 = 4ax$ gives:

L.H.S. $y^2 = (2at)^2 = 4a^2t^2$
R.H.S. $4ax = 4a(at^2) = 4a^2t^2$

Since L.H.S. = R.H.S., $x = at^2$, $y = 2at$ are suitable parametric equations. To plot the graph for values of t between, say, -3 and 3, the values of x and y are calculated from the parametric equations. Thus, when

$t =$	-3	-2	-1	0	1	2	3
$x = at^2 =$	$9a$	$4a$	a	0	a	$4a$	$9a$
$y = 2at =$	$-6a$	$-4a$	$-2a$	0	$2a$	$4a$	$6a$

The parabola represented by $x = at^2$, $t = 2at$, over a range $-3 \leqslant t \leqslant 3$ is shown in Fig. 12.2

There can be more than one parametric form of the co-ordinates used to represent a particular curve. For example, the circle having parameter θ and parametric equations $x = a \cos \theta$, $y = a \sin \theta$ may be expressed in terms of

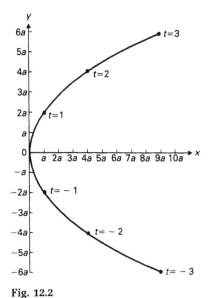

Fig. 12.2

a different parameter t, as

$$x = \frac{a(1 - t^2)}{1 + t^2} \qquad y = \frac{2at}{1 + t^2}$$

where the parameter t is given by $t = \tan \dfrac{\theta}{2}$.

The list below shows some of the more common parametric equations used in mathematics.

Parabola	$x = at^2$,	$y = 2at$
Ellipse	$x = a \cos \theta$,	$y = b \sin \theta$
Hyperbola	$x = a \sec \theta$,	$y = b \tan \theta$
(also	$x = \pm a \cosh u$,	$y = b \sinh u$)
Rectangular $\Big\}$ Hyperbola	$x = ct$,	$y = \dfrac{c}{t}$
Astroid	$x = a \cos^3 \theta$,	$y = a \sin^3 \theta$
Cardioid	$x = a(2 \cos \theta - \cos 2\theta)$,	$y = a(2 \sin \theta - \sin 2\theta)$
Cycloid	$x = a(\theta - \sin \theta)$,	$y = a(1 - \cos \theta)$

Sketches showing the approximate shapes of these curves are shown in Fig. 12.3.

12.2 Differentiation in parameters

When x and y are given in terms of a parameter, say, t, then by the chain rule of differentiation:

$$\frac{dy}{dx} = \frac{dy}{dt} \times \frac{dt}{dx}$$

Also, if y is a function of x, then x must be a function of y. The differential coefficient, $\dfrac{dx}{dy}$, is defined as

$$\frac{dx}{dy} = \lim_{\delta y \to 0} \frac{\delta x}{\delta y}$$

Since δx and δy are finite, measurable quantities, the basic rules of fractions apply, and $\dfrac{dx}{dy}$ may be written as

$$\frac{dx}{dy} = \lim_{\delta y \to 0} \frac{1}{\dfrac{\delta y}{\delta x}}$$

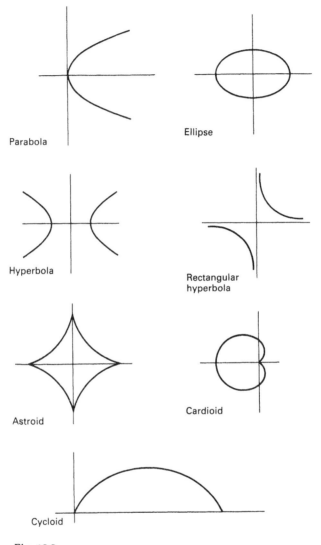

Parabola

Ellipse

Hyperbola

Rectangular
hyperbola

Astroid

Cardioid

Cycloid

Fig. 12.3

As δy approaches zero, δx also approaches zero, hence

$$\frac{dx}{dy} = \lim_{\delta x \to 0} \frac{1}{\dfrac{\delta x}{\delta y}} = \frac{1}{\dfrac{dy}{dx}}, \text{ i.e. } \frac{dy}{dx} = \frac{1}{\dfrac{dx}{dy}}$$

Thus the chain rule can be written as:

$$\frac{dy}{dx} = \frac{\dfrac{dy}{dt}}{\dfrac{dx}{dt}} \qquad (1)$$

The second derivative of y with respect to x can also be expressed in terms of a parameter t as follows:

$$\frac{d^2 y}{dx^2} = \frac{d}{dx}\left(\frac{dy}{dx}\right)$$

Let $u = \dfrac{dy}{dx}$. Then by the chain rule of differentiation and equation (1) above,

$$\frac{du}{dx} = \frac{du}{dt} \times \frac{dt}{dx} = \frac{\dfrac{du}{dt}}{\dfrac{dx}{dt}}$$

Since $u = \dfrac{dy}{dx}$, then $\dfrac{d}{dx}\dfrac{dy}{dx} = \dfrac{d^2 y}{dx^2} = \dfrac{\dfrac{d}{dt}\left(\dfrac{dy}{dx}\right)}{\dfrac{dx}{dt}} \qquad (2)$

Worked problems on differentiation in parameters

Problem 1. If $x = 3t + 1$ and $y = 3t(t - 2)$, find $\dfrac{dy}{dx}$ in terms of parameter t.

$x = 3t + 1$, hence $\dfrac{dx}{dt} = 3$

$y = 3t(t - 2)$, hence $\dfrac{dy}{dt} = 6(t - 1)$

From equation (1), $\dfrac{dy}{dx} = \dfrac{\dfrac{dy}{dt}}{\dfrac{dx}{dt}} = \dfrac{6(t - 1)}{3} = \mathbf{2(t - 1)}$

Problem 2. The equation of a tangent drawn to a curve at point (x_1, y_1) is given by:

$$y - y_1 = \frac{dy_1}{dx_1}(x - x_1)$$

Determine the equation of the tangent drawn to the parabola $x = 3t^2$, $y = 6t$, at the point t

At point t, $x_1 = 3t^2$ and $y_1 = 6t$

$$\frac{dx_1}{dt} = 6t \quad \frac{dy_1}{dt} = 6$$

From equation (1), $\dfrac{dy_1}{dx_1} = \dfrac{\dfrac{dy_1}{dt}}{\dfrac{dx_1}{dt}} = \dfrac{6}{6t} = \dfrac{1}{t}$

Hence, the equation of the tangent is: $y - 6t = \dfrac{1}{t}(x - 3t^2)$

Problem 3. The equation of the normal drawn to a curve at point (x_1, y_1) is given by:

$$y - y_1 = \frac{-1}{\dfrac{dy_1}{dx_1}}(x - x_1)$$

Determine the equation of the normal drawn to the astroid $x = a \cos^3 \theta$, $y = a \sin^3 \theta$ at the point $\theta = \dfrac{\pi}{4}$

$$x = a \cos^3 \theta, \frac{dx}{d\theta} = -3a \cos^2 \theta \sin \theta$$

$$y = a \sin^3 \theta, \frac{dy}{d\theta} = 3a \sin^2 \theta \cos \theta$$

From equation (1), $\dfrac{dy}{dx} = \dfrac{\dfrac{dy}{d\theta}}{\dfrac{dx}{d\theta}} = \dfrac{3a \sin^2 \theta \cos \theta}{-3a \cos^2 \theta \sin \theta}$

$$= -\tan \theta$$

When $\theta = \dfrac{\pi}{4}, \dfrac{dy_1}{dx_1} = -1$

$$x_1 = a\cos^3\frac{\pi}{4} = a\left(\frac{1}{\sqrt{2}}\right)^3 = \frac{a}{2\sqrt{2}}$$

$$y_1 = a\sin^3\frac{\pi}{4} = a\left(\frac{1}{\sqrt{2}}\right)^3 = \frac{a}{2\sqrt{2}}$$

Hence the equation of the normal is:

$$y - \frac{a}{2\sqrt{2}} = \frac{-1}{-1}\left(x - \frac{a}{2\sqrt{2}}\right)$$

i.e. $x = y$

Problem 4. Determine the values of $\dfrac{dy}{dx}$ and $\dfrac{d^2y}{dx^2}$ for the cycloid

$x = 3(\theta - \sin\theta)$, $y = 3(1 - \cos\theta)$ at the point $\theta = \dfrac{\pi}{3}$.

$$x = 3(\theta - \sin\theta), \quad \frac{dx}{d\theta} = 3(1 - \cos\theta)$$

$$y = 3(1 - \cos\theta), \quad \frac{dy}{d\theta} = 3\sin\theta$$

From equation (1) $\dfrac{dy}{dx} = \dfrac{\dfrac{dy}{d\theta}}{\dfrac{dx}{d\theta}} = \dfrac{3\sin\theta}{3(1 - \cos\theta)} = \dfrac{\sin\theta}{1 - \cos\theta}$

When $\theta = \dfrac{\pi}{3}, \dfrac{dy}{dx} = \dfrac{\sin\dfrac{\pi}{3}}{1 - \cos\dfrac{\pi}{3}} = \dfrac{\dfrac{\sqrt{3}}{2}}{\left(1 - \dfrac{1}{2}\right)} = \sqrt{3}$

From equation (2) $\dfrac{d^2y}{dx^2} = \dfrac{\dfrac{d}{d\theta}\left(\dfrac{dy}{dx}\right)}{\dfrac{dx}{d\theta}} = \dfrac{\dfrac{d}{d\theta}\left(\dfrac{\sin\theta}{1 - \cos\theta}\right)}{3(1 - \cos\theta)}$

$$= \frac{(1 - \cos\theta)\cos\theta - \sin\theta(\sin\theta)}{3(1 - \cos\theta)^3}$$

$$= \frac{\cos\theta - (\cos^2\theta + \sin^2\theta)}{3(1 - \cos\theta)^3} = -\frac{4}{3}$$

Problem 5. When determining the surface tension of a liquid, the radius of curvature, ρ, of part of the surface is given by

$$\rho = \frac{\left[1 + \left(\dfrac{dy}{dx}\right)^2\right]^{\frac{3}{2}}}{\dfrac{d^2 y}{dx^2}}$$

Determine the radius of curvature of the part of the surface having the parametric equations $x = 2t^2$, $y = 4t$ at the point $t = 1$.

$$x = 2t^2, \quad \frac{dx}{dt} = 4t$$

$$y = 4t, \quad \frac{dy}{dt} = 4$$

From equation (1), $\dfrac{dy}{dx} = \dfrac{\dfrac{dy}{dt}}{\dfrac{dx}{dt}} = \dfrac{4}{4t} = \dfrac{1}{t}$

From equation (2), $\dfrac{d^2 y}{dx^2} = \dfrac{\dfrac{d}{dt}\left(\dfrac{dy}{dx}\right)}{\dfrac{dx}{dt}} = \dfrac{\dfrac{d}{dt}\left(\dfrac{1}{t}\right)}{\dfrac{dx}{dt}}$

$$= \dfrac{-\dfrac{1}{t^2}}{4t} = -\dfrac{1}{4t^3}$$

$$\rho = \frac{\left[1 + \left(\dfrac{dy}{dx}\right)^2\right]^{\frac{3}{2}}}{\dfrac{d^2 y}{dx^2}} = \frac{\left[1 + \left(\dfrac{1}{t}\right)^2\right]^{\frac{3}{2}}}{-\dfrac{1}{4t^3}}$$

When $t = 1$, $\rho = \dfrac{\sqrt{8}}{-\dfrac{1}{4}} = -4\sqrt{8} = -8\sqrt{2}$

Problem 6. Determine the radius of curvature, ρ, of the cardioid $x = 3(2 \cos \theta - \cos 2\theta)$, $y = 3(2 \sin \theta - \sin 2\theta)$ at the point $\theta = \dfrac{\pi}{6}$

radians. Express the result correct to four decimal places.

$$\left(\rho = \left[1 + \left(\frac{dy}{dx}\right)^2\right]^{\frac{3}{2}} \Big/ \frac{d^2y}{dx^2}\right)$$

$$x = 3(2\cos\theta - \cos 2\theta), \quad y = 3(2\sin\theta - \sin 2\theta)$$

$$\frac{dx}{d\theta} = -6\sin\theta + 6\sin 2\theta, \quad \frac{dy}{d\theta} = 6\cos\theta - 6\cos 2\theta$$

From equation (1), $\dfrac{dy}{dx} = \dfrac{\dfrac{dy}{d\theta}}{\dfrac{dx}{d\theta}} = \dfrac{6\cos\theta - 6\cos 2\theta}{-6\sin\theta + 6\sin 2\theta}$

$$= \frac{\cos\theta - \cos 2\theta}{\sin 2\theta - \sin\theta}$$

From equation (2), $\dfrac{d^2y}{dx^2} = \dfrac{\dfrac{d}{d\theta}\left(\dfrac{dy}{dx}\right)}{\dfrac{dx}{d\theta}}$

$$\frac{d}{d\theta}\left(\frac{dy}{dx}\right) = \frac{\begin{array}{c}(\sin 2\theta - \sin\theta)(-\sin\theta + 2\sin 2\theta) \\ - (\cos\theta - \cos 2\theta)(2\cos 2\theta - \cos\theta)\end{array}}{(\sin 2\theta - \sin\theta)^2}$$

$$= \frac{\left[\begin{array}{c}(-\sin 2\theta\sin\theta + 2\sin^2 2\theta + \sin^2\theta - 2\sin\theta\sin 2\theta) \\ - (2\cos\theta\cos 2\theta - \cos^2\theta - 2\cos^2 2\theta + \cos\theta\cos 2\theta)\end{array}\right]}{(\sin 2\theta - \sin\theta)^2}$$

Since $\sin^2\theta + \cos^2\theta = 1$ and $\sin^2 2\theta + \cos^2 2\theta = 1$

$$\frac{d^2y}{dx^2} = \frac{3 - 3\sin\theta\sin 2\theta - 3\cos\theta\cos 2\theta}{(\sin 2\theta - \sin\theta)^2(6)(\sin 2\theta - \sin\theta)}$$

$$= \frac{1 - \sin\theta\sin 2\theta - \cos\theta\cos 2\theta}{2(\sin 2\theta - \sin\theta)^3}$$

Radius of curvature, $\rho = \left[1 + \left(\dfrac{dy}{dx}\right)^2\right]^{\frac{3}{2}} \Big/ \dfrac{d^2y}{dx^2}$

$$= \frac{\left[1 + \left(\dfrac{\cos\theta - \cos 2\theta}{\sin 2\theta - \sin\theta}\right)^2\right]^{\frac{3}{2}}}{(1 - \sin\theta\sin 2\theta - \cos\theta\cos 2\theta)/2(\sin 2\theta - \sin\theta)^3}$$

When $\theta = \dfrac{\pi}{6}$, $\sin \theta = \dfrac{1}{2}$, $\cos \theta = \dfrac{\sqrt{3}}{2}$, $\sin 2\theta = \dfrac{\sqrt{3}}{2}$ and $\cos 2\theta = \dfrac{1}{2}$

Hence $\rho = \dfrac{\left[1 + \left(\dfrac{\dfrac{\sqrt{3}}{2} - \dfrac{1}{2}}{\dfrac{\sqrt{3}}{2} - \dfrac{1}{2}}\right)^2\right]^{\frac{3}{2}} \left[2\left(\dfrac{\sqrt{3}}{2} - \dfrac{1}{2}\right)^3\right]}{\left(1 - \dfrac{1}{2} \cdot \dfrac{\sqrt{3}}{2} - \dfrac{\sqrt{3}}{2} \cdot \dfrac{1}{2}\right)}$

$= \dfrac{(\sqrt{8})(0.098\,08)}{0.133\,97} = \mathbf{2.070\,7}$, **correct to four decimal places.**

Further problems on differentiating in parameters may be found in the following Section 12.3 (Problems 1 to 20).

12.3   Further problems

Differentiation in parameters

In Problems 1 to 5, determine $\dfrac{dy}{dx}$ in terms of the parameter given

and hence find the value of $\dfrac{dy}{dx}$ at the point stated.

1. The ellipse $x = 2 \cos \theta$, $y = 3 \sin \theta$ at $\theta = \dfrac{\pi}{6}$. $\left[-\dfrac{3\sqrt{3}}{2}\right]$

2. The parabola $x = \dfrac{1}{2}t^2$, $y = t$ at $t = \ln 1.3$, correct to four significant figures.   [3.811]

3. The rectangular hyperbola $x = 4t$, $y = \dfrac{4}{t}$ at $t = 0.25$.   [−16]

4. The hyperbola $x = 2 \sec \theta$, $y = 5 \tan \theta$ at $\theta = 1.4$ radians, correct to four decimal places.   [2.536 9]

5. The cycloid $x = 3(\theta - \cos \theta)$, $y = 3(1 - \cos \theta)$ at $\theta = 0.57$ radians, correct to four decimal places.   [0.350 5]

In Problems 6 to 10, given that the equation of the tangent drawn to a curve at point (x_1, y_1) is $y - y_1 = \dfrac{dy_1}{dx_1}(x - x_1)$ and that the equation of the normal drawn to a curve at point (x_1, y_1) is

$y - y_1 = \dfrac{-1}{\dfrac{dy_1}{dx_1}}(x - x_1)$, determine the equation stated, expressing it

in the form $y = mx + c$.

6. The tangent drawn to the ellipse $x = 2 \cos \theta$, $y = 3 \sin \theta$ at
$\theta = \dfrac{\pi}{3}$. $\left[y = -\dfrac{\sqrt{3}}{2}x + 2\sqrt{3} \right]$

7. The normal to the parabola $x = \dfrac{1}{2}t^2$, $y = t$ at $t = \dfrac{1}{4}$.
$\left[y = -\dfrac{1}{4}x + \dfrac{33}{128} \right]$

8. The tangent to the rectangular hyperbola $x = 4t$, $y = \dfrac{4}{t}$ at $t = 3$.
$\left[y = -\dfrac{1}{9}x + 2\dfrac{2}{3} \right]$

9. The normal to the hyperbola $x = 2 \sec \theta$, $y = 5 \tan \theta$ at
$\theta = \dfrac{7\pi}{6}$ radians, correct to four decimal places.
$[y = 0.200\,0x + 3.348\,6]$

10. The equations of the tangent and normal to the cycloid
$x = 3(\theta - \sin \theta)$, $y = 3(1 - \cos \theta)$ at $\theta = 1.3$ radians, correct
to four decimal places. $\left[\begin{array}{l} \text{Tangent } y = 1.315\,4x + 0.869\,8, \\ \text{normal } \ \ y = 0.760\,2x + 2.964\,8 \end{array} \right]$

In Problems 11 to 15, find the values of $\dfrac{dy}{dx}$ and $\dfrac{d^2y}{dx^2}$ at the points
stated, for the curves represented by the parametric equations
given.

11. The hyperbola $x = 2.3 \sec \theta$, $y = 3.4 \tan \theta$ at $\theta = 1$ radian,
correct to four decimal places. $[1.756\,8, \ -0.391\,3]$
12. The parabola $x = \sqrt{5t^2}$, $y = 2\sqrt{5t}$ at $t = 2.83$, correct to four
significant figures. $[0.353\,4, \ \pm 0.009\,866]$
13. The ellipse $x = 3.7 \cos \theta$, $y = 4.4 \sin \theta$ at $\theta = 1.43$ radians,
correct to four decimal places. $[-0.168\,5, \ -0.331\,1]$
14. The rectangular hyperbola $x = \sqrt{3t}$, $y = \dfrac{\sqrt{3}}{t}$ at $t = 0.18$,
correct to four significant figures. $[-30.86, \ \pm 198.0]$
15. The cardioid $x = 4(2 \cos \theta - \cos 2\theta)$, $y = 4(2 \sin \theta - \sin 2\theta)$
at $\theta = 3.5$ radians, correct to four decimal
places. $[-1.677\,3, \ 7.228\,2]$

16. If $x = \sqrt{2}(1 - \cos\theta)$ and $y = \sqrt{2}\sin\theta$, show that $\dfrac{dy}{dx} = \cot\theta$

and that $\dfrac{d^2y}{dx^2} = \dfrac{-\operatorname{cosec}^3\theta}{\sqrt{2}}$.

17. A point on a curve is given by $x = 7\cot t + 3.5\cos 2t$,

$y = 7\sin t - 3.5\sin 2t$. Express $\dfrac{d^2y}{dx^2}$ in terms of t.

$$\left[\frac{1}{28}\cos^3\left(\frac{t}{2}\right)\sin\frac{3t}{2} \right]$$

18. For the curve described by $x = 2\sqrt{2}t^2$, $y = 2\sqrt{2}t^3$, find $\dfrac{d^2y}{dx^2}$

in terms of t. $\left[\dfrac{3}{8\sqrt{2}t} \text{ or } \dfrac{3\sqrt{2}}{16t} \right]$

In Problems 19 and 20, determine the radius of curvature, ρ, for the curves stated in terms of their parameters, given that

$$\rho = \frac{\left[1 + \left(\dfrac{dy}{dx} \right)^2 \right]^{\frac{3}{2}}}{\dfrac{d^2y}{dx^2}}$$

19. The astroid, $x = 5\cos^3 t$, $y = 5\sin^3 t$.
     $[15\sin t\cos t \text{ or } 7.5\sin 2t]$

20. The curve, $x = 3\sin 2\theta(1 + \cos 2\theta)$, $y = 3\cos 2\theta(1 - \cos 2\theta)$.
     $[12\cos 3\theta]$

Chapter 13

Logarithmic differentiation

13.1 The laws of logarithms applied to functions

The laws of logarithms may be used to change problems involving products to additions, quotients to subtractions and powers to products. They are stated mathematically as:

$$\log(A \cdot B) = \log A + \log B$$

$$\log\left(\frac{A}{B}\right) = \log A - \log B$$

and $\qquad \log A^n = n \log A,$

where 'log' means the logarithm to any base. In calculus, logarithms to the base of 'e' are invariably used. Also, the constants A and B can be changed to functions of x, and the laws stated above are still true. Thus for two functions of x, $f(x)$ and $g(x)$, the laws may be expressed as:

$$\ln(f(x) \cdot g(x)) = \ln f(x) + \ln g(x) \qquad (1)$$

$$\ln\left(\frac{f(x)}{g(x)}\right) = \ln f(x) - \ln g(x) \qquad (2)$$

and $\qquad \ln[f(x)]^n = n \ln f(x) \qquad (3)$

If $y = (x^3 + 4) \sin 2x$, let $(x^3 + 4) = f(x)$ and $\sin 2x = g(x)$. Taking logarithms to the base of e of each side of the equation gives:

$\ln y = \ln[(x^3 + 4) \cdot \sin 2x]$, and by applying law (1) above:

$\ln y = \ln(x^3 + 4) + \ln(\sin 2x)$

Functions may be made up of a mixture of products and quotients. When logarithms are applied to numbers,

$$\log\left(\frac{A \cdot B}{C}\right) = \log(A \cdot B) - \log C$$

$$= \log A + \log B - \log C$$

197

Applying this principle to functions of x, say $f(x)$, $g(x)$ and $F(x)$ gives:

$$y = \frac{f(x) \cdot g(x)}{F(x)}$$

Then, $\ln y = \ln f(x) + \ln g(x) - \ln F(x)$

For a more complicated expression such as

$$y = \frac{f(x) \cdot [g(x)]^n}{F(x) \cdot [G(x)]^m}$$

the three laws of logarithms may be applied as follows:

$$\ln y = \ln \left\{ \frac{f(x) \cdot [g(x)]^n}{F(x) \cdot [G(x)]^m} \right\}$$

Applying law (2),

$$\ln y = \ln\{f(x) \cdot [g(x)]^n\} - \ln\{F(x) \cdot [G(x)]^m\}$$

Applying law (1),

$$\ln y = \{\ln f(x) + \ln[g(x)]^n\} - \{\ln F(x) + \ln[G(x)]^m\}$$
$$= \ln f(x) + \ln[g(x)]^n - \ln F(x) - \ln[G(x)]^m$$

Applying law (3),

$$\ln y = \ln f(x) + n \ln g(x) - \ln F(x) - m \ln G(x)$$

These principles are used in the worked problems following, and also form the basis of logarithmic differentiation introduced in Section 13.2.

Worked problems on the laws of logarithms applied to functions

In Worked Problems 1 and 2 below, by taking Napierian logarithms of each side of the equations given, convert the functions from 'product, quotient, power' form to 'addition, subtraction, product' form.

Problem 1. (a) $y = (x^2 + 3) \cos 2x$ (b) $p = \dfrac{3 \tan 2q}{4 \, e^{(q/2) - 3}}$

(c) $\alpha = 4 \sin^3 \dfrac{\theta}{2}$

(a) $y = (x^2 + 3) \cos 2x$

Taking logarithms to the base of e of each side of the equation gives:

$\ln y = \ln[(x^2 + 3) \cdot \cos 2x]$

Applying law (1), **$\ln y = \ln(x^2 + 3) + \ln \cos 2x$**

(b) $p = \dfrac{3 \tan 2q}{4\, e^{(q/2) - 3}}$

Taking Napierian logarithms of each side of the equation gives:

$\ln p = \ln\left\{\dfrac{3 \tan 2q}{4\, e^{(q/2) - 3}}\right\}$

Applying law (2), $\ln p = \ln(3 \tan 2q) - \ln(4\, e^{(q/2) - 3})$

Applying law (1), $\ln p = \ln 3 + \ln \tan 2q - \ln 4 - \ln e^{(q/2) - 3}$

Applying law (3), $\ln p = \ln 3 + \ln \tan 2q - \ln 4 - \left(\dfrac{q}{2} - 3\right)\ln e$.

From the definition of a logarithm to any base, a, if

$\qquad y = a^x$ \hfill (1)

then $x = \log_a y$ \hfill (2)

When $a = y$, giving $\log_a a$ in equation (2), then from equation (1),

$\qquad y = y^x$, i.e. $x = 1$

It follows that $\log_a a = 1$, where a is any value, thus

$\qquad \ln e = \log_e e = 1$

Hence, **$\ln p = \ln 3 + \ln \tan 2q - \ln 4 - \left(\dfrac{q}{2} - 3\right)$**

(c) $\alpha = 4 \sin^3 \dfrac{\theta}{2}$

$\ln \alpha = \ln\left(4 \sin^3 \dfrac{\theta}{2}\right)$

$\quad\ \ = \ln\left[4\left(\sin \dfrac{\theta}{2}\right)^3\right]$

Applying law (1), $\ln \alpha = \ln 4 + \ln\left(\sin \dfrac{\theta}{2}\right)^3$

Applying law (3), **$\ln \alpha = \ln 4 + 3 \ln \sin \dfrac{\theta}{2}$**

Problem 2. (a) $(2x^2 - 3x)^{\frac{1}{2}} \sec^3 x$ (b) $\dfrac{x^2 \sin^2 x}{e^x(3 - 4x^{\frac{1}{2}})^5}$

(c) $\dfrac{e^{3x} \ln(x^2)}{\ln(x^3) \operatorname{cosec}^3 x}$

(a) Let $y = (2x^2 - 3x)^{\frac{1}{2}} \sec^3 x$

Taking logarithms to the base of e of each side of the equation gives:

$$\ln y = \ln[(2x^2 - 3x)^{\frac{1}{2}} \sec^3 x]$$

Applying law (1), $\ln y = \ln(2x^2 - 3x)^{\frac{1}{2}} + \ln \sec^3 x$

Applying law (3), $\ln y = \dfrac{1}{2}\ln(2x^2 - 3x) + 3\ln \sec x$

i.e. $\ln y = \dfrac{1}{2}\ln x(2x - 3) + 3\ln \sec x$

Applying law (1) to the first term,

$$\ln y = \frac{1}{2}\ln x + \frac{1}{2}\ln(2x - 3) + 3\ln \sec x$$

(b) Let $y = \dfrac{x^2 \sin^2 x}{e^x(3 - 4x^{\frac{1}{2}})^5}$

Taking logarithms to the base of e of each side of the equation gives:

$$\ln y = \ln\left[\frac{x^2 \sin^2 x}{e^x(3 - 4x^{\frac{1}{2}})^5}\right]$$

Applying law (2), $\ln y = \ln(x^2 \sin^2 x) - \ln[e^x(3 - 4x^{\frac{1}{2}})^5$

Applying law (1), $\ln y = (\ln x^2 + \ln \sin^2 x)$
$\qquad\qquad\qquad\qquad - [\ln e^x + \ln(3 - 4x^{\frac{1}{2}})^5]$

Applying law (3), $\ln y = 2\ln x + 2\ln \sin x$
$\qquad\qquad\qquad\qquad - x\ln e - 5\ln(3 - 4x^{\frac{1}{2}})$

But $\ln e = 1$,

hence $\ln y = 2(\ln x + \ln \sin x) - x - 5\ln(3 - 4x^{\frac{1}{2}})$

(c) Let $y = \dfrac{e^{3x} \ln(x^2)}{\ln(x^3) \operatorname{cosec}^3 x}$

Taking logarithms to the base of e of each side of the equation gives:

$$\ln y = \ln\left[\frac{e^{3x} \ln(x^2)}{\ln(x^3) \operatorname{cosec}^3 x}\right]$$

Applying law (3), $\ln y = \ln\left[\dfrac{e^{3x} \cdot 2 \, \cancel{\ln x}}{3 \, \cancel{\ln x} \cdot \text{cosec}^3 x}\right]$

$= \ln\left[\dfrac{2 \, e^{3x}}{3 \, \text{cosec}^3 x}\right]$

Applying law (2), $\ln y = \ln 2 + \ln e^{3x} - \ln 3 - \ln \text{cosec}^3 x$

Applying law (3), $\ln y = \ln 2 + 3x \ln e - \ln 3 - 3 \ln \text{cosec} \, x$

But $\ln e = 1$,

Hence, $\ln y = \ln 2 + 3x - \ln 3 - 3 \ln \text{cosec} \, x$

Applying law (2), $\mathbf{\ln y = \ln \dfrac{2}{3} + 3x - 3 \ln \text{cosec} \, x}$

Further problems on the laws of logarithms applied to functions may be found in Section 13.3 (Problems 1 to 5), page 205.

13.2   Logarithmic differentiation

To differentiate a function which contains products and/or quotients and/or powers, it is often easier to firstly take logarithms to a base of e and to express the function in a different form. For example, a simple function such as, say, $y = x^2 \sin x$ may be differentiated using the product rule of differentiation, giving:

$$\frac{dy}{dx} = x^2 \cos x + 2x \sin x$$

An alternative method of differentiating this function is by logarithmic differentiation as follows:

Taking Napierian logarithms of each side of the equation gives:

$\ln y = \ln(x^2 \sin x)$

i.e. $\ln y = 2 \ln x + \ln \sin x$

from the laws of logarithms introduced in Section 13.1.

In Chapter 11, dealing with implicit functions, it is shown that

$$\frac{d}{dx}(f(y)) = \frac{d}{dy}(f(y)) \times \frac{dy}{dx}$$

Thus the left-hand side, i.e. $\ln y$, becomes $\dfrac{1}{y}\dfrac{dy}{dx}$ when differentiated with respect to x. With logarithmic differentiation, the right-hand side of the equation is of the form $\ln f(x) \pm \ln g(x) \pm \ldots$. To differentiate, say, $\ln f(x)$ with respect to x, a substitution method is used.

Thus, let $u = f(x)$, then $\dfrac{du}{dx} = f'(x)$

Hence, $\dfrac{d}{dx}(\ln f(x)) = \dfrac{d}{du}(\ln u) \times \dfrac{du}{dx}$

$$= \dfrac{1}{u}\dfrac{du}{dx}$$

i.e. $\dfrac{d}{dx}(\ln f(x)) = \dfrac{f'(x)}{f(x)}$ (4)

Applying equation (4) to the right-hand side, i.e. $2\ln x + \ln \sin x$ gives: $\dfrac{2}{x} + \dfrac{\cos x}{\sin x}$ when it is differentiated with respect to x. Thus, applying logarithmic differentiation to the equation $y = x^2 \sin x$, gives:

$$\dfrac{1}{y}\dfrac{dy}{dx} = \dfrac{2}{x} + \dfrac{\cos x}{\sin x}$$

i.e. $\dfrac{dy}{dx} = y\left(\dfrac{2}{x} + \dfrac{\cos x}{\sin x}\right)$

But $y = x^2 \sin x$, hence

$$\dfrac{dy}{dx} = x^2 \sin x\left(\dfrac{2}{x} + \dfrac{\cos x}{\sin x}\right)$$

$$= 2x \sin x + x^2 \cos x,$$

as obtained by applying the product rule of differentiation.

In this case, using a comparatively simple function, the process of logarithmic differentiation is longer than that of applying the product rule of differentiation. However, for more complicated functions, logarithmic differentiation is usually the simplest method of differentiating the function.

Procedure for logarithmic differentiation

The procedure for differentiating a function of the form, say, $y = \dfrac{f(x)[g(x)]^n}{h(x)}$ is shown below.

(i) Take Napierian logarithms of each side of the equation, giving

$$\ln y = \left\{\dfrac{f(x)[g(x)]^n}{h(x)}\right\}$$

(ii) Apply the laws of logarithms to change products to addition, etc. This gives:

$$\ln y = \ln f(x) + n \ln g(x) - \ln h(x)$$

(iii) Differentiate each side with respect to the variable, giving:

$$\frac{1}{y}\frac{dy}{dx} = \frac{f'(x)}{f(x)} + \frac{ng'(x)}{g(x)} - \frac{h'(x)}{h(x)}$$

(iv) Multiply throughout by y, giving

$$\frac{dy}{dx} = y\left\{\frac{f'(x)}{f(x)} + \frac{ng'(x)}{g(x)} - \frac{h'(x)}{h(x)}\right\}$$

(v) Substitute for y in terms of x, giving:

$$\frac{dy}{dx} = \frac{f(x)[g(x)]^n}{h(x)}\left\{\frac{f'(x)}{f(x)} + \frac{ng'(x)}{g(x)} - \frac{h'(x)}{h(x)}\right\}$$

(vi) Simplify the expression obtained in (v) where possible.

This procedure is used to determine the differential coefficients in the worked problems following.

Worked problems on logarithmic differentiation

Problem 1. Use logarithmic differentiation to differentiate

$$y = \frac{3(4 - x^2)}{\tan x}$$

With reference to the procedure for logarithmic differentiation given above:

(i) $\ln y = \ln\left\{\dfrac{3(4 - x^2)}{\tan x}\right\}$

(ii) $\ln y = \ln 3 + \ln(4 - x^2) - \ln \tan x$

(iii) $\dfrac{1}{y}\dfrac{dy}{dx} = 0 + \dfrac{(-2x)}{4 - x^2} - \dfrac{\sec^2 x}{\tan x}$

(iv) $\dfrac{dy}{dx} = y\left\{\dfrac{-2x}{4 - x^2} - \dfrac{\sec^2 x}{\tan x}\right\}$

(v) $\dfrac{dy}{dx} = \dfrac{3(4 - x^2)}{\tan x}\left\{\dfrac{-2x}{4 - x^2} - \dfrac{\sec^2 x}{\tan x}\right\}$

(vi) $\dfrac{dy}{dx} = \dfrac{-6x}{\tan x} - \dfrac{3(4 - x^2)\sec^2 x}{\tan^2 x}$

$\qquad = -\dfrac{3}{\tan x}\left\{2x + \dfrac{(4 - x^2)\sec^2 x}{\tan x}\right\}$

$\qquad = -3\cot x\left\{2x + \dfrac{4 - x^2}{\sin x \cos x}\right\}$

Problem 2. Use logarithmic differentiation to differentiate

(a) $y = \dfrac{4\,e^{-2x}\sec x}{\left(x^2 + \dfrac{1}{2}\right)^{\frac{3}{2}}}$ (b) $p = \dfrac{q^3 \ln 2q}{(2 - q)^{\frac{1}{3}}\operatorname{cosec} 2q}$

(a) $y = \dfrac{4\,e^{-2x}\sec x}{\left(x^2 + \dfrac{1}{2}\right)^{\frac{3}{2}}}$

Applying the laws of logarithms:

$\ln y = \ln 4 + (-2x)\ln e + \ln \sec x - \dfrac{3}{2}\ln\left(x^2 + \dfrac{1}{2}\right)$

Differentiating each term gives:

$\dfrac{1}{y}\dfrac{dy}{dx} = -2 + \dfrac{\sec x \tan x}{\sec x} - \dfrac{3(2x)}{2\left(x^2 + \dfrac{1}{2}\right)}$

$\dfrac{dy}{dx} = \dfrac{4\,e^{-2x}\sec x}{\left(x^2 + \dfrac{1}{2}\right)^{\frac{3}{2}}}\left[-2 + \tan x - \dfrac{3x}{\left(x^2 + \dfrac{1}{2}\right)}\right]$

(b) $p = \dfrac{q^3 \ln 2q}{(2 - q)^{\frac{1}{3}}\operatorname{cosec} 2q}$

$\ln p = 3\ln q + \ln(\ln 2q) - \dfrac{1}{3}\ln(2 - q) - \ln \operatorname{cosec} 2q$

$\dfrac{1}{p}\dfrac{dp}{dq} = \dfrac{3}{q} + \dfrac{\dfrac{2}{2q}}{\ln 2q} - \dfrac{(-1)}{3(2 - q)} - \dfrac{-2\operatorname{cosec} 2q \cot 2q}{\operatorname{cosec} 2q}$

$\dfrac{dp}{dq} = \dfrac{q^3 \ln 2q}{(2 - q)^{\frac{1}{3}}\operatorname{cosec} 2q}\left\{\dfrac{3}{q} + \dfrac{1}{q \ln 2q} + \dfrac{1}{3(2 - q)} + 2\cot 2q\right\}$

Further problems on logarithmic differentiation may be found in the following Section 13.3 (Problems 6 to 17).

13.3   Further problems

Laws of logarithms applied to functions

In Problems 1 to 5, by taking Napierian logarithms of each side of the equation, convert the functions from 'product, quotient, power' form to 'addition, subtraction, product' form.

1. $y = (3x - 4) \tan 2x$ $[\ln y = \ln(3x - 4) + \ln(\tan 2x)]$

2. $p = \dfrac{3 \cot 2q}{e^{3-q}}$ $[\ln p = \ln 3 + \ln \cot 2q - (3 - q)]$

3. $m = \dfrac{n^3 e^{2n}}{\operatorname{cosec} \dfrac{n}{2}}$ $\left[\ln m = 3 \ln n + 2n - \ln \operatorname{cosec} \dfrac{n}{2} \right]$

4. $y = \dfrac{2(1 + x^{\frac{3}{2}})^{\frac{1}{2}} \cos(x - 2)}{3\, e^{3x+4} \ln 2x}$

$\left[\ln y = \ln 2 + \dfrac{1}{2}\ln(1 + x^{\frac{3}{2}}) + \ln \cos(x - 2) \right.$

$\left. \qquad\qquad\qquad - \ln 3 - (3x + 4) - \ln(\ln 2x) \right]$

5. $\alpha = \dfrac{3\, \theta^{\frac{3}{2}} \sin^3 \theta}{(2 - \theta^{\frac{1}{2}})^{\frac{1}{3}} \sec^3 \theta}$

$\left[\ln \alpha = \ln 3 + \dfrac{3}{2} \ln \theta + 3 \ln \sin \theta - \dfrac{1}{3}\ln(2 - \theta^{\frac{1}{2}}) - 3 \ln \sec \theta \right]$

Logarithmic differentiation

In Problems 6 to 11, use logarithmic differentiation to differentiate the functions given.

6. $y = 3\sqrt{x} \sin 2x$ $\left[\dfrac{dy}{dx} = 3\sqrt{x}\left(\dfrac{\sin 2x}{2x} + 2 \cos 2x \right) \right]$

7. $l = 4\, e^{2 - 3m} \operatorname{cosec} \dfrac{m}{2}$

$\left[\dfrac{dl}{dm} = -4\, e^{2 - 3m} \operatorname{cosec} \dfrac{m}{2}\left(3 + \dfrac{1}{2} \cot \dfrac{m}{2} \right) \right]$

8. $p = \dfrac{2(1 - q^2)}{3 \sec(\pi - q)}$

$\left[\dfrac{dp}{dq} = \dfrac{2}{3 \sec(\pi - q)}\left\{ (1 - q^2) \tan(\pi - q) - 2q \right\} \right]$

9. $r = \dfrac{2 \ln\left(4 - \dfrac{s}{3}\right)}{3 \cot(s^2)}$

$$\left[\frac{dr}{ds} = \frac{2}{3 \cot(s^2)}\left\{-\frac{1}{12 - s} + \frac{2s \ln\left(4 - \dfrac{s}{3}\right)}{\sin s^2 \cos s^2}\right\}\right]$$

10. $u = 1.8\sqrt{\sin^3 v}$ $\left[\dfrac{du}{dv} = 2.7\sqrt{(\sin v)}\cos v\right]$

11. $\alpha = \dfrac{4}{9}[3 - (2\theta)^{\frac{1}{3}}]^{\frac{1}{2}}$ $\left[\dfrac{d\alpha}{d\theta} = -\dfrac{4}{27}\left\{\dfrac{1}{(2\theta)^{\frac{2}{3}}(3 - (2\theta)^{\frac{1}{3}})^{\frac{1}{2}}}\right\}\right]$

12. If $\dfrac{y}{e^{2x}} = \sin^3(4 - 2x)$, find $\dfrac{dy}{dx}$.

$$\left[\frac{dy}{dx} = 2\,e^{2x}\sin^3(4 - 2x)\{1 - 3\cot(4 - 2x)\}\right]$$

13. If $\dfrac{r}{\sec^2 \dfrac{x}{2}} = \ln 3x \tan 3x$, show that

$$\frac{dr}{dx} = \ln 3x \cdot \sec^2 \frac{x}{2}\tan 3x\left\{\frac{1}{x \ln 3x} + \tan\frac{x}{2} + \frac{3\sec^2 3x}{\tan 3x}\right\}$$

14. If $\dfrac{p}{4\left(3 - \dfrac{q}{4}\right)^{\frac{1}{4}}} = e^{2q-4}\sqrt{(\cos(3 - 2q))}$, find $\dfrac{dp}{dq}$.

$$\left[\frac{dp}{dq} = 4\left(3 - \frac{q}{4}\right)^{\frac{1}{4}} e^{2q-4}\sqrt{(\cos(3 - 2q))}\right.$$
$$\left.\times \left\{2 + \tan(3 - 2q) - \frac{1}{4(12 - q)}\right\}\right]$$

15. If $y\left(\ln\dfrac{x}{4}\right)^2 = (x \sin x)^3$, show that

$$\frac{dy}{dx} = \frac{(x \sin x)^3}{\left(\ln\dfrac{x}{4}\right)^2}\left\{\frac{3}{x} + 3\cot x - \frac{2}{x \ln\dfrac{x}{4}}\right\}$$

16. If $u(v - 1)^{\frac{1}{2}} \cot^3 v = v^3 \sec^2 v$, find $\dfrac{du}{dv}$.

$$\left[\frac{du}{dv} = \frac{v^3 \sec^2 v}{(v - 1)^{\frac{1}{2}} \cot^3 v} \left\{ \frac{3}{v} + 2 \tan v - \frac{1}{2(v - 1)} + \frac{3}{\sin v \cos v} \right\} \right]$$

17. If $p \, e^{q/3} \sqrt{\left(\dfrac{q}{2}\right)} = 16 \cot^2 q$, show that

$$\frac{dp}{dq} = \frac{-16 \cot^2 q}{\sqrt{\left(\dfrac{q}{2}\right)} e^{q/3}} \left\{ \frac{2}{\sin q \cos q} + \frac{1}{2q} + \frac{1}{3} \right\}$$

Chapter 14

Differentiation of inverse trigonometric and inverse hyperbolic functions

14.1 Inverse functions

If y is a function of x, i.e. $y = f(x)$, it is often possible to transpose the equation to find x in terms of y.

For example, if $y = 2x - 1$ then $x = \dfrac{y + 1}{2}$

The two functions, i.e. $y = 2x - 1$ and $x = \dfrac{y + 1}{2}$, are called **inverse functions**.

Further examples of inverse functions are:

(i)   if $y = x^2$ then $x = \sqrt{y}$,
(ii) if $y = 10^x$ then $x = \log_{10} y$,
(iii) if $y = \sin x$ then x is the inverse sine of y, which is written as $x = \arcsin y$,
(iv) if $y = \cosh x$ then x is the inverse hyperbolic cosine of y, which is written as $x = \operatorname{arcosh} y$.

The inverse circular functions are denoted by prefixing the function with 'arc'. For example, $\arcsin x$, $\operatorname{arcsec} x$, etc. The former notation $\sin^{-1} x$, $\tan^{-1} x$, etc., is discouraged because of possible confusion with $(\sin x)^{-1}$, $(\tan x)^{-1}$, etc.

The inverse hyperbolic functions are denoted by prefixing the function with 'ar'. For example, $\operatorname{arsinh} x$, $\operatorname{arcosech} x$, etc. The former notation $\sinh^{-1} x$, $\coth^{-1} x$, etc., is discouraged because of possible confusion with $(\sinh x)^{-1}$, $(\coth x)^{-1}$, etc.

14.2 Differentiation of inverse trigonometric functions

(a) $y = \arcsin x$

If $y = \arcsin x$ then $x = \sin y$.

Differentiating x with respect to y gives $\dfrac{dx}{dy} = \cos y$

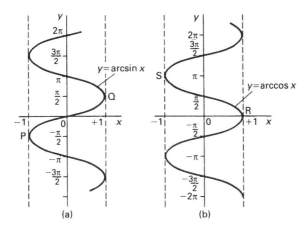

Fig. 14.1

Now $\cos^2 y + \sin^2 y = 1$. Hence $\cos y = \sqrt{(1 - \sin^2 y)} = \sqrt{(1 - x^2)}$.

Hence $\dfrac{dx}{dy} = \sqrt{(1 - x^2)}$.

It may be shown that $\dfrac{dy}{dx} = \dfrac{1}{dx/dy}$ (see Chapter 12, page 188).

Thus when $y = \arcsin x$, $\dfrac{dy}{dx} = \dfrac{1}{\sqrt{(1 - x^2)}}$ and there are two possible values, one positive and one negative.

A sketch of part of the graph $y = \arcsin x$ is shown in Fig. 14.1(a) and it is seen that there are an infinite number of values for y for a given value of x. When y has more than one value for a given value of x, such as the relationship $y = \arcsin x$, the numerically least of these values is called the **principal value** of y. If there are two numerically equal least values, the positive one is called the principal value.

The principal value of $\arcsin x$ is defined as the value between $-\dfrac{\pi}{2}$ and $+\dfrac{\pi}{2}$, shown between points P and Q in Fig. 14.1(a). This range of values covers every possible value of x that can occur and when adopting principal values, if given a particular value of x, only one value of y is possible. The gradient of the curve between P and Q is positive for all values of x, and hence if $\arcsin x$ is understood to mean the principal value of $\arcsin x$ then the differential coefficient will be positive and only the positive value is taken when evaluating $\sqrt{(1 - x^2)}$.

Hence if $y = \arcsin x$ then $\dfrac{dy}{dx} = \dfrac{1}{\sqrt{(1 - x^2)}}$

If $y = \arcsin \dfrac{x}{a}$ then $\dfrac{x}{a} = \sin y$

i.e. $x = a \sin y$

$$\frac{dx}{dy} = a \cos y = a\sqrt{(1 - \sin^2 y)} = a\sqrt{\left[1 - \left(\frac{x}{a}\right)^2\right]}$$

i.e. $$\frac{dx}{dy} = a\sqrt{\left(\frac{a^2 - x^2}{a^2}\right)}$$

$$= a\frac{\sqrt{(a^2 - x^2)}}{a} = \sqrt{(a^2 - x^2)}$$

Hence $\dfrac{dy}{dx} = \dfrac{1}{dx/dy} = \dfrac{1}{\sqrt{(a^2 - x^2)}}$

This latter form is important when integrating functions of the form

$$\frac{1}{\sqrt{(a^2 - x^2)}}, \text{ i.e. } \int \frac{1}{\sqrt{(a^2 - x^2)}}\, dx = \arcsin \frac{x}{a} + c$$

(This is discussed in Chapter 18, page 282.)

The most general form of the differential coefficient of arcsin $f(x)$ is obtained by using the 'chain rule' of differentiation for 'functions of a function'. That is, given $y = \arcsin f(x)$, let $u = f(x)$

then $\dfrac{du}{dx} = f'(x)$

$$\frac{dy}{du} = \frac{1}{\sqrt{(1 - u^2)}}$$

Thus $\dfrac{dy}{dx} = \dfrac{dy}{du} \cdot \dfrac{du}{dx} = \dfrac{1}{\sqrt{(1 - u^2)}} f'(x),$

i.e. $\dfrac{dy}{dx} = \dfrac{1}{\sqrt{\{1 - [f(x)]^2\}}} \cdot f'(x)$

(b) $y = \arccos x$

If $y = \arccos x$ then $x = \cos y$

Differentiating x with respect to y gives:

$$\frac{dx}{dy} = -\sin y = -\sqrt{(1 - \cos^2 y)}$$

$$= -\sqrt{(1 - x^2)}$$

Hence $\dfrac{dy}{dx} = \dfrac{1}{dx/dy} = -\dfrac{1}{\sqrt{(1-x^2)}}$

A sketch of part of the graph $y = \arccos x$ is shown in Fig. 14.1(b) and $\arccos x$ is defined as that angle which lies between 0 and π shown between points R and S, i.e. the principal value of $y = \arccos x$ is between 0 and π. (If the range $-\dfrac{\pi}{2}$ to $+\dfrac{\pi}{2}$ were used, as in the case of $\arcsin x$, then the cosine of an angle would always be positive. Thus $\arccos\left(-\dfrac{1}{3}\right)$, for example, would have no meaning.) The gradient of the curve between R and S is negative for all values of x and hence the differential coefficient is negative as shown above.

It may be shown that if $y = \arccos \dfrac{x}{a}$ then $\dfrac{dy}{dx} = -\dfrac{1}{\sqrt{(a^2 - x^2)}}$

The most general form of the differential coefficient of $\arccos f(x)$ is:

$$\dfrac{d}{dx}[\arccos f(x)] = -\dfrac{1}{\sqrt{\{1 - [f(x)]^2\}}}[f'(x)]$$

(c) $y = \arctan x$

A sketch of part of the graph of $y = \arctan x$ is shown in Fig. 14.2(a). By definition the principal value of $y = \arctan x$ is defined as that angle which lies between $-\dfrac{\pi}{2}$ and $+\dfrac{\pi}{2}$. Hence the gradient $\left(\text{i.e. } \dfrac{dy}{dx}\right)$ is always positive.

If $y = \arctan x$, then $x = \tan y$

$$\dfrac{dx}{dy} = \sec^2 y$$

Now $1 + \tan^2 y = \sec^2 y$

Hence $\dfrac{dx}{dy} = 1 + \tan^2 y = 1 + x^2$

$$\dfrac{dy}{dx} = \dfrac{1}{dx/dy} = \dfrac{1}{1 + x^2}$$

If $y = \arctan \dfrac{x}{a}$, then $\dfrac{x}{a} = \tan y$, i.e. $x = a \tan y$

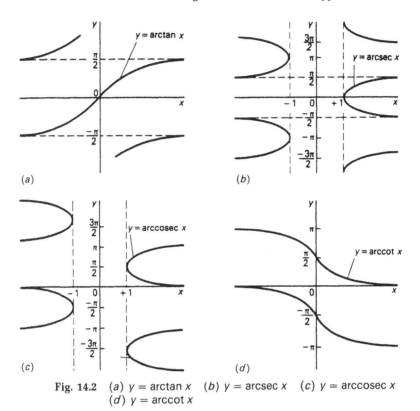

Fig. 14.2   (a) $y = \arctan x$ (b) $y = \text{arcsec } x$ (c) $y = \text{arccosec } x$
(d) $y = \text{arccot } x$

$$\frac{dx}{dy} = a \sec^2 y = a(1 + \tan^2 y) = a\left[1 + \left(\frac{x}{a}\right)^2\right]$$

$$= a\left(\frac{a^2 + x^2}{a^2}\right) = \frac{a^2 + x^2}{a}$$

Hence $\dfrac{dy}{dx} = \dfrac{a}{a^2 + x^2}$

The latter form is important when integrating functions of the form

$$\frac{a}{a^2 + x^2},$$

i.e. $\displaystyle\int \frac{a}{a^2 + x^2}\,dx = \arctan\frac{x}{a} + c$ or $\displaystyle\int \frac{1}{a^2 + x^2}\,dx = \frac{1}{a}\arctan\frac{x}{a} + c$

(This is discussed in Chapter 18, page 284.)

The most general form of the differential coefficient of arctan $f(x)$ is:

$$\frac{d}{dx}[\arctan f(x)] = \frac{1}{\{1 + [f(x)]^2\}}[f'(x)]$$

(d) $y = \operatorname{arcsec} x$

A sketch of part of the graph of $y = \operatorname{arcsec} x$ is shown in Fig. 14.2(b). It is a discontinuous curve with no part of it between $x = -1$ and $x = +1$. By definition, the principal value of $y = \operatorname{arcsec} x$ is defined as that angle which lies between 0 and π. In this range the gradient of the curve is always positive.

If $y = \operatorname{arcsec} x$ then $x = \sec y$

$$\frac{dx}{dy} = \sec y \tan y$$

Now $1 + \tan^2 y = \sec^2 y$. Hence $\tan y = \sqrt{(\sec^2 y - 1)} = \sqrt{(x^2 - 1)}$

Thus $\quad \dfrac{dx}{dy} = \sec y \tan y = x\sqrt{(x^2 - 1)}$

$$\frac{dy}{dx} = \frac{1}{dx/dy} = \frac{1}{x\sqrt{(x^2 - 1)}}$$

By a similar method it may be shown that if $y = \operatorname{arcsec} \dfrac{x}{a}$, then

$$\frac{dy}{dx} = \frac{a}{x\sqrt{(x^2 - a^2)}}$$

The most general form of the differential coefficient of arcsec $f(x)$ is:

$$\frac{d}{dx}[\operatorname{arcsec} f(x)] = \frac{1}{f(x)\sqrt{\{[f(x)]^2 - 1\}}}[f'(x)]$$

(e) $y = \operatorname{arccosec} x$

A sketch of the part of the graph of $y = \operatorname{arccosec} x$ is shown in Fig. 14.2(c). It is a discontinuous curve with no part of it between $x = -1$ and $x = +1$. By definition, the principal value of $y = \operatorname{arccosec} x$ is defined as that angle which lies between $-\dfrac{\pi}{2}$ and $+\dfrac{\pi}{2}$. In this range the gradient of the curve is always negative.

If $y = \operatorname{arccosec} x$ then $x = \operatorname{cosec} y$

$$\frac{dx}{dy} = -\operatorname{cosec} y \cot y$$

Now $\cot^2 y + 1 = \operatorname{cosec}^2 y$. Hence $\cot y = \sqrt{(\operatorname{cosec}^2 y - 1)} = \sqrt{(x^2 - 1)}$

Thus $\quad \dfrac{dx}{dy} = -\operatorname{cosec} y \cot y = -x\sqrt{(x^2 - 1)}$

$$\frac{dy}{dx} = \frac{1}{dx/dy} = -\frac{1}{x\sqrt{(x^2 - 1)}}$$

By a similar method it may be shown that if $y = \text{arccosec } \dfrac{x}{a}$, then

$$\frac{dy}{dx} = \frac{-a}{x\sqrt{(x^2 - a^2)}}$$

The most general form of the differential coefficient of arccosec $f(x)$ is:

$$\frac{d}{dx}[\text{arccosec } f(x)] = -\frac{1}{f(x)\sqrt{\{[f(x)]^2 - 1\}}}[f'(x)]$$

(f) $y = \text{arccot } x$

A sketch of part of the graph of $y = \text{arccot } x$ is shown in Fig. 14.2(d). By definition, the principal value of $y = \text{arccot } x$ is defined as that angle which

lies between $-\dfrac{\pi}{2}$ and $+\dfrac{\pi}{2}$

In this range the gradient of the curve is always negative.
If $y = \text{arccot } x$ then $x = \cot y$

$$\frac{dx}{dy} = -\text{cosec}^2 y$$

Now $\cot^2 y + 1 = \text{cosec}^2 y$. Hence $\dfrac{dx}{dy} = -(\cot^2 y + 1) = -(x^2 + 1)$

Thus $\quad \dfrac{dy}{dx} = \dfrac{1}{dx/dy} = -\dfrac{1}{(1 + x^2)}$

By a similar method it may be shown that if $y = \text{arccot } \dfrac{x}{a}$, then

$$\frac{dy}{dx} = \frac{-a}{(a^2 + x^2)}$$

The most general form of the differential coefficient of arccot $f(x)$ is:

$$\frac{d}{dx}[\text{arccot } f(x)] = -\frac{1}{\{1 + [f(x)]^2\}}[f'(x)]$$

Summary of differential coefficients of inverse trigonometrical functions

y or $f(x)$	$\dfrac{dy}{dx}$ or $f'(x)$
(a) $\arcsin \dfrac{x}{a}$	$\dfrac{1}{\sqrt{(a^2 - x^2)}}$
$\arcsin f(x)$	$\dfrac{1}{\sqrt{\{1 - [f(x)]^2\}}} [f'(x)]$
(b) $\arccos \dfrac{x}{a}$	$\dfrac{-1}{\sqrt{(a^2 - x^2)}}$
$\arccos f(x)$	$\dfrac{-1}{\sqrt{\{1 - [f(x)]^2\}}} [f'(x)]$
(c) $\arctan \dfrac{x}{a}$	$\dfrac{a}{a^2 + x^2}$
$\arctan f(x)$	$\dfrac{1}{\{1 + [f(x)]^2\}} [f'(x)]$
(d) $\operatorname{arcsec} \dfrac{x}{a}$	$\dfrac{a}{x\sqrt{(x^2 - a^2)}}$
$\operatorname{arcsec} f(x)$	$\dfrac{1}{f(x)\sqrt{\{[f(x)]^2 - 1\}}} [f'(x)]$
(e) $\operatorname{arccosec} \dfrac{x}{a}$	$\dfrac{-a}{x\sqrt{(x^2 - a^2)}}$
$\operatorname{arccosec} f(x)$	$\dfrac{-1}{f(x)\sqrt{\{[f(x)]^2 - 1\}}} [f'(x)]$
(f) $\operatorname{arccot} \dfrac{x}{a}$	$\dfrac{-a}{a^2 + x^2}$
$\operatorname{arccot} f(x)$	$\dfrac{-1}{\{1 + [f(x)]^2\}} [f'(x)]$

Worked problems on differentiating inverse trigonometric functions

Problem 1. Find the differential coefficient of $y = \arcsin 3x^2$.

$$\frac{dy}{dx} = \frac{1}{\sqrt{\{1 - [f(x)]^2\}}} [f'(x)] = \frac{1}{\sqrt{\{1 - [3x^2]^2\}}} [6x] = \frac{6x}{\sqrt{(1 - 9x^4)}}$$

Problem 2. Differentiate $y = \ln(\arccos 2t)$.

Let $u = \arccos 2t$

then $y = \ln u$

$$\frac{dy}{dt} = \frac{dy}{du} \cdot \frac{du}{dt} = \frac{1}{u} \cdot \frac{d}{dt} (\arccos 2t)$$

$$= \left(\frac{1}{\arccos 2t}\right)\left(-\frac{2}{\sqrt{\{1 - (2t)^2\}}}\right)$$

Hence

$$\frac{d}{dt} (\ln \arccos 2t) = \frac{-2}{\sqrt{(1 - 4t^2)} \arccos 2t}$$

Problem 3. Find $\dfrac{d}{d\theta}\left(\arctan \dfrac{2}{\theta^2}\right)$.

$$\frac{d}{d\theta}\left(\arctan \frac{2}{\theta^2}\right) = \left(\frac{1}{1 + \left(\dfrac{2}{\theta^2}\right)^2}\right)\left(\frac{-4}{\theta^3}\right) = \left(\frac{1}{\dfrac{\theta^4 + 4}{\theta^4}}\right)\left(\frac{-4}{\theta^3}\right) = \frac{-4\theta}{\theta^4 + 4}$$

Problem 4. Find the differential coefficient of $f(x) = x \operatorname{arcsec} x$.

$$f'(x) = (x)\left(\frac{1}{x\sqrt{(x^2 - 1)}}\right) + (\operatorname{arcsec} x)(1), \text{ by the product rule of differentiation}$$

$$= \frac{1}{\sqrt{(x^2 - 1)}} + \operatorname{arcsec} x$$

Problem 5. Differentiate $y = \dfrac{\operatorname{arccot} x}{(1 + x^2)}$.

$$\frac{dy}{dx} = \frac{(1 + x^2)\left(\dfrac{-1}{1 + x^2}\right) - (\operatorname{arccot} x)(2x)}{(1 + x^2)^2}, \text{ by the quotient rule of differentiation}$$

$$= \frac{-(1 + 2x \operatorname{arccot} x)}{(1 + x^2)^2}$$

Problem 6. Show that if $y = \text{arccot}\left(\dfrac{\cos \theta}{1 - \sin \theta}\right)$ then $\dfrac{dy}{d\theta} = -\dfrac{1}{2}$.

$$\frac{dy}{d\theta} = \frac{-1}{\{1 + [f(x)]^2\}} [f'(x)]$$

$$= -\frac{1}{\left\{1 + \dfrac{\cos \theta}{1 - \sin \theta}\right\}^2} \left\{\frac{(1 - \sin \theta)(-\sin \theta) - (\cos \theta)(-\cos \theta)}{(1 - \sin \theta)^2}\right\}$$

$$= \left\{\frac{-1}{\dfrac{(1 - \sin \theta)^2 + \cos^2 \theta}{(1 - \sin \theta)^2}}\right\}\left\{\frac{-\sin \theta + \sin^2 \theta + \cos^2 \theta}{(1 - \sin \theta)^2}\right\}$$

$$= \left\{\frac{-(1 - \sin \theta)^2}{(1 - \sin \theta)^2 + \cos^2 \theta}\right\}\left\{\frac{1 - \sin \theta}{(1 - \sin \theta)^2}\right\}$$

$$= \left\{\frac{-1}{1 - 2 \sin \theta + \sin^2 \theta + \cos^2 \theta}\right\}(1 - \sin \theta)$$

$$= \frac{-1}{(2 - 2 \sin \theta)}(1 - \sin \theta) = \frac{-(1 - \sin \theta)}{2(1 - \sin \theta)} = -\frac{1}{2}$$

Further problems on differentiating inverse trigonometric functions may be found in Section 14.5 (Problems 1 to 16), page 229.

14.3   Logarithmic forms of the inverse hyperbolic functions

(i)   arsinh $\dfrac{x}{a}$

If $y = \text{arsinh} \dfrac{x}{a}$ then $\dfrac{x}{a} = \sinh y$.

From Chapter 9, $e^y = \cosh y + \sinh y$. Also $\cosh^2 y - \sinh^2 y = 1$, from which $\cosh y = \sqrt{(1 + \sinh^2 y)}$, which is positive since $\cosh y$ is always positive (see Fig. 9.1).

Hence $e^y = \sqrt{(1 + \sinh^2 y)} + \sinh y$

i.e. $e^y = \sqrt{\left[1 + \left(\dfrac{x}{a}\right)^2\right]} + \dfrac{x}{a} = \dfrac{\sqrt{(a^2 + x^2)}}{a} + \dfrac{x}{a}$

Taking Napierian logarithms of both sides gives:

$$y = \ln\left\{\frac{x + \sqrt{(a^2 + x^2)}}{a}\right\}$$

Hence $\text{arsinh}\dfrac{x}{a} = \ln\left\{\dfrac{x + \sqrt{(a^2 + x^2)}}{a}\right\}$

(ii) $\text{arcosh}\dfrac{x}{a}$

If $y = \text{arcosh}\dfrac{x}{a}$ then $\dfrac{x}{a} = \cosh y$

$e^y = \cosh y + \sinh y$

$\quad = \cosh y \pm \sqrt{(\cosh^2 y - 1)}$, since $\sinh y$ may be positive or
negative (see Fig. 9.2)

i.e. $e^y = \dfrac{x}{a} \pm \sqrt{\left[\left(\dfrac{x}{a}\right)^2 - 1\right]} = \dfrac{x}{a} \pm \dfrac{\sqrt{(x^2 - a^2)}}{a}$

Taking Napierian logarithms of both sides gives:

$$y = \ln\left\{\frac{x \pm \sqrt{(x^2 - a^2)}}{a}\right\}$$

The two values obtained are $y = \ln\left\{\dfrac{x + \sqrt{(x^2 - a^2)}}{a}\right\}$ and

$$y = \ln\left\{\frac{x - \sqrt{(x^2 - a^2)}}{a}\right\}$$

Adding these two values of y gives:

$\ln\left\{\dfrac{x + \sqrt{(x^2 - a^2)}}{a}\right\} + \ln\left\{\dfrac{x - \sqrt{(x^2 - a^2)}}{a}\right\}$

$\quad = \ln\left\{\dfrac{x + \sqrt{(x^2 - a^2)}}{a}\right\}\left\{\dfrac{x - \sqrt{(x^2 - a^2)}}{a}\right\}$ from the laws of logarithms

$\quad = \ln\left\{\dfrac{x^2 - (x^2 - a^2)}{a^2}\right\} = \ln 1 = 0$

Hence the two values of y are equal but opposite in sign.

Assuming the principal value, $\text{arcosh}\dfrac{x}{a} = \ln\left\{\dfrac{x + \sqrt{(x^2 - a^2)}}{a}\right\}$

(iii) $\text{artanh}\dfrac{x}{a}$

If $y = \text{artanh}\dfrac{x}{a}$ then $\dfrac{x}{a} = \tanh y$

Now $\tanh y = \dfrac{\sinh y}{\cosh y} = \dfrac{\frac{1}{2}(e^y - e^{-y})}{\frac{1}{2}(e^y + e^{-y})} = \dfrac{e^{2y} - 1}{e^{2y} + 1}$

Then $\dfrac{x}{a} = \dfrac{e^{2y} - 1}{e^{2y} + 1}$ and $x(e^{2y} + 1) = a(e^{2y} - 1)$

Hence
$$x + a = a\,e^{2y} - x\,e^{2y}$$
$$e^{2y} = \frac{a + x}{a - x}$$

Taking Napierian logarithms of both sides gives:

$$2y = \ln\left(\frac{a + x}{a - x}\right)$$

$$y = \tfrac{1}{2}\ln\left(\frac{a + x}{a - x}\right)$$

Hence artanh $\dfrac{x}{a} = \tfrac{1}{2}\ln\left(\dfrac{a + x}{a - x}\right)$

Worked problems on evaluating inverse hyperbolic functions

Problem 1. Evaluate, correct to four decimal places;

(a) arsinh $\dfrac{4}{3}$ (b) arcosh 3   (c) artanh 0.3

(a) In logarithmic form, arsinh $\dfrac{x}{a} = \ln\left\{\dfrac{x + \sqrt{(a^2 + x^2)}}{a}\right\}$

If $x = 4$ and $a = 3$ then arsinh $\dfrac{4}{3} = \ln\left\{\dfrac{4 + \sqrt{(3^2 + 4^2)}}{3}\right\}$

$$= \ln\left(\frac{4 + 5}{3}\right) = \textbf{ln 3 or 1.098\,6}$$

(b) In logarithmic form, arcosh $\dfrac{x}{a} = \ln\left\{\dfrac{x + \sqrt{(x^2 - a^2)}}{a}\right\}$

If $x = 3$ and $a = 1$ then arcosh $\dfrac{3}{1} = \ln\left\{\dfrac{3 + \sqrt{(3^2 - 1^2)}}{1}\right\}$

$$= \ln(3 + \sqrt{8}) = \ln 5.8284 = \textbf{1.762\,7}$$

(c) In logarithmic form, artanh $\dfrac{x}{a} = \tfrac{1}{2}\ln\left(\dfrac{a + x}{a - x}\right)$

artanh $0.3 =$ artanh $\dfrac{3}{10}$

$$\text{If } x = 3 \text{ and } a = 10 \text{ then artanh } \frac{3}{10} = \frac{1}{2} \ln\left(\frac{10+3}{10-3}\right) = \frac{1}{2} \ln\frac{13}{7}$$

$$= 0.309\,5$$

Further problems on evaluating inverse hyperbolic functions may be found in Section 14.5 (Problems 17 to 19), page 231.

14.4   Differentiation of inverse hyperbolic functions

(*a*) $y = \text{arsinh } \dfrac{x}{a}$

If $y = \text{arsinh } \dfrac{x}{a}$ then $\dfrac{x}{a} = \sinh y$ and $x = a \sinh y$

$\dfrac{dx}{dy} = a \cosh y$. (For differential coefficients of hyperbolic functions see Chapter 9).

Now $\cosh^2 y - \sinh^2 y = 1$. Thus $\cosh y = \sqrt{(1 + \sinh^2 y)} = \sqrt{\left\{1 + \left(\dfrac{x}{a}\right)^2\right\}}$

Hence $\dfrac{dx}{dy} = a \cosh y = (a)\left[\dfrac{\sqrt{(a^2 + x^2)}}{a}\right] = \sqrt{(a^2 + x^2)}$

$\dfrac{dy}{dx} = \dfrac{1}{dx/dy} = \dfrac{1}{\sqrt{(a^2 + x^2)}}$, and there are two possible values, one positive and one negative, due to the square root sign. A sketch of part of the graph $y = \text{arsinh } x$ is shown in Fig. 14.3(*a*) where it is seen that the gradient $\left(\text{i.e. } \dfrac{dy}{dx}\right)$ is always positive.

Hence if $y = \text{arsinh } \dfrac{x}{a}$ **then** $\dfrac{dy}{dx} = \dfrac{1}{\sqrt{(a^2 + x^2)}}$

Alternatively, since $y = \text{arsinh } \dfrac{x}{a} = \ln\left\{\dfrac{x + \sqrt{(a^2 + x^2)}}{a}\right\}$ then

$$\frac{d}{dx}\left(\text{arsinh } \frac{x}{a}\right) = \frac{d}{dx}\left[\ln\left\{\frac{x + \sqrt{(a^2 + x^2)}}{a}\right\}\right]$$

$$= \left(\frac{1}{\dfrac{x + \sqrt{(a^2 + x^2)}}{a}}\right)\left(\frac{1}{a}\right)[1 + \tfrac{1}{2}(a^2 + x^2)^{-\frac{1}{2}}2x]$$

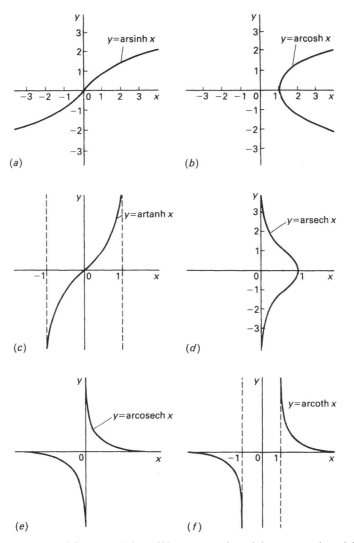

Fig. 14.3   (a) $y = $ arsinh x; (b) $y = $ arcosh x; (c) $y = $ artanh x; (d) $y = $ arsech x; (e) $y = $ arcosech x; (f) $y = $ arcoth x

$$= \left(\frac{a}{x + \sqrt{(a^2 + x^2)}}\right)\left(\frac{1}{a}\right)\left(1 + \frac{x}{\sqrt{(a^2 + x^2)}}\right)$$

$$= \left(\frac{1}{x + \sqrt{(a^2 + x^2)}}\right)\left(\frac{\sqrt{(a^2 + x^2)} + x}{\sqrt{(a^2 + x^2)}}\right)$$

$$= \frac{1}{\sqrt{(a^2 + x^2)}}, \text{ as above}$$

When $a = 1$, $\dfrac{d}{dx}(\text{arsinh } x) = \dfrac{1}{\sqrt{(1 + x^2)}}$

The most general form of the differential coefficient of arsinh $f(x)$ is:

$$\frac{d}{dx}[\text{arsinh } f(x)] = \frac{1}{\sqrt{\{1 + [f(x)]^2\}}}[f'(x)]$$

(b) $y = \text{arcosh } \dfrac{x}{a}$

If $y = \text{arcosh } \dfrac{x}{a}$ then $\dfrac{x}{a} = \cosh y$ and $x = a \cosh y$

$\dfrac{dx}{dy} = a \sinh y = (a)[\sqrt{(\cosh^2 y - 1)}]$, since $\cosh^2 y - \sinh^2 y = 1$

$\dfrac{dx}{dy} = (a)\sqrt{\left[\left(\dfrac{x}{a}\right)^2 - 1\right]} = (a)\left(\dfrac{\sqrt{(x^2 - a^2)}}{a}\right) = \sqrt{(x^2 - a^2)}$

$\dfrac{dy}{dx} = \dfrac{1}{dx/dy} = \dfrac{1}{\sqrt{(x^2 - a^2)}}$ and there are are two possible values, one positive and one negative, due to the square root sign. A sketch of part of the graph $y = \text{arcosh } x$ is shown in Fig. 14.3(b), where it is seen that when x is greater than $+1$ there are two values of y which correspond to a particular value of x. The positive value of y is defined as the principal value of $\text{arcosh } \dfrac{x}{a}$.

Thus if $y = \text{arcosh } \dfrac{x}{a}$ **then** $\dfrac{dy}{dx} = \dfrac{1}{\sqrt{(x^2 - a^2)}}$

Alternatively, the logarithmic equivalent of $\text{arcosh } \dfrac{x}{a}$, that is

$$\ln\left\{\frac{x + \sqrt{(x^2 - a^2)}}{a}\right\}$$

may be differentiated with the same result. When $a = 1$,

$$\frac{d}{dx}(\text{arcosh } x) = \frac{1}{\sqrt{(x^2 - 1)}}$$

The most general form of the differential coefficient of arcosh $f(x)$ is:

$$\frac{d}{dx}[\text{arcosh } f(x)] = \frac{1}{\sqrt{\{[f(x)]^2 - 1\}}}[f'(x)]$$

(c) $y = \text{artanh } \dfrac{x}{a}$

If $y = \text{artanh } \dfrac{x}{a}$, then $\dfrac{x}{a} = \tanh y$ and $x = a \tanh y$

$$\frac{dx}{dy} = a \operatorname{sech}^2 y$$

Now $1 - \operatorname{sech}^2 y = \tanh^2 y$ thus $\operatorname{sech}^2 y = 1 - \tanh^2 y$

Hence $\dfrac{dx}{dy} = a(1 - \tanh^2 y) = a\left[1 - \left(\dfrac{x}{a}\right)^2\right] = a\left(\dfrac{a^2 - x^2}{a^2}\right)$

i.e. $\dfrac{dx}{dy} = \dfrac{a^2 - x^2}{a}$

Hence $\dfrac{dy}{dx} = \dfrac{1}{dx/dy} = \dfrac{a}{a^2 - x^2}$

Alternatively, the logarithmic equivalent of $\operatorname{artanh} \dfrac{x}{a}$, that is, $\frac{1}{2} \ln\left(\dfrac{a + x}{a - x}\right)$, may be differentiated with the same result. When $a = 1$,

$$\frac{d}{dx}(\text{artanh } x) = \frac{1}{1 - x^2}$$

The most general form of the differential coefficient of artanh $f(x)$ is:

$$\frac{d}{dx}[\text{artanh } f(x)] = \frac{1}{\{1 - [f(x)]^2\}}[f'(x)]$$

A sketch of part of the graph $y = \operatorname{artanh} x$ is shown in Fig. 14.3(c), where the gradient $\left(\text{i.e. } \dfrac{dy}{dx}\right)$ is seen to be always positive.

(d) $y = \text{arsech} \dfrac{x}{a}$

If $y = \operatorname{arsech} \dfrac{x}{a}$, then $\dfrac{x}{a} = \operatorname{sech} y$ and $x = a \operatorname{sech} y$

$$\frac{dx}{dy} = -a \operatorname{sech} y \tanh y$$

Now $1 - \tanh^2 y = \operatorname{sech}^2 y$, thus $\tanh y = \sqrt{(1 - \operatorname{sech}^2 y)}$

$$= \sqrt{\left[1 - \left(\frac{x}{a}\right)^2\right]} = \frac{\sqrt{(a^2 - x^2)}}{a},$$

the positive value being taken, since the positive value is defined as the principal value.

$$\frac{dx}{dy} = -a \operatorname{sech} y \tanh y = -a\left(\frac{x}{a}\right)\left[\frac{\sqrt{(a^2 - x^2)}}{a}\right]$$

Hence $$\frac{dy}{dx} = \frac{1}{dx/dy} = \frac{-a}{x\sqrt{(a^2 - x^2)}}$$

When $a = 1$, $$\frac{d}{dx}(\operatorname{arsech} x) = \frac{-1}{x\sqrt{(1 - x^2)}}$$

A sketch of part of the graph $y = \operatorname{arsech} x$ is shown in Fig. 14.3(d), where it is seen that the gradient is negative for principal values of y.

The most general form of the differential coefficient of arsech $f(x)$ is:

$$\frac{d}{dx}[\operatorname{arsech} f(x)] = \frac{-1}{f(x)\sqrt{\{1 - [f(x)]^2\}}}[f'(x)]$$

(e) $y = \operatorname{arcosech} \dfrac{x}{a}$

If $y = \operatorname{arcosech} \dfrac{x}{a}$, then $\dfrac{x}{a} = \operatorname{cosech} y$ and $x = a \operatorname{cosech} y$

$$\frac{dx}{dy} = -a \operatorname{cosech} y \coth y$$

Now $\coth^2 y - 1 = \operatorname{cosech}^2 y$, thus $\coth y = \sqrt{(\operatorname{cosech}^2 y + 1)}$

$$= \sqrt{\left[\left(\frac{x}{a}\right)^2 + 1\right]}$$

$$= \frac{\sqrt{(x^2 + a^2)}}{a}$$

the positive value being taken since the positive value is defined as the principal value.

Hence $$\frac{dx}{dy} = -a \operatorname{cosech} y \coth y = -a\left(\frac{x}{a}\right)\left[\frac{\sqrt{(x^2 + a^2)}}{a}\right] = -\frac{x}{a}\sqrt{(x^2 + a^2)}$$

$$\frac{dy}{dx} = \frac{1}{dx/dy} = \frac{-a}{x\sqrt{(x^2 + a^2)}}$$

When $a = 1$, $$\frac{d}{dx}(\operatorname{arcosech} x) = \frac{-1}{x\sqrt{(x^2 + 1)}}$$

A sketch of part of the graph $y = \text{arcosech } x$ is shown in Fig. 14.3(e), where it is seen that the gradient is negative for principal values of y.

The most general form of the differential coefficient of arcosech $f(x)$ is:

$$\frac{d}{dx} [\text{arcosech } f(x)] = \frac{-1}{f(x)\sqrt{\{[f(x)]^2 + 1\}}} [f'(x)]$$

(f) $y = \text{arcoth } \dfrac{x}{a}$

If $y = \text{arcoth } \dfrac{x}{a}$ then $\dfrac{x}{a} = \coth y$ and $x = a \coth y$

$$\frac{dx}{dy} = -a \cosech^2 y$$

Now $\coth^2 y - 1 = \cosech^2 y$. Hence $\dfrac{dx}{dy} = -a(\coth^2 y - 1)$

$$= -a\left[\left(\frac{x}{a}\right)^2 - 1\right]$$

$$= -\frac{(x^2 - a^2)}{a}$$

$$\frac{dy}{dx} = \frac{1}{dx/dy} = \frac{-a}{x^2 - a^2} = \frac{a}{a^2 - x^2}$$

$\left(\text{i.e. the same result as obtained for artanh } \dfrac{x}{a}\right)$

When $a = 1$, $\dfrac{d}{dx} (\text{arcoth } x) = \dfrac{1}{1 - x^2}$

A sketch of part of the graph $y = \text{arcoth } x$ is shown in Fig. 14.3(f), where it is seen that the gradient is negative for positive values of y.

The most general form of the differential coefficient of arcoth $f(x)$ is:

$$\frac{d}{dx} [\text{arcoth } f(x)] = \frac{1}{\{1 - [f(x)]^2\}} [f'(x)]$$

From the above results it may be seen that:

$$\int \frac{1}{\sqrt{(x^2 + a^2)}} dx = \text{arsinh } \frac{x}{a} + c \text{ or } \ln\left\{\frac{x + \sqrt{(x^2 + a^2)}}{a}\right\} + c,$$

$$\int \frac{1}{\sqrt{(x^2 - a^2)}} dx = \text{arcosh } \frac{x}{a} + c \text{ or } \ln\left\{\frac{x + \sqrt{(x^2 - a^2)}}{a}\right\} + c,$$

Summary of differential coefficients of inverse hyperbolic functions

y or $f(x)$	$\dfrac{\mathrm{d}y}{\mathrm{d}x}$ or $f'(x)$
(a) $\operatorname{arsinh} \dfrac{x}{a}$	$\dfrac{1}{\sqrt{(x^2 + a^2)}}$
$\operatorname{arsinh} f(x)$	$\dfrac{1}{\sqrt{\{[f(x)]^2 + 1\}}}\,[f'(x)]$
(b) $\operatorname{arcosh} \dfrac{x}{a}$	$\dfrac{1}{\sqrt{(x^2 - a^2)}}$
$\operatorname{arcosh} f(x)$	$\dfrac{1}{\sqrt{\{[f(x)]^2 - 1\}}}\,[f'(x)]$
(c) $\operatorname{artanh} \dfrac{x}{a}$	$\dfrac{a}{(a^2 - x^2)}$
$\operatorname{artanh} f(x)$	$\dfrac{1}{\{1 - [f(x)]^2\}}\,[f'(x)]$
(d) $\operatorname{arsech} \dfrac{x}{a}$	$\dfrac{-a}{x\sqrt{(a^2 - x^2)}}$
$\operatorname{arsech} f(x)$	$\dfrac{-1}{f(x)\sqrt{\{1 - [f(x)]^2\}}}\,[f'(x)]$
(e) $\operatorname{arcosech} \dfrac{x}{a}$	$\dfrac{-a}{x\sqrt{(x^2 + a^2)}}$
$\operatorname{arcosech} f(x)$	$\dfrac{-1}{f(x)\sqrt{\{[f(x)]^2 + 1\}}}\,[f'(x)]$
(f) $\operatorname{arcoth} \dfrac{x}{a}$	$\dfrac{a}{a^2 - x^2}$
$\operatorname{arcoth} f(x)$	$\dfrac{1}{\{1 - [f(x)]^2\}}\,[f'(x)]$

and

$$\int \frac{1}{(a^2 - x^2)} \, dx = \frac{1}{a} \operatorname{artanh} \frac{x}{a} + c \text{ or } \frac{1}{2a} \ln\left(\frac{a + x}{a - x}\right) + c$$

The method of determining such integrals is discussed in Section 18.3.

Worked problems on differentiating inverse hyperbolic functions

Problem 1. Find the differential coefficient of (a) arsinh $3x$ and (b) arcosh $\sqrt{(1 + t^2)}$.

(a) $\dfrac{d}{dx} [\operatorname{arsinh} f(x)] = \dfrac{1}{\sqrt{\{[f(x)]^2 + 1\}}} [f'(x)]$

Hence $\dfrac{d}{dx} [\operatorname{arsinh} 3x] = \dfrac{1}{\sqrt{\{(3x)^2 + 1\}}} [3] = \dfrac{3}{\sqrt{(9x^2 + 1)}}$

(b) $\dfrac{d}{dt} [\operatorname{arcosh} f(t)] = \dfrac{1}{\sqrt{\{[f(t)]^2 - 1\}}} [f'(t)]$

Hence $\dfrac{d}{dt} [\operatorname{arcosh} \sqrt{(1 + t^2)}] = \dfrac{1}{\sqrt{\{[\sqrt{(1 + t^2)}]^2 - 1\}}} [\tfrac{1}{2}(1 + t^2)^{-\frac{1}{2}} 2t]$

$$= \frac{1}{\sqrt{(1 + t^2 - 1)}} \left[\frac{t}{\sqrt{(1 + t^2)}}\right]$$

$$= \frac{1}{t} \left[\frac{t}{\sqrt{(1 + t^2)}}\right] = \frac{1}{\sqrt{(1 + t^2)}}$$

Problem 2. Differentiate (a) artanh $\dfrac{3x}{4}$, (b) arcosech(sinh x).

(a) $\dfrac{d}{dx} [\operatorname{artanh} f(x)] = \dfrac{1}{\{1 - [f(x)]^2\}} [f'(x)]$

Hence $\dfrac{d}{dx} \left[\operatorname{artanh} \dfrac{3x}{4}\right] = \dfrac{1}{\left\{1 - \left(\dfrac{3x}{4}\right)^2\right\}} \left[\dfrac{3}{4}\right] = \dfrac{1}{\left(1 - \dfrac{9x^2}{16}\right)} \left[\dfrac{3}{4}\right]$

$$= \frac{1}{\left(\dfrac{16 - 9x^2}{16}\right)} \left[\frac{3}{4}\right] = \frac{16}{(16 - 9x^2)} \left[\frac{3}{4}\right]$$

$$= \frac{12}{16 - 9x^2}$$

(b) $\dfrac{d}{dx}[\text{arcosech } f(x)] = \dfrac{-1}{f(x)\sqrt{\{[f(x)]^2 + 1\}}}[f'(x)]$

Hence $\dfrac{d}{dx}[\text{arcosech}(\sinh x)] = \dfrac{-1}{\sinh x\sqrt{\{\sinh^2 x + 1\}}}[\cosh x]$

Now $\cosh^2 x - \sinh^2 x = 1$, hence $\sinh^2 x + 1 = \cosh^2 x$

Therefore $\dfrac{d}{dx}[\text{arcosech}(\sinh x)] = \dfrac{-1}{\sinh x\sqrt{(\cosh^2 x)}}[\cosh x]$

$$= \frac{-1}{\sinh x} = -\text{cosech } x$$

Problem 3. Find (a) $\dfrac{d}{dx}[\text{arsech}(3x - 1)]$ and (b) $\dfrac{d}{d\theta}[\text{arcoth}\sqrt{(1 - 2\theta^2)}]$.

(a) $\dfrac{d}{dx}[\text{arsech } f(x)] = \dfrac{-1}{f(x)\sqrt{\{1 - [f(x)]^2\}}}[f'(x)]$

Hence $\dfrac{d}{dx}[\text{arsech}(3x - 1)] = \dfrac{-1}{(3x - 1)\sqrt{\{1 - (3x - 1)^2\}}}[3]$

$$= \frac{-3}{(3x - 1)\sqrt{\{1 - (9x^2 - 6x + 1)\}}}$$

$$= \frac{-3}{(3x - 1)\sqrt{(6x - 9x^2)}}$$

$$= \frac{-3}{(3x - 1)\sqrt{[3x(2 - 3x)]}}$$

(b) $\dfrac{d}{d\theta}[\text{arcoth } f(\theta)] = \dfrac{1}{\{1 - [f(\theta)]^2\}}[f'(\theta)]$

Hence $\dfrac{d}{d\theta}[\text{arcoth}\sqrt{(1 - 2\theta^2)}]$

$$= \frac{1}{\{1 - [\sqrt{(1 - 2\theta^2)}]^2\}}[\tfrac{1}{2}(1 - 2\theta^2)^{-\frac{1}{2}}(-4\theta)]$$

$$= \frac{1}{\{1 - (1 - 2\theta^2)\}}\left[\frac{-2\theta}{\sqrt{(1 - 2\theta^2)}}\right]$$

$$= \frac{1}{2\theta^2}\left[\frac{-2\theta}{\sqrt{(1-2\theta^2)}}\right] = \frac{-1}{\theta\sqrt{(1-2\theta^2)}}$$

Problem 4. Differentiate: (a) $(1-t^2)$ arcoth t, (b) $\dfrac{\text{arcosh}(\sec x)}{x}$.

(a) $f(t) = (1-t^2)$ arcoth t

Hence $f'(t) = (1-t^2)\dfrac{1}{(1-t^2)} + (\text{arcoth } t)(-2t)$, by the product
rule of differentiation

$\qquad\qquad = 1 - 2t$ **arcoth** t

(b) $f(x) = \dfrac{\text{arcosh}(\sec x)}{x}$

Hence $f'(x) = \dfrac{(x)\left[\dfrac{1}{\sqrt{(\sec^2 x - 1)}}(\sec x \tan x)\right] - [\text{arcosh}(\sec x)](1)}{x^2}$,

by the quotient rule of differentiation

$$= \frac{(x)\left(\dfrac{\sec x \tan x}{\tan x}\right) - \text{arcosh}(\sec x)}{x^2}, \text{ since } \sqrt{(\sec^2 x - 1)} = \tan x$$

$$= \frac{x \sec x - \text{arcosh}(\sec x)}{x^2}$$

---

*Further problems on differentiating inverse hyperbolic functions may be found
in the following Section 14.5 (Problems 20 to 35), page 232.*

14.5   Further problems

Differentiation of inverse trigonometrical functions

In Problems 1 to 8, differentiate with respect to the variable.

1.   (a) arcsin $3x$ (b) arcsin $\dfrac{x}{5}$

  (a) $\left[\dfrac{3}{\sqrt{(1-9x^2)}}\right]$ (b) $\left[\dfrac{1}{\sqrt{(25-x^2)}}\right]$

2.   (a) 2 arccos $4x$ (b) 3 arccos $\dfrac{2x}{3}$

  (a) $\left[\dfrac{-8}{\sqrt{(1-16x^2)}}\right]$ (b) $\left[\dfrac{-6}{\sqrt{(9-4x^2)}}\right]$

3.   (a) $2 \arctan 2x$ (b) $\arctan \sqrt{(x-1)}$

     (a) $\left[\dfrac{4}{1 + 4x^2} \right]$ (b) $\left[\dfrac{1}{2x\sqrt{(x-1)}} \right]$

4.   (a) $\operatorname{arcsec} 4\theta$ (b) $2 \operatorname{arcsec} \sqrt{x}$

     (a) $\left[\dfrac{1}{\theta\sqrt{(16\theta^2 - 1)}} \right]$ (b) $\left[\dfrac{1}{x\sqrt{(x-1)}} \right]$

5.   (a) $\operatorname{arccosec} \dfrac{x}{3}$ (b) $\operatorname{arccosec}(x^2 + 1)$

     (a) $\left[\dfrac{-3}{x\sqrt{(x^2 - 9)}} \right]$ (b) $\left[\dfrac{-2}{(x^2 + 1)\sqrt{(x^2 + 2)}} \right]$

6.   (a) $\operatorname{arccot} x^2$ (b) $3 \operatorname{arccot} \sqrt{(x^2 - 1)}$

     (a) $\left[\dfrac{-2x}{1 + x^4} \right]$ (b) $\left[\dfrac{-3}{x\sqrt{(x^2 - 1)}} \right]$

7.   (a) $\ln(\arcsin x)$ (b) $3 e^{\operatorname{arccot} x}$

     (a) $\left[\dfrac{1}{\sqrt{(1 - x^2)} \arcsin x} \right]$ (b) $\left[\dfrac{-3 e^{\operatorname{arccot} x}}{1 + x^2} \right]$

8.   (a) $\arctan\left(\dfrac{2t}{1 - t^2} \right)$ (b) $\dfrac{1}{2} \operatorname{arcsec}\left(\dfrac{x^2 + 1}{x^2 - 1} \right)$

     (a) $\left[\dfrac{2}{1 + t^2} \right]$ (b) $\left[\dfrac{-1}{(x^2 + 1)} \right]$

9.   Show that (a) $\dfrac{d}{dx}\left[\arcsin\left(\dfrac{2x^2 + 3}{5} \right) \right] = \dfrac{2x}{\sqrt{(4 - 3x^2 - x^4)}}$

     (b) $\dfrac{d}{dt}\left[\operatorname{arcsec} \sqrt{\left(\dfrac{t}{2} \right)} \right] = \dfrac{1}{t\sqrt{[2(t - 2)]}}$

10.   Show that (a) if $y = \arctan\left(\dfrac{\sin \theta}{1 - \cos \theta} \right)$ then $\dfrac{dy}{d\theta} = -\dfrac{1}{2}$,

     (b) if $y = 2 \operatorname{arcsec} \sqrt{(\cos t)}$ then $\dfrac{dy}{dt} = \dfrac{-\tan t}{\sqrt{(\cos t - 1)}}$

In Problems 11 to 16 differentiate with respect to the variable.

11.   (a) $x \arcsin \dfrac{x}{2}$ (b) $x^2 \operatorname{arccosec} 2x$

     (a) $\left[\dfrac{x}{\sqrt{(4 - x^2)}} + \arcsin \dfrac{x}{2} \right]$

     (b) $\left[x\left(2 \operatorname{arccosec} 2x - \dfrac{1}{\sqrt{(4x^2 - 1)}} \right) \right]$

12. (a) $t \arccos(2t^2 - 1)$ (b) $3t^3 \operatorname{arccot} t$

(a) $\left[\dfrac{-2t}{\sqrt{(1 - t^2)}} + \arccos(2t^2 - 1) \right]$

(b) $\left[3t^2 \left(3 \operatorname{arccot} t - \dfrac{t}{1 + t^2} \right) \right]$

13. (a) $(1 + \theta^2) \arctan \theta$ (b) $\sqrt{(1 - x^2)} \arcsin x$

(a) $[1 + 2\theta \arctan \theta]$ (b) $\left[1 - \dfrac{x \arcsin x}{\sqrt{(1 - x^2)}} \right]$

14. (a) $\sin \left(\operatorname{arcsec} \dfrac{1}{x} \right)$ (b) $2t^2 \arccos(t - 1)$

(a) $\left[\dfrac{-\cos(\operatorname{arcsec} 1/x)}{\sqrt{(1 - x^2)}} \right]$

(b) $\left[4t \arccos(t - 1) - \dfrac{2t^2}{\sqrt{[t(2 - t)]}} \right]$

15. (a) $\dfrac{\arcsin 2x}{x}$ (b) $\dfrac{\arccos x}{\sqrt{(1 - x^2)}}$

(a) $\left[\dfrac{1}{x^2} \left(\dfrac{2x}{\sqrt{(1 - 4x^2)}} - \arcsin 2x \right) \right]$

(b) $\left[\dfrac{-1}{(1 - x^2)} + \dfrac{x \arccos x}{\sqrt{(1 - x^2)^3}} \right]$

16. (a) $\dfrac{\operatorname{arcsec} \sqrt{(x^2 + 1)}}{\sqrt{(x^2 + 1)}}$ (b) $\dfrac{\operatorname{arccot} x}{(1 + x^2)}$

(a) $\left[\dfrac{1 - x \operatorname{arcsec} \sqrt{(x^2 + 1)}}{\sqrt{(x^2 + 1)^3}} \right]$

(b) $\left[\dfrac{-(1 + 2x \operatorname{arccot} x)}{(1 + x^2)^2} \right]$

Evaluation of inverse hyperbolic functions

In Problems 17 to 19 use logarithmic equivalents of inverse hyperbolic functions to evaluate correct to four decimal places.

17. (a) $\operatorname{arsinh} \frac{2}{3}$ (b) $\operatorname{arsinh} \frac{3}{4}$ (c) $\operatorname{arsinh} 0.5$

    (a) $[0.625\,1]$ (b) $[0.693\,1]$ (c) $[0.481\,2]$

18. (a) $\operatorname{arcosh} \frac{5}{3}$ (b) $\operatorname{arcosh} 2$ (c) $\operatorname{arcosh} \dfrac{13}{12}$

    (a) $[1.098\,6]$ (b) $[1.317\,0]$ (c) $[0.405\,5]$

19. (a) artanh $\frac{1}{2}$ (b) artanh $\frac{3}{5}$ (c) artanh 0.3

    (a) [0.549 3] (b) [0.693 1] (c) [0.309 5]

Differentiation of inverse hyperbolic functions

In Problems 20 to 34 differentiate with respect to the variable.

20. (a) arsinh $\frac{x}{2}$ (b) arsinh 5x

    (a) $\left[\frac{1}{\sqrt{(x^2 + 4)}} \right]$ (b) $\left[\frac{5}{\sqrt{(25x^2 + 1)}} \right]$

21. (a) arcosh $\frac{x}{3}$ (b) arcosh 4x

    (a) $\left[\frac{1}{\sqrt{(x^2 - 9)}} \right]$ (b) $\left[\frac{4}{\sqrt{(16x^2 - 1)}} \right]$

22. (a) artanh $\frac{2x}{3}$ (b) artanh 3x

    (a) $\left[\frac{6}{9 - 4x^2} \right]$ (b) $\left[\frac{3}{1 - 9x^2} \right]$

23. (a) arsech $\frac{3t}{4}$ (b) arsech 3t

    (a) $\left[\frac{-4}{t\sqrt{(16 - 9t^2)}} \right]$ (b) $\left[\frac{-1}{t\sqrt{(1 - 9t^2)}} \right]$

24. (a) arcosech $\frac{x}{5}$ (b) arcosech 2x

    (a) $\left[\frac{-5}{x\sqrt{(x^2 + 25)}} \right]$ (b) $\left[\frac{-1}{x\sqrt{(4x^2 + 1)}} \right]$

25. (a) arcoth $\frac{2\theta}{5}$ (b) arcoth 4θ

    (a) $\left[\frac{10}{25 - 4\theta^2} \right]$ (b) $\left[\frac{4}{1 - 16\theta^2} \right]$

26. (a) arsinh $\sqrt{(t^2 - 1)}$ (b) arsech(2x − 1)

    (a) $\left[\frac{1}{\sqrt{(t^2 - 1)}} \right]$ (b) $\left[\frac{-1}{(2x - 1)\sqrt{x(1 - x)}} \right]$

27. (a) arcosh $\sqrt{(1 + x^2)}$ (b) arcosh(cosh x)

    (a) $\left[\frac{1}{\sqrt{(1 + x^2)}} \right]$ (b) [1]

28. (a) artanh $\sqrt{(2t^2 + 1)}$ (b) arcoth$(\sin x)$

(a) $\left[\dfrac{-1}{t\sqrt{(2t^2 + 1)}} \right]$ (b) $[\sec x]$

29. (a) arcosech $\sqrt{(x^2 - 1)}$ (b) arcoth $\sqrt{(1 - 3t^2)}$

(a) $\left[\dfrac{-1}{(x^2 - 1)} \right]$ (b) $\left[\dfrac{-1}{t\sqrt{(1 - 3t^2)}} \right]$

30. (a) x arsinh x (b) $(1 - x^2)$ artanh x

(a) $\left[\dfrac{x}{\sqrt{(x^2 + 1)}} + \text{arsinh } x \right]$ (b) $[1 - 2x \text{ artanh } x]$

31. (a) $3x$ arsech $2x$ (b) x^2 arcoth $\sqrt{(1 - x^2)}$

(a) $\left[\dfrac{-3}{\sqrt{(1 - 4x^2)}} + 3 \text{ arsech } 2x \right]$

(b) $\left[2x \text{ arcoth } \sqrt{(1 - x^2)} \dfrac{-x}{\sqrt{(1 - x^2)}} \right]$

32. (a) $\sqrt{(x^2 - 1)}$ arcosh x (b) $\sqrt{(x^2 - 1)}$ arcosech $\sqrt{(x^2 - 1)}$

(a) $\left[1 + \dfrac{x \text{ arcosh } x}{\sqrt{(x^2 - 1)}} \right]$

(b) $\left[\dfrac{1}{\sqrt{(x^2 - 1)}} (x \text{ arcosech } \sqrt{(x^2 - 1)} - 1) \right]$

33. (a) $\dfrac{\text{arsinh } \sqrt{(x^2 - 1)}}{x}$ (b) $\dfrac{\text{artanh}(2x - 1)}{2x}$

(a) $\left[\dfrac{1}{x^2} \left\{ \dfrac{x}{\sqrt{(x^2 - 1)}} - \text{arsinh } \sqrt{(x^2 - 1)} \right\} \right]$

(b) $\left[\dfrac{1}{4x^2} \left\{ \dfrac{1}{(1 - x)} - 2 \text{ artanh}(2x - 1) \right\} \right]$

34. (a) $\dfrac{\text{arsech } x}{\sqrt{(1 - x^2)}}$ (b) $\dfrac{\text{arcoth } x/2}{4 - x^2}$

(a) $\left[\dfrac{1}{(1 - x^2)} \left\{ -\dfrac{1}{x} + \dfrac{x \text{ arsech } x}{\sqrt{(1 - x^2)}} \right\} \right]$ (b) $\left[\dfrac{2(1 + x \text{ arcoth } x/2)}{(4 - x^2)^2} \right]$

35. Show that $\dfrac{d}{dx} [x \text{ arcosh}(\sec x)] = x \sec x + \text{arcosh}(\sec x)$.

Chapter 15

Partial differentiation

15.1 Differentiating a function having two variables

When differentiating functions previously, problems have always been associated with one variable only. For a given function of x, say $y = f(x)$, the problem has been to find $\dfrac{dy}{dx}$, i.e. $f'(x)$. When applying the product or quotient rules, $y = uv$ or $y = \dfrac{u}{v}$, y, u, and v are all functions of x only. However, there are many formulae and occurrences in mathematics, engineering and science, in which the variation of one function may depend on changes taking place in two or more variables. Some examples include:

(i)   The volume of a cylinder, $V = \pi r^2 h$. The volume of the cylinder will change if either radius r or height h is changed.

(ii) Resonant frequency, $fr = \dfrac{1}{2\pi\sqrt{(LC)}}$. The resonant frequency will change if either inductance L or capacitance C is changed.

(iii) The pressure of an ideal gas, $p = \dfrac{mRT}{V}$. The pressure will change if either the thermodynamic temperature T or the volume V is changed.

(iv) Torque, $T = I\alpha$. The torque on a shaft will change if either the moment of inertia, I, or the angular acceleration, α, is changed, and so on.

For the formula giving the volume of a right circular cylinder, the value of V depends on the values of r and h and this is expressed mathematically as $V = f(r, h)$, called 'V is some function of r and h'. A statement such as '$V = f(r, h)$ and $V = \pi r^2 h$' indicates that both r and h are variable quantities.

When differentiating a function having two variables, one variable is kept constant and the differential coefficient is found of the other variable, with respect to that variable. The differential coefficient obtained is called a **partial derivative** of the function. Thus, for a right circular cylinder, the increase in volume with respect to radius, $\dfrac{dV}{dr}$, when h remains constant, is the partial derivative of V with respect to r, and is written $\dfrac{\partial V}{\partial r}$. The 'curly' dee, ∂, is used to denote a differential coefficient in an expression containing more than one variable. Alternatively,

$$\left[\frac{dV}{dh}\right]_{r \text{ constant}} = \frac{\partial V}{\partial h}$$

the partial derivative of V with respect to h, with r kept constant.

15.2   First order partial derivatives

By applying the principles given in Section 15.1 to the volume of a right circular cylinder, the partial derivatives $\frac{\partial V}{\partial r}$ and $\frac{\partial V}{\partial h}$ may be found. These are called **first order partial derivatives** since $n = 1$ when they are written in the form $\frac{\partial^n V}{\partial r^n}$. Second order partial derivatives contain at least one term in which $n = 2$, i.e. a term of the form $\frac{\partial^2 V}{\partial r^2}$.

The partial derivative of V with respect to r is found by keeping h constant, i.e. since $V = \pi r^2 h$, then (πh) becomes the constant term. Thus $\frac{\partial V}{\partial r} = (\pi h)\frac{d(r^2)}{dr} = 2\pi rh$. The partial derivative of V with respect to h is found by keeping r constant, i.e. since $V = \pi r^2 h$ then (πr^2) becomes the constant term. Thus $\frac{\partial V}{\partial h} = (\pi r^2)\frac{d(h)}{dh} = \pi r^2$.

Worked problems on first order partial derivatives

Problem 1. The pressure p of a given mass of gas is given by $pV = kT$, where k is a constant, V is the volume and T is the thermodynamic temperature. Determine equations for $\frac{\partial p}{\partial V}$ and $\frac{\partial p}{\partial T}$.

Since $pV = kT$, $p = \dfrac{kT}{V}$

To find $\dfrac{\partial p}{\partial V}$, T is kept constant and $p = (kT)V^{-1}$.

Hence, $\dfrac{\partial p}{\partial V} = (kT)\dfrac{d(V^{-1})}{dV} = kT(-1)V^{-2} = -\dfrac{kT}{V^2}$

To find $\dfrac{\partial V}{\partial T}$, V is kept constant and $p = \left(\dfrac{k}{V}\right) \cdot T$

Hence, $\dfrac{\partial V}{\partial T} = \left(\dfrac{k}{V}\right) \cdot \dfrac{\mathrm{d}(T)}{\mathrm{d}T} = \left(\dfrac{k}{V}\right) \cdot 1 = \dfrac{k}{V}$

Problem 2. If $z = (x + y)\ln\dfrac{x}{y}$, prove that $x\dfrac{\partial z}{\partial x} + y\dfrac{\partial z}{\partial y} = z$.

As there is a product of two functions having x in each, $\dfrac{\partial z}{\partial x}$ is found by keeping y constant and applying the product rule of differentiation. The partial differential coefficient of the first function with respect to x is:

$$\frac{\partial(x + y)}{\partial x} = 1,$$

since y is a constant. The partial differential coefficient of the second function with respect to x is:

$$\frac{\partial(\ln x/y)}{\partial x} = \frac{1}{x/y} \times \frac{1}{y} = \frac{1}{x}$$

Alternatively, $\dfrac{\partial(\ln x/y)}{\partial x} = \dfrac{\partial(\ln x - \ln y)}{\partial x} = \dfrac{1}{x}$

since $\ln y$ is a constant.

Hence, by the product rule, $\dfrac{\partial z}{\partial x} = (x + y) \cdot \dfrac{1}{x} + \left(\ln\dfrac{x}{y}\right) \cdot 1$

$\dfrac{\partial z}{\partial y}$ can be found by keeping x constant.

As there are two functions of y in each, the product rule of differentiation is again used. The partial differential coefficient of the first function with respect to y is: $\dfrac{\partial(x + y)}{\partial y} = 1$, since x is a constant. The partial differential coefficient of the second function with respect to y is:

$$\frac{\partial(\ln x/y)}{\partial y} = \frac{\partial(\ln x - \ln y)}{\partial y} = -\frac{1}{y}.$$

Thus, $\dfrac{\partial z}{\partial y} = (x + y)\left(-\dfrac{1}{y}\right) + \left(\ln\dfrac{x}{y}\right) \cdot 1.$

Taking the L.H.S. of the identity and substituting for $\dfrac{\partial z}{\partial x}$ and $\dfrac{\partial z}{\partial y}$ gives:

$$x\left[\frac{1}{x}(x+y)+\ln\frac{x}{y}\right]+y\left[-\frac{1}{y}(x+y)+\ln\frac{x}{y}\right]$$

i.e. L.H.S. $= (x+y) + x\ln\dfrac{x}{y} - (x+y) + y\ln\dfrac{x}{y}$

$$= (x+y)\ln\frac{x}{y} = z = \text{R.H.S.}$$

Problem 3. If $z = f(\theta, \phi)$, find $\dfrac{\partial z}{\partial \theta}$ and $\dfrac{\partial z}{\partial \phi}$, if $z = 3\sin 2\theta \cos 3\phi$.

$\dfrac{\partial z}{\partial \theta}$ may be found by keeping ϕ constant, i.e. $(3\cos 3\phi)$ is the constant term.

Thus, $\dfrac{\partial z}{\partial \theta} = 3\cos 3\phi \dfrac{d(\sin 2\theta)}{d\theta}$,

i.e. $\dfrac{\partial z}{\partial \theta} = 3\cos 3\phi\, 2\cos 2\theta$

$$= \textbf{6}\cos 2\theta \cos 3\phi$$

$\dfrac{\partial z}{\partial \phi}$ may be found by keeping θ constant, i.e. $(3\sin 2\theta)$ is the constant term.

Thus, $\dfrac{\partial z}{\partial \phi} = 3\sin 2\theta \cdot \dfrac{d(\cos 3\phi)}{d\phi}$

$$= 3\sin 2\theta \cdot (-3\sin 3\phi)$$

$$= -\textbf{9}\sin 2\theta \sin 3\phi$$

Problem 4. If $z = f(x^2 + y^2)$, show that $x\dfrac{\partial z}{\partial y} - y\dfrac{\partial z}{\partial x} = 0$.

The actual function of z is not known and z could be equal to, say, $\sin(x^2 + y^2)$ or $\ln(x^2 + y^2)$ or $(x^2 + y^2)^{\frac{3}{2}}$ or any other function of $(x^2 + y^2)$. However, proofs based on functions are often possible, even though the actual function is not known. Such techniques are sometimes used in solving partial differential equations. Making a substitution such as let $u = (x^2 + y^2)$ usually helps in proving the identity. If such a substitution is made in this case, then $z = f(u)$. Also, since $u = x^2 + y^2$, $\dfrac{\partial u}{\partial x} = 2x$ (y is kept constant) and $\dfrac{\partial u}{\partial y} = 2y$ (x is kept constant). The chain rule of differentiation applies equally to partial derivatives, hence

$$\frac{\partial z}{\partial x} = \frac{\partial z}{\partial u} \times \frac{\partial u}{\partial x} = f'(u) \times 2x$$

Similarly,

$$\frac{\partial z}{\partial y} = \frac{\partial z}{\partial u} \times \frac{\partial u}{\partial y} = f'(u) \times 2y$$

Substituting for $\dfrac{\partial z}{\partial x}$ and $\dfrac{\partial z}{\partial y}$ in the L.H.S. of the equation gives:

$$x \cdot 2y\ f'(u) - y \cdot 2x\ f'(u) = 0 = \text{R.H.S.}$$

Further problems on first order partial derivatives may be found in Section 15.4 (Problems 1 to 7), page 242.

15.3   Second order partial derivatives

It is shown in Section 15.2 that if $z = f(x, y)$, then it is possible to find $\dfrac{\partial z}{\partial x}$ and $\dfrac{\partial z}{\partial y}$. The partial derivatives $\dfrac{\partial z}{\partial x}$ and $\dfrac{\partial z}{\partial y}$ are themselves functions of x and y and hence can be differentiated again with respect to x or with respect to y. When this is done, four **second order partial derivatives** are possible for a function containing two variables. These are:

(i) differentiating $\dfrac{\partial z}{\partial x}$ with respect to x (keeping y constant) gives

$$\frac{\partial\left(\dfrac{\partial z}{\partial x}\right)}{\partial x}, \text{ written as } \frac{\partial^2 z}{\partial x^2},$$

(ii) differentiating $\dfrac{\partial z}{\partial x}$ with respect to y (keeping x constant) gives

$$\frac{\partial\left(\dfrac{\partial z}{\partial x}\right)}{\partial y}, \text{ written as } \frac{\partial^2 z}{\partial y\, \partial x},$$

(iii) differentiating $\dfrac{\partial z}{\partial y}$ with respect to x (keeping y constant) gives

$\dfrac{\partial\left(\dfrac{\partial z}{\partial y}\right)}{\partial x}$, written as $\dfrac{\partial^2 z}{\partial x\,\partial y}$, and

(iv) differentiating $\dfrac{\partial z}{\partial y}$ with respect to y (keeping x constant) gives

$\dfrac{\partial\left(\dfrac{\partial z}{\partial y}\right)}{\partial y}$, written as $\dfrac{\partial^2 z}{\partial y^2}$.

To illustrate these four possibilities, consider the function:

$$z = \sin xy$$

$\dfrac{\partial z}{\partial x} = y \cos xy$ and $\dfrac{\partial z}{\partial y} = x \cos xy$. Then,

(i) $\dfrac{\partial^2 z}{\partial x^2} = \dfrac{\partial\left(\dfrac{\partial z}{\partial x}\right)}{\partial x} = \dfrac{\partial(y\cos xy)}{\partial x} = -y^2 \sin xy$

(ii) $\dfrac{\partial^2 z}{\partial y\,\partial x} = \dfrac{\partial\left(\dfrac{\partial z}{\partial x}\right)}{\partial y} = \dfrac{\partial(y\cos xy)}{\partial y}$

$$= y(-x \sin xy) + 1(\cos xy)$$
$$= \cos xy - xy \sin xy$$

(iii) $\dfrac{\partial^2 z}{\partial x\,\partial y} = \dfrac{\partial\left(\dfrac{\partial z}{\partial y}\right)}{\partial x} = \dfrac{\partial(x\cos xy)}{\partial x}$

$$= x(-y \sin xy) + 1(\cos xy)$$
$$= \cos xy - xy \sin xy$$

(iv) $\dfrac{\partial^2 z}{\partial y^2} = \dfrac{\partial\left(\dfrac{\partial z}{\partial y}\right)}{\partial y} = \dfrac{\partial(x\cos xy)}{\partial y} = -x^2 \sin xy$

It can be seen that the values of $\dfrac{\partial^2 z}{\partial y\,\partial x}$ and $\dfrac{\partial^2 z}{\partial x\,\partial y}$ are the same in this case. This result is true for all functions which are continuous. (A function which is continuous means that the graph of the function has no sudden jumps or breaks.)

Since $\dfrac{\partial^2 z}{\partial y\, \partial x}$ is always equal to $\dfrac{\partial^2 z}{\partial x\, \partial y}$ for functions being considered at this level, there are in fact only three second order partial derivatives for functions of two variables, namely:

$$\frac{\partial^2 z}{\partial x^2}, \quad \frac{\partial^2 z}{\partial y^2} \quad \text{and either} \quad \frac{\partial^2 z}{\partial x\, \partial y} \quad \text{or} \quad \frac{\partial^2 z}{\partial y\, \partial x}.$$

Second order partial derivatives will be met in such topic areas as entropy and the continuity theorem in thermodynamics, the waveguide theory in electrical engineering and in the solution of partial differential equations.

Worked problems on second order partial derivatives

Problem 1. If $u = f(x, y)$ and $u = \dfrac{x - y}{x + y}$, find

$$\frac{\partial^2 u}{\partial x^2}, \frac{\partial^2 u}{\partial y^2}, \frac{\partial^2 u}{\partial x\, \partial y} \quad \text{and} \quad \frac{\partial^2 u}{\partial y\, \partial x}.$$

To find $\dfrac{\partial u}{\partial x}$, the quotient rule of differentiation is used, y being treated as a constant. Hence

$$\frac{\partial u}{\partial x} = \frac{(x + y)(1) - (x - y)(1)}{(x + y)^2} = \frac{2y}{(x + y)^2} = 2y(x + y)^{-2}$$

$$\frac{\partial^2 u}{\partial x^2} = \frac{\partial \left(\dfrac{\partial u}{\partial x}\right)}{\partial x} = \frac{\partial [2y(x + y)^{-2}]}{\partial x}$$

$$= (2y)(-2)(x + y)^{-3}(1) = \frac{-4y}{(x + y)^3}$$

To find $\dfrac{\partial u}{\partial y}$, the quotient rule of differentiation is used, x being treated as a constant. Hence,

$$\frac{\partial u}{\partial y} = \frac{(x + y)(-1) - (x - y)(1)}{(x + y)^2} = \frac{-2x}{(x + y)^2}$$

$$= -2x(x + y)^{-2}$$

$$\frac{\partial^2 u}{\partial y^2} = \frac{\partial\left(\frac{\partial u}{\partial y}\right)}{\partial y} = \frac{\partial(-2x(x+y)^{-2})}{\partial y}$$

$$= (-2x)(-2)(x+y)^{-3}(1) = \frac{4x}{(x+y)^3}$$

$$\frac{\partial^2 u}{\partial x \, \partial y} = \frac{\partial\left(\frac{\partial u}{\partial y}\right)}{\partial x} = \frac{\partial\left(\frac{-2x}{(x+y)^2}\right)}{\partial x}$$

$$= \frac{(x+y)^2(-2) - (-2x)(2)(x+y)(1)}{(x+y)^4} = \frac{2(x-y)}{(x+y)^3}$$

$$\frac{\partial^2 u}{\partial y \, \partial x} = \frac{\partial\left(\frac{\partial u}{\partial x}\right)}{\partial y} = \frac{\partial\left(\frac{2y}{(x+y)^2}\right)}{\partial y}$$

$$= \frac{(x+y)^2(2) - 2y(2)(x+y)(1)}{(x+y)^4} = \frac{2(x-y)}{(x+y)^3}$$

The results obtained for $\dfrac{\partial^2 u}{\partial x \, \partial y}$ and $\dfrac{\partial^2 u}{\partial y \, \partial x}$ verify that in this case

the relationship stated in the text that $\dfrac{\partial^2 u}{\partial x \, \partial y} = \dfrac{\partial^2 u}{\partial y \, \partial x}$ is true.

Problem 2. If $z = f(x, y)$ and $z = x \cos(x+y)$, find $\dfrac{\partial^2 z}{\partial x^2}$ and $\dfrac{\partial^2 z}{\partial y^2}$.
Show that $\dfrac{\partial^2 z}{\partial x \, \partial y} = \dfrac{\partial^2 z}{\partial y \, \partial x}$.

Applying the product rule gives:

$$\frac{\partial z}{\partial x} = x(-\sin(x+y)) + 1 \cdot \cos(x+y)$$

$$= -x \sin(x+y) + \cos(x+y)$$

Differentiating $\dfrac{\partial z}{\partial x}$ with respect to x gives:

$$\frac{\partial^2 z}{\partial x^2} = -x \cos(x+y) + (-1)\sin(x+y) - \sin(x+y)$$

$$= -x \cos(x+y) - 2 \sin(x+y)$$

Differentiating $z = x \cos(x + y)$ with respect to y gives:

$$\frac{\partial z}{\partial y} = -x \sin(x + y).$$

Differentiating $\dfrac{\partial z}{\partial y}$ with respect to y gives:

$$\frac{\partial^2 z}{\partial y^2} = -x \cos(x + y)$$

$$\frac{\partial^2 z}{\partial x\, \partial y} = \frac{\partial\left(\dfrac{\partial z}{\partial y}\right)}{\partial x} = \frac{\partial(-x \sin(x + y))}{\partial x}$$

$$= -x \cos(x + y) + (-1) \sin(x + y)$$
$$= -x \cos(x + y) - \sin(x + y) \tag{1}$$

$$\frac{\partial^2 z}{\partial y\, \partial x} = \frac{\partial\left(\dfrac{\partial z}{\partial x}\right)}{\partial y} = \frac{\partial(-x \sin(x + y) + \cos(x + y))}{\partial y}$$

$$= -x \cos(x + y) - \sin(x + y) \tag{2}$$

Since equation (1) is equal to equation (2), then

$$\frac{\partial^2 z}{\partial x\, \partial y} = \frac{\partial^2 z}{\partial y\, \partial x}$$

Further problems on second order partial derivatives may be found in the following Section 15.4 (Problems 8 to 19), page 244.

15.4   Further problems

First order partial derivatives

1. If $z = f(x, y)$, find $\dfrac{\partial z}{\partial x}$ and $\dfrac{\partial z}{\partial y}$ for the following equations:

   (a) $z = x^2 + 3xy + y^3$ (b) $z = (3x - 2y)^2$ (c) $z = e^y \ln(x + y)$

   (a) $\left[\dfrac{\partial z}{\partial x} = 2x + 3y, \dfrac{\partial z}{\partial y} = 3(x + y^2) \right]$

   (b) $\left[\dfrac{\partial z}{\partial x} = 6(3x - 2y), \dfrac{\partial z}{\partial y} = -4(3x - 2y) \right]$

   (c) $\left[\dfrac{\partial z}{\partial x} = \dfrac{e^y}{x + y}, \dfrac{\partial z}{\partial y} = e^y\left(\dfrac{1}{x + y} + \ln(x + y)\right) \right]$

2. If $u = f(p, q)$, find $\dfrac{\partial u}{\partial p}$ and $\dfrac{\partial u}{\partial q}$ for the following equations:

(a) $u = \dfrac{1}{\sqrt{(p^2 + q^2)}}$ (b) $u = e^{(4p + q/2)} \tan(3p + 2q)$

(c) $u = pe^q + \dfrac{1}{p} \ln q$

(a) $\left[\dfrac{\partial u}{\partial p} = -\dfrac{p}{\sqrt{(p^2 + q^2)^3}}, \dfrac{\partial u}{\partial q} = -\dfrac{q}{\sqrt{(p^2 + q^2)^3}} \right]$

(b) $\left[\dfrac{\partial u}{\partial p} = e^{(4p + q/2)} \{3 \sec^2(3p + 2q) + 4 \tan(3p + 2q)\} \right.$

$\left. \dfrac{\partial u}{\partial q} = e^{(4p + q/2)} \{2 \sec^2(3p + 2q) + \dfrac{1}{2} \tan(3p + 2q)\} \right]$

(c) $\left[\dfrac{\partial u}{\partial p} = e^q - \dfrac{\ln q}{p^2}, \dfrac{\partial u}{\partial q} = p e^q + \dfrac{1}{pq} \right]$

3. If $m = f(u, v)$ and $m = \cos uv$, prove that

$$\dfrac{1}{v} \dfrac{\partial m}{\partial u} = \dfrac{1}{u} \dfrac{\partial m}{\partial v}$$

4. The volume V of a right circular cone of height h and base radius r is given by:

$$V = \dfrac{1}{3} \pi r^2 h$$

Find $\dfrac{\partial V}{\partial r}$ and $\dfrac{\partial V}{\partial h}$ $\left[\dfrac{\partial V}{\partial r} = \dfrac{2}{3} \pi r h, \dfrac{\partial V}{\partial h} = \dfrac{1}{3} \pi r^2 \right]$

5. The time of oscillation T of a simple pendulum is given by $T = 2\pi \sqrt{\left(\dfrac{l}{g}\right)}$, where l is the length of the pendulum and g is the free fall acceleration due to gravity. Find $\dfrac{\partial T}{\partial l}$ and $\dfrac{\partial T}{\partial g}$.

$$\left[\dfrac{\partial T}{\partial l} = \dfrac{\pi}{\sqrt{(lg)}}, \dfrac{\partial T}{\partial g} = -\pi \sqrt{\left(\dfrac{l}{g^3}\right)} \right]$$

6. If $z = f(p - q)$, prove that

$$\dfrac{\partial z}{\partial p} + \dfrac{\partial z}{\partial q} = 0$$

7. If $z = m^2 f(m - n)$, show that

$$\dfrac{\partial z}{\partial m} + \dfrac{\partial z}{\partial n} = \dfrac{2z}{m}$$

Second order partial derivatives

In Problems 8 to 10, find (a) $\dfrac{\partial^2 z}{\partial x^2}$, (b) $\dfrac{\partial^2 z}{\partial y\,\partial x}$, (c) $\dfrac{\partial^2 z}{\partial x\,\partial y}$ and

(d) $\dfrac{\partial^2 z}{\partial y^2}$, given that $z = f(x, y)$.

8. $z = \ln xy$ (a) $\left[-\dfrac{1}{x^2}\right]$ (b) $[0]$ (c) $[0]$ (d) $\left[-\dfrac{1}{y^2}\right]$

9. $z = x\cos y - y\cos x$ (a) $[y\cos x]$ (b) $[\sin x - \sin y]$
   (c) $[\sin x - \sin y]$ (d) $[-x\cos y]$

10. $z = \dfrac{x}{x + y}$ (a) $\left[-\dfrac{2y}{(x+y)^3}\right]$ (b) $\left[\dfrac{x-y}{(x+y)^3}\right]$

   (c) $\left[\dfrac{x-y}{(x+y)^3}\right]$ (d) $\left[\dfrac{2x}{(x+y)^3}\right]$

11.   If $z = f(x, y)$ and $z = e^{-y}\sin x$, show that

$$\frac{\partial^2 z}{\partial x^2} + \frac{\partial^2 z}{\partial y^2} = 0$$

12.   If $z = f(x, t)$ and $z = \sqrt{\left(\dfrac{x}{4t}\right)}$, find $\dfrac{\partial^2 z}{\partial x^2}$ and show that

$$\frac{\partial^2 z}{\partial x\,\partial t} = \frac{\partial^2 z}{\partial t\,\partial x}\qquad\left[-\frac{1}{8\sqrt{(tx^3)}}\right]$$

13.   If $z = f(p, q)$ and $z = p^3\tan\dfrac{p}{q}$, prove that

$$\frac{\partial^2 z}{\partial p\,\partial q} = \frac{\partial^2 z}{\partial q\,\partial p} = -\frac{2p^3}{q^2}\left(\sec^2\frac{p}{q}\right)\left\{\frac{p}{q}\tan\left(\frac{p}{q}\right) + 2\right\}$$

14.   If $u = f(p, q)$ and $u = \dfrac{2p}{q^2}\ln q$, prove that

$$\frac{\partial^2 u}{\partial q\,\partial p} - \frac{\partial^2 u}{\partial p\,\partial q} = 0$$

15.   If $\phi = f(r, \theta)$ and $\phi = (Ar^n + Br^{-n})\sin(n\theta + \alpha)$ where A, B, n and α are constants, show that

$$\frac{\partial^2 \phi}{\partial r^2} + \frac{1}{r}\frac{\partial \phi}{\partial r} + \frac{1}{r^2}\frac{\partial^2 \phi}{\partial \theta^2} = 0$$

16. An equation used in thermodynamics is the Benedict–Webb–Rubine equation of state for the expansion of a gas. The equation is:

$$p = \frac{RT}{V} + \left(B_0 RT - A_0 - \frac{C_0}{T^2}\right)\frac{1}{V^2} + (bRT - a)\frac{1}{V^3} + \frac{A\alpha}{V^6}$$
$$+ \frac{C(1 + \gamma/V^2)}{T^2}\left(\frac{1}{V^3}\right)e^{-\gamma/V^2}$$

Show that $\dfrac{\partial^2 P}{\partial T^2} = \dfrac{6}{V^2 T^4}\left\{\dfrac{C}{V}\left(1 + \dfrac{\gamma}{V^2}\right)e^{-\gamma/V^2} - C_0\right\}.$

15. An equation resulting from plucking a string is

$$y = \sin\frac{n\pi x}{l}\left\{k\cos\frac{n\pi b}{l}t + c\sin\frac{n\pi b}{l}t\right\}$$

Determine $\dfrac{\partial y}{\partial t}$ and $\dfrac{\partial y}{\partial x}$.

$$\left[\begin{array}{l}\dfrac{\partial y}{\partial t} = \dfrac{bn\pi}{l}\sin\dfrac{n\pi}{l}x\left\{c\cos\dfrac{n\pi b}{l}t - k\sin\dfrac{n\pi b}{l}t\right\} \\[3mm] \dfrac{\partial y}{\partial x} = \dfrac{n\pi}{l}\cos\dfrac{n\pi}{l}x\left\{k\cos\dfrac{n\pi b}{l}t + c\sin\dfrac{n\pi b}{l}t\right\}\end{array}\right]$$

18. The magnetic field vector H due to a steady current I flowing around a circular wire of radius a and at a distance z from its centre is given by:

$$H = \pm\frac{I}{2}\frac{\partial}{\partial z}\left(\frac{z}{\sqrt{(a^2 + z^2)}}\right)$$

Determine H. $\left[H = \pm\dfrac{a^2 I}{2\sqrt{(a^2 + z^2)^3}}\right]$

19. In a thermodynamic system
$k = A\,e^{(T\Delta S - \Delta H)/RT}$,

where R, k and A are constants.

Find $\dfrac{\partial k}{\partial T}, \dfrac{\partial A}{\partial T}, \dfrac{\partial(\Delta S)}{\partial T}$ and $\dfrac{\partial(\Delta H)}{\partial T}$.

$$\left[\begin{array}{l}\dfrac{\partial k}{\partial T} = \dfrac{A\Delta H}{RT^2}e^{(T\Delta S - \Delta H)/RT} \\[3mm] \dfrac{\partial A}{\partial T} = \dfrac{-k\Delta H}{RT^2}e^{(\Delta H - T\Delta S)/RT} \\[3mm] \dfrac{\partial(\Delta S)}{\partial T} = \dfrac{-\Delta H}{T^2}, \dfrac{\partial(\Delta H)}{\partial T} = \Delta S - R\ln\left(\dfrac{k}{A}\right)\end{array}\right]$$

Chapter 16

Total differential, rates of change and small changes

16.1 Total differential

In Chapter 15 the principles of partial differentiation are introduced for the case where only one variable changes at a time, the other variables being kept constant. In practice, variables may all be changing at the same time. Consider a right cylindrical block of metal being heated for a time δt. Let the initial volume of the block, V, be $\pi r^2 h$, where r is the radius of the base and h is the height of the cylinder. If the metal of the block has a positive temperature coefficient then both r and h will increase as the block is heated. Let the increase in r be δr, the increase in h be δh and the resultant increase in volume be δV. Then, after time δt:

the new volume,

$$V + \delta V = \pi (r + \delta r)^2 (h + \delta h)$$

The change in volume,

$$\delta V = \pi (r + \delta r)^2 (h + \delta h) - V$$
$$= \pi (r^2 + 2r\delta r + \delta r^2)(h + \delta h) - \pi r^2 h$$

i.e.
$$\delta V = \pi r^2 h + 2\pi r h \delta r + \pi h \delta r^2 + \pi r^2 \delta h + 2\pi r \delta r \delta h$$
$$+ \pi \delta r^2 \delta h - \pi r^2 h$$

i.e.
$$\delta V = 2\pi r h \delta r + \pi h \delta r^2 + \pi r^2 \delta h + 2\pi r \delta r \delta h + \pi \delta r^2 \delta h \qquad (1)$$

Change of volume in unit time, $\dfrac{\delta V}{\delta t}$, is given by:

$$\frac{\delta V}{\delta t} = 2\pi r h \frac{\delta r}{\delta t} + \pi h \delta r \frac{\delta r}{\delta t} + \pi r^2 \frac{\delta h}{\delta t} + 2\pi r \delta r \frac{\delta h}{\delta t} + \pi \delta r^2 \frac{\delta h}{\delta t} \qquad (2)$$

As $\delta t \to 0$, the ratios $\dfrac{\delta V}{\delta t}, \dfrac{\delta r}{\delta t}$ and $\dfrac{\delta h}{\delta t}$ become $\dfrac{dV}{dt}, \dfrac{dr}{dt}$ and $\dfrac{dh}{dt}$ respectively, and terms containing the factors δr and δr^2 become zero. Hence rate of increase in volume, $\dfrac{dV}{dt} = 2\pi r h \dfrac{dr}{dt} + \pi r^2 \dfrac{dh}{dt}$. But it is shown in Chapter 15 that if $V = \pi r^2 h$, then $\dfrac{\partial V}{\partial r} = 2\pi r h$ and $\dfrac{\partial V}{\partial h} = \pi r^2$

Thus, $\qquad \dfrac{\mathrm{d}V}{\mathrm{d}t} = \dfrac{\partial V}{\partial r}\dfrac{\mathrm{d}r}{\mathrm{d}t} + \dfrac{\partial V}{\partial h}\dfrac{\mathrm{d}h}{\mathrm{d}t}$ $\qquad\qquad$ (3)

Equation (3) may be made independent of time by multiplying throughout by $\mathrm{d}t$, giving

$$\frac{\mathrm{d}V}{\mathrm{d}t}\,\mathrm{d}t = \frac{\partial V}{\partial r}\frac{\mathrm{d}r}{\mathrm{d}t}\,\mathrm{d}t + \frac{\partial V}{\partial h}\frac{\mathrm{d}h}{\mathrm{d}t}\,\mathrm{d}t$$

i.e. $\qquad \mathbf{d}V = \dfrac{\partial V}{\partial r}\,\mathbf{d}r + \dfrac{\partial V}{\partial h}\,\mathbf{d}h$ $\qquad\qquad$ (4)

In this form, $\mathrm{d}V$ is called the **total differential** of V and $\dfrac{\partial V}{\partial r}\,\mathrm{d}r$ and $\dfrac{\partial V}{\partial h}\,\mathrm{d}h$ are called the **partial differentials** of V. Thus the total differential is the sum of the separate partial differentials. The total differential can be used as a basis for finding the change of volume with respect to some other quantity. For example, to find the change of volume with respect to, say, thermodynamic temperature, T, each term is divided by $\mathrm{d}T$, giving

$$\frac{\mathrm{d}V}{\mathrm{d}T} = \frac{\partial V}{\partial r}\frac{\mathrm{d}r}{\mathrm{d}T} + \frac{\partial V}{\partial h}\frac{\mathrm{d}h}{\mathrm{d}T}$$

The total differential is also used as a basis for solving partial differential equations.

The result for the total differential of a right circular cylindrical block can be generalised. If $z = f(u, v, w)$, then from equation (4),

$$\mathrm{d}z = \frac{\partial z}{\partial u}\,\mathrm{d}u + \frac{\partial z}{\partial v}\,\mathrm{d}v + \frac{\partial z}{\partial w}\,\mathrm{d}w + \cdots \qquad\qquad (5)$$

Worked problems on total differentials

Problem 1. If $z = f(x, y)$ and $z = x^3 y^2 + \dfrac{y}{x}$, find the total differential, $\mathrm{d}z$.

$\dfrac{\partial z}{\partial x} = y^2(3x^2) - \dfrac{y}{x^2}$ (y is kept constant)

$\dfrac{\partial z}{\partial y} = x^3(2y) + \dfrac{1}{x}$ (x is kept constant)

The total differential is the sum of the partial differentials,

i.e. $dz = \dfrac{\partial z}{\partial x} dx + \dfrac{\partial z}{\partial y} dy$

Thus, $\mathbf{dz} = \left(3x^2y^2 - \dfrac{y}{x^2}\right)\mathbf{dx} + \left(2x^3y + \dfrac{1}{x}\right)\mathbf{dy}$

Problem 2. The pressure p, volume V and temperature T of unit mass of a gas are related by the formula $pV = RT$, where R is a constant.

Prove that $dp = \dfrac{p}{T} dT - \dfrac{p}{V} dV$, and that

$$dT = \dfrac{T}{V} dV + \dfrac{T}{p} dp$$

Since $pV = RT$, $p = \dfrac{RT}{V}$, $\dfrac{\partial p}{\partial T} = \dfrac{R}{V}$ and $\dfrac{\partial p}{\partial V} = -\dfrac{RT}{V^2}$

From equation (5), $dp = \dfrac{\partial p}{\partial T} dT + \dfrac{\partial p}{\partial V} dV$

$$= \dfrac{R}{V} dT + \left(-\dfrac{RT}{V^2}\right) dV$$

But from the initial equation, $\dfrac{R}{V} = \dfrac{p}{T}$, hence substituting for $\dfrac{R}{V}$ gives:

$$dp = \dfrac{p}{T} dT - \left(\dfrac{p}{T} \cdot \dfrac{T}{V}\right) dV$$

$$= \dfrac{p}{T} dT - \dfrac{p}{V} dV$$

Since $pV = RT$, $T = \dfrac{pV}{R}$, $\dfrac{\partial T}{\partial p} = \dfrac{V}{R}$ and $\dfrac{\partial T}{\partial V} = \dfrac{p}{R}$

From equation (5), $dT = \dfrac{\partial T}{\partial V} dV + \dfrac{\partial T}{\partial p} dp$

$$= \dfrac{p}{R} dV + \dfrac{V}{R} dp$$

But from the initial equation, $\dfrac{p}{R} = \dfrac{T}{V}$ and $\dfrac{V}{R} = \dfrac{T}{p}$,

hence, $\mathbf{d}T = \dfrac{T}{V}\mathbf{d}V + \dfrac{T}{p}\mathbf{d}p$.

Further problems on total differentials may be found in Section 16.4 (Problems 1 to 5), page 254.

16.2   Rates of change

It is shown in Section 16.1, equation (3), that for a right circular cylindrical block being heated, the rate of change of volume is given by

$$\frac{dV}{dt} = \frac{\partial V}{\partial r}\frac{dr}{dt} + \frac{\partial V}{\partial h}\frac{dh}{dt}$$

This result may be generalised and if $z = f(u, v, w, \ldots)$ then

$$\frac{dz}{dt} = \frac{\partial z}{\partial u}\frac{du}{dt} + \frac{\partial z}{\partial v}\frac{dv}{dt} + \frac{\partial z}{\partial w}\frac{dw}{dt} + \cdots \tag{6}$$

Equation (6) may be used to solve problems in which different quantities have different rates of change. The principles used are shown in the worked problems following.

Worked problems on rates of change

Problem 1. If $z = f(u, v)$ and $z = e^u \sin 2v$, find the rate of change of z, correct to four decimal places, when u is increasing at 3 units per second, v is decreasing at 5 units per second, u is 0.1 units and v is 0.5 units.

Since $z = e^u \sin 2v$, $\dfrac{\partial z}{\partial u} = e^u \sin 2v$ and $\dfrac{\partial z}{\partial v} = 2\,e^u \cos 2v$

If u is increasing at 3 units/s, then $\dfrac{du}{dt} = 3$. Similarly, $\dfrac{dv}{dt} = -5$

From equation (6), $\dfrac{dz}{dt} = \dfrac{\partial z}{\partial u}\dfrac{du}{dt} + \dfrac{\partial z}{\partial v}\dfrac{dv}{dt}$

$$= (e^u \sin 2v)(3) + (2\,e^u \cos 2v)(-5)$$

But $u = 0.1$ and $v = 0.5$ units, hence

$$\frac{dz}{dt} = 3\,e^{0.1}\sin 1 - 10\,e^{0.1}\cos 1$$

Sin 1 means the sine of 1 radian, i.e. $\sin\left(\frac{180}{\pi}\right)^{\circ}$

Also cos 1 means $\cos\left(\frac{180}{\pi}\right)^{\circ}$

Hence, $\frac{dz}{dt} = e^{0.1}\left[3\sin\left(\frac{180}{\pi}\right)^{\circ} - 10\cos\left(\frac{180}{\pi}\right)^{\circ}\right]$, i.e. $\frac{dz}{dt} = -3.181\,4$, correct to four decimal places.

That is, z **is decreasing at 3.181 4 units per second**, correct to four decimal places.

Problem 2. The radius of a right circular cylinder is increasing at a rate of 2 cm/s and the height is decreasing at a rate of 3 cm/s. Find the rate at which the volume is changing when the radius is 8 cm and the height is 5 cm.

The volume of a right cylinder, $V = \pi r^2 h$

From equation (6), $\frac{dV}{dt} = \frac{\partial V}{\partial r}\frac{dr}{dt} + \frac{\partial V}{\partial h}\frac{dh}{dt}$

$$\frac{\partial V}{\partial r} = 2\pi r h \text{ and } \frac{\partial V}{\partial h} = \pi r^2$$

Since the radius is increasing at 2 cm/s, $\frac{dr}{dt} = 2$

Since the height is decreasing at 3 cm/s, $\frac{dh}{dt} = -3$

Thus, rate of change of volume, $\frac{dV}{dt} = (2\,\pi r h)(2) + \pi r^2(-3)$

When $r = 8$ and $h = 5$, $\frac{dV}{dt} = 4\,\pi(8)(5) - 3\,\pi(8)^2$

$$= 160\,\pi - 192\,\pi$$

$$= -32\,\pi$$

i.e. **the volume is decreasing at a rate of 32 π cm^3/s.**

Problem 3. A rectangular box has sides of length x mm, y mm and z mm. Sides x and y are expanding at rates of 1.5 and 2.5 mm/s respectively and side z is contracting at a rate of 0.25 mm/s. Find the rate of change of volume when x is 15 mm, y is 12 mm and z is 10 mm.

Volume of box, $V = xyz$

From equation (6),

$$\frac{dV}{dt} = \frac{\partial V}{\partial x}\frac{dx}{dt} + \frac{\partial V}{\partial y}\frac{dy}{dt} + \frac{\partial V}{\partial z}\frac{dz}{dt}$$

$$\frac{\partial V}{\partial x} = yz, \quad \frac{\partial V}{\partial y} = xz \text{ and } \frac{\partial V}{\partial z} = xy$$

Also, $\dfrac{dx}{dt} = 1.5, \dfrac{dy}{dt} = 2.5$ and $\dfrac{dz}{dt} = -0.25$

Hence, $\dfrac{dV}{dt} = yz(1.5) + xz(2.5) + xy(-0.25)$

When $x = 15, y = 12$ and $z = 10$,

$$\frac{dV}{dt} = 12 \times 10 \times 1.5 + 15 \times 10 \times 2.5 - 15 \times 12 \times 0.25$$

$$= +510,$$

i.e. **the rate of increase of volume is 510 mm^3/s.**

Further problems on rates of change may be found in Section 16.4 (Problems 6 to 10), page 255.

16.3   Small changes

It is shown in Section 16.1, equation (1), that for a right circular cylinder being heated, the change of volume is given by

$$\delta V = 2\pi r h \delta r + \pi h \delta r^2 + \pi r^2 \delta h + 2\pi r \delta r \delta h + \pi \delta r^2 \delta h$$

If δr and δh are small compared with r and h, then products of these small terms such as δr^2, $\delta r \delta h$ and $\delta r^2 \delta h$ will be very small and may be neglected. Thus the change of volume simplifies to:

$$\delta V \approx 2\pi r h \delta r + \pi r^2 \delta h$$

But, $V = \pi r^2 h$, so $\dfrac{\partial V}{\partial r} = 2\pi r h$ and $\dfrac{\partial V}{\partial h} = \pi r^2$

Hence, $\delta V \approx \dfrac{\partial V}{\partial r}\delta r + \dfrac{\partial V}{\partial h}\delta h$

This result may be generalised and if $z = f(u, v, w, \ldots)$, then the approximate change in z, δz, for small changes in u, v and w, denoted by δu, δv and δw respectively, is given by

$$\delta z \approx \frac{\partial z}{\partial u} \delta u + \frac{\partial z}{\partial v} \delta v + \frac{\partial z}{\partial w} \delta w + \cdots \tag{7}$$

Worked problems on small changes

Problem 1. The current in an electrical circuit is given by $i = \dfrac{v}{R}$ amperes. Determine the approximate change in current if the voltage falls from 240 V to 238 V and the value of the resistance is increased from 100 Ω to 100.5 Ω.

The relationship is $i = f(v, R)$ and $i = \dfrac{v}{R}$.

From equation (7), the change in current $\delta i \approx \dfrac{\partial i}{\partial v} \delta v + \dfrac{\partial i}{\partial R} \delta R$.

Since $i = \dfrac{v}{R}$, $\dfrac{\partial i}{\partial v} = \dfrac{1}{R}$ and $\dfrac{\partial i}{\partial R} = -\dfrac{v}{R^2}$.

The small change in voltage δv is -2 volts and change in resistance δR is $+0.5$ ohms.

Thus, $\delta i = \dfrac{1}{R}(-2) + \left(-\dfrac{v}{R^2}\right)(0.5)$.

But $R = 100 \, \Omega$ and $v = 240$ volts (the initial values),

hence, $\delta i = -\dfrac{2}{100} - \dfrac{240}{2 \times 100^2} = -0.032$.

That is, the approximate change in current is a decrease of 0.032 A.

Problem 2. The pressure p and volume V of a gas are related by the equation $pV^{1.4} = C$.

Find the approximate percentage change in C when the pressure is increased by 2.3 per cent and the volume is decreased by 0.84 per cent.

Let p, V and C refer to the initial values.

From equation (7), $\delta C = \dfrac{\partial C}{\partial p} \delta p + \dfrac{\partial C}{\partial V} \delta V$.

Since $C = f(p, V)$ and $C = pV^{1.4}$, then

$$\frac{\partial C}{\partial p} = V^{1.4} \text{ and } \frac{\partial C}{\partial V} = 1.4\, pV^{0.4}.$$

The pressure is increased by 2.3 per cent, i.e. the change in pressure

$$\delta p = \frac{2.3}{100} \cdot p = 0.023\, p$$

The volume is decreased by 0.84 per cent, i.e. the change in volume

$$\delta V = -\frac{0.84}{100} \cdot V = -0.008\,4\, V.$$

Hence, the approximate change in C,

$$\delta C \approx V^{1.4}\,(0.023\, p) + 1.4\, pV^{0.4}(-0.008\,4\, V)$$

$$\approx pV^{1.4}(0.023 - 1.4(0.008\,4))$$

$$\approx 0.011\,2\, pV^{1.4}$$

i.e. $\delta C \approx \dfrac{1.12}{100} \cdot C$

That is, the approximate change in C is a 1.12 per cent increase.

Problem 3. The side a of a triangle is calculated from:
$a^2 = b^2 + c^2 - 2\, bc \cos A$. If b, c and A are measured as 2 mm,
4 mm and $60°$ respectively and the measurement errors which
occur are $+0.1$ mm, $+0.15$ mm and $+2°$ respectively, determine
the error in the calculated value of a.

Since $a^2 = b^2 + c^2 - 2bc \cos A$

$$a = (b^2 + c^2 - 2bc \cos A)^{\frac{1}{2}}$$

From equation (7), $\delta a \approx \dfrac{\partial a}{\partial b}\, \delta b + \dfrac{\partial a}{\partial c}\, \delta c + \dfrac{\partial a}{\partial A}\, \delta A$

$$\frac{\partial a}{\partial b} = \frac{1}{2}(b^2 + c^2 - 2bc \cos A)^{-\frac{1}{2}}(2b - 2c \cos A)$$

$$\frac{\partial a}{\partial c} = \frac{1}{2}(b^2 + c^2 - 2bc \cos A)^{-\frac{1}{2}}(2c - 2b \cos A)$$

and $\dfrac{\partial a}{\partial A} = \dfrac{1}{2}(b^2 + c^2 - 2bc \cos A)^{-\frac{1}{2}}(2bc \sin A)$

Since $b = 2, c = 4$ and $A = 60° = \dfrac{\pi}{3}$ rad.

$$b^2 + c^2 - 2bc \cos A = 4 + 16 - 2 \times 8 \times \dfrac{1}{2}$$

$$= 12$$

and $\dfrac{1}{2}(b^2 + c^2 - 2bc \cos A)^{-\frac{1}{2}} = \dfrac{1}{2\sqrt{12}} \approx 0.144\,3$

Hence $\dfrac{\partial a}{\partial b} \approx 0.144\,3(4 - 4) = 0$

$\dfrac{\partial a}{\partial c} \approx 0.144\,3(8 - 2) \approx 0.866$

$\dfrac{\partial a}{\partial A} \approx 0.144\,3\left(16\,\dfrac{\sqrt{3}}{2}\right) \approx 1.999$

Also $\delta b = 0.1,\ \delta c = 0.15$ and $\delta A = 2° = \dfrac{2 \times \pi}{180} = 0.035$ rad

Hence, $\delta a \approx 0 + 0.866 \times 0.15 + 1.999 \times 0.035$

$$\approx 0.199 \text{ mm}$$

i.e. **the approximate error in the calculated value of a is $+0.20$ mm.**

Further problems on small changes may be found in the following Section 16.4 (Problems 11 to 22), page 256.

16.4   Further problems

Total differential

1. Determine the total differential, dz, given $z = f(x, y)$ and
    $z = 3xy^2 - \dfrac{\sqrt{x}}{y^3}$.
    $$\left[dz = \left(3y^2 - \dfrac{1}{2\sqrt{xy^3}} \right) dx + \left(6xy + \dfrac{3\sqrt{x}}{y^4} \right) dy \right]$$

2. Find the total differential, dz, if $z = f(r, \theta)$ and
    $z = e^r(\cos \theta + j \sin \theta)$.
    $[dz = e^r(\cos \theta + j \sin \theta)\, dr + e^r(-\sin \theta + j \cos \theta)\, d\theta]$

3. If $z = f(u, v, w)$ and $z = 2u^3 + 3v^2 - 4w$, find the total
    differential, dz. $[dz = 6u^2\, du + 6v\, dv - 4\, dw]$

4. If $z = p\,e^q + \dfrac{1}{p}\ln q$, where both p and q are variables, show that

$$dz = \left(e^q - \frac{1}{p^2}\ln q\right)dp + \left(p\,e^q + \frac{1}{pq}\right)dq$$

5. If $xyz = k$, where k is a constant, prove that

$$dz = -z\left(\frac{dx}{x} + \frac{dy}{y}\right)$$

Rates of change

6. If $z = 2p^3\tan\dfrac{q}{2}$, p is decreasing at 5 units per second and q is decreasing at 0.04 units per second, find the rate of change of z when $p = 1$ and $q = \pi/4$, correct to two decimal places.
   [z is decreasing at 12.47 units per second]

7. Determine the rate of change of m, correct to four significant figures, given the following data:

   $m = f(u, v, w)$

   $m = 3\ln\dfrac{v}{3} + w^3\sec u$

   v is decreasing at 0.4 cm/s
   u is increasing at 0.2 cm/s
   w is increasing at 0.7 cm/s
   $u = 3$ cm, $v = 3.7$ cm and $w = 4.4$ cm
   [m is decreasing at 38.94 cm/s, correct to four significant figures]

8. If $z = f(x, y)$ and $z = e^{y/2}\ln(2x + 3y)$, find the rate of increase of z, correct to four significant figures, when $x = 2$ cm, $y = 3$ cm, x is increasing at 5 cm/s and y is increasing at 4 cm/s.   [30.58 cm/s]

9. The radius of a right circular cone is increasing at a rate of 1 mm/s and its height is increasing at 2 mm/s. Determine, correct to three decimal places, the rate at which the volume is increasing in cm³/s when the radius is 1.2 cm and the height is 3.6 cm.   [1.206 cm³/s]

10. The area of a triangle is given by: area $= \dfrac{1}{2}bc\sin A$, where A is the angle between sides b and c. If b is increasing at 0.4 units/s, c is decreasing at 0.25 units/s and A is increasing at 0.1 units/s, find the rate of change of area of the triangle, correct to four decimal places, when b is 2 units, c is 3 units and A is 0.5 units.   [Increasing at 0.4311 units²/s]

Small changes

11. An error of 3.5 per cent too large is made when measuring
     the radius of a sphere. Determine the approximate error
     in calculating: (a) the volume and (b) the surface area
     when they are calculated using the incorrect radius
     measurement.   (a) [10.5 per cent too large]
     (b) [7 per cent too large]

12. The area of a triangle is given by $A = \frac{1}{2} ab \sin C$, where C is

     the angle between sides a and b of the triangle. Calculate the
     approximate change in area when: (a) both a and b are
     increased by 2 per cent and (b) when a is increased by
     2 per cent and b is reduced by 2 per cent.
     (a) [4 per cent increase]   (b) [no change]

13. The moment of inertia of a body about an axis is given
     by $I = kbd^3$ where k is a constant and b and d are
     the dimensions of the body. If b and d are measured as
     2 m and 0.8 m respectively and the measurement errors are
     10 cm in b and -8 mm in d, determine the error in the
     calculated value of the moment of inertia using the measured
     values, in terms of k.   [$0.02k$ m^4]

14. The radius of a cone is reduced from 10 cm to 9.55 cm and
     its height is increased from 20 cm to 20.3 cm. Determine the
     approximate percentage change in its volume.
     [7.5 per cent reduction]

15. The power developed by an engine is given by $I = b\,PLAN$,
     where b is a constant. Find the approximate change in
     power in terms of b (Pa m^3/min units) when:

     P is increased from 2×10^5 to 2.1×10^5 Pa
     L is reduced from 8.3 to 8.1 cm
     A is reduced from 2.7 cm^2 to 2.6 cm^2, and
     N is increased from 300 to 302 rev/min

     [A reduction of $6b$ (Pa m^3/min) units]

16. The modulus of rigidity G is given by $G = \dfrac{R^4\theta}{L}$, where R is

     the radius, θ the angle of twist and L the length. Find the
     approximate percentage change in G when R is increased by
     1.5 per cent and θ is reduced by 5 per cent.
     [An increase of 1 per cent]

17. In triangle ABC, AB = 50 mm, AC = 60 mm and angle BAC = 45°. If length can be measured correct to the nearest 0.1 mm, and angles correct to the nearest 0.5°, determine the approximate value of maximum error which can occur in the calculated value of BC based on the measured results. [0.5 mm]

16. Side b of a triangle is calculated using the cosine rule. Determine the approximate error in b using the measured values if the measured values of a, c and B are 120 mm, 180 mm and 32° respectively, and the actual values are 121 mm, 179 mm and 32.25° respectively. [−0.6 mm]

19. The power consumed in an electrical resistor is given by $P = \dfrac{E^2}{R}$ watts, where E is the voltage drop across the resistor in volts and R the resistance of the resistor in ohms. Determine the approximate change in power when E increases by 4 per cent and R decreases by 0.05 per cent, if the original values of E and R are 100 volts and 10 ohms respectively. [An increase of 80.5 watts]

20. The resonant frequency of a series-connected electrical circuit is given by $f = \dfrac{1}{2\pi\sqrt{(LC)}}$ Hz, where L is the inductance in henrys and C the capacitance in farads. Determine the approximate percentage change in the resonant frequency when L is increased by 1.7 per cent and C is reduced by 3.4 per cent. [An increase of 0.85 per cent]

21. The rate of flow of gas in a pipe is given by $v = Cd^{\frac{1}{2}}T^{-\frac{5}{6}}$, where C is a constant, d is the diameter of the pipe and T is the thermodynamic temperature of the gas. When determining the rate of flow experimentally, d is measured and subsequently found to be in error by +1.6 per cent, and T has an error of −1.2 per cent. Determine the percentage error in the rate of flow based on the measured values of d and T. [1.8 per cent too large]

22. The volume (V) of a liquid of viscosity coefficient (η) delivered after time, t, when passed through a tube of length l and diameter d by a pressure p is given by:

$$V = \frac{pd^4t}{128\,\eta l}$$

If the errors in V, p and l are 1 per cent, 2 per cent and 3 per cent respectively, determine the error in η. [−2 per cent]

Chapter 17

Revision of basic integration

17.1 Introduction

The process of integration reverses the process of differentiation. In differentiation, if $f(x) = 3x^2$ then $f'(x) = 6x$. Since integration reverses the process of moving from $f(x)$ to $f'(x)$, it follows that the integral of $6x$ is $3x^2$, i.e., it is the process of moving from $f'(x)$ to $f(x)$. Similarly, if $y = 4x^3$ then $\dfrac{dy}{dx} = 12x^2$. Reversing the process shows that the integral of $12x^2$ is $4x^3$.

Integration is also a process of summation or adding parts together and an elongated S, shown as $\int$, is used to replace the words 'the integral of'. Hence, from above, $\int 6x = 3x^2$ and $\int 12x^2$ is $4x^3$.

In differentiation, the differential coefficient $\dfrac{dy}{dx}$ or $\dfrac{d}{dx}[f(x)]$ indicates that a function of x is being differentiated with respect to x, the dx indicating this. In integration, the variable of integration is shown by adding d(the variable) after the function to be integrated. Thus $\int 6x\, dx$ means 'the integral of $6x$ with respect to x', and $\int 5t^2\, dt$ means 'the integral of $5t^2$ with respect to t'.

Arbitrary constant of integration

The differential coefficient of $3x^2$ is $6x$, hence $\int 6x\, dx = 3x^2$. However, the differential coefficient of $(3x^2 + 2)$ is also $6x$, hence $\int (3x^2 + 2)\, dx = 6x$. Since the differential coefficient of any constant is zero, it follows that the differential coefficient of $3x^2 + c$, where c is any constant, is $6x$. To allow for the possible presence of this constant, whenever the process of integration is performed the constant should be added to the result.

Hence $\int 6x\, dx = 3x^2 + c$ and $\int 12x^2\, dx = 4x^3 + c$

'c' is called the **arbitrary constant of integration**.

17.2   The general solution of integrals of the form ax^n

The general solution of integrals of the form $\int ax^n \, dx$, where a and n are constants, is given by:

$$\int ax^n \, dx = \frac{ax^{n+1}}{n+1} + c$$

This rule is true when n is fractional, zero, or a positive or negative integer, with the exception of $n = -1$.

Using the general rule:

$$\int 6x \, dx = \frac{6x^{1+1}}{1+1} + c = \frac{6x^2}{2} + c = 3x^2 + c \text{ (as obtained above)}$$

$$\int \frac{1}{x^2} \, dx = \int x^{-2} \, dx = \frac{x^{-2+1}}{-2+1} + c = \frac{x^{-1}}{-1} + c = \frac{-1}{x} + c \text{ and}$$

$$\int 3\sqrt{x} \, dx = \int 3x^{\frac{1}{2}} \, dx = \frac{3x^{\frac{1}{2}+1}}{\frac{1}{2}+1} + c = \frac{3x^{\frac{3}{2}}}{\frac{3}{2}} + c = 2\sqrt{x^3} + c$$

Each of these results may be checked by differentiation. The integral of a constant k with respect to x is $kx + c$. For example, $\int 5 \, dx = 5x + c$.

When a sum of several terms is integrated, the result is the sum of the integrals of the separate terms. For example,

$$\int (2x + 7x^2 - 3) \, dx = \int 2x \, dx + \int 7x^2 \, dx - \int 3 \, dx$$

$$= \frac{2x^2}{2} + \frac{7x^3}{3} - 3x + c = x^2 + \frac{7}{3}x^3 - 3x + c$$

17.3   Standard integrals

Since integration is the reverse process of differentiation the **standard integrals** listed in Table 17.1 on page 260 may be deduced and readily checked by differentiation (see Tables 1.1 and 1.2).

Worked problems on standard integrals

Problem 1. Determine (a) $\int 3x^2 \, dx$ (b) $\int 5t^3 \, dt$

The standard integral of $\int ax^n \, dx = \frac{ax^{n+1}}{n+1} + c$

Table 17.1 Standard integrals

(1)	$\displaystyle\int ax^n \, dx$	$= \dfrac{ax^{n+1}}{n+1} + c$ (except when $n = -1$)
(2)	$\displaystyle\int \cos ax \, dx$	$= \dfrac{1}{a} \sin ax + c$
(3)	$\displaystyle\int \sin ax \, dx$	$= -\dfrac{1}{a} \cos ax + c$
(4)	$\displaystyle\int \sec^2 ax \, dx$	$= \dfrac{1}{a} \tan ax + c$
(5)	$\displaystyle\int \mathrm{cosec}^2 ax \, dx$	$= -\dfrac{1}{a} \cot ax + c$
(6)	$\displaystyle\int \mathrm{cosec}\, ax \cot ax \, dx$	$= -\dfrac{1}{a} \mathrm{cosec}\, ax + c$
(7)	$\displaystyle\int \sec ax \tan ax \, dx$	$= \dfrac{1}{a} \sec ax + c$
(8)	$\displaystyle\int e^{ax} \, dx$	$= \dfrac{1}{a} e^{ax} + c$
(9)	$\displaystyle\int \dfrac{1}{x} \, dx$	$= \ln x + c$

(a) When $a = 3$ and $n = 2$, $\displaystyle\int 3x^2 \, dx = \dfrac{3x^{2+1}}{2+1} + c = \dfrac{3x^3}{3} + c$

$$= x^3 + c$$

(b) When $a = 5$ and $n = 3$, $\displaystyle\int 5t^3 \, dt = \dfrac{5t^{3+1}}{3+1} + c = \dfrac{5}{4} t^4 + c$

Problem 2. Find (a) $\displaystyle\int 5 \, dx$,   (b) $\displaystyle\int \dfrac{2}{3} x \, dx$.

(a) $\int 5 \, dx$ is the same as $\int 5x^0 \, dx$, and using the standard integral $\int ax^n \, dx$, where $a = 5$ and $n = 0$, gives:

$$\int 5x^0 \, dx = \dfrac{5x^{0+1}}{0+1} + c = 5x + c$$

(In general, if k is a constant, then $\int k \, dx = kx + c$)

(b) When $a = \dfrac{2}{3}$ and $n = 1$, then $\displaystyle\int \dfrac{2}{3} x\,dx = \left(\dfrac{2}{3}\right) \dfrac{x^{1+1}}{1+1} + c$

$$= \dfrac{2x^2}{3(2)} + c$$

$$= \dfrac{1}{3} x^2 + c$$

Problem 3. Determine $\displaystyle\int \left(5 + \dfrac{1}{4} x - 2x^2\right) dx$.

$\displaystyle\int \left(5 + \dfrac{1}{4} x - 2x^2\right) dx$ may be written as

$\displaystyle\int 5\,dx + \int \dfrac{1}{4} x\,dx - \int 2x^2\,dx$

i.e. each term is integrated separately.

Hence $\displaystyle\int \left(5 + \dfrac{1}{4} x - 2x^2\right) dx = 5x + \left(\dfrac{1}{4}\right) \dfrac{x^{1+1}}{1+1} - (2) \dfrac{x^{2+1}}{2+1} + c$

$$= 5x + \dfrac{1}{8} x^2 - \dfrac{2}{3} x^3 + c$$

Note that when an integral contains more than one term there is no need to have an arbitrary constant for each; just a single constant at the end is sufficient.

Problem 4. Find (a) $\displaystyle\int \left(\dfrac{3x^4 - 2x}{5x}\right) dx$, (b) $\displaystyle\int (1 - x)^2\,dx$.

(a) Rearranging into standard integral form gives:

$\displaystyle\int \left(\dfrac{3x^4 - 2x}{5x}\right) dx = \int \left(\dfrac{3x^4}{5x} - \dfrac{2x}{5x}\right) dx = \int \left(\dfrac{3}{5} x^3 - \dfrac{2}{5}\right) dx$

$$= \left(\dfrac{3}{5}\right) \dfrac{x^{3+1}}{3+1} - \dfrac{2}{5} x + c = \dfrac{3}{20} x^4 - \dfrac{2}{5} x + c$$

(b) $\displaystyle\int (1 - x)^2\,dx = \int (1 - 2x + x^2)\,dx$

$$= x - \dfrac{2x^{1+1}}{1+1} + \dfrac{x^{2+1}}{2+1} + c$$

$$= x - x^2 + \dfrac{1}{3} x^3 + c$$

This problem shows that functions often have to be rearranged into the standard form of ax^n before it is possible to integrate them.

Problem 5. Determine (a) $\int \dfrac{2}{x^2}\,dx,$ (b) $-\dfrac{2}{3}\int \dfrac{1}{x^5}\,dx.$

(a) $\displaystyle\int \frac{2}{x^2}\,dx = \int 2x^{-2}\,dx = (2)\frac{x^{-2+1}}{-2+1} + c = -2x^{-1} + c$

$$= -\frac{2}{x} + c$$

(b) $\displaystyle -\frac{2}{3}\int \frac{1}{x^5}\,dx = -\frac{2}{3}\int x^{-5}\,dx = \left(-\frac{2}{3}\right)\frac{x^{-5+1}}{-5+1} + c$

$$= \left(-\frac{2}{3}\right)\frac{x^{-4}}{-4} + c$$

$$= \frac{1}{6}x^{-4} + c = \frac{1}{6x^4} + c$$

Problem 6. Find (a) $\displaystyle\int 2\sqrt{x^3}\,dx,$ (b) $\displaystyle\int \frac{3}{5}\sqrt[4]{x^5}\,dx.$

(a) $\displaystyle\int 2\sqrt{x^3}\,dx = \int 2x^{3/2}\,dx = (2)\frac{x^{(\frac{3}{2})+1}}{\frac{3}{2}+1} + c = \frac{2x^{\frac{5}{2}}}{\frac{5}{2}} + c$

$$= \frac{4}{5}\sqrt{x^5} + c$$

(b) $\displaystyle\int \frac{3}{5}\sqrt[4]{x^5}\,dx = \int \frac{3}{5}x^{\frac{5}{4}}\,dx = \left(\frac{3}{5}\right)\frac{x^{(\frac{5}{4})+1}}{5/4+1} + c$

$$= \left(\frac{3}{5}\right)\frac{x^{\frac{9}{4}}}{9/4} + c$$

$$= \frac{4}{15}x^{\frac{9}{4}} + c = \frac{4}{15}\sqrt[4]{x^9} + c$$

Problem 7. Determine (a) $\displaystyle\int \frac{3}{\sqrt{x}}\,dx,$ (b) $\displaystyle\int \frac{-3}{7\sqrt[3]{x^4}}\,dx.$

(a) $\displaystyle\int \frac{3}{\sqrt{x}}\,dx = \int \frac{3}{x^{\frac{1}{2}}}\,dx = \int 3x^{-\frac{1}{2}}\,dx = \frac{3x^{-\frac{1}{2}+1}}{-\frac{1}{2}+1} + c$

$$= \frac{3x^{\frac{1}{2}}}{\frac{1}{2}} + c = 6\sqrt{x} + c$$

(b) $\displaystyle\int \frac{-3}{7\sqrt[3]{x^4}}\,dx = \int \frac{-3}{7x^{\frac{4}{3}}}\,dx = \int -\frac{3x^{-\frac{4}{3}}}{7}\,dx$

$$= \left(-\frac{3}{7}\right)\frac{x^{-\frac{4}{3}+1}}{-\frac{4}{3}+1} + c = \left(-\frac{3}{7}\right)\frac{x^{-\frac{1}{3}}}{-\frac{1}{3}} + c$$

$$= \frac{9}{7}x^{-\frac{1}{3}} + c = \frac{9}{7x^{\frac{1}{3}}} + c = \frac{9}{7\sqrt[3]{x}} + c$$

Problem 8. Find (a) $\displaystyle\int 3t^{1.2}\,dt$, (b) $\displaystyle\int \frac{6}{x^{2.4}}\,dx$.

(a) $\displaystyle\int 3t^{1.2}\,dt = \frac{3t^{1.2+1}}{1.2+1} + c$

$$= \frac{3}{2.2}t^{2.2} + c \text{ or } \mathbf{1.364}t^{2.2} + c$$

correct to four significant figures

(b) $\displaystyle\int \frac{6}{x^{2.4}}\,dx = \int 6x^{-2.4}\,dx = \frac{6x^{-2.4+1}}{-2.4+1} + c$

$$= \frac{6x^{-1.4}}{-1.4} + c = \frac{-\mathbf{4.286}}{x^{1.4}} + c$$

correct to four significant figures

Problem 9. Determine $\displaystyle\int \frac{(2+x)^2}{\sqrt{x}}\,dx$.

$$\int \frac{(2+x)^2}{\sqrt{x}}\,dx = \int \left(\frac{4+4x+x^2}{x^{\frac{1}{2}}}\right)\,dx = \int \left(\frac{4}{x^{\frac{1}{2}}} + \frac{4x}{x^{\frac{1}{2}}} + \frac{x^2}{x^{\frac{1}{2}}}\right)\,dx$$

$$= \int (4x^{-\frac{1}{2}} + 4x^{\frac{1}{2}} + x^{\frac{3}{2}})\,dx, \text{ by the laws of indices}$$

$$= \frac{4x^{-\frac{1}{2}+1}}{-\frac{1}{2}+1} + \frac{4x^{\frac{1}{2}+1}}{\frac{1}{2}+1} + \frac{x^{\frac{3}{2}+1}}{\frac{3}{2}+1} + c$$

$$= 8x^{\frac{1}{2}} + \frac{4}{3/2} x^{\frac{3}{2}} + \frac{x^{\frac{5}{2}}}{5/2} + c$$

$$= 8\sqrt{x} + \frac{8}{3}\sqrt{x^3} + \frac{2}{5}\sqrt{x^5} + c$$

Problem 10. Find (a) $\displaystyle\int 5 \cos 2x \, dx$,   (b) $\displaystyle\int 3 \sin 2\theta \, d\theta$.

(a) From Table 17.1, result (2),

$$\int 5 \cos 2x \, dx = (5)\left(\frac{1}{2} \sin 2x\right) + c = \frac{5}{2} \sin 2x + c$$

(b) From Table 17.1, result (3),

$$\int 3 \sin 2\theta \, d\theta = (3)\left(-\frac{1}{2} \cos 2\theta\right) + c$$

$$= -\frac{3}{2} \cos 2\theta + c$$

Problem 11. Determine (a) $\displaystyle\int 5 \sec^2 3t \, dt$,   (b) $\displaystyle\int 2 \operatorname{cosec}^2 4\theta \, d\theta$.

(a) From Table 17.1, result (4),

$$\int 5 \sec^2 3t \, dt = (5)\left(\frac{1}{3} \tan 3t\right) + c$$

$$= \frac{5}{3} \tan 3t + c$$

(b) From Table 17.1, result (5),

$$\int 2 \operatorname{cosec}^2 4\theta \, d\theta = (2)\left(-\frac{1}{4} \cot 4\theta\right) + c$$

$$= -\frac{1}{2} \cot 4\theta + c$$

Problem 12. Find (a) $\displaystyle\int 3 \operatorname{cosec} 2t \cot 2t \, dt$,   (b) $\displaystyle\int 5 \sec 4\theta \tan 4\theta \, d\theta$.

(a) From Table 17.1, result (6),

$$\int 3 \operatorname{cosec} 2t \cot 2t \, dt = (3)\left(-\frac{1}{2}\operatorname{cosec} 2t\right) + c$$

$$= -\frac{3}{2}\operatorname{cosec} 2t + c$$

(b) From Table 17.1, result (7),

$$\int 5 \sec 4\theta \tan 4\theta \, d\theta = (5)\left(\frac{1}{4}\sec 4\theta\right) + c$$

$$= \frac{5}{4}\sec 4\theta + c$$

Problem 13. Determine (a) $\int 4e^{2x}\,dx$, (b) $\int \frac{3}{4e^{3t}}\,dt$.

(a) From Table 17.1, result (8),

$$\int 4e^{2x}\,dx = (4)\left(\frac{1}{2}e^{2x}\right) + c = 2e^{2x} + c$$

(b) $\int \frac{3}{4e^{3t}}\,dt = \int \frac{3}{4}e^{-3t}\,dt = \left(\frac{3}{4}\right)\left(-\frac{1}{3}e^{-3t}\right) + c$

$$= -\frac{1}{4}e^{-3t} + c = -\frac{1}{4e^{3t}} + c$$

Problem 14. Find (a) $\int \frac{2}{3x}\,dx$, (b) $\int \left(\frac{3z^2 - 1}{z}\right)dz$.

(a) From Table 17.1, result (9),

$$\int \frac{2}{3x}\,dx = \int \frac{2}{3}\left(\frac{1}{x}\right)dx = \frac{2}{3}\ln x + c$$

(b) $\int \left(\frac{3z^2 - 1}{z}\right)dz = \int \left(\frac{3z^2}{z} - \frac{1}{z}\right)dz = \int \left(3z - \frac{1}{z}\right)dz$

$$= \frac{3z^2}{2} - \ln z + c$$

Further problems on standard integrals may be found in Section 17.5 (Problems 1 to 14), page 269.

17.4   Definite integrals

Integrals containing an arbitrary constant c in their results are called **indefinite integrals** since their precise value cannot be determined without further information.

Definite integrals are those in which limits are applied. If an expression is written as $[x]_a^b$, 'b' is called the upper limit and 'a' the lower limit.

By definition $[x]_a^b = (b - a)$

The increase in the value of the integral $(x^2 - 1)$ as x increases from 1 to 3

is written as $\displaystyle\int_1^3 (x^2 - 1)\, dx$

Applying the limits gives:

$$\int_1^3 (x^2 - 1)\, dx = \left[\frac{x^3}{3} - x + c\right]_1^3 = \left(\frac{(3)^3}{3} - (3) + c\right) - \left(\frac{(1)^3}{3} - (1) + c\right)$$

$$= (9 - 3 + c) - \left(\frac{1}{3} - 1 + c\right)$$

$$= 9 - 3 + c - \frac{1}{3} + 1 - c = 6\,\frac{2}{3}$$

Note that the 'c' term always cancels out when limits are applied and the arbitrary constant need not be shown with definite integrals.

Worked problems on definite integrals

Problem 1. Evaluate (a) $\displaystyle\int_1^2 4x\, dx$, (b) $\displaystyle\int_{-2}^4 (3 - x^2)\, dx$.

(a) $\displaystyle\int_1^2 4x\, dx = \left[\frac{4x^2}{2}\right]_1^2 = 2[x^2]_1^2 = 2(2^2 - 1^2) = 2(4 - 1)$

$$= 6$$

(b) $\displaystyle\int_{-2}^4 (3 - x^2)\, dx = \left[3x - \frac{x^3}{3}\right]_{-2}^4$

$$= \left(3(4) - \frac{(4)^3}{3} \right) - \left(3(-2) - \frac{(-2)^3}{3} \right)$$

$$= \left(12 - 21\frac{1}{3} \right) - \left(-6 + 2\frac{2}{3} \right)$$

$$= 12 - 21\frac{1}{3} + 6 - 2\frac{2}{3} = -6$$

Problem 2. Evaluate $\displaystyle\int_1^4 \left(\frac{t+1}{\sqrt{t}} \right) dt$, taking positive square roots only.

$$\int_1^4 \left(\frac{t+1}{\sqrt{t}} \right) dt = \int_1^4 \left(\frac{t}{t^{\frac{1}{2}}} + \frac{1}{t^{\frac{1}{2}}} \right) dt = \int_1^4 (t^{\frac{1}{2}} + t^{-\frac{1}{2}}) \, dt$$

$$= \left[\frac{t^{\frac{1}{2}+1}}{\frac{1}{2}+1} + \frac{t^{-\frac{1}{2}+1}}{-\frac{1}{2}+1} \right]_1^4 = \left[\frac{t^{\frac{3}{2}}}{\frac{3}{2}} + \frac{t^{\frac{1}{2}}}{\frac{1}{2}} \right]_1^4$$

$$= \left[\frac{2}{3}\sqrt{t^3} + 2\sqrt{t} \right]_1^4 = \left(\frac{2}{3}\sqrt{4^3} + 2\sqrt{4} \right) - \left(\frac{2}{3}\sqrt{1^3} + 2\sqrt{1} \right)$$

$$= \left(\frac{2}{3}(8) + 2(2) \right) - \left(\frac{2}{3} + 2 \right)$$

$$= 5\frac{1}{3} + 4 - 2\frac{2}{3} = 6\frac{2}{3}$$

Problem 3. $\displaystyle\int_0^{\pi/2} 5 \sin 2x \, dx.$

$$\int_0^{\pi/2} 5 \sin 2x \, dx = \left[5 \left(-\frac{1}{2} \cos 2x \right) \right]_0^{\pi/2} = \left[-\frac{5}{2} \cos 2x \right]_0^{\pi/2}$$

$$= -\frac{5}{2} [\cos 2x]_0^{\pi/2} = -\frac{5}{2} \left[\cos 2\frac{\pi}{2} - \cos 2(0) \right]$$

$$= -\frac{5}{2} [\cos \pi - \cos 0] = -\frac{5}{2}(-1 - 1)$$

$$= -\frac{5}{2}(-2) = 5$$

Problem 4. Find the value of $\int_1^3 2\cos 3\theta \, d\theta$, correct to four significant figures.

$$\int_1^3 2\cos 3\theta \, d\theta = \left[(2)\left(\frac{1}{3}\sin 3\theta\right) \right]_1^3 = \left[\frac{2}{3}\sin 3\theta \right]_1^3$$

$$= \frac{2}{3}[\sin 3\theta]_1^3 = \frac{2}{3}[\sin 9 - \sin 3]$$

Limits of trigonometric functions are expressed in radians. Thus sin 9 means the sine of 9 radians. It is important to make sure that a calculator is set on the radian mode.

Hence $\int_1^3 2\cos 3\theta \, d\theta = \frac{2}{3}[0.412\,118\,48\cdots - 0.141\,120\,00\cdots]$

$$= \textbf{0.180 7}, \text{ correct to four significant figures}$$

Problem 5. Evaluate (a) $\int_1^2 2\,e^{5x}\,dx$, (b) $\int_3^4 \frac{3}{x}\,dx$, each correct to five significant figures.

(a) $\int_1^2 2\,e^{5x}\,dx = \left[\frac{2}{5}e^{5x} \right]_1^2 = \frac{2}{5}[e^{5x}]_1^2 = \frac{2}{5}[e^{10} - e^5]$

$$= \frac{2}{5}[22\,026.465\,7\cdots - 148.413\,159\cdots]$$

$$= \textbf{8 751.2}, \text{ correct to five significant figures}$$

(b) $\int_3^4 \frac{3}{x}\,dx = [3\ln x]_3^4 = 3[\ln x]_3^4$

$$= 3[\ln 4 - \ln 3] = \textbf{0.863 05}, \text{ correct to five significant}$$
$$\text{figures}$$

Problem 6. Evaluate $\int_{\pi/4}^{\pi/3} (3\sec^2 x - 2\cos 2x)\,dx$, correct to three decimal places.

$$\int_{\pi/4}^{\pi/3} (3\sec^2 x - 2\cos 2x)\,dx = \left[3\tan x - \frac{2}{2}\sin 2x \right]_{\pi/4}^{\pi/3}$$

$$= \left(3 \tan \frac{\pi}{3} - \sin 2 \frac{\pi}{3}\right) - \left(3 \tan \frac{\pi}{4} - \sin 2 \frac{\pi}{4}\right)$$

$$= (5.196\,15 \cdots - \mathbf{0.866\,026} \cdots) - (3 - 1)$$

$$= \mathbf{2.330}, \text{ correct to three decimal places}$$

Further problems on definite integrals may be found in the following Section 17.5 (Problems 15 to 24), page 271.

17.5   Further problems

Standard integrals

In Problems 1 to 14, determine the indefinite integrals.

1. (a) $\int 3x \, dx$ (b) $\int 9x \, dx$ (a) $\left[\dfrac{3x^2}{2} + c\right]$ (b) $\left[\dfrac{9x^2}{2} + c\right]$

2. (a) $\int \dfrac{3}{5} x^2 \, dx$ (b) $\int \dfrac{4}{3} x^3 \, dx$ (a) $\left[\dfrac{x^3}{5} + c\right]$ (b) $\left[\dfrac{1}{3} x^4 + c\right]$

3. (a) $\int (3 - 2\theta + \theta^2) \, d\theta$ (b) $3 \int (x - 2x^2) \, dx$

    (a) $\left[3\theta - \theta^2 + \dfrac{\theta^3}{3} + c\right]$ (b) $\left[\dfrac{3}{2} x^2 - 2x^3 + c\right]$

4. (a) $\int \left(\dfrac{5x^2 - 2x}{x}\right) \, dx$ (b) $\int (3 + t)^2 \, dt$

    (a) $\left[\dfrac{5x^2}{2} - 2x + c\right]$ (b) $\left[9t + 3t^2 + \dfrac{t^3}{3} + c\right]$

5. (a) $\int \dfrac{4}{3x^2} \, dx$ (b) $\int \dfrac{2}{5x^4} \, dx$ (a) $\left[\dfrac{-4}{3x} + c\right]$ (b) $\left[\dfrac{-2}{15x^3} + c\right]$

6. (a) $\int 3\sqrt{x^3} \, dx$ (b) $\int \dfrac{1}{3} \sqrt[4]{t^5} \, dt$

    (a) $\left[\dfrac{6}{5}\sqrt{x^5} + c\right]$ (b) $\left[\dfrac{4}{27} \sqrt[4]{t^9} + c\right]$

7. (a) $\int \dfrac{-2}{\sqrt{x^3}} \, dx$ (b) $\int \dfrac{1}{2} \sqrt[4]{\theta^3} \, d\theta$

    (a) $\left[\dfrac{4}{\sqrt{x}} + c\right]$ (b) $\left[\dfrac{2}{7} \sqrt[4]{\theta^7} + c\right]$

8. (a) $\int \frac{1}{2} x^{2.4}\, dx$ (b) $\int \frac{3}{x^{1.3}}\, dx$

    (a) $\left[\frac{1}{6.8} x^{3.4} + c \right]$ (b) $\left[\frac{-10}{x^{0.3}} + c \right]$

9. (a) $\int 7 \cos 3x\, dx$ (b) $\int 2 \sin 5t\, dt$

    (a) $\left[\frac{7}{3} \sin 3x + c \right]$ (b) $\left[-\frac{2}{5} \cos 5t + c \right]$

10. (a) $\int \frac{1}{2} \sec^2 7x\, dx$ (b) $\int 3 \operatorname{cosec}^2 2\theta\, d\theta$

    (a) $\left[\frac{1}{14} \tan 7x + c \right]$ (b) $\left[-\frac{3}{2} \cot 2\theta + c \right]$

11. (a) $\int 3 \cot 5t \operatorname{cosec} 5t\, dt$ (b) $\int \frac{1}{2} \sec 3x \tan 3x\, dx$

    (a) $\left[-\frac{3}{5} \operatorname{cosec} 5t + c \right]$ (b) $\left[\frac{1}{6} \sec 3x + c \right]$

12. (a) $\int 2 e^{3x}\, dx$ (b) $\frac{3}{5} \int \frac{dt}{e^{2t}}$

    (a) $\left[\frac{2}{3} e^{3x} + c \right]$ (b) $\left[\frac{-3}{10\, e^{2t}} + c \right]$

13. (a) $\int \frac{3}{4x}\, dx$ (b) $\int \left(\frac{t^2 - 1}{t} \right) dt$

    (a) $\left[\frac{3}{4} \ln x + c \right]$ (b) $\left[\frac{t^2}{2} - \ln t + c \right]$

14. (a) $\int \frac{(2 - 3x)^2}{\sqrt{x}}\, dx$ (b) $\int \left(\frac{1}{\theta} - 2\theta \right)^2 d\theta$

    (a) $\left[8\sqrt{x} - 8\sqrt{x^3} + \frac{18}{5} \sqrt{x^5} + c \right]$

    (b) $\left[-\frac{1}{\theta} - 4\theta + \frac{4}{3} \theta^3 + c \right]$

Definite integrals

In Problems 15 to 23, evaluate the definite integrals (where necessary correct to four significant figures).

15. (a) $\int_1^4 3 \, dx$ (b) $\int_{-1}^2 2x \, dx$ (a) [9] (b) [3]

16. (a) $\int_1^4 5t^2 \, dt$ (b) $\int_{-2}^1 -\dfrac{3}{5}\theta^2 \, d\theta$ (a) [105] (b) $\left[-1\dfrac{4}{5}\right]$

17. (a) $\int_1^2 (4 - x^2) \, dx$ (b) $\int_1^3 (x^2 - 2x + 3) \, dx$

(a) $\left[1\dfrac{2}{3}\right]$ (b) $\left[6\dfrac{2}{3}\right]$

18. (a) $\int_0^\pi \dfrac{1}{2} \cos 2t \, dt$ (b) $\int_0^{\pi/2} 3 \cos \theta \, d\theta$ (a) [0] (b) [3]

19. (a) $\int_{\pi/6}^{\pi/3} \sin 2\theta \, d\theta$ (b) $\int_1^2 5 \sin \theta \, d\theta$ (a) $\left[\dfrac{1}{2}\right]$ (b) [4.782]

20. (a) $\int_0^1 3 \cos 2x \, dx$ (b) $\int_0^{\pi/4} 4 \sec^2 3x \, dx$

(a) [1.364]   (b) $\left[-1\dfrac{1}{3}\right]$

21. (a) $\int_1^3 2 \operatorname{cosec}^2 3t \, dt$

(b) $\int_{\pi/4}^{\pi/2} (2 \sin 3x - 3 \cos 2x) \, dx$

(a) [−3.203]   (b) [1.971]

22. (a) $\int_0^2 \dfrac{1}{2} e^{3\theta} \, d\theta$ (b) $\int_{-3}^2 \dfrac{4}{5e^{2t}} \, dt$ (a) [67.07] (b) [161.4]

23. (a) $\int_2^3 \dfrac{4}{3x} \, dx$ (b) $\int_2^3 \left(\dfrac{2t^2 - 1}{t}\right) \, dt$

(a) [0.540 6]   (b) [4.595]

24. Evaluate $\int_1^2 \dfrac{(3x - 1)(x - 4)}{x} \, dx$, correct to three decimal places.

[−5.727]

Chapter 18

Integration using substitutions

18.1 Introduction

Functions which require integrating are often not in the 'standard form' of the list shown in Table 17.1, page 260. In such cases it may be possible by one of four methods to change the function into a form which can be readily integrated.

The methods available are:

(a) by using algebraic substitutions (see Section 18.2),
(b) by using trigonometrical and hyperbolic identities and substitutions (see Section 18.3),
(c) by using partial fractions (see Chapter 19), and
(d) integration by parts (see Chapter 20).

It should be realised that many mathematical functions cannot be integrated by analytical means at all and approximate methods have then to be used (see *Technician mathematics 3*, Chapter 27 on Numerical integration).

18.2 Integration using algebraic substitutions

An integral may often be reduced to a standard integral by making an algebraic substitution. The substitution usually made is to let u be equal to $f(x)$, such that $f(u)\,\mathrm{d}u$ is a standard integral.

A most important point in the use of substitution is that once it has been made the original variable must be removed completely, because a variable can only be integrated with respect to itself, i.e. we cannot integrate, for example, a function of t with respect to x.

A concept that $\dfrac{\mathrm{d}u}{\mathrm{d}x}$ is a single entity (indicating the change of u with respect to x) has been established in the work done on differentiation. Frequently, in work on integration and on differential equations, $\dfrac{\mathrm{d}u}{\mathrm{d}x}$ is split. Provided that when this is done the original differential coefficient can be re-formed by applying the rules of algebra, then it is in order to do so. For example, if $\dfrac{dy}{dx} = x$, then it is in order to write $dy = x\,dx$ since dividing both sides by dx re-forms the original differential coefficient. This principle is shown in the following worked problems.

Worked problems on integration by algebraic substitutions

Type 1

Problem 1. Find (a) $\int \sin(6x + 5)\,dx$, (b) $\int (3t - 2)^8\,dt$

(c) $\int \dfrac{1}{(9x + 4)}\,dx$

(a) $\int \sin(6x + 5)\,dx$

Let $u = 6x + 5$, then $\dfrac{du}{dx} = 6$, i.e. $dx = \dfrac{du}{6}$

Hence

$$\int \sin(6x + 5)\,dx = \int \sin u\,\frac{du}{6} = \frac{1}{6}\int \sin u\,du = \frac{1}{6}(-\cos u) + c$$

Since the original integral is given in terms of x the result should be stated in terms of x.

Since $u = 6x + 5$, $\quad\displaystyle\int \sin(6x + 5)\,dx = -\frac{1}{6}\cos(6x + 5) + c$

(b) $\displaystyle\int (3t - 2)^8\,dt$

Let $u = 3t - 2$, then $\dfrac{du}{dt} = 3$, i.e. $dt = \dfrac{du}{3}$

Hence

$$\int (3t - 2)^8\,dt = \int u^8\,\frac{du}{3} = \frac{1}{3}\int u^8\,du = \frac{1}{3}\left(\frac{u^9}{9}\right) + c = \frac{1}{27}u^9 + c$$

Since $u = 3t - 2$, $\displaystyle\int (3t - 2)^8\,dt = \frac{1}{27}(3t - 2)^9 + c$

(c) $\displaystyle\int \dfrac{1}{(9x + 4)}\,dx$

Let $u = 9x + 4$, then $\dfrac{du}{dx} = 9$, i.e. $dx = \dfrac{du}{9}$

Hence $\displaystyle\int \frac{1}{(9x + 4)}\,dx = \int \frac{1}{u}\frac{du}{9} = \frac{1}{9}\ln u + c = \frac{1}{9}\ln(9x + 4) + c$

It may be seen from Problem 1 that:

If 'x' in a standard integral is replaced by $(ax + b)$, where a and b

are constants, then $(ax + b)$ is written for x in the result and the result is multiplied by $\dfrac{1}{a}$.

In general, $\displaystyle\int (ax + b)^n \, dx = \dfrac{1}{a(n + 1)} (ax + b)^{n+1} + c$

[except when $n = -1$ (see (c) above)]

With practice integrals of this type may be determined by inspection. A check can always be made by differentiating the result.

Problem 2. Determine the following by inspection:

(a) $\displaystyle\int 2 \sin(4x - 1) \, dx$ (b) $\displaystyle\int 3 \, e^{7t-2} \, dt$ (c) $\displaystyle\int \dfrac{4}{6\theta + 7} \, d\theta$

(a) $\displaystyle\int 2 \sin (4x - 1) \, dx = \dfrac{2}{4}[-\cos(4x-1)] + c = -\dfrac{1}{2}\cos(4x-1) + c$

(b) $\displaystyle\int 3 \, e^{7t-2} \, dt = 3(e^{7t-2})\left(\dfrac{1}{7}\right) + c = \dfrac{3}{7} e^{7t-2} + c$

(c) $\displaystyle\int \dfrac{4}{6\theta + 7} \, d\theta = 4[\ln(6\theta + 7)]\left(\dfrac{1}{6}\right) + c = \dfrac{2}{3}\ln(6\theta + 7) + c$

Type 2

Problem 3. Find (a) $\displaystyle\int 2x(3x^2 + 5)^6 \, dx$, (b) $\displaystyle\int \sin^3 \theta \cos \theta \, d\theta$.

(a) $\displaystyle\int 2x(3x^2 + 5)^6 \, dx$

There appear to be two possible choices for the substitution, i.e. either to let $u = 2x$ or to let $u = (3x^2 + 5)$. Choosing the latter substitution, if $u = (3x^2 + 5)$, then $\dfrac{du}{dx} = 6x$, i.e. $dx = \dfrac{du}{6x}$

Hence $\displaystyle\int 2x(3x^2 + 5)^6 \, dx = \int 2x(u)^6 \dfrac{du}{6x} = \int \dfrac{2u^6}{6} \, du = \dfrac{1}{3} \int u^6 \, du$

The original variable, x, has been removed completely and the integral is now only in terms of u.

Hence $\dfrac{1}{3}\displaystyle\int u^6 \, du = \dfrac{1}{3}\left(\dfrac{u^7}{7}\right) + c = \dfrac{1}{21}(3x^2 + 5)^7 + c$

(b) $\displaystyle\int \sin^3 \theta \cos \theta \, d\theta$

Let $u = \sin \theta$, then $\dfrac{du}{d\theta} = \cos \theta$, i.e. $d\theta = \dfrac{du}{\cos \theta}$

Hence $\displaystyle\int \sin^3 \theta \cos \theta \, d\theta = \int u^3 \cos \theta \, \dfrac{du}{\cos \theta} = \int u^3 \, du = \dfrac{u^4}{4} + c$

$$= \frac{1}{4} \sin^4 \theta + c$$

Whenever a power of a sine is multiplied by a cosine of power 1, or vice versa, the integral may be determined by inspection.

For example, $\displaystyle\int \cos^5 x \sin x \, dx = -\frac{1}{6} \cos^6 x + c,$

since $\dfrac{d}{dx}\left\{ -\dfrac{1}{6} \cos^6 x + c \right\} = \cos^5 x \sin x$

It may be seen from Problem 3 that:

Integrals of the form $k \displaystyle\int [f(x)]^n f'(x) \, dx$ (where k and n are constants) can be integrated by substituting u for $f(x)$.

Type 3

Problem 4. Find (a) $\displaystyle\int \frac{x}{1 + x^2} \, dx$, (b) $\displaystyle\int \frac{4x}{\sqrt{(5x^2 - 1)}} \, dx$,

(c) $\displaystyle\int \tan \theta \, d\theta$.

(a) $\displaystyle\int \frac{x}{1 + x^2} \, dx$

Let $u = 1 + x^2$, then $\dfrac{du}{dx} = 2x$, i.e. $dx = \dfrac{du}{2x}$

Hence $\displaystyle\int \frac{x}{1 + x^2} \, dx = \int \frac{x \, du}{u \, 2x} = \frac{1}{2} \int \frac{1}{u} \, du = \frac{1}{2} \ln u + c$

$$= \frac{1}{2} \ln(1 + x^2) + c$$

(b) $\displaystyle\int \frac{4x}{\sqrt{(5x^2 - 1)}}\,dx$

Let $u = 5x^2 - 1$, then $\dfrac{du}{dx} = 10x.$, i.e., $dx = \dfrac{du}{10x}$

Hence $\displaystyle\int \frac{4x}{\sqrt{(5x^2 - 1)}}\,dx = \int \frac{4x}{\sqrt{u}}\frac{du}{10x} = \frac{2}{5}\int u^{-\frac{1}{2}}\,du = \frac{2}{5}\frac{u^{\frac{1}{2}}}{1/2} + c$

$$= \frac{4}{5}\sqrt{u} + c$$

$$= \frac{4}{5}\sqrt{(5x^2 - 1)} + c$$

(c) $\displaystyle\int \tan\theta\,d\theta = \int \frac{\sin\theta}{\cos\theta}\,d\theta$

Let $u = \cos\theta$, then $\dfrac{du}{d\theta} = -\sin\theta$, i.e. $d\theta = -\dfrac{du}{\sin\theta}$

Hence $\displaystyle\int \frac{\sin\theta}{\cos\theta}\,d\theta = \int \frac{\sin\theta}{u}\left(\frac{-du}{\sin\theta}\right) = -\int \frac{1}{u}\,du = -\ln u + c$

$$= -\ln(\cos\theta) + c$$

$$= \ln(\cos\theta)^{-1} + c$$

i.e. $\displaystyle\int \tan\theta\,d\theta = \ln(\sec\theta) + c$

It may be seen from Problem 4 that:

Integrals of the form $k \displaystyle\int \dfrac{f'(x)}{[f(x)]^n}\,dx$ (where k and n are constants) can be integrated by substituting u for $f(x)$.

If, in an integral, it is recognised that the numerator is the differential coefficient of the denominator (times a constant) then it may be integrated by inspection.

General problem on definite integrals

Problem 5. Evaluate the following: (a) $\displaystyle\int_1^3 3\sec^2(2\theta - 1)\,d\theta$

(b) $\displaystyle\int_0^3 3x\sqrt{(3x^2 + 9)}\,dx$, taking positive values of roots only

(c) $\displaystyle\int_1^2 \frac{2e^t}{3 + e^t}\,dt$

(a) $\int_{1}^{3} 3 \sec^2(2\theta - 1)\, d\theta = \left[\frac{3}{2} \tan(2\theta - 1)\right]_{1}^{3} = \frac{3}{2}[\tan 5 - \tan 1]$

where 'tan 5' means 'the tangent of 5 radians' and 'tan 1'
means 'the tangent of 1 radian'.

$$= \frac{3}{2}[-3.380\,5 - 1.557\,4] = -7.406\,9$$

(b) $\int_{0}^{3} 3x\sqrt{(3x^2 + 9)}\, dx$

Let $u = 3x^2 + 9$, then $\dfrac{du}{dx} = 6x$, i.e. $dx = \dfrac{du}{6x}$

Then $\int_{0}^{3} 3x\sqrt{(3x^2 + 9)}\, dx = \int 3x\sqrt{u}\,\dfrac{du}{6x}$

$$= \frac{1}{2}\int \sqrt{u}\, du = \frac{1}{3}\sqrt{u^3}$$

$$= \frac{1}{3}[\sqrt{(3x^2 + 9)^3}]_{0}^{3}$$

$$= \frac{1}{3}[\sqrt{(36)^3} - \sqrt{(9)^3}]$$

$$= \frac{1}{3}[216 - 27],\ \text{taking positive}$$
values of roots only

$$= 63$$

(c) $\int_{1}^{2} \dfrac{2e^t}{3 + e^t}\, dt = [2\ln(3 + e^t)]_{1}^{2}$ (by inspection, or by using the
substitution $u = 3 + e^t$)

$$= 2[\ln(3 + e^2) - \ln(3 + e^1)] = 2\ln\left(\frac{3 + e^2}{3 + e^1}\right)$$

$$= 1.194\,2$$

---

*Further problems on integration by algebraic substitutions may be found in
Section 18.5 (Problems 1 to 38), page 293.*

18.3 Integration using trigonometric and hyperbolic identities and substitutions

(A) Integration of $\cos^2 x$, $\sin^2 x$, $\tan^2 x$ and $\cot^2 x$

It may be shown from Chapter 16 of *Technician mathematics 3* that the
compound angle addition formula $\cos(A + B)$ is given by:

$$\cos(A + B) = \cos A \cos B - \sin A \sin B$$

If $B = A$, then $\cos 2A = \cos^2 A - \sin^2 A$ (1)

Since $\cos^2 A + \sin^2 A = 1$ then $\sin^2 A = 1 - \cos^2 A$
  Hence $\cos 2A = \cos^2 A - (1 - \cos^2 A)$

i.e. $\cos 2A = 2\cos^2 A - 1$ (2)

Also $\cos^2 A = 1 - \sin^2 A$
  Hence $\cos 2A = (1 - \sin^2 A) - \sin^2 A$

i.e. $\cos 2A = 1 - 2\sin^2 A$ (3)

From equation (2), $\cos^2 A = \frac{1}{2}(1 + \cos 2A)$ (4)

From equation (3), $\sin^2 A = \frac{1}{2}(1 - \cos 2A)$ (5)

Thus $\int \cos^2 x \, dx = \int \frac{1}{2}(1 + \cos 2x) \, dx = \frac{1}{2}\left(x + \frac{\sin 2x}{2}\right) + c$

Similarly, $\int \sin^2 x \, dx = \int \frac{1}{2}(1 - \cos 2x) \, dx = \frac{1}{2}\left(x - \frac{\sin 2x}{2}\right) + c$

From $1 + \tan^2 x = \sec^2 x$, $\tan^2 x = \sec^2 x - 1$

  Thus $\int \tan^2 x \, dx = \int (\sec^2 x - 1) \, dx = \tan x - x + c$

From $\cot^2 x + 1 = \operatorname{cosec}^2 x$, $\cot^2 x = \operatorname{cosec}^2 x - 1$

  Thus $\int \cot^2 x \, dx = \int (\operatorname{cosec}^2 x - 1) \, dx = -\cot x - x + c$

Problem 1. Find (a) $\int \cos^2 5t \, dt$, (b) $\int \tan^2 3\theta \, d\theta$.

(a) $\int \cos^2 5t \, dt = \int \frac{1}{2}(1 + \cos 10t) \, dt = \frac{1}{2}\left(t + \frac{\sin 10t}{10}\right) + c$

(b) $\int \tan^2 3\theta \, d\theta = \int (\sec^2 3\theta - 1) \, d\theta = \frac{1}{3}\tan 3\theta - \theta + c$

Problem 2. Determine (a) $\int 3\sin^2 2x \, dx$, (b) $\int 2\cot^2 4\theta \, d\theta$.

(a) $\int 3\sin^2 2x \, dx = 3\int \frac{1}{2}(1 - \cos 4x) \, dx$

$= \frac{3}{2}\left(x - \frac{\sin 4x}{4}\right) + c$

(b) $\displaystyle\int 2\cot^2 4\theta\, d\theta = 2\int (\text{cosec}^2\, 4\theta - 1)\, d\theta$

$$= 2\left(-\frac{\cot 4\theta}{4} - \theta\right) + c$$

$$= -\tfrac{1}{2}\cot 4\theta - 2\theta + c$$

(B) Powers of sines and cosines of the form $\int \cos^m x \sin^n x\, dx$

(i) *To evaluate $\int \cos^m x \sin^n x\, dx$ when either m or n is **odd** (but not both), use is made of the trigonometric identity $\cos^2 x + \sin^2 x = 1$ as shown in Problems 1 and 2*

Problem 1. Find $\int \cos^5 x\, dx$.

$$\int \cos^5 x\, dx = \int \cos x\, (\cos^2 x)^2\, dx$$

$$= \int \cos x\, (1 - \sin^2 x)^2\, dx = \int \cos x\, (1 - 2\sin^2 x + \sin^4 x)\, dx$$

$$= \int \cos x - 2\sin^2 x \cos x + \sin^4 x \cos x\, dx$$

Hence the integral has been reduced to one which may be determined by inspection (see Section 18.2, type 2).

Hence $\displaystyle\int \cos^5 x\, dx = \sin x - \tfrac{2}{3}\sin^3 x + \tfrac{1}{5}\sin^5 x + c$

Problem 2. Find $\int \sin^3 x \cos^2 x\, dx$.

$$\int \sin^3 x \cos^2 x\, dx = \int \sin x\, (\sin^2 x)(\cos^2 x)\, dx$$

$$= \int \sin x\, (1 - \cos^2 x)(\cos^2 x)\, dx$$

$$= \int (\sin x \cos^2 x - \sin x \cos^4 x)\, dx$$

$$= -\tfrac{1}{3}\cos^3 x + \tfrac{1}{5}\cos^5 x + c$$

(ii) *To evaluate $\int \cos^m x \sin^n x\, dx$ when m **and** n are both **even** use is made of equations (4) and (5) of Section 18.3(A), i.e.*

$\cos^2 x = \tfrac{1}{2}(1 + \cos 2x)$ *and* $\sin^2 x = \tfrac{1}{2}(1 - \cos 2x)$

(It follows from these equations that $\cos^2 2x = \frac{1}{2}(1 + \cos 4x)$, $\sin^2 3x = \frac{1}{2}(1 - \cos 6x)$, and so on)

Problem 3. Find $\int \cos^4 x \, dx$.

$$\int \cos^4 x \, dx = \int (\cos^2 x)^2 \, dx = \int [\frac{1}{2}(1 + \cos 2x)]^2 \, dx$$

$$= \frac{1}{4} \int (1 + 2 \cos 2x + \cos^2 2x) \, dx$$

$$= \frac{1}{4} \int 1 + 2 \cos 2x + \frac{1}{2}(1 + \cos 4x) \, dx$$
$$\text{(since } \cos 4x = 2 \cos^2 2x - 1)$$

$$= \frac{1}{4} \int [\frac{3}{2} + 2 \cos 2x + \frac{1}{2} \cos 4x] \, dx$$

Hence $\qquad \int \cos^4 x \, dx = \frac{1}{4}[\frac{3}{2}x + \sin 2x + \frac{1}{8} \sin 4x] + c$

Problem 4. Find $\int \sin^4 x \cos^2 x \, dx$.

$$\int \sin^4 x \cos^2 x \, dx = \int (\sin^2 x)^2 \cos^2 x \, dx$$

$$= \int \left(\frac{1 - \cos 2x}{2} \right)^2 \left(\frac{1 + \cos 2x}{2} \right) dx$$

$$- \frac{1}{8} \int (1 - 2 \cos 2x + \cos^2 2x)(1 + \cos 2x) \, dx$$

$$= \frac{1}{8} \int (1 - 2 \cos 2x + \cos^2 2x + \cos 2x$$
$$- 2 \cos^2 2x + \cos^3 2x) \, dx$$

$$= \frac{1}{8} \int (1 - \cos 2x - \cos^2 2x + \cos^3 2x) \, dx$$

$$= \frac{1}{8} \int 1 - \cos 2x - \left(\frac{1 + \cos 4x}{2} \right)$$
$$+ \cos 2x(1 - \sin^2 2x) \, dx$$

$$= \frac{1}{8} \int [\frac{1}{2} - \cos 2x - \frac{1}{2} \cos 4x + \cos 2x$$
$$- \cos 2x \sin^2 2x] \, dx$$

$$= \frac{1}{8} \int [\frac{1}{2} - \frac{1}{2} \cos 4x - \cos 2x \sin^2 2x] \, dx$$

$$= \frac{1}{8} \left[\frac{x}{2} - \frac{\sin 4x}{8} - \frac{\sin^3 2x}{6} \right] + c$$

Hence $\int \sin^4 x \cos^2 x \, dx = \frac{1}{16}(x - \frac{1}{4} \sin 4x - \frac{1}{3} \sin^3 2x) + c$

(C) Products of sines and cosines

It is shown in *Technician mathematics 3*, Section 16.5, that:

$$\sin A \cos B = \tfrac{1}{2}[\sin(A + B) + \sin (A - B)] \tag{1}$$

$$\cos A \sin B = \tfrac{1}{2}[\sin(A + B) - \sin(A - B)] \tag{2}$$

$$\cos A \cos B = \tfrac{1}{2}[\cos(A + B) + \cos(A - B)] \tag{3}$$

$$\sin A \sin B = -\tfrac{1}{2}[\cos(A + B) - \cos(A - B)] \tag{4}$$

These formulae are used when integrating products of sines and cosines and are particularly useful for Fourier series calculations (see Chapter 27).

Problem 1. Find $\int \sin 4\theta \cos 3\theta \, d\theta$.

$$\int \sin 4\theta \cos 3\theta \, d\theta = \int \tfrac{1}{2}(\sin 7\theta + \sin \theta) \, d\theta \text{ from equation (1)}$$

$$= \frac{1}{2}\left(-\frac{\cos 7\theta}{7} - \cos \theta\right) + c$$

$$= -\tfrac{1}{14}(\cos 7\theta + 7 \cos \theta) + c$$

Problem 2. Find $\int \cos 6x \sin 2x \, dx$.

$$\int \cos 6x \sin 2x \, dx = \int \tfrac{1}{2}(\sin 8x - \sin 4x) \, dx, \text{ from equation (2)}$$

$$= \frac{1}{2}\left(\frac{-\cos 8x}{8} + \frac{\cos 4x}{4}\right) + c$$

$$= \tfrac{1}{16}(2 \cos 4x - \cos 8x) + c$$

Problem 3. Find $\int \cos 3t \cos t \, dt$.

$$\int \cos 3t \cos t \, dt = \int \tfrac{1}{2}(\cos 4t + \cos 2t) \, dt, \text{ from equation (3)}$$

$$= \frac{1}{2}\left(\frac{\sin 4t}{4} + \frac{\sin 2t}{2}\right) + c$$

$$= \tfrac{1}{8}(\sin 4t + 2 \sin 2t) + c$$

Problem 4. Find $\int \sin 6x \sin 3x \, dx$.

$$\int \sin 6x \sin 3x \, dx = \int -\tfrac{1}{2}(\cos 9x - \cos 3x) \, dx, \text{ from equation (4)}$$

$$= -\frac{1}{2}\left(\frac{\sin 9x}{9} - \frac{\sin 3x}{3}\right) + c$$

$$= -\tfrac{1}{18}(\sin 9x - 3 \sin 3x) + c$$

(D) Integrals containing $\sqrt{(a^2 - x^2)}$ – the 'sine θ' substitution

When an integral contains a term $\sqrt{(a^2 - x^2)}$, the substitution $x = a \sin \theta$ is used. The reasons for this are made obvious by the following worked problems.

Problem 1. Find $\displaystyle\int \frac{1}{\sqrt{(a^2 - x^2)}} \, dx$.

Let $x = a \sin \theta$, then $\dfrac{dx}{d\theta} = a \cos \theta$, i.e. $dx = a \cos \theta \, d\theta$

Hence $\displaystyle\int \frac{1}{\sqrt{(a^2 - x^2)}} \, dx = \int \frac{1}{\sqrt{(a^2 - a^2 \sin^2 \theta)}} \, a \cos \theta \, d\theta$

$$= \int \frac{a \cos \theta \, d\theta}{\sqrt{[(a^2(1 - \sin^2 \theta)]}}$$

$$= \int \frac{a \cos \theta}{a \cos \theta} \, d\theta, \text{ since } 1 - \sin^2 \theta = \cos^2 \theta$$

$$= \int d\theta = \theta + c$$

Since $x = a \sin \theta$ then $\sin \theta = \dfrac{x}{a}$ and $\theta = \arcsin \dfrac{x}{a}$

Hence $\displaystyle\int \frac{1}{\sqrt{(a^2 - x^2)}} \, dx = \arcsin \frac{x}{a} + c$

Problem 2. Evaluate $\displaystyle\int_0^2 \frac{1}{\sqrt{(16 - x^2)}} \, dx$.

Using the result of Problem 1, $\displaystyle\int_0^2 \frac{1}{\sqrt{(16 - x^2)}} \, dx = \left[\arcsin \frac{x}{4} \right]_0^2$

$$= (\arcsin \tfrac{1}{2} - \arcsin 0)$$

$$= \frac{\pi}{6} \text{ or } 0.523\,6$$

Problem 3. Find $\int \sqrt{(a^2 - x^2)} \, dx$.

Let $x = a \sin \theta$, then $\dfrac{dx}{d\theta} = a \cos \theta$, i.e. $dx = a \cos \theta \, d\theta$.

Thus $\int \sqrt{(a^2 - x^2)}\, dx = \int \sqrt{(a^2 - a^2 \sin^2 \theta)}(a \cos \theta \, d\theta)$

$$= \int \sqrt{[a^2(1 - \sin^2 \theta)]}(a \cos \theta \, d\theta)$$

$$= \int (a \cos \theta)(a \cos \theta) \, d\theta$$

$$= \int a^2 \cos^2 \theta \, d\theta$$

$$= a^2 \int \tfrac{1}{2}(1 + \cos 2\theta) \, d\theta$$

$$= \frac{a^2}{2}\left(\theta + \frac{\sin 2\theta}{2}\right) + c$$

In the compound angle addition formula
$\sin(A + B) = \sin A \cos B + \cos A \sin B$. If $B = A$ then
$\sin 2A = 2 \sin A \cos A$

Hence $\int \sqrt{(a^2 - x^2)}\, dx = \dfrac{a^2}{2}\left(\theta + \dfrac{2 \sin \theta \cos \theta}{2}\right) + c$

$$= \frac{a^2}{2}(\theta + \sin \theta \cos \theta) + c$$

Since $x = a \sin \theta$ then $\sin \theta = \dfrac{x}{a}$ and $\theta = \arcsin \dfrac{x}{a}$

Also $\cos^2 \theta + \sin^2 \theta = 1$ from which $\cos \theta = \sqrt{(1 - \sin^2 \theta)}$

i.e. $\cos \theta = \sqrt{\left[1 - \left(\dfrac{x}{a}\right)^2\right]} = \sqrt{\left(\dfrac{a^2 - x^2}{a^2}\right)} = \dfrac{\sqrt{(a^2 - x^2)}}{a}$

Hence $\int \sqrt{(a^2 - x^2)}\, dx = \dfrac{a^2}{2}\left[\arcsin \dfrac{x}{a} + \left(\dfrac{x}{a}\right)\dfrac{\sqrt{(a^2 - x^2)}}{a}\right] + c$

$$= \frac{a^2}{2}\arcsin \frac{x}{a} + \frac{x}{2}\sqrt{(a^2 - x^2)} + c$$

Problem 4. Evaluate $\int_0^3 \sqrt{(9 - x^2)}\, dx$.

Using the result of Problem 3:

$$\int_0^3 \sqrt{(9 - x^2)}\, dx = \left[\frac{9}{2}\arcsin \frac{x}{3} + \frac{x}{2}\sqrt{(9 - x^2)}\right]_0^3$$

$$= \left[\frac{9}{2}\arcsin 1 + \frac{3}{2}\sqrt{(9 - 9)}\right] - \left[\frac{9}{2}\arcsin 0 + 0\right]$$

$$= \frac{9}{2}\arcsin 1$$

'arcsin 1' means 'the angle whose sine is equal to 1', i.e. $\dfrac{\pi}{2}$ radians

Hence $\displaystyle\int_0^3 \sqrt{(9 - x^2)}\,dx = \dfrac{9}{2} \times \dfrac{\pi}{2} = \dfrac{9\pi}{4}$ or **7.068 6**

Problem 5. Find $\displaystyle\int \dfrac{1}{\sqrt{(5 + 4x - x^2)}}\,dx.$

$$\int \frac{1}{\sqrt{(5 + 4x - x^2)}}\,dx = \int \frac{1}{\sqrt{[-(x^2 - 4x - 5)]}}\,dx$$

$$= \int \frac{1}{\sqrt{[-\{(x - 2)^2 - 9\}]}}\,dx$$

$$= \int \frac{1}{\sqrt{[(3)^2 - (x - 2)^2]}}\,dx = \mathbf{arcsin}\left(\frac{x - 2}{3}\right) + c$$

(E) $\displaystyle\int \dfrac{1}{a^2 + x^2}\,dx$ – the 'tan θ' substitution

When an integral is of the form $\dfrac{1}{a^2 + x^2}$ the substitution $x = a\tan\theta$ is used. The reason for this is made obvious by the following worked problem.

Problem 1. Find $\displaystyle\int \dfrac{1}{a^2 + x^2}\,dx.$

Let $x = a\tan\theta$, then $\dfrac{dx}{d\theta} = a\sec^2\theta$, i.e. $dx = a\sec^2\theta\,d\theta$

$$\int \frac{1}{a^2 + x^2}\,dx = \int \frac{1}{a^2 + a^2\tan^2\theta}\,a\sec^2\theta\,d\theta = \int \frac{a\sec^2\theta\,d\theta}{a^2(1 + \tan^2\theta)}$$

$$= \int \frac{a\sec^2\theta}{a^2\sec^2\theta}\,d\theta, \text{ since } 1 + \tan^2\theta = \sec^2\theta$$

$$= \int \frac{1}{a}\,d\theta = \frac{1}{a}(\theta) + c$$

Hence $\displaystyle\int \dfrac{1}{a^2 + x^2}\,dx = \dfrac{1}{a}\arctan\dfrac{x}{a} + c$, since $x = a\tan\theta$

Problem 2. Evaluate $\displaystyle\int_0^1 \dfrac{1}{9 + 4x^2}\,dx$, correct to four significant figures.

$$\int_0^1 \frac{1}{9 + 4x^2} \, dx = \int_0^1 \frac{1}{4(\frac{9}{4} + x^2)} \, dx = \frac{1}{4} \int_0^1 \frac{1}{(\frac{3}{2})^2 + x^2} \, dx$$

$$= \frac{1}{4} \left[\frac{1}{(\frac{3}{2})} \arctan \frac{x}{(\frac{3}{2})} \right]_0^1$$

from the result of Problem 1

$$= \frac{1}{4} \left[\arctan \frac{2x}{3} \right]_0^1$$

$$= \frac{1}{6} \left[\arctan \frac{2}{3} - \arctan 0 \right]$$

$$= \frac{1}{6} (0.588\,0) = \mathbf{0.098\,0}$$

(F) Integrals containing $\sqrt{(x^2 + a^2)}$ – the 'sinh θ' substitution

When an integral contains a term $\sqrt{(x^2 + a^2)}$ the substitution $x = a \sinh \theta$ is used. The reason for this is made obvious by the following worked problems.

Problem 1. Find (a) $\int \dfrac{1}{\sqrt{(x^2 + a^2)}} \, dx$, and

(b) evaluate $\int_0^3 \dfrac{1}{\sqrt{(x^2 + 9)}} \, dx$.

(a)   Let $x = a \sinh \theta$ then $\dfrac{dx}{d\theta} = a \cosh \theta$, i.e. $dx = a \cosh \theta \, d\theta$

$$\int \frac{1}{\sqrt{(x^2 + a^2)}} \, dx = \int \frac{1}{\sqrt{(a^2 \sinh^2 \theta + a^2)}} a \cosh \theta \, d\theta$$

$$= \int \frac{a \cosh \theta \, d\theta}{\sqrt{[a^2(\sinh^2 \theta + 1)]}}$$

$$= \int \frac{a \cosh \theta}{a \cosh \theta} \, d\theta, \text{ since } \sinh^2 \theta + 1 = \cosh^2 \theta$$

$$= \int d\theta = \theta = \mathbf{arsinh} \, \frac{x}{a} + c \text{ since } x = a \sinh \theta$$

From Chapter 14, $\text{arsinh} \, \dfrac{x}{a} = \ln \left\{ \dfrac{x + \sqrt{(x^2 + a^2)}}{a} \right\}$ which

gives an alternative solution to $\int \dfrac{1}{\sqrt{(x^2 + a^2)}} \, dx$.

(b) $\int_0^3 \dfrac{1}{\sqrt{(x^2 + 9)}}\, dx = \left[\text{arsinh}\, \dfrac{x}{3}\right]_0^3$ or $\left[\ln\left\{\dfrac{x + \sqrt{(x^2 + 9)}}{3}\right\}\right]_0^3$

from part (a)

$\left[\text{arsinh}\, \dfrac{x}{3}\right]_0^3 = (\text{arsinh}\, 1 - \text{arsinh}\, 0)$

To evaluate arsinh 1: If $x = \text{arsinh}\, 1$

then $\sinh x = 1$

$\dfrac{e^x - e^{-x}}{2} = 1$

$e^x - e^{-x} = 2$

Multiplying each term by e^x gives: $(e^x)^2 - 2\, e^x - 1 = 0$

$e^x = \dfrac{-(-2) \pm \sqrt{[(-2)^2 - 4(1)(-1)]}}{2(1)}$

$= \dfrac{2 \pm \sqrt{8}}{2}$

Taking natural logarithms of both sides gives:

$x = \ln\left(\dfrac{2 + \sqrt{8}}{2}\right)$, since the logarithm of a negative number cannot be determined in real terms

$= 0.881\, 4$

Hence $\text{arsinh}\, 1 = 0.881\, 4$

Hence $\int_0^3 \dfrac{1}{\sqrt{(x^2 + 9)}}\, dx = \mathbf{0.881\, 4}$

Using the logarithmic form:

$\int_0^3 \dfrac{1}{\sqrt{(x^2 + 9)}}\, dx = \left[\ln\left\{\dfrac{x + \sqrt{(x^2 + 9)}}{3}\right\}\right]_0^3$

$= \ln\left(\dfrac{3 + \sqrt{18}}{3}\right) = \mathbf{0.881\, 4}$, as above

The logarithmic form is thus seen to be easier for evaluating definite integrals.

Problem 2. Find $\int \sqrt{(x^2 + a^2)}\, dx$.

Let $x = a \sinh \theta$, then $dx = a \cosh \theta\, d\theta$.

$$\int \sqrt{(x^2 + a^2)}\, dx = \int \sqrt{(a^2 \sinh^2 \theta + a^2)}\, a \cosh \theta\, d\theta$$

$$= \int \sqrt{[a^2(\sinh^2 \theta + 1)]}\, a \cosh \theta\, d\theta$$

$$= \int (a \cosh \theta)(a \cosh \theta)\, d\theta,$$
$$\text{since } \sinh^2 \theta + 1 = \cosh^2 \theta$$

$$= \int a^2 \cosh^2 \theta\, d\theta = a^2 \int \left(\frac{1 + \cosh 2\theta}{2}\right) d\theta$$
$$\text{since } \cosh 2\theta = 2 \cosh^2 \theta - 1 \text{ (see Chapter 9)}$$

$$= \frac{a^2}{2}\left[\theta + \frac{\sinh 2\theta}{2}\right] + c$$

Hence $\int \sqrt{(x^2 + a^2)}\, dx = \dfrac{a^2}{2}[\theta + \sinh \theta \cosh \theta] + c$, since
$$\sinh 2\theta = 2 \sinh \theta \cosh \theta$$

Since $x = a \sinh \theta$ then $\sinh \theta = \dfrac{x}{a}$ and $\theta = \text{arsinh}\, \dfrac{x}{a}$

Since $\cosh^2 \theta - \sinh^2 \theta = 1$, then

$$\cosh \theta = \sqrt{(1 + \sinh^2 \theta)} = \sqrt{\left[1 + \left(\frac{x}{a}\right)^2\right]} = \frac{1}{a}\sqrt{(a^2 + x^2)}$$

Hence $\int \sqrt{(x^2 + a^2)}\, dx = \dfrac{a^2}{2}\left[\text{arsinh}\, \dfrac{x}{a} + \left(\dfrac{x}{a}\right)\left(\dfrac{1}{a}\sqrt{(a^2 + x^2)}\right)\right] + c$

$$= \frac{a^2}{2}\, \text{arsinh}\, \frac{x}{a} + \frac{x}{2}\sqrt{(x^2 + a^2)} + c$$

Problem 3. Evaluate $\displaystyle\int_{-1}^{1} \dfrac{1}{x^2 \sqrt{(1 + x^2)}}\, dx$.

Since the integral contains a term in $\sqrt{(a^2 + x^2)}$, let $x = \sinh \theta$,
then $dx = \cosh \theta\, d\theta$

Hence $\displaystyle\int \dfrac{1}{x^2 \sqrt{(1 + x^2)}}\, dx = \int \dfrac{\cosh \theta\, d\theta}{(\sinh^2 \theta)\sqrt{(1 + \sinh^2 \theta)}}$

$$= \int \text{cosech}^2 \theta\, d\theta = -\coth \theta + c$$

Since $\cosh^2 \theta - \sinh^2 \theta = 1$ then
$\cosh \theta = \sqrt{(1 + \sinh^2 \theta)} = \sqrt{(1 + x^2)}$

Hence $-\coth \theta = \dfrac{-\cosh \theta}{\sinh \theta} = \dfrac{-\sqrt{(1 + x^2)}}{x}$

Therefore

$$\int_{-1}^{1} \frac{1}{x^2\sqrt{(1+x^2)}}\,dx = -\left[\frac{\sqrt{(1+x^2)}}{x}\right]_{-1}^{1} = -\left[\frac{\sqrt{2}}{1} - \frac{\sqrt{2}}{-1}\right]$$

$$= -2\sqrt{2} = -2.828\,4$$

(G) Integrals containing $\sqrt{(x^2 - a^2)}$ – the 'cosh θ' substitution

When an integral contains a term $\sqrt{(x^2 - a^2)}$ the substitution $x = a \cosh \theta$ is used. The reason for this is made obvious by the following worked problems.

Problem 1. Find $\displaystyle\int \frac{1}{\sqrt{(x^2 - a^2)}}\,dx.$

Let $x = a \cosh \theta$, then $dx = a \sinh \theta\, d\theta$

$$\int \frac{1}{\sqrt{(x^2 - a^2)}}\,dx = \int \frac{1}{\sqrt{(a^2 \cosh^2 \theta - a^2)}}\, a \sinh \theta\, d\theta$$

$$= \int \frac{a \sinh \theta\, d\theta}{\sqrt{[a^2(\cosh^2 \theta - 1)]}} = \int \frac{a \sinh \theta\, d\theta}{a \sinh \theta}$$

$$\text{since } \cosh^2 \theta - 1 = \sinh^2 \theta$$

$$= \int d\theta = \theta + c$$

Hence $\displaystyle\int \frac{1}{\sqrt{(x^2 - a^2)}}\,dx = \text{arcosh}\,\frac{x}{a} + c$, since $x = a \cosh \theta$

From Chapter 14, $\text{arcosh}\,\dfrac{x}{a} = \ln\left\{\dfrac{x + \sqrt{(x^2 - a^2)}}{a}\right\}$ which gives

an alternative solution to $\displaystyle\int \frac{1}{\sqrt{(x^2 - a^2)}}\,dx$

Problem 2. Find $\int \sqrt{(x^2 - a^2)}\,dx.$

Let $x = a \cosh \theta$, then $dx = a \sinh \theta\, d\theta$

$$\int \sqrt{(x^2 - a^2)}\,dx = \int \sqrt{(a^2 \cosh^2 \theta - a^2)}\, a \sinh \theta\, d\theta$$

$$= \int (a \sinh \theta)(a \sinh \theta)\, d\theta = \int a^2 \sinh^2 \theta\, d\theta$$

$$= a^2 \int \left(\frac{\cosh 2\theta - 1}{2} \right) d\theta$$

since $\cosh 2\theta = 1 + 2 \sinh^2 \theta$

$$= \frac{a^2}{2} \left[\frac{\sinh 2\theta}{2} - \theta \right] + c$$

$$= \frac{a^2}{2} [\sinh \theta \cosh \theta - \theta] + c$$

since $\sinh 2\theta = 2 \sinh \theta \cosh \theta$

Since $x = a \cosh \theta$ then $\cosh \theta = \dfrac{x}{a}$ and $\theta = \operatorname{arcosh} \dfrac{x}{a}$

Since $\cosh^2 \theta - \sinh^2 \theta = 1$ then

$$\sinh \theta = \sqrt{(\cosh^2 \theta - 1)} = \sqrt{\left[\left(\frac{x}{a} \right)^2 - 1 \right]}$$

$$= \frac{1}{a} \sqrt{(x^2 - a^2)}$$

Hence $\displaystyle\int \sqrt{(x^2 - a^2)}\, dx = \frac{a^2}{2} \left[\frac{1}{a} \sqrt{(x^2 - a^2)} \left(\frac{x}{a} \right) - \operatorname{arcosh} \frac{x}{a} \right] + c$

$$= \frac{x}{2} \sqrt{(x^2 - a^2)} - \frac{a^2}{2} \operatorname{arcosh} \frac{x}{a} + c$$

Problem 3. Find $\displaystyle\int \frac{4x + 1}{\sqrt{(x^2 + 4x - 12)}}\, dx.$

With this type, two fractions are created, one of the form

$k \displaystyle\int \dfrac{f'(x)}{f(x)}\, dx$, and the other to keep the ratio the same.

Thus $\displaystyle\int \frac{4x + 1}{\sqrt{(x^2 + 4x - 12)}}\, dx$

$$= \int \left\{ \frac{2(2x + 4)}{\sqrt{(x^2 + 4x - 12)}} - \frac{7}{\sqrt{(x^2 + 4x - 12)}} \right\} dx$$

$$= 4\sqrt{(x^2 + 4x - 12)} - 7 \int \frac{1}{\sqrt{[(x + 2)^2 - (4)^2]}}\, dx$$

$$= 4\sqrt{(x^2 + 4x - 12)} - 7 \operatorname{arcosh} \left(\frac{x + 2}{4} \right) + c$$

Further problems on integration using trigonometric and hyperbolic identities and substitutions may be found in Section 18.5 (Problems 39 to 58), page 295.

Summary

$f(x)$	$\int f(x)\,dx$	Method
$\cos^2 x$	$\dfrac{1}{2}\left(x + \dfrac{\sin 2x}{2}\right) + c$	Use $\cos 2x = 2\cos^2 x - 1$
$\sin^2 x$	$\dfrac{1}{2}\left(x - \dfrac{\sin 2x}{2}\right) + c$	Use $\cos 2x = 1 - 2\sin^2 x$
$\tan^2 x$	$\tan x - x + c$	Use $1 + \tan^2 x = \sec^2 x$
$\cot^2 x$	$-\cot x - x + c$	Use $\cot^2 x + 1 = \operatorname{cosec}^2 x$
$\cos^m x \sin^n x$	If m or n is odd, use $\cos^2 x + \sin^2 x = 1$ If both m and n are even, use either $\cos 2x = 2\cos^2 x - 1$ or $\cos 2x = 1 - 2\sin^2 x$	
$\sin A \cos B$		Use $\tfrac{1}{2}[\sin(A + B) + \sin(A - B)]$
$\cos A \sin B$		Use $\tfrac{1}{2}[\sin(A + B) - \sin(A - B)]$
$\cos A \cos B$		Use $\tfrac{1}{2}[\cos(A + B) + \cos(A - B)]$
$\sin A \sin B$		Use $-\tfrac{1}{2}[\cos(A + B) - \cos(A - B)]$
$\dfrac{1}{\sqrt{(a^2 - x^2)}}$	$\arcsin \dfrac{x}{a} + c$	Use $x = a \sin \theta$ substitution
$\sqrt{(a^2 - x^2)}$	$\dfrac{a^2}{2}\arcsin \dfrac{x}{a} + \dfrac{x}{2}\sqrt{(a^2 - x^2)} + c$	Use $x = a \sin \theta$ substitution
$\dfrac{1}{a^2 + x^2}$	$\dfrac{1}{a}\arctan \dfrac{x}{a} + c$	Use $x = a \tan \theta$ substitution
$\dfrac{1}{\sqrt{(x^2 + a^2)}}$	$\begin{cases} \operatorname{arsinh} \dfrac{x}{a} + c \\[2ex] \text{or } \ln\left\{\dfrac{x + \sqrt{(x^2 + a^2)}}{a}\right\} + c \end{cases}$	Use $x = a \sinh \theta$ substitution
$\sqrt{(x^2 + a^2)}$	$\dfrac{a^2}{2}\operatorname{arsinh} \dfrac{x}{a} + \dfrac{x}{2}\sqrt{(x^2 + a^2)} + c$	Use $x = a \sinh \theta$ substitution
$\dfrac{1}{\sqrt{(x^2 - a^2)}}$	$\begin{cases} \operatorname{arcosh} \dfrac{x}{a} + c \\[2ex] \text{or } \ln\left\{\dfrac{x + \sqrt{(x^2 - a^2)}}{a}\right\} + c \end{cases}$	Use $x = a \cosh \theta$ substitution
$\sqrt{(x^2 - a^2)}$	$\dfrac{x}{2}\sqrt{(x^2 - a^2)} - \dfrac{a^2}{2}\operatorname{arcosh} \dfrac{x}{a} + c$	Use $x = a \cosh \theta$ substitution

18.4   Change of limits of integration by a substitution

When evaluating definite integrals involving substitutions it is often easier to change the limits of the integral as shown in the following worked problems.

Problem 1. Evaluate $\displaystyle\int_0^2 \frac{t}{\sqrt{(2t^2 + 1)}}\, dt$, taking positive values of square roots only.

Let $u = 2t^2 + 1$, then $\dfrac{du}{dt} = 4t$, i.e. $dt = \dfrac{du}{4t}$

Hence $\displaystyle\int_0^2 \frac{t}{\sqrt{(2t^2 + 1)}}\, dt = \int_{t=0}^{t=2} \frac{t}{\sqrt{u}}\,\frac{du}{4t} = \frac{1}{4}\int_{t=0}^{t=2} u^{-\frac{1}{2}}\, du$

However if $u = 2t^2 + 1$ then: when $t = 2$, $u = 9$ and when $t = 0$, $u = 1$

Hence $\dfrac{1}{4}\displaystyle\int_{t=0}^{t=2} u^{-\frac{1}{2}}\, du = \frac{1}{4}\int_{u=1}^{u=9} u^{-\frac{1}{2}}\, du$, i.e. the limits have been changed

$$= \frac{1}{4}\left[\frac{u^{-\frac{1}{2}}}{\frac{1}{2}}\right]_1^9 = \tfrac{1}{2}(\sqrt{9} - \sqrt{1}) = 1$$

taking positive values of square roots only.

When the limits are changed it makes it unnecessary to change back to the original variable (in this case, t) after integration.

Problem 2. Evaluate $\displaystyle\int_0^5 \frac{1}{\sqrt{(25 - x^2)}}\, dx$.

Let $x = 5 \sin \theta$, then $dx = 5 \cos \theta\, d\theta$

When $x = 5$, $\sin \theta = 1$ and $\theta = \dfrac{\pi}{2}$

When $x = 0$, $\sin \theta = 0$ and $\theta = 0$

Hence $\displaystyle\int_{x=0}^{x=5} \frac{1}{\sqrt{(25 - x^2)}}\, dx = \int_{\theta=0}^{\theta=\pi/2} \frac{5 \cos \theta\, d\theta}{\sqrt{[25(1 - \sin^2 \theta)]}}$

$$= [\theta]_0^{\pi/2} = \frac{\pi}{2} \text{ or } 1.570\,8$$

Problem 3. Evaluate $\displaystyle\int_0^\infty \frac{1}{(x^2 + 4)}\,dx.$

Let $x = 2 \tan \theta$, then $dx = 2 \sec^2 \theta \, d\theta$

When $x = +\infty$, $\theta = \dfrac{\pi}{2}$ and when $x = 0$, $\theta = 0$

Hence $\displaystyle\int_{x=0}^{x=\infty} \frac{1}{(x^2 + 4)}\,dx = \int_{\theta=0}^{\theta=\pi/2} \frac{2 \sec^2 \theta \, d\theta}{4(\tan^2 \theta + 1)} = \int_0^{\pi/2} \frac{1}{2}\,d\theta$

$$= \left[\frac{\theta}{2}\right]_0^{\pi/2} = \frac{1}{2}\left(\frac{\pi}{2} - 0\right) = \frac{\pi}{4} \text{ or } 0.785\,4$$

Problem 4. Evaluate $\displaystyle\int_0^1 \sqrt{(4 - x^2)}\,dx$

Let $x = 2 \sin \theta$, then $dx = 2 \cos \theta \, d\theta$

When $x = 1$, $\sin \theta = \dfrac{1}{2}$ and $\theta = \dfrac{\pi}{6}$

When $x = 0$, $\sin \theta = 0$ and $\theta = 0$

Hence $\displaystyle\int_{x=0}^{x=1} \sqrt{(4 - x^2)}\,dx = \int_{\theta=0}^{\theta=\pi/6} \sqrt{[4(1 - \sin^2 \theta)]}\, 2 \cos \theta \, d\theta$

$$= \int_0^{\pi/6} 4 \cos^2 \theta \, d\theta$$

$$= \int_0^{\pi/6} 4\left(\frac{1 + \cos 2\theta}{2}\right) d\theta$$

$$= 2\left[\theta + \frac{\sin 2\theta}{2}\right]_0^{\pi/6}$$

$$= 2\left[\left(\frac{\pi}{6} + \frac{\sin \pi/3}{2}\right) - (0)\right]$$

$$= \frac{\pi}{3} + \frac{\sqrt{3}}{2} \text{ or } 1.913\,2$$

Problem 5. Evaluate $\displaystyle\int_1^5 \frac{dx}{2x^2 + 4x + 26}.$

$$\int_1^5 \frac{dx}{2x^2 + 4x + 26} = \frac{1}{2} \int_1^5 \frac{dx}{x^2 + 2x + 13} = \frac{1}{2} \int_1^5 \frac{dx}{(x+1)^2 + 12}$$

$$= \frac{1}{2} \int_1^5 \frac{dx}{(x+1)^2 + (2\sqrt{3})^2}$$

Let $x + 1 = 2\sqrt{3} \tan \theta$, then $dx = 2\sqrt{3} \sec^2 \theta \, d\theta$

When $x = 5$, $\theta = \dfrac{\pi}{3}$ and when $x = 1$, $\theta = \dfrac{\pi}{6}$

Hence $\dfrac{1}{2} \displaystyle\int_1^5 \dfrac{dx}{(x+1)^2 + (2\sqrt{3})^2} = \dfrac{1}{2} \int_{\pi/6}^{\pi/3} \dfrac{2\sqrt{3} \sec^2 \theta \, d\theta}{(2\sqrt{3} \tan \theta)^2 + (2\sqrt{3})^2}$

$$= \frac{1}{2} \int_{\pi/6}^{\pi/3} \frac{2\sqrt{3} \sec^2 \theta \, d\theta}{(2\sqrt{3})^2 (\tan^2 \theta + 1)}$$

$$= \frac{1}{2} \int_{\pi/6}^{\pi/3} \frac{1}{2\sqrt{3}} \, d\theta = \frac{1}{4\sqrt{3}} [\theta]_{\pi/6}^{\pi/3}$$

$$= \frac{\pi}{24\sqrt{3}} \text{ or } 0.075\,57$$

Further problems on evaluating definite integrals may be found in the following Section 18.5 (Problems 59 to 69), page 298.

18.5   Further problems

Integration using algebraic substitution

In Problems 1 to 24 integrate with respect to the variable.

1. $\sin(5x - 3)$ $[-\frac{1}{5} \cos(5x - 3) + c]$

2. $4 \cos(3t - 1)$ $[\frac{4}{3} \sin(3t - 1) + c]$

3. $5 \sec^2(7\theta - 2)$ $[\frac{5}{7} \tan(7\theta - 2) + c]$

4. $(3x + 2)^9$ $[\frac{1}{30}(3x + 2)^{10} + c]$

5. $14(7t + 4)^6$ $[\frac{2}{7}(7t + 4)^7 + c]$

6. $\dfrac{1}{3d + 1}$ $[\frac{1}{3} \ln(3d + 1) + c]$

7. $\dfrac{-2}{8x + 3}$ $[-\frac{1}{4} \ln(8x + 3) + c]$

8. $3 e^{4x-5}$ $[\frac{3}{4} e^{4x-5} + c]$

9. $6 e^{3-2t}$ $[-3 e^{3-2t} + c]$

10. $4x(2x^2 - 4)^6$ $[\frac{1}{7}(2x^2 - 4)^7 + c]$

11. $18t(1 - 3t^2)^5$ $[-\frac{1}{2}(1 - 3t^2)^6 + c]$

12. $\frac{2}{3} \sin^3 \theta \cos \theta$ $[\frac{1}{6} \sin^4 \theta + c]$

13. $3 \cos^4 t \sin t$ $[-\frac{3}{5} \cos^5 t + c]$

14. $\sec^2 2x \tan 2x$ $[\frac{1}{4} \tan^2 2x + c]$

15. $4x\sqrt{(4x^2 + 1)}$ $[\frac{1}{3}\sqrt{(4x^2 + 1)^3} + c]$

16. $(9x^2 - 4x)\sqrt{(3x^3 - 2x^2 + 1)}$ $[\frac{2}{3}\sqrt{(3x^3 - 2x^2 + 1)^3} + c]$

17. $\dfrac{2 \ln t}{t}$ $[(\ln t)^2 + c]$

18. $\dfrac{4x + 3}{(2x^2 + 3x - 1)^4}$ $\left[-\dfrac{1}{3(2x^2 + 3x - 1)^3} + c \right]$

19. $\dfrac{t}{\sqrt{(t^2 - 3)}}$ $[\sqrt{(t^2 - 3)} + c]$

20. $\dfrac{x^2 - 1}{\sqrt{(x^3 - 3x + 4)}}$ $[\frac{2}{3}\sqrt{(x^3 - 3x + 4)} + c]$

21. $\dfrac{4 e^t}{\sqrt{(2 + e^t)}}$ $[8\sqrt{(2 + e^t)} + c]$

22. $5 \tan 2\alpha$ $[\frac{5}{2} \ln(\sec 2\alpha) + c]$

23. $(5\theta + 2) \sec^2(5\theta^2 + 4\theta)$ $[\frac{1}{2} \tan(5\theta^2 + 4\theta) + c]$

24. $2t e^{3t^2 - 2}$ $[\frac{1}{3} e^{3t^2 - 2} + c]$

In Problems 25 to 35 evaluate the definite integrals.

25. $\int_0^1 (2x - 1)^6 \, dx$ $[\frac{1}{7}]$

26. $\int_0^2 t\sqrt{(2t^2 + 1)} \, dt$ $[4\frac{1}{3}]$

27. $\displaystyle\int_0^{\pi/2} \sin\left(2\theta + \frac{\pi}{3}\right) d\theta$ $[\frac{1}{2}]$

28. $\int_1^3 2 \cos(2t - 1) \, dt$ $[-1.8004]$

29. $\int_0^2 (2x^2 - 1)\sqrt{(2x^3 - 3x)} \, dx$ $[7.0273]$

30. $\displaystyle\int_1^2 \frac{5 \ln x}{x} \, dx$ $[1.2011]$

31. $\displaystyle\int_1^3 \frac{(x - 2)}{(x^2 - 4x + 1)^4} \, dx$ $[0]$

32. $\int_0^1 2t \, e^{4t^2-3} \, dt$ [0.667 1]

33. $\int_0^{\pi/3} 2 \sin^5 \theta \cos \theta \, d\theta$ [0.140 6]

34. $\int_0^1 \theta \sec^2(5\theta^2) \, d\theta$ [-0.338 1]

35. $\int_1^2 \left[\dfrac{e^{(2x-1)} - e^{(-2x+1)}}{2} \right] dx$ [4.262 3]

36. The entropy change for an ideal gas is given by the equation

$$\Delta S = \int_{T_1}^{T_2} C_V \frac{dT}{T} - R \int_{V_1}^{V_2} \frac{dV}{V}$$

where T is the thermodynamic temperature, V is the volume and $R = 8.314$.

Determine the entropy change when a gas expands from 2 litres to 3 litres for a temperature rise from 200 K to 400 K given that

$$C_V = 45 + 6 \times 10^{-3}T + 8 \times 10^{-6}T^2 \qquad [29.5]$$

37. The electrostatic potential on all parts of a conducting circular disc of radius r is given by the equation

$$V = 2\pi\sigma \int_0^9 \frac{R \, dR}{\sqrt{(R^2 + r^2)}}$$

Solve the equation by determining the integral.

$$[V = 2\pi\sigma[\sqrt{(9^2 + r^2)} - r]]$$

38. In the study of a rigid rotor the following integration occurs:

$$Z_r = \int_0^\infty (2J + 1) \, e^{-J(J+1)h^2/8\pi^2 Ikt} \, dJ$$

Determine Z_r for constant temperature T assuming h, I, and k are constants. $\left[\dfrac{8\pi^2 IkT}{h^2} \right]$

Integration using trigonometric and hyperbolic identities and substitutions

In Problems 39 to 58 integrate with respect to the variable.

39. (a) $\cos^2 2x$ (b) $\sin^2 3x$ (c) $\tan^2 4x$ (d) $\cot^2 5x$

(a) $\left[\dfrac{1}{2}\left(x + \dfrac{\sin 4x}{4}\right) + c\right]$ (b) $\left[\dfrac{1}{2}\left(x - \dfrac{\sin 6x}{6}\right) + c\right]$

(c) $\left[\dfrac{1}{4}\tan 4x - x + c\right]$ (d) $\left[-\dfrac{1}{5}\cot 5x - x + c\right]$

Powers of sines and cosines

40. (a) $\cos^3 \theta$ (b) $\sin^3 2\theta$

(a) $\left[\sin \theta - \dfrac{\sin^3 \theta}{3} + c\right]$ (b) $\left[\dfrac{1}{6}\cos^3 2\theta - \dfrac{1}{2}\cos 2\theta + c\right]$

41. (a) $\cos^3 t \sin^2 t$ (b) $\sin^4 x \cos^3 x$

(a) $[\frac{1}{3}\sin^3 t - \frac{1}{5}\sin^5 t + c]$ (b) $[\frac{1}{5}\sin^5 x - \frac{1}{7}\sin^7 x + c]$

42. (a) $\sin^4 2x$ (b) $\cos^2 x \sin^2 x$

(a) $\left[\dfrac{3x}{8} - \dfrac{1}{8}\sin 4x + \dfrac{1}{64}\sin 8x + c\right]$ (b) $\left[\dfrac{x}{8} - \dfrac{1}{32}\sin 4x + c\right]$

43. (a) $3 \cos^4 3t \sin^2 3t$ (b) $\frac{1}{2}\sin^4 2\theta \cos^2 2\theta$

(a) $[\frac{3}{16}(t - \frac{1}{12}\sin 12t + \frac{1}{9}\sin^3 6t) + c]$

(b) $[\frac{1}{32}(\theta - \frac{1}{8}\sin 8\theta - \frac{1}{6}\sin^3 4\theta) + c]$

Products of sines and cosines

44. (a) $\sin 3t \cos t$ (b) $2 \cos 4t \cos 2t$

(a) $[-\frac{1}{8}(\cos 4t + 2 \cos 2t) + c]$ (b) $[\frac{1}{6}\sin 6t + \frac{1}{2}\sin 2t + c]$

45. (a) $3 \cos 6x \cos 2x$ (b) $4 \sin 5x \sin 2x$

(a) $[\frac{3}{16}(\sin 8x + 2 \sin 4x) + c]$ (b) $[2(\frac{1}{3}\sin 3x - \frac{1}{7}\sin 7x) + c]$

46. (a) $\frac{1}{2}\sin 9t \sin 3t$ (b) $2 \sin 4\theta \cos 3\theta$

(a) $[\frac{1}{48}(2 \sin 6t - \sin 12t) + c]$ (b) $[-\frac{1}{7}\cos 7\theta - \cos \theta + c]$

47. (a) $9 \cos 5t \cos 4t$ (b) $\frac{3}{2}\cos 2x \sin x$

(a) $[\frac{9}{2}(\frac{1}{9}\sin 9t + t) + c]$ (b) $[\frac{3}{4}(\cos x - \frac{1}{3}\cos 3x) + c]$

'Sine θ' substitution

48. (a) $\dfrac{2}{\sqrt{(4 - x^2)}}$ (b) $\dfrac{1}{\sqrt{(9 - 4x^2)}}$

(a) $\left[2 \arcsin \dfrac{x}{2} + c\right]$ (b) $\left[\dfrac{1}{2}\arcsin \dfrac{2x}{3} + c\right]$

49. (a) $\sqrt{(16 - x^2)}$ (b) $\sqrt{(16 - 9x^2)}$

(a) $\left[8 \arcsin \dfrac{x}{4} + \dfrac{x}{2} \sqrt{(16 - x^2)} + c \right]$

(b) $\left[\dfrac{8}{3} \arcsin \dfrac{3x}{4} + \dfrac{x}{2} \sqrt{(16 - 9x^2)} + c \right]$

50. (a) $\dfrac{1}{\sqrt{(4 + 2x - x^2)}}$ (b) $\sqrt{(4 + 2x - x^2)}$

(a) $\left[\arcsin \dfrac{(x - 1)}{\sqrt{5}} + c \right]$

(b) $\left[\dfrac{5}{2} \arcsin \dfrac{(x - 1)}{\sqrt{5}} + \left(\dfrac{x - 1}{2} \right) \sqrt{(4 + 2x - x^2)} + c \right]$

'Tan θ' substitution

51. (a) $\dfrac{2}{1 + x^2}$ (b) $\dfrac{3}{16 + x^2}$

(a) $[2 \arctan x + c]$ (b) $\left[\dfrac{3}{4} \arctan \dfrac{x}{4} + c \right]$

52. (a) $\dfrac{1}{9 + 16x^2}$ (b) $\dfrac{1}{2x^2 + 4x + 18}$

(a) $\left[\dfrac{1}{12} \arctan \dfrac{4x}{3} + c \right]$ (b) $\left[\dfrac{1}{4\sqrt{2}} \arctan \dfrac{x + 1}{2\sqrt{2}} + c \right]$

'Sinh θ' substitution

53. (a) $\dfrac{1}{\sqrt{(x^2 + 4)}}$ (b) $\dfrac{1}{\sqrt{(9 + 2x^2)}}$

(a) $\left[\operatorname{arsinh} \dfrac{x}{2} + c \quad \text{or} \quad \ln \left\{ \dfrac{x + \sqrt{(x^2 + 4)}}{2} \right\} + c \right]$

(b) $\left[\dfrac{1}{\sqrt{2}} \operatorname{arsinh} \dfrac{\sqrt{2}x}{3} + c \text{ or } \dfrac{1}{\sqrt{2}} \ln \left\{ \dfrac{x + \sqrt{(\frac{9}{2} + x^2)}}{3/\sqrt{2}} \right\} + c \right]$

54. (a) $\sqrt{(x^2 + 16)}$ (b) $\sqrt{(4x^2 + 25)}$

(a) $\left[8 \operatorname{arsinh} \dfrac{x}{4} + \dfrac{x}{2} \sqrt{(x^2 + 16)} + c \right]$

(b) $\left[\dfrac{25}{4} \operatorname{arsinh} \dfrac{2x}{5} + x \sqrt{\left(x^2 + \dfrac{25}{4} \right)} + c \right]$

55. (a) $\sqrt{(x^2 + 2x + 5)}$ (b) $\dfrac{2}{3x^2\sqrt{(4 + x^2)}}$ (by letting $x = 2\sinh\theta$)

(a) $\left[\,2\,\text{arsinh}\,\dfrac{x+1}{2} + \left(\dfrac{x+1}{2}\right)\sqrt{(x^2 + 2x + 5)} + c\,\right]$

(b) $\left[\,\dfrac{-\sqrt{(4 + x^2)}}{6x} + c\,\right]$

'Cosh θ' substitution

56. (a) $\dfrac{1}{\sqrt{(x^2 - 9)}}$ (b) $\dfrac{1}{\sqrt{(4x^2 - 9)}}$

(a) $\left[\,\text{arcosh}\,\dfrac{x}{3} + c \ \text{ or } \ \ln\left\{\dfrac{x + \sqrt{(x^2 - 9)}}{3}\right\} + c\,\right]$

(b) $\left[\,\dfrac{1}{2}\,\text{arcosh}\,\dfrac{2x}{3} + c \ \text{ or } \ \dfrac{1}{2}\ln\left\{\dfrac{x + \sqrt{(x^2 - \frac{9}{4})}}{\frac{3}{2}}\right\} + c\,\right]$

57. (a) $\sqrt{(x^2 - 4)}$ (b) $\sqrt{(9x^2 - 16)}$

(a) $\left[\,\dfrac{x}{2}\sqrt{(x^2 - 4)} - 2\,\text{arcosh}\,\dfrac{x}{2} + c\,\right]$

(b) $\left[\,3\left\{\dfrac{x}{2}\sqrt{(x^2 - \frac{16}{9})} - \dfrac{8}{9}\,\text{arcosh}\,\dfrac{3x}{4}\right\} + c\,\right]$

58. (a) $\dfrac{1}{\sqrt{(x^2 + 4x - 5)}}$ (b) $\sqrt{(x^2 + 6x + 5)}$

(a) $\left[\,\text{arcosh}\,\dfrac{x+2}{3} + c \ \text{ or } \ \ln\left\{\dfrac{(x+2) + \sqrt{(x^2 + 4x - 5)}}{3}\right\} + c\,\right]$

(b) $\left[\,\left(\dfrac{x+3}{2}\right)\sqrt{(x^2 + 6x + 5)} - 2\,\text{arcosh}\left(\dfrac{x+3}{2}\right) + c\,\right]$

Definite integrals

In Problems 59 to 69 evaluate the definite integrals.

59. (a) $\int_0^{\pi/2} \sin^2\theta\,d\theta$ (b) $\int_0^1 2\cos^2\theta\,d\theta$

(a) $\left[\,\dfrac{\pi}{4} \ \text{ or } \ 0.785\,4\,\right]$ (b) $[1.454\,6]$

60. (a) $\int_0^{\pi/4} \sin^3 t\cos t\,dt$ (b) $\int_0^{\pi/3} \cos^3 2t\,dt$

(a) $[0.062\,5]$ (b) $[0.324\,8]$

61. (a) $\int_{\pi/4}^{\pi/2} 4 \sin^2 \theta \cos^2 \theta \, d\theta$ (b) $\int_0^{\pi/2} 3 \sin^2 2t \cos^4 2t \, dt$

(a) $\left[\dfrac{\pi}{8} \text{ or } 0.392\,7 \right]$ (b) $\left[\dfrac{3\pi}{32} \text{ or } 0.294\,5 \right]$

62. (a) $\int_0^{\pi/4} \sin 5\theta \cos 3\theta \, d\theta$ (b) $\int_0^{\pi/3} 4 \cos 6t \sin 3t \, dt$

(a) $[\frac{1}{4}]$ (b) $[-\frac{8}{9}]$

63. (a) $\int_0^1 2 \cos 7x \cos 2x \, dx$ (b) $\int_0^{\pi/2} \sin 8\alpha \sin 5\alpha \, d\alpha$

(a) $[-0.146\,0]$ (b) $[-0.205\,1]$

64. (a) $\displaystyle\int_0^3 \dfrac{1}{\sqrt{(9 - x^2)}} \, dx$ (b) $\int_0^3 \sqrt{(9 - x^2)} \, dx$

(a) $\left[\dfrac{\pi}{2} \text{ or } 1.570\,8 \right]$ (b) $\left[\dfrac{9\pi}{4} \text{ or } 7.068\,6 \right]$

65. (a) $\displaystyle\int_0^{1/2} \dfrac{2}{\sqrt{(1 - x^2)}} \, dx$ (b) $\int_0^{3/4} \sqrt{(9 - 16x^2)} \, dx$

(a) $\left[\dfrac{\pi}{3} \text{ or } 1.047\,2 \right]$ (b) $\left[\dfrac{9\pi}{16} \text{ or } 1.767\,1 \right]$

66. (a) $\displaystyle\int_0^1 \dfrac{1}{1 + x^2} \, dx$ (b) $\displaystyle\int_0^2 \dfrac{3}{4 + x^2} \, dx$

(a) $\left[\dfrac{\pi}{2} \text{ or } 0.785\,4 \right]$ (b) $\left[\dfrac{3\pi}{8} \text{ or } 1.178\,1 \right]$

67. (a) $\displaystyle\int_{-1}^2 \dfrac{1}{2x^2 + 4x + 20} \, dx$ (b) $\displaystyle\int_{-3}^0 \dfrac{5x - 2}{x^2 + 6x + 13} \, dx$

(a) $\left[\dfrac{\pi}{24} \text{ or } 0.130\,9 \right]$ (b) $[-5.407\,1]$

68. (a) $\displaystyle\int_0^2 \dfrac{1}{\sqrt{(x^2 + 4)}} \, dx$ (b) $\int_0^1 \sqrt{(x^2 + 9)} \, dx$

(a) $[0.881\,4]$ (b) $[3.054\,7]$

69. (a) $\displaystyle\int_1^2 \dfrac{1}{\sqrt{(x^2 - 1)}} \, dx$ (b) $\int_2^3 \sqrt{(x^2 - 4)} \, dx$

(a) $[1.317\,0]$ (b) $[1.429\,3]$

Chapter 19

Integration using partial fractions

19.1   Introduction

The process of expressing a fraction in terms of simpler fractions – called partial fractions – was explained in Chapter 3. There are certain functions that can only be integrated when they have been resolved into partial fractions.

This is demonstrated in the following worked problems.

Worked problems on integration using partial fractions

(a)   *Denominator containing linear factors*

Problem 1. Find $\int \dfrac{x-8}{x^2 - x - 2} \, dx$.

It was shown on page 38 that

$$\frac{x-8}{x^2 - x - 2} \equiv \frac{3}{x+1} - \frac{2}{x-2}$$

Hence $\displaystyle\int \frac{x-8}{x^2 - x - 2} \, dx = \int \left(\frac{3}{x+1} - \frac{2}{x-2} \right) dx$

$$= 3 \ln(x+1) - 2 \ln(x-2) + c$$

$$\text{or } \ln\left\{ \frac{x+1)^3}{(x-2)^2} \right\} + c$$

Problem 2. Determine $\int \dfrac{6x^2 + 7x - 25}{(x-1)(x+2)(x-3)} \, dx$.

It was shown on page 40 that

$$\frac{6x^2 + 7x - 25}{(x-1)(x+2)(x-3)} \equiv \frac{2}{(x-1)} - \frac{1}{(x+2)} + \frac{5}{(x-3)}$$

Hence $\displaystyle\int \frac{6x^2 + 7x - 25}{(x-1)(x+2)(x-3)} \, dx = \int \left(\frac{2}{(x-1)} - \frac{1}{(x+2)} + \frac{5}{(x-3)} \right) dx$

$= 2 \ln(x - 1) - \ln(x + 2) + 5 \ln(x - 3) + c$

or $\ln \left\{ \dfrac{(x-1)^2(x-3)^5}{(x+2)} \right\} + c$

Problem 3. Evaluate $\displaystyle\int_3^4 \frac{x^3 - x^2 - 5x}{x^2 - 3x + 2} \, dx.$

It was shown on page 40 that

$$\frac{x^3 - x^2 - 5x}{x^2 - 3x + 2} \equiv x + 2 + \frac{5}{(x-1)} - \frac{6}{(x-2)}$$

Hence $\displaystyle\int_3^4 \frac{x^3 - x^2 - 5x}{x^2 - 3x + 2} \, dx = \int_3^4 \left(x + 2 + \frac{5}{(x-1)} - \frac{6}{(x-2)} \right) dx$

$$= \left[\frac{x^2}{2} + 2x + 5 \ln(x-1) - 6 \ln(x-2) \right]_3^4$$

$$= (8 + 8 + 5 \ln 3 - 6 \ln 2)$$
$$- (\tfrac{9}{2} + 6 + 5 \ln 2 - 6 \ln 1)$$

$$= \mathbf{3.368\,4}$$

(b) Denominator containing repeated linear factors

Problem 4. Find $\displaystyle\int \frac{x + 5}{(x + 3)^2} \, dx.$

It was shown on page 41 that $\dfrac{x + 5}{(x + 3)^2} \equiv \dfrac{1}{(x + 3)} + \dfrac{2}{(x + 3)^2}$

Hence $\displaystyle\int \frac{x + 5}{(x + 3)^2} \, dx = \int \left(\frac{1}{(x + 3)} + \frac{2}{(x + 3)^2} \right) dx$

$$= \ln(x + 3) - \frac{2}{(x + 3)} + c$$

Problem 5. Find $\displaystyle\int \frac{5x^2 - 19x + 3}{(x - 2)^2(x + 1)} \, dx.$

It was shown on page 42 that

$$\frac{5x^2 - 19x + 3}{(x - 2)^2(x + 1)} \equiv \frac{2}{(x - 2)} - \frac{5}{(x - 2)^2} + \frac{3}{(x + 1)}$$

Hence $\displaystyle\int \frac{5x^2 - 19x + 3}{(x - 2)^2(x + 1)}\,dx = \int\left(\frac{2}{(x - 2)} - \frac{5}{(x - 2)^2} + \frac{3}{(x + 1)}\right)dx$

$$= 2\ln(x - 2) + \frac{5}{(x - 2)} + 3\ln(x + 1) + c$$

$$= \ln(x - 2)^2(x + 1)^3 + \frac{5}{(x - 2)} + c$$

Problem 6. Evaluate $\displaystyle\int_5^6 \frac{2x^2 - 13x + 13}{(x - 4)^3}\,dx.$

It was shown on page 43 that

$$\frac{2x^2 - 13x + 13}{(x - 4)^3} \equiv \frac{2}{(x - 4)} + \frac{3}{(x - 4)^2} - \frac{7}{(x - 4)^3}$$

Hence $\displaystyle\int_5^6 \frac{2x^2 - 13x + 13}{(x - 4)^3} = \int_5^6\left(\frac{2}{(x-4)} + \frac{3}{(x-4)^2} - \frac{7}{(x-4)^3}\right)dx$

$$= \left[2\ln(x - 4) - \frac{3}{(x - 4)} + \frac{7}{2(x - 4)^2}\right]_5^6$$

$$= (2\ln 2 - \tfrac{3}{2} + \tfrac{7}{8}) - (2\ln 1 - \tfrac{3}{1} + \tfrac{7}{2})$$

$$= 0.261\,3$$

(c)   Denominator containing a quadratic factor

Problem 7. Find $\displaystyle\int \frac{8x^2 - 3x + 19}{(x^2 + 3)(x - 1)}\,dx.$

It was shown on page 44 that $\displaystyle\frac{8x^2 - 3x + 19}{(x^2 + 3)(x - 1)} \equiv \frac{2x - 1}{(x^2 + 3)} + \frac{6}{(x - 1)}$

Hence $\displaystyle\int \frac{8x^2 - 3x + 19}{(x^2 + 3)(x - 1)}\,dx = \int\left(\frac{2x - 1}{(x^2 + 3)} + \frac{6}{(x - 1)}\right)dx$

$$= \int\left(\frac{2x}{(x^2 + 3)} - \frac{1}{(x^2 + 3)} + \frac{6}{(x - 1)}\right)dx$$

$$= \ln(x^2 + 3) - \frac{1}{\sqrt{3}}\arctan\frac{x}{\sqrt{3}} + 6\ln(x - 1) + c$$

$$= \ln(x^2 + 3)(x - 1)^6 - \frac{1}{\sqrt{3}}\arctan\frac{x}{\sqrt{3}} + c$$

Problem 8. Find $\displaystyle\int \frac{2 + x + 6x^2 - 2x^3}{x^2(x^2 + 1)}\,dx.$

It was shown on page 44 that

$$\frac{2 + x + 6x^2 - 2x^3}{x^2(x^2 + 1)} \equiv \frac{1}{x} + \frac{2}{x^2} + \frac{4 - 3x}{x^2 + 1}$$

Hence $\displaystyle\int \frac{2 + x + 6x^2 - 2x^3}{x^2(x^2 + 1)}\,dx = \int \left(\frac{1}{x} + \frac{2}{x^2} + \frac{4 - 3x}{x^2 + 1}\right)dx$

$$= \int \left(\frac{1}{x} + \frac{2}{x^2} + \frac{4}{x^2 + 1} - \frac{3x}{x^2 + 1}\right)dx$$

$$= \ln x - \frac{2}{x} + 4 \arctan x - \frac{3}{2}\ln(x^2 + 1) + c$$

$$= \ln\left\{\frac{x}{(x^2 + 1)^{\frac{3}{2}}}\right\} - \frac{2}{x} + 4 \arctan x + c$$

Problem 9. Find $\displaystyle\int \frac{5(x^2 + x + 3)}{x(x^2 + 2x + 5)}\,dx.$

Let $\displaystyle\frac{5(x^2 + x + 3)}{x(x^2 + 2x + 5)} \equiv \frac{A}{x} + \frac{Bx + C}{x^2 + 2x + 5}$

$$\equiv \frac{A(x^2 + 2x + 5) + (Bx + C)(x)}{x(x^2 + 2x + 5)}$$

Hence $5x^2 + 5x + 15 \equiv A(x^2 + 2x + 5) + (Bx + C)(x)$ by equating numerators.

Let $x = 0$, then $A = 3$

Equating the coefficients of x^2 gives $5 = A + B$, i.e. $B = 2$

Equating the coefficients of x gives: $5 = 2A + C$, i.e. $C = -1$

Hence $\displaystyle\int \frac{5(x^2 + x + 3)}{x(x^2 + 2x + 5)}\,dx = \int \left(\frac{3}{x} + \frac{2x - 1}{x^2 + 2x + 5}\right)dx$

Now $\displaystyle\int \frac{2x - 1}{x^2 + 2x + 5}\,dx = \int \frac{2x + 2}{(x^2 + 2x + 5)}\,dx - \int \frac{3}{(x^2 + 2x + 5)}\,dx$

i.e. the numerator of the first integral on the right-hand side has deliberately been made equal to the differential coefficient of the denominator so that the integral will integrate as $\ln(x^2 + 2x + 5)$.

$$\int \frac{3}{(x^2 + 2x + 5)}\,dx = \int \frac{3}{(x + 1)^2 + 4}\,dx = \int \frac{3}{(x + 1)^2 + (2)^2}\,dx$$

$$= \frac{3}{2}\arctan \frac{(x + 1)}{2}$$

Hence $\int \dfrac{5(x^2 + x + 3)}{x(x^2 + 2x + 5)}\,dx = 3\ln x + \ln(x^2 + 2x + 5)$

$$-\frac{3}{2}\arctan\frac{(x + 1)}{2} + c$$

$$= \ln x^3(x^2 + 2x + 5) - \frac{3}{2}\arctan\frac{(x + 1)}{2} + c$$

(d)   *Integrals of the form* $\displaystyle\int \dfrac{1}{x^2 - a^2}\,dx$ *and* $\displaystyle\int \dfrac{1}{a^2 - x^2}\,dx$

Problem 10. Determine (a) $\displaystyle\int \dfrac{1}{x^2 - a^2}\,dx$, (b) $\displaystyle\int_1^2 \dfrac{1}{(9 - x^2)}\,dx$.

(a)   Let $\dfrac{1}{x^2 - a^2} \equiv \dfrac{A}{(x - a)} + \dfrac{B}{(x + a)} \equiv \dfrac{A(x + a) + B(x - a)}{(x - a)(x + a)}$

Hence $1 = A(x + a) + B(x - a)$ by equating numerators

Let $x = a$, then $A = \dfrac{1}{2a}$. Let $x = -a$, then $B = -\dfrac{1}{2a}$

Hence $\displaystyle\int \dfrac{1}{x^2 - a^2}\,dx = \dfrac{1}{2a}\int \left(\dfrac{1}{(x - a)} - \dfrac{1}{(x + a)}\right)\,dx$

$$= \frac{1}{2a}\left[\ln(x - a) - \ln(x + a)\right] + c$$

$$= \frac{1}{2a}\ln\left(\frac{x - a}{x + a}\right) + c$$

Similarly, it may be shown that $\displaystyle\int \dfrac{1}{a^2 - x^2}\,dx = \dfrac{1}{2a}\ln\left(\dfrac{a + x}{a - x}\right) + c$

or $\dfrac{1}{a}\mathbf{artanh}\,\dfrac{x}{a} + c$ (by using the substitution $x = a\tanh\theta$)

(b) $\displaystyle\int_1^2 \dfrac{1}{9 - x^2}\,dx = \int_1^2 \dfrac{1}{3^2 - x^2}\,dx = \dfrac{1}{2(3)}\left[\ln\left(\dfrac{3 + x}{3 - x}\right)\right]_1^2$ from (a)

$$= \frac{1}{6}\ln\frac{5}{2} \text{ or } 0.1527$$

19.2   Further problems

In Problems 1 to 20 integrate after resolving into partial fractions.

1. $\dfrac{8}{x^2 - 4}$

$$\left[2\ln(x-2) - 2\ln(x+2) + c \text{ or } 2\ln\left(\frac{x-2}{x+2}\right) + c \right]$$

2. $\dfrac{3x+5}{x^2 + 2x - 3}$

$$[2\ln(x-1) + \ln(x+3) + c \text{ or } \ln\{(x-1)^2(x+3)\} + c]$$

3. $\dfrac{y-13}{y^2 - y - 6}$

$$\left[3\ln(y+2) - 2\ln(y-3) + c \text{ or } \ln\left\{\frac{(y+2)^3}{(y-3)^2}\right\} + c \right]$$

4. $\dfrac{17x^2 - 21x - 6}{x(x+1)(x-3)}$

$$[2\ln x + 8\ln(x+1) + 7\ln(x-3) + c \text{ or}$$
$$\ln\{x^2(x+1)^8(x-3)^7\} + c]$$

5. $\dfrac{6x^2 + 7x - 49}{(x-4)(x+1)(2x-3)}$

$$\left[3\ln(x-4) - 2\ln(x+1) + 2\ln(2x-3) + c \text{ or}\right.$$
$$\left.\ln\left\{\frac{(x-4)^3(2x-3)^2}{(x+1)^2}\right\} + c \right]$$

6. $\dfrac{x^2 + 2}{(x+4)(x-2)}$

$$\left[x - 3\ln(x+4) + \ln(x-2) + c \text{ or } x + \ln\left\{\frac{(x-2)}{(x+4)^3}\right\} + c \right]$$

7. $\dfrac{2x^2 + 4x + 19}{2(x-3)(x+4)}$

$$\left[x + \tfrac{7}{2}\ln(x-3) - \tfrac{5}{2}\ln(x+4) + c \text{ or } x + \ln\left\{\frac{(x-3)^{\frac{7}{2}}}{(x+4)^{\frac{5}{2}}}\right\} + c \right]$$

8. $\dfrac{2x^3 + 7x^2 - 2x - 27}{(x-1)(x+4)}$

$\left[x^2 + x - 4 \ln(x-1) + 7 \ln(x+4) + c \text{ or} \right.$

$$\left. x^2 + x + \ln\left\{\dfrac{(x+4)^7}{(x-1)^4}\right\} + c \right]$$

9. $\dfrac{2t-1}{(t+1)^2}$ $\left[2 \ln(t+1) + \dfrac{3}{(t+1)} + c \right]$

10. $\dfrac{8x^2 + 12x - 3}{(x+2)^3}$ $\left[8 \ln(x+2) + \dfrac{20}{(x+2)} - \dfrac{5}{2(x+2)^2} + c \right]$

11. $\dfrac{6x+1}{(2x+1)^2}$ $\left[\tfrac{3}{2} \ln(2x+1) + \dfrac{1}{(2x+1)} + c \right]$

12. $\dfrac{1}{x^2(x+2)}$

$$\left[-\dfrac{1}{2x} - \dfrac{1}{4}\ln x + \dfrac{1}{4}\ln(x+2) + c \text{ or } \dfrac{1}{4}\ln\left(\dfrac{x+2}{x}\right) - \dfrac{1}{2x} + c \right]$$

13. $\dfrac{9x^2 - 73x + 150}{(x-7)(x-3)^2}$

$\left[5 \ln(x-7) + 4 \ln(x-3) + \dfrac{3}{(x-3)} + c \text{ or} \right.$

$$\left. \ln\{(x-7)^5(x-3)^4\} + \dfrac{3}{x-3} + c \right]$$

14. $\dfrac{-(9x^2 + 4x + 4)}{x^2(x^2 - 4)}$

$\left[\ln x - \dfrac{1}{x} + 2 \ln(x+2) - 3 \ln(x-2) + c \text{ or} \right.$

$$\left. \ln\left\{\dfrac{x(x+2)^2}{(x-2)^3}\right\} - \dfrac{1}{x} + c \right]$$

15. $\dfrac{-(a^2 + 5a + 13)}{(a^2 + 5)(a - 2)}$

$\left[\ln(a^2+5) - \dfrac{1}{\sqrt{5}}\arctan\dfrac{a}{\sqrt{5}} - 3\ln(a-2) + c \text{ or} \right.$

$$\left. \ln\left\{\dfrac{(a^2+5)}{(a-2)^3}\right\} - \dfrac{1}{\sqrt{5}}\arctan\dfrac{a}{\sqrt{5}} + c \right]$$

16. $\dfrac{3 - x}{(x^2 + 3)(x + 3)}$

$$\left[\dfrac{1}{2\sqrt{3}} \arctan \dfrac{x}{\sqrt{3}} - \dfrac{1}{4} \ln(x^2 + 3) + \dfrac{1}{2} \ln(x + 3) + c \text{ or} \right.$$

$$\left. \ln \left\{ \dfrac{(x + 3)^{\frac{1}{2}}}{(x^2 + 3)^{\frac{1}{4}}} \right\} + \dfrac{1}{2\sqrt{3}} \arctan \dfrac{x}{\sqrt{3}} + c \right]$$

17. $\dfrac{12 - 2x - 5x^2}{(x^2 + x + 1)(3 - x)}$

$$\left[\ln(x^2 + x + 1) + \dfrac{8}{\sqrt{3}} \arctan \dfrac{2(x + \frac{1}{2})}{\sqrt{3}} + 3 \ln(3 - x) + c \right]$$

18. $\dfrac{x^3 + 7x^2 + 8x + 10}{x(x^2 + 2x + 5)}$

$$\left[x + 2 \ln x + \tfrac{3}{2} \ln(x^2 + 2x + 5) - 2 \arctan \left(\dfrac{x + 1}{2} \right) + c \right]$$

19. $\dfrac{5x^3 - 3x^2 + 41x - 64}{(x^2 + 6)(x - 1)^2}$

$$\left[\dfrac{2}{\sqrt{6}} \arctan \dfrac{x}{\sqrt{6}} - \dfrac{3}{2} \ln(x^2 + 6) + 8 \ln(x - 1) + \dfrac{3}{(x - 1)} + c \right]$$

20. $\dfrac{6x^3 + 5x^2 + 4x + 3}{(x^2 + x + 1)(x^2 - 1)}$

$$\left[\ln(x^2 + x + 1) - \dfrac{4}{\sqrt{3}} \arctan \dfrac{2(x + \frac{1}{2})}{\sqrt{3}} + 3 \ln(x - 1) + \ln(x + 1) + c \right]$$

In Problems 21 to 30 evaluate the definite integrals correct to four decimal places.

21. $\displaystyle\int_4^6 \dfrac{x - 7}{x^2 - 2x - 3} \, dx \qquad [-0.425\,7]$

22. $\displaystyle\int_2^3 \dfrac{x^3 - 2x^2 - 3x - 2}{(x + 2)(x - 1)} \, dx \qquad [-0.993\,7]$

23. $\displaystyle\int_3^4 \dfrac{4x^2 + 15x - 1}{(x + 1)(x - 2)(x + 3)} \, dx \qquad (2.371\,6]$

24. $\displaystyle\int_3^5 \dfrac{x^2 + 1}{x^2 + x - 6} \, dx \qquad [2.523\,2]$

25. $\int_6^8 \dfrac{1}{(x^2 - 25)}\,dx$ [0.093 2]

26. $\int_2^3 \dfrac{1}{(16 - x^2)}\,dx$ [0.105 9]

27. $\int_1^2 \dfrac{2 + x + 6x^2 - 2x^3}{x^2(x^2 + 1)}\,dx$ [1.605 7]

28. $\int_2^3 \dfrac{2x^2 - x - 2}{(x - 1)^3}\,dx$ [2.511 3]

29. $\int_3^4 \dfrac{2x^3 + 4x^2 - 8x - 4}{x^2(x^2 - 4)}\,dx$ [1.041 8]

30. $\int_0^1 \dfrac{-(4x^2 + 9x + 8)}{(x + 1)^2(x + 2)}\,dx$ [−2.546 5]

31. The velocity constant k of a given chemical reaction is given by

$$kt = \int \frac{dx}{(3 - 0.4x)(2 - 0.6x)}$$

where $x = 0$ when $t = 0$. Determine kt

$$\left[\ln\left\{\frac{2(3 - 0.4x)}{3(2 - 0.6x)}\right\}\right]$$

Chapter 20

The $t = \tan \theta/2$ substitution

20.1 Introduction

In the right-angled triangle shown in Fig. 20.1, let angle B be equal to $\frac{\theta}{2}$, with side AB = 1, and side AC = t

Fig. 20.1

Then $\tan \frac{\theta}{2} = \frac{t}{1} = t$, and by Pythagoras's theorem, side BC = $\sqrt{(1 + t^2)}$

Hence $\sin \frac{\theta}{2} = \frac{t}{\sqrt{(1 + t^2)}}$ and $\cos \frac{\theta}{2} = \frac{t}{\sqrt{(1 + t^2)}}$

From double-angle formulae (see Chapter 16 of *Technician mathematics 3*),

$$\sin 2x = 2 \sin x \cos x$$

and it follows that $\sin \theta = 2 \sin \frac{\theta}{2} \cos \frac{\theta}{2}$

Hence, from above, $\sin \theta = 2 \left(\frac{t}{\sqrt{(1 + t^2)}} \right) \left(\frac{1}{\sqrt{(1 + t^2)}} \right)$

i.e. $\qquad \sin \theta = \frac{2t}{(1 + t^2)}$ \hfill (1)

Also from double-angle formulae, $\cos 2x = \cos^2 x - \sin^2 x$

and it follows that $\cos \theta = \cos^2 \frac{\theta}{2} - \sin^2 \frac{\theta}{2}$

Hence, from above, $\cos \theta = \left[\dfrac{1}{\sqrt{(1 + t^2)}}\right]^2 - \left[\dfrac{t}{\sqrt{(1 + t^2)}}\right]^2$

$$= \dfrac{1}{(1 + t^2)} - \dfrac{t^2}{(1 + t^2)}$$

i.e. $$\cos \theta = \dfrac{(1 - t^2)}{(1 + t^2)} \qquad (2)$$

Since $t = \tan \dfrac{\theta}{2}$, then $\dfrac{dt}{d\theta} = \dfrac{1}{2} \sec^2 \dfrac{\theta}{2} = \dfrac{1}{2}\left(1 + \tan^2 \dfrac{\theta}{2}\right)$ from trigonometric identities

i.e. $$\dfrac{dt}{d\theta} = \dfrac{1}{2}(1 + t^2)$$

from which,

$$d\theta = \dfrac{2}{(1 + t^2)}\, dt \qquad (3)$$

Equations (1), (2) and (3) are used to determine integrals of the form $\displaystyle\int \dfrac{d\theta}{a \cos \theta + b \sin \theta + c}$, where a, b and c are constants (and may be zero) and are known as the $t = \tan \dfrac{\theta}{2}$ substitution. This is demonstrated in the worked problems following.

Worked problems on the $t = \tan \theta/2$ substitution

Problem 1. Find $\displaystyle\int \dfrac{d\theta}{\sin \theta}$.

If $t = \tan \dfrac{\theta}{2}$ then, from equation (1), $\sin \theta = \dfrac{2t}{1 + t^2}$ and from equation (3), $d\theta = \dfrac{2}{(1 + t^2)}\, dt$

Thus $\displaystyle\int \dfrac{d\theta}{\sin \theta} = \int \dfrac{1}{\left(\dfrac{2t}{1 + t^2}\right)} \cdot \dfrac{2}{(1 + t^2)}\, dt = \int \dfrac{1}{t}\, dt = \ln t + c$

Hence $\displaystyle\int \dfrac{d\theta}{\sin \theta} = \ln\left(\tan \dfrac{\theta}{2}\right) + c$

Problem 2. Determine $\int \dfrac{d\theta}{1 + \cos \theta}$.

If $t = \tan \dfrac{\theta}{2}$ then, from equation (2), $\cos \theta = \dfrac{(1 - t^2)}{(1 + t^2)}$ and from

equation (3), $d\theta = \dfrac{2}{(1 + t^2)} \, dt$

Thus $\displaystyle\int \frac{d\theta}{1 + \cos \theta} = \int \frac{1}{1 + \dfrac{(1 - t^2)}{(1 + t^2)}} \cdot \frac{2}{(1 + t^2)} \, dt$

$$= \int \frac{1}{\dfrac{(1 + t^2) + (1 - t^2)}{(1 + t^2)}} \cdot \frac{2}{(1 + t^2)} \, dt$$

$$= \int dt = t + c$$

Hence $\displaystyle\int \frac{d\theta}{1 + \cos \theta} = \tan \frac{\theta}{2} + c$

Problem 3. Determine $\int \dfrac{d\theta}{4 + 3 \cos \theta}$.

If $t = \tan \dfrac{\theta}{2}$, then $\cos \theta = \dfrac{(1 - t^2)}{(1 + t^2)}$ and $d\theta = \dfrac{2}{(1 + t^2)} \, dt$, from

equations (2) and (3)

Thus $\displaystyle\int \frac{d\theta}{4 + 3 \cos \theta} = \int \frac{1}{4 + 3\dfrac{(1 - t^2)}{(1 + t^2)}} \cdot \frac{2}{(1 + t^2)} \, dt$

$$= \int \frac{1}{\dfrac{4(1 + t^2) + 3(1 - t^2)}{(1 + t^2)}} \cdot \frac{2}{(1 + t^2)} \, dt = \int \frac{2}{7 + t^2} \, dt$$

$$= 2 \int \frac{1}{(\sqrt{7})^2 + t^2} \, dt = 2\left(\frac{1}{\sqrt{7}} \arctan \frac{t}{\sqrt{7}} \right) + c$$

Hence $\displaystyle\int \frac{d\theta}{4 + 3 \cos \theta} = \frac{2}{\sqrt{7}} \arctan\left(\frac{1}{\sqrt{7}} \tan \frac{\theta}{2} \right) + c$

Problem 4. Determine $\int \dfrac{d\theta}{\cos \theta}$.

If $t = \tan \dfrac{\theta}{2}$, then from equation (2), $\cos \theta = \dfrac{(1 - t^2)}{(1 + t^2)}$ and from

equation (3), $d\theta = \dfrac{2}{(1 + t^2)}\, dt$

Thus $\displaystyle\int \dfrac{d\theta}{\cos \theta} = \int \dfrac{1}{\dfrac{(1 - t^2)}{(1 + t^2)}} \cdot \dfrac{2}{(1 + t^2)}\, dt = \int \dfrac{2}{(1 - t^2)}\, dt$

$\dfrac{2}{(1 - t^2)}$ may be resolved into partial fractions (see Chapter 3)

Let $\dfrac{2}{(1 - t^2)} = \dfrac{2}{(1 - t)(1 + t)} \equiv \dfrac{A}{(1 - t)} + \dfrac{B}{(1 + t)}$

$$= \dfrac{A(1 + t) + B(1 - t)}{(1 - t)(1 + t)}$$

Hence $2 = A(1 + t) + B(1 - t)$
When $t = 1$, $2 = 2A$, from which $A = 1$
When $t = -1$, $2 = 2B$, from which $B = 1$

Hence $\displaystyle\int \dfrac{2}{(1 - t^2)}\, dt = \int \dfrac{1}{(1 - t)} + \dfrac{1}{(1 + t)}\, dt$

$$= -\ln(1 - t) + \ln(1 + t) + c$$

$$= \ln \dfrac{(1 + t)}{(1 - t)} + c, \text{ by the laws of logarithms}$$

Thus $\displaystyle\int \dfrac{d\theta}{\cos \theta} = \ln \left\{ \dfrac{1 + \tan \dfrac{\theta}{2}}{1 - \tan \dfrac{\theta}{2}} \right\} + c$

[Note that since $\tan \dfrac{\pi}{4} = 1$, the above result may be re-written as

$$\int \dfrac{d\theta}{\cos \theta} = \ln \left\{ \dfrac{\tan \dfrac{\pi}{4} + \tan \dfrac{\theta}{2}}{\tan \dfrac{\pi}{4} - \tan \dfrac{\theta}{2}} \right\} + c$$

$$= \ln \left\{ \tan \left(\dfrac{\pi}{4} + \dfrac{\theta}{2} \right) \right\} + c$$

from compound angles]

Problem 5. Find $\displaystyle\int \frac{d\theta}{\sin\theta + \cos\theta}$.

If $t = \tan\dfrac{\theta}{2}$, then $\sin\theta = \dfrac{2t}{(1+t^2)}$, $\cos\theta = \dfrac{(1-t^2)}{(1+t^2)}$ and $d\theta = \dfrac{2}{(1+t^2)} \, dt$ from equations (1), (2) and (3).

Thus $\displaystyle\int \frac{d\theta}{\sin\theta + \cos\theta} = \int \frac{\dfrac{2}{(1+t^2)}\,dt}{\dfrac{2t}{(1+t^2)} + \dfrac{(1-t^2)}{(1+t^2)}}$

$$= \int \frac{\dfrac{2}{(1+t^2)}\,dt}{\dfrac{2t+1-t^2}{(1+t^2)}} = \int \frac{2\,dt}{1+2t-t^2}$$

$$= \int \frac{-2\,dt}{t^2 - 2t - 1} = \int \frac{-2\,dt}{(t-1)^2 - 2}$$

$$= \int \frac{2\,dt}{2-(t-1)^2} = \int \frac{2\,dt}{(\sqrt{2})^2 - (t-1)^2}$$

$$= 2\left[\frac{1}{2\sqrt{2}} \ln\left\{\frac{\sqrt{2}+(t-1)}{\sqrt{2}-(t-1)}\right\}\right] + c$$

from Problem 10(b), page 304, i.e.

$$\int \frac{d\theta}{\sin\theta + \cos\theta} = \frac{1}{\sqrt{2}} \ln\left\{\frac{\sqrt{2}-1+\tan\dfrac{\theta}{2}}{\sqrt{2}+1-\tan\dfrac{\theta}{2}}\right\} + c$$

Problem 6. Determine $\displaystyle\int \frac{dx}{3 - 3\sin x + 2\cos x}$.

Using equations (1), (2) and (3),

$$\int \frac{dx}{3 - 3\sin x + 2\cos x} = \int \frac{\dfrac{2}{(1+t^2)}\,dt}{3 - 3\dfrac{(2t)}{(1+t^2)} + \dfrac{2(1-t^2)}{(1+t^2)}}$$

$$= \int \frac{\dfrac{2 \, dt}{(1+t^2)}}{\dfrac{3(1+t^2) - 3(2t) + 2(1-t^2)}{(1+t^2)}} = \int \frac{2 \, dt}{3 + 3t^2 - 6t + 2 - 2t^2}$$

$$= \int \frac{2 \, dt}{t^2 - 6t + 5} = \int \frac{2 \, dt}{(t-3)^2 - (2)^2}$$

$$= 2 \left[\frac{1}{2(2)} \ln \left\{ \frac{(t-3) - 2}{(t-3) + 2} \right\} \right] + c, \text{ from Problem 10(a), page 304,}$$

$$= \frac{1}{2} \ln \left(\frac{t-5}{t-1} \right) + c$$

i.e. $\displaystyle \int \frac{dx}{3 - 3 \sin x + 2 \cos x} = \frac{1}{2} \ln \left\{ \frac{\tan \dfrac{x}{2} - 5}{\tan \dfrac{x}{2} - 1} \right\} + c$

Problem 7. Determine $\displaystyle \int \frac{d\theta}{3 \sin \theta - 4 \cos \theta}$.

From equations (1), (2) and (3),

$$\int \frac{d\theta}{3 \sin \theta - 4 \cos \theta} = \int \frac{\dfrac{2}{(1+t^2)} \, dt}{3 \left(\dfrac{2t}{1+t^2} \right) - 4 \left(\dfrac{1-t^2}{1+t^2} \right)}$$

$$= \int \frac{\dfrac{2}{(1+t^2)} \, dt}{\dfrac{6t - 4 + 4t^2}{(1+t^2)}} = \int \frac{2 \, dt}{4t^2 + 6t - 4}$$

$$= \int \frac{dt}{2t^2 + 3t - 2} = \frac{1}{2} \int \frac{dt}{t^2 + \dfrac{3}{2}t - 1}$$

$$= \frac{1}{2} \int \frac{dt}{\left(t + \dfrac{3}{4} \right)^2 - \dfrac{25}{16}}$$

$$= \frac{1}{2} \int \frac{dt}{\left(t + \dfrac{3}{4} \right)^2 - \left(\dfrac{5}{4} \right)^2}$$

$$= \frac{1}{2} \left[\frac{1}{2\left(\frac{5}{4}\right)} \ln\left\{ \frac{\left(t+\frac{3}{4}\right) - \frac{5}{4}}{\left(t+\frac{3}{4}\right) + \frac{5}{4}} \right\} \right] + c$$

$$= \frac{1}{5} \ln\left(\frac{t - \frac{1}{2}}{t + 2} \right) + c$$

i.e. $\displaystyle\int \frac{d\theta}{3\sin\theta - 4\cos\theta} = \frac{1}{5} \ln\left\{ \frac{\tan\dfrac{\theta}{2} - \dfrac{1}{2}}{\tan\dfrac{\theta}{2} + 2} \right\} + c$

or

$$\frac{1}{5} \ln\left\{ \frac{2\tan\dfrac{\theta}{2} - 1}{2\tan\dfrac{\theta}{2} + 4} \right\} + c$$

Further problems on the $t = \tan\dfrac{\theta}{2}$ *substitution may be found in the following Section 20.2 (Problems 1 to 12).*

20.2   Further problems

In Problems 1 to 9, integrate with respect to the variable.

1. $\displaystyle\int \frac{dx}{1 + \sin x}$ $\qquad \left[\dfrac{-2}{1 + \tan\dfrac{x}{2}} + c \right]$

2. $\displaystyle\int \frac{d\theta}{5 + 4\cos\theta}$ $\qquad \left[\dfrac{2}{3} \arctan\left(\dfrac{1}{3} \tan\dfrac{\theta}{2} \right) + c \right]$

3. $\displaystyle\int \frac{d\theta}{7 - 3\sin\theta + 6\cos\theta}$ $\qquad \left[\arctan\left(\dfrac{\tan\dfrac{\theta}{2} - 3}{2} \right) + c \right]$

4. $\displaystyle\int \frac{dx}{3 \sin x + 4 \cos x}$ $\left[\dfrac{1}{5} \ln \left\{ \dfrac{1 + 2 \tan \dfrac{x}{2}}{4 - 2 \tan \dfrac{x}{2}} \right\} + c \right]$

5. $\displaystyle\int \frac{d\theta}{1 + \sin \theta - \cos \theta}$ $\left[\ln \left\{ \dfrac{\tan \dfrac{\theta}{2}}{1 + \tan \dfrac{\theta}{2}} \right\} + c \right]$

6. $\displaystyle\int \frac{dx}{2 + \cos x}$ $\left[\dfrac{2}{\sqrt{3}} \arctan \left(\dfrac{\tan \dfrac{x}{3}}{\sqrt{3}} \right) + c \right]$

7. $\displaystyle\int \frac{d\theta}{4 \sin \theta - 3 \cos \theta}$ $\left[\dfrac{1}{5} \ln \left\{ \dfrac{\tan \dfrac{\theta}{2} - \dfrac{1}{3}}{\tan \dfrac{\theta}{2} + 3} \right\} + c \right]$

8. $\displaystyle\int \frac{d\theta}{6 + 5 \sin \theta}$ $\left[\dfrac{2}{\sqrt{11}} \arctan \left(\dfrac{6 \tan \dfrac{\theta}{2} + 5}{\sqrt{11}} \right) + c \right]$

9. $\displaystyle\int \frac{d\alpha}{2 - 3 \sin \alpha}$ $\left[\dfrac{1}{\sqrt{5}} \ln \left\{ \dfrac{2 \tan \dfrac{\alpha}{2} - 3 - \sqrt{5}}{2 \tan \dfrac{\alpha}{2} - 3 + \sqrt{5}} \right\} + c \right]$

10. Show that $\displaystyle\int_0^{\pi/6} \frac{3}{\cos x} \, dx = 1.648$, correct to four significant figures.

11. Show that $\displaystyle\int_{\pi/6}^{\pi/4} \frac{2}{\sin \theta} \, d\theta = 0.871\,2,$ correct to four decimal places.

12. Show that $\displaystyle\int \frac{d\theta}{3 + 4 \cos \theta} = \dfrac{1}{\sqrt{7}} \ln \left\{ \dfrac{\sqrt{7} + \tan \dfrac{\theta}{2}}{\sqrt{7} - \tan \dfrac{\theta}{2}} \right\} + c$

Chapter 21

Integration by parts

21.1   Introduction

When differentiating the product uv, where u and v are both functions of x, then:

$$\frac{\mathrm{d}}{\mathrm{d}x}(uv) = v\frac{\mathrm{d}u}{\mathrm{d}x} + u\frac{\mathrm{d}v}{\mathrm{d}x}$$

This is known as the **product rule for differentiation**.
Rearranging this formula gives:

$$u\frac{\mathrm{d}v}{\mathrm{d}x} = \frac{\mathrm{d}}{\mathrm{d}x}(uv) - v\frac{\mathrm{d}u}{\mathrm{d}x}$$

Integrating both sides with respect to x gives:

$$\int u\frac{\mathrm{d}v}{\mathrm{d}x}\,\mathrm{d}x = \int \frac{\mathrm{d}}{\mathrm{d}x}(uv)\,\mathrm{d}x - \int v\frac{\mathrm{d}u}{\mathrm{d}x}\,\mathrm{d}x$$

Since integration is the reversal of the differentiation process this becomes:

$$\int u\,\mathrm{d}v = uv - \int v\,\mathrm{d}u$$

This formula enables products of certain simple functions to be integrated in cases where it is possible to evaluate $\int v\,\mathrm{d}u$. This is known as the **integration by parts formula** and is a useful method of integration enabling such integrals as $\int x\,e^x\,\mathrm{d}x$, $\int x\cos x\,\mathrm{d}x$, $\int t^2\sin t\,\mathrm{d}t$, $\int x^3\ln x\,\mathrm{d}x$, $\int \ln x\,\mathrm{d}x$, $\int e^{ax}\sin bx\,\mathrm{d}x$, etc., to be determined.

21.2   Application of the integration by parts formula

Problem 1. Find $\int x\,e^x\,\mathrm{d}x$.
In the integration by parts formula we must let one function of our product be equal to u and the other be equal to dv.

Let $u = x$ and $\mathrm{d}v = e^x\,\mathrm{d}x$

Then $\dfrac{du}{dx} = 1$, i.e. $du = dx$ and $v = \int e^x\,dx = e^x$

There are now expressions for u, du, dv and v which are substituted into the formula:

$$\int u \quad dv \quad = u\cdot v - \int v \quad du$$

$$\int x \quad e^x\,dx = x\cdot e^x - \int e^x\,dx$$

$$= x\cdot e^x - e^x + c$$

Hence, $\displaystyle\int x\,e^x\,dx = e^x(x-1) + c$

The following four points should be noted:

(i)   The above result may be checked by differentiation.

Thus $\dfrac{d}{dx}\,[e^x(x-1) + c] = e^x(1) + (x-1)e^x + 0$

$$= e^x + x\,e^x - e^x = x\,e^x$$

(ii) Given that $dv = e^x\,dx$ then $v = \int e^x\,dx$, which is strictly equal to $e^x +$ a constant. However, the constant is omitted at this stage. If a constant, say k, were included, then:

$$\int x\,e^x\,dx = x(e^x + k) - \int (e^x + k)\,dx$$

$$= x\,e^x + xk - (e^x + kx + c)$$

$$= x\,e^x - e^x + c, \text{ as before}$$

Thus the constant k is an unnecessary addition. A constant is added only after the final integration (i.e. c in the above problem).

(iii) If instead of choosing to let $u = x$ and $dv = e^x\,dx$ we let $u = e^x$ and $dv = x\,dx$ then $du = e^x\,dx$ and $v = \displaystyle\int x\,dx = \dfrac{x^2}{2}$

Hence $\displaystyle\int x\,e^x\,dx = (e^x)\!\left(\dfrac{x^2}{2}\right) - \int\!\left(\dfrac{x^2}{2}\right)e^x\,dx$

The integral on the far right-hand side is seen to be more complicated than the original, thus the choice of letting $u = e^x$ and $dv = x\,dx$ was wrong. The choice must be such that the 'u part' becomes a constant after differentiation and the 'dv part' can be integrated easily. (It will be seen later that for the 'u part' to become a constant often requires more than one differentiation (see Problem 5).)

(iv)   If a product to be integrated contains an 'x' term then this term is chosen as the 'u part' and the other function as the 'dv part' except where a logarithmic term is involved (see Problems 6 and 7).

Problem 2. Determine $\int x \sin x \, dx$.

Let $u = x$ and $dv = \sin x \, dx$
Then $du = dx$ and $v = \int \sin x \, dx = -\cos x$
Substituting into $\int u \, dv = uv - \int v \, du$ gives:

$$\int x \sin x \, dx = (x)(-\cos x) - \int (-\cos x) \, dx$$

$$= -x \cos x + \int \cos x \, dx$$

$$= -x \cos x + \sin x + c$$

This result can be checked by differentiating.

Thus

$$\frac{d}{dx}(-x \cos x + \sin x + c) = (-x)(-\sin x) + (\cos x)(-1) + \cos x$$

$$= x \sin x - \cos x + \cos x = x \sin x$$

Problem 3. Evaluate $\int_0^1 2x \, e^{3x} \, dx$.

Let $u = 2x$ and $dv = e^{3x} \, dx$

Then $\dfrac{du}{dx} = 2$, i.e. $du = 2 \, dx$ and $v = \int e^{3x} \, dx = \dfrac{e^{3x}}{3}$

Substituting into $\int u \, dv = uv - \int v \, du$ gives:

$$\int 2x \, e^{3x} \, dx = (2x)\left(\frac{e^{3x}}{3}\right) - \int \left(\frac{e^{3x}}{3}\right)(2 \, dx)$$

$$= \frac{2}{3} x \, e^{3x} - \frac{2}{3} \int e^{3x} \, dx$$

$$= \frac{2}{3} x \, e^{3x} - \frac{2}{3}\left(\frac{e^{3x}}{3}\right) + c$$

$$= \frac{2}{3} x \, e^{3x} - \frac{2}{9} e^{3x} + c$$

Hence $\int_0^1 2x \, e^{3x} \, dx = \left[\dfrac{2}{3} x \, e^{3x} - \dfrac{2}{9} e^{3x} \right]_0^1$

$$= \left(\dfrac{2}{3} e^3 - \dfrac{2}{9} e^3 \right) - \left(0 - \dfrac{2}{9} e^0 \right)$$

$$= \dfrac{4}{9} e^3 + \dfrac{2}{9} = 8.926 \, 9 + 0.222 \, 2$$

$$= \textbf{9.149 correct to three decimal places}$$

Problem 4. Evaluate $\displaystyle \int_0^{\pi/2} 3t \cos 2t \, dt$.

Let $u = 3t$ and $dv = \cos 2t \, dt$

Then $du = 3 \, dt$ and $v = \displaystyle \int \cos 2t \, dt = \dfrac{1}{2} \sin 2t$

Substituting into $\int u \, dv = uv - \int v \, du$ gives:

$$\int 3t \cos 2t \, dt = (3t)\left(\dfrac{\sin 2t}{2} \right) - \int \left(\dfrac{\sin 2t}{2} \right)(3 \, dt)$$

$$= \dfrac{3t}{2} \sin 2t - \dfrac{3}{2} \int \sin 2t \, dt$$

$$= \dfrac{3t}{2} \sin 2t - \dfrac{3}{2} \left(\dfrac{-\cos 2t}{2} \right) + c$$

$$= \dfrac{3t}{2} \sin 2t + \dfrac{3}{4} \cos 2t + c$$

Hence $\displaystyle \int_0^{\pi/2} 3t \cos 2t \, dt = \left[\dfrac{3t}{2} \sin 2t + \dfrac{3}{4} \cos 2t \right]_0^{\pi/2}$

$$= \left[\dfrac{3}{2}\left(\dfrac{\pi}{2} \right) \sin \pi + \dfrac{3}{4} \cos \pi \right] - \left[0 + \dfrac{3}{4} \cos 0 \right]$$

$$= \left(0 + \dfrac{3}{4}(-1) \right) - \left(0 + \dfrac{3}{4} \right) = -\dfrac{3}{4} - \dfrac{3}{4} = -1\tfrac{1}{2}$$

Problem 5. Determine $\displaystyle \int x^2 \cos x \, dx$.

Let $u = x^2$ and $dv = \cos x \, dx$
Then $du = 2x \, dx$ and $v = \int \cos x \, dx = \sin x$
Substituting into $\int u \, dv = uv - \int v \, du$ gives:

$$\int x^2 \cos x \, dx = (x^2)(\sin x) - \int (\sin x)(2x \, dx)$$

$$= x^2 \sin x \quad - 2\left[\int x \sin x \, dx\right]$$

The integral on the right-hand side in the bracket is not a standard and cannot be determined 'on sight'. Since it is a product of two simple functions we may use integration by parts again.

Now $\int x \sin x \, dx = -x \cos x + \sin x$ (from Problem 2)

Thus $\int x^2 \cos x \, dx = x^2 \sin x - 2[-x \cos x + \sin x] + c$

$$= x^2 \sin x + 2x \cos x - 2 \sin x + c$$

$$= (x^2 - 2) \sin x + 2x \cos x + c$$

Generally, if the term in x is of power n then the integration by parts formula is applied n times, provided one of the functions is not ln x, as in Problem 6 following.

Problem 6. $\int x^2 \ln x \, dx.$

Whenever a product consists of a term in x and a logarithmic function, as in this problem, it is always the logarithmic function that is chosen as the 'u part'. The reason for this is that $\dfrac{d}{dx}(\ln x)$ is $\dfrac{1}{x}$, but $\int \ln x \, dx$ is not normally remembered as a standard integral.

Thus if $u = \ln x$ and $dv = x^2 \, dx$

then $du = \dfrac{1}{x} \, dx$ and $v = \int x^2 \, dx = \dfrac{x^3}{3}$

Substituting into $\int u \, dv = uv - \int v \, du$ gives:

$$\int x^2 \ln x \, dx = (\ln x)\left(\frac{x^3}{3}\right) - \int \left(\frac{x^3}{3}\right)\left(\frac{1}{x} \, dx\right)$$

$$= \frac{x^3}{3} \ln x - \frac{1}{3} \int x^2 \, dx$$

$$= \frac{x^3}{3} \ln x - \frac{1}{3}\left(\frac{x^3}{3}\right) + c$$

$$= \frac{x^3}{3} \ln x - \frac{x^3}{9} + c$$

$$= \frac{x^3}{9} (3 \ln x - 1) + c$$

Problem 7. Determine $\int \ln x \, dx$.

In each of the previous problems the components of the product have been obvious. However, $\int \ln x \, dx$ is a special case for initially it appears not to be a product. However, $\int \ln x \, dx$ is the same as $\int 1 \times \ln x \, dx$

Let $u = \ln x$ and $dv = 1 \, dx$

Then $du = \frac{1}{x} dx$ and $v = \int 1 \, dx = x$

Hence $\int \ln x \, dx = (\ln x)(x) - \int (x) \left(\frac{1}{x} dx \right)$

$$= x \ln x - \int dx$$

$$= x \ln x - x + c$$

$$= x(\ln x - 1) + c$$

Problem 8. Find $\int e^{ax} \sin bx \, dx$.

With an integral of a product of an exponential function and a sine or cosine function it does not matter which function is made equal to 'u'.
Thus let $u = e^{ax}$ and $dv = \sin bx \, dx$

Then $du = a e^{ax} dx$ and $v = \int \sin bx \, dx = \frac{-\cos bx}{b}$

Thus $\int e^{ax} \sin bx \, dx = (e^{ax}) \left(\frac{-\cos bx}{b} \right) - \int \left(\frac{-\cos bx}{b} \right) (a e^{ax} dx)$

$$= -\frac{1}{b} e^{ax} \cos bx + \frac{a}{b} \left[\int e^{ax} \cos bx \, dx \right] \qquad (1)$$

It would seem that we are no nearer a solution of the initial integral. However, the integration by parts formula may be applied to the integral in the bracket.

Let $u = e^{ax}$ and $dv = \cos bx\, dx$

Then $du = a\, e^{ax}\, dx$ and $v = \int \cos bx\, dx = \dfrac{\sin bx}{b}$

Thus $\int e^{ax} \cos bx\, dx = (e^{ax})\left(\dfrac{\sin bx}{b}\right) - \int \left(\dfrac{\sin bx}{b}\right)(a\, e^{ax}\, dx)$

$$= \dfrac{1}{b} e^{ax} \sin bx - \dfrac{a}{b} \int e^{ax} \sin bx\, dx$$

Substituting this result into equation (1) gives:

$$\int e^{ax} \sin bx\, dx = -\dfrac{1}{b} e^{ax} \cos bx + \dfrac{a}{b}\left[\dfrac{1}{b} e^{ax} \sin bx - \dfrac{a}{b} \int e^{ax} \sin bx\, dx\right]$$

$$= \dfrac{-1}{b} e^{ax} \cos bx + \dfrac{a}{b^2} e^{ax} \sin bx - \dfrac{a^2}{b^2}\left[\int e^{ax} \sin bx\, dx\right]$$

The integral in the bracket on the right-hand side is the same as the integral on the left-hand side, thus they may be combined on the left-hand side of the equation. Thus:

$$\int e^{ax} \sin bx\, dx + \dfrac{a^2}{b^2} \int e^{ax} \sin bx\, dx = \dfrac{-1}{b} e^{ax} \cos bx + \dfrac{a}{b^2} e^{ax} \sin bx$$

i.e. $\left(1 + \dfrac{a^2}{b^2}\right) \int e^{ax} \sin bx\, dx = \dfrac{e^{ax}}{b^2}(a \sin bx - b \cos bx)$

$\left(\dfrac{b^2 + a^2}{b^2}\right) \int e^{ax} \sin bx\, dx = \dfrac{e^{ax}}{b^2}(a \sin bx - b \cos bx)$

Hence $\int e^{ax} \sin bx\, dx = \left(\dfrac{b^2}{a^2 + b^2}\right) \dfrac{e^{ax}}{b^2}(a \sin bx - b \cos bx)$

$$= \left(\dfrac{e^{ax}}{a + b^2}\right)(a \sin bx - b \cos bx) + c$$

A product of an exponential function and a sine or cosine function thus involves integration by parts twice. If, in the above problem, the exponential function is made equal to dv instead of u the same result is obtained. It is left as a student exercise to prove this.

By a similar method to above it may be shown that:

$$\int e^{ax} \cos bx\, dx = \left(\dfrac{e^{ax}}{a^2 + b^2}\right)(b \sin bx + a \cos bx) + c$$

Further problems on integration by parts may be found in the following Section 21.3 (Problems 1 to 27).

21.3   Further problems

Find the following integrals using integration by parts.

1. $\int x e^{3x} \, dx$ $\left[\dfrac{e^{3x}}{9} (3x - 1) + c \right]$

2. $\int 3x \, e^{2x} \, dx$ $\left[\dfrac{3}{4} e^{2x}(2x - 1) + c \right]$

3. $\int \dfrac{5x}{e^{4x}} \, dx$ $\left[\dfrac{-5}{16} e^{-4x}(4x + 1) + c \right]$

4. $\int x \cos x \, dx$ $[x \sin x + \cos x + c]$

5. $\int x \ln x \, dx$ $\left[\dfrac{x^2}{4} (2 \ln x - 1) + c \right]$

6. $\int 3x \sin 2x \, dx$ $\left[\dfrac{-3x}{2} \cos 2x + \dfrac{3}{4} \sin 2x + c \right]$

7. $\int \ln 4x \, dx$ $[x(\ln 4x - 1) + c]$

8. $\int \dfrac{\ln y \, dy}{y^2}$ $\left[-\dfrac{1}{y} (\ln y + 1) + c \right]$

9. $\int 2x \cos 5x \, dx$ $\left[\dfrac{2}{25} (5x \sin 5x + \cos 5x) + c \right]$

10. $\int 2x^2 e^x \, dx$ $[2e^x(x^2 - 2x + 2) + c]$

11. $\int x^2 \sin 2x \, dx$ $\left[\left(\dfrac{1}{4} - \dfrac{x^2}{2} \right) \cos 2x + \dfrac{x}{2} \sin 2x + c \right]$

12. $\int e^{2x} \cos x \, dx$ $\left[\dfrac{e^{2x}}{5} (\sin x + 2 \cos x) + c \right]$

13. $\int 4 \, \theta^2 \cos 3\theta \, d\theta$ $\left[4 \left(\dfrac{\theta^2}{3} - \dfrac{2}{27} \right) \sin 3\theta + \dfrac{8}{9} \theta \cos 3\theta + c \right]$

14. $\int 3e^x \sin 2x \, dx$ $\left[\dfrac{3}{5} e^x(\sin 2x - 2 \cos 2x) + c \right]$

15. $\int 4x \sec^2 x \, dx$ $\left[4[x \tan x - \ln(\sec x)] + c \right]$

Evaluate the following integrals correct to three decimal places.

16. $\displaystyle\int_0^2 x \, e^x \, dx$ $[8.389]$

17. $\displaystyle\int_0^{\pi/2} x \cos x \, dx$ [0.571]

18. $\displaystyle\int_0^1 t \, e^{2t} \, dt$ [2.097]

19. $\displaystyle\int_0^{\pi/4} \phi \sin 2\phi \, d\phi$ [0.250]

20. $\displaystyle\int_1^2 \ln x^2 \, dx$ [0.773]

21. $\displaystyle\int_0^{\pi/2} 3x^2 \cos x \, dx$ [1.402]

22. $\displaystyle\int_0^1 x^2 \, e^x \, dx$ [0.718]

23. $\displaystyle\int_1^4 \sqrt{t} \ln t \, dt$ [4.282]

24. $\displaystyle\int_0^1 2e^t \cos 2t \, dt$ [1.125]

25. $\displaystyle\int_0^{\pi/2} e^\theta \sin \theta \, d\theta$ [2.905]

26. In the study of damped oscillations integrations of the following type are important:

$$C = \int_0^1 e^{-0.4\theta} \cos 1.2\theta \, d\theta$$

and $\displaystyle S = \int_0^1 e^{-0.4\theta} \sin 1.2\theta \, d\theta$

Determine C and S. [$C = 0.66$, $S = 0.41$]

27. If a string is plucked at a point $x = \dfrac{l}{3}$ with an amplitude a and released, the equation of motion is

$$K = \frac{2}{l}\left\{\int_0^{l/3} \frac{3a}{l} x \sin \frac{n\pi}{l} x \, dx + \int_{l/3}^l \frac{3a}{2l}(l - x) \sin \frac{n\pi}{l} x \, dx\right\}$$

where n is a constant.

Show that $K = \dfrac{9a}{\pi^2 n^2} \sin \dfrac{n\pi}{3}$.

First order differential equations by separation of the variables

22.1 Introduction

Definition. A differential equation is one that contains differential coefficients. Examples include:

(i) $\dfrac{dy}{dx} = 5x$

(ii) $\dfrac{d^2y}{dx^2} + 3\dfrac{dy}{dx} + 4y = 0$

(iii) $\left(\dfrac{d^2s}{dt^2}\right)^3 + 2\left(\dfrac{ds}{dt}\right)^4 = 4$

Order. Differential equations are classified according to the highest derivative which occurs in them. Hence example (i) above is a **first order differential equation** and examples (ii) and (iii) are both **second order differential equations**.

Degree. The degree of a differential equation is that of the highest power of the highest differential which the equation contains after any necessary simplification. Thus example (i) above is of first order, first degree, example (ii) is of second order, first degree, and example (iii) is of second order, third degree.

Solutions of a differential equation

Starting with a differential equation it is possible, by integration and by being given enough data to determine the constants. to obtain the original function. The process of determining the original relationship between the two variables is called 'solving the differential equation'. A solution to a differential equation which contains one or more arbitrary constants of integration is called the **general solution** of the differential equation. When finding the general solution of a first order differential equation one arbitrary constant results; when finding the general solution of a second order differential equation two arbitrary constants result, and so on. When additional information is given so that constants may be calculated the **particular solution** of the differential equation is obtained. The additional information is called the **boundary conditions**.

In this text only differential equations of the first and second order and first degree are discussed and the method of solution of five types of first order and two types of second order are considered. The first order types discussed are:

(a) differential equations of the form $\dfrac{dy}{dx} = f(x)$,

(b) differential equations of the form $\dfrac{dy}{dx} = f(y)$,

(c) 'variable separable' types of differential equations,

(d) differential equations which are homogeneous in x and y, and

(e) differential equations of the form $\dfrac{dy}{dx} + Py = Q$, where P and Q are functions of x

Types (a), (b) and (c) are considered in this chapter, type (d) in Chapter 23 and type (e) in Chapter 24. Two second order types of differential equation are discussed in Chapters 25 and 26.

22.2   Solution of differential equations of the form $\dfrac{dy}{dx} = f(x)$

An equation of the form $\dfrac{dy}{dx} = f(x)$ may be solved immediately by integration.

The solution is $\boxed{y = \int f(x)\, dx.}$

Problem 1. Find the general solutions of the following differential equations.

(a) $\dfrac{dy}{dx} = 5x^2 + \cos 3x$ (b) $x\dfrac{dy}{dx} = 3 - 2x^2$

(a) If $\dfrac{dy}{dx} = 5x^2 + \cos 3x$, then $y = \int (5x^2 + \cos 3x)\, dx$

i.e. $y = \dfrac{5x^3}{3} + \dfrac{1}{3}\sin 3x + c$

(b) If $x\dfrac{dy}{dx} = 3 - 2x^2$, then $\dfrac{dy}{dx} = \dfrac{3 - 2x^2}{x} = \dfrac{3}{x} - 2x$

Hence $y = \int \left(\dfrac{3}{x} - 2x \right) dx$

i.e. $y = 3 \ln x - x^2 + c$

Problem 2. Find the particular solutions of the following differential equations satisfying the given boundary conditions:

(a) $3 \dfrac{dy}{dx} + x = 6$ and $y = 5\frac{1}{2}$ when $x = 3$

(b) $2 \dfrac{dr}{d\theta} + \sin 2\theta = 0$ and $r = 2$ when $\theta = \dfrac{\pi}{2}$

(a) If $3 \dfrac{dy}{dx} + x = 6$, then $\dfrac{dy}{dx} = \dfrac{6 - x}{3} = 2 - \dfrac{x}{3}$

Hence $y = \int \left(2 - \dfrac{x}{3} \right) dx$

i.e. $y = 2x - \dfrac{x^2}{6} + c$

(This is the general solution)
Substituting the boundary conditions $y = 5\frac{1}{2}$ when $x = 3$ to evaluate c gives:

$5\frac{1}{2} = 6 - 1\frac{1}{2} + c$

i.e. $c = 1$

Hence the particular solution is $y = 2x - \dfrac{x^2}{6} + 1$

(b) If $2 \dfrac{dr}{d\theta} + \sin 2\theta = 0$, then $\dfrac{dr}{d\theta} = \dfrac{-\sin 2\theta}{2}$

Hence $r = \int \dfrac{-\sin 2\theta}{2} \, d\theta$

i.e. $r = \dfrac{1}{4} \cos 2\theta + c$

Substituting the boundary conditions $r = 2$ when $\theta = \dfrac{\pi}{2}$ to evaluate c gives:

$2 = \dfrac{1}{4} \cos \pi + c$

i.e. $c = 2 - \left(-\dfrac{1}{4} \right) = 2\dfrac{1}{4}$

Hence the particular solution is

$$r = \dfrac{1}{4} \cos 2\theta + 2\dfrac{1}{4} = \dfrac{1}{4} (\cos 2\theta + 9)$$

Further problems on solving differential equations of the form $\dfrac{dy}{dx} = f(x)$
may be found in Section 22.5 (Problems 1 to 14), page 334.

22.3 Solution of differential equations of the form $\dfrac{dy}{dx} = f(y)$

An equation of the form $\dfrac{dy}{dx} = f(y)$ may be rearranged to give:

$$\frac{dx}{dy} = \frac{1}{f(y)}$$

i.e. $dx = \dfrac{dy}{f(y)}$

Integrating both sides gives: $\boxed{\displaystyle\int dx = \int \frac{dy}{f(y)}}$

Hence the solution may be obtained by direct integration.

Problem 1. Solve the differential equations:

(a) $\dfrac{dy}{dx} = 2 + y$ (b) $3\dfrac{dy}{dx} = \sec 2y$

(a)   Rearranging $\dfrac{dy}{dx} = 2 + y$ gives $dx = \dfrac{dy}{(2 + y)}$

Integrating both sides gives: $\displaystyle\int dx = \int \frac{dy}{(2 + y)}$

i.e. $x = \ln(2 + y) + c$ (1)

The general solutions of differential equations can sometimes be rearranged. In this case, for example, if $c = \ln D$, where D is a constant, then:

$x = \ln(2 + y) + \ln D$

i.e. $x = \ln D(2 + y)$, from the law of logarithms,

from which $e^x = D(2 + y)$ (2)

Equations (1) and (2) are both acceptable general solutions of the differential equation $\dfrac{dy}{dx} = 2 + y$

(b)   Rearranging $3\dfrac{dy}{dx} = \sec 2y$ gives

$$dx = \frac{3}{\sec 2y}\, dy = 3\cos 2y\, dy$$

Integrating both sides gives: $\int dx = \int 3\cos 2y\, dy$

Hence the general solution is $x = \dfrac{3}{2}\sin 2y + c$

Problem 2. The rate at which a body cools is given by the equation $\dfrac{d\theta}{dt} = -k\theta$, where θ is the temperature of the body above its surroundings and k is a constant. Solve the equation for θ given that at $t = 0$, $\theta = \theta_0$

$\dfrac{d\theta}{dt} = -k\theta$ is of the form $\dfrac{dy}{dx} = f(y)$

Rearranging gives $dt = \dfrac{-1}{k\theta}\, d\theta$

Integrating both sides gives: $\displaystyle\int dt = \frac{-1}{k}\int \frac{d\theta}{\theta}$

i.e. $t = \dfrac{-1}{k}\ln\theta + c$ (1)

Substituting the boundary conditions $t = 0$ and $\theta = \theta_0$ to find c gives:

$$0 = \frac{-1}{k}\ln\theta_0 + c$$

i.e. $c = \dfrac{1}{k}\ln\theta_0$

Substituting $c = \dfrac{1}{k}\ln\theta_0$ in equation (1) gives:

$$t = -\frac{1}{k}\ln\theta + \frac{1}{k}\ln\theta_0$$

$$t = \frac{1}{k}(\ln\theta_0 - \ln\theta) = \frac{1}{k}\ln\left(\frac{\theta_0}{\theta}\right)$$

$$kt = \ln\left(\frac{\theta_0}{\theta}\right)$$

$$e^{kt} = \frac{\theta_0}{\theta}$$

$$e^{-kt} = \frac{\theta}{\theta_0}$$

Hence $\theta = \theta_0 \, e^{-kt}$

Further problems on solving differential equations of the form $\dfrac{dy}{dx} = f(y)$ *may be found in Section 22.5 (Problems 15 to 29), page 335.*

22.4   Solution of 'variables separable' type of differential equations

An equation of the form $\dfrac{dy}{dx} = f(x) \cdot g(y)$, where $f(x)$ is a function of x only and $g(y)$ is a function of y only, may be rearranged thus:

$$\frac{dy}{g(y)} = f(x)\,dx$$

Integrating both sides gives $\boxed{\displaystyle\int \frac{dy}{g(y)} = \int f(x)\,dx,}$ i.e. the left-hand side is

the integral of a function of y with respect to y and the right-hand side is the integral of a function of x with respect to x.

When two variables can be rearranged into two separate groups as shown above, each consisting of only one variable, the variables are said to be separable.

The equations of the type $\dfrac{dy}{dx} = f(x)$ and $\dfrac{dy}{dx} = f(y)$ discussed in Sections 22.2 and 22.3 are, in fact, merely special simple cases of 'separating the variables'.

Problem 1. Solve the differential equations:

(a) $\dfrac{dy}{dx} = \dfrac{3x^2 - 2}{2y - 1}$ (b) $2xy\,\dfrac{dy}{dx} = 1 + y^2$

(a) $\dfrac{dy}{dx} = \dfrac{3x^2 - 2}{2y - 1}$

Separating the variables gives $(2y - 1)\,dy = (3x^2 - 2)\,dx$
Integrating both sides gives $\int (2y - 1)\,dy = \int (3x^2 - 2)\,dx$
Hence the general solution is $y^2 - y = x^3 - 2x + C$

Note that when integrating both sides of an equation there is no need to put an arbitrary constant on both sides of the result. In this case, if this was done, then:

$$y^2 - y + A = x^3 - 2x + B$$

and $y^2 - y = x^3 - 2x + C$, where $C = B - A$

(b) $2xy \dfrac{dy}{dx} = 1 + y^2$

Separating the variables gives: $\dfrac{2y}{1 + y^2} dy = \dfrac{1}{x} dx$

Integrating both sides gives: $\displaystyle\int \dfrac{2y}{(1 + y^2)} dy = \int \dfrac{1}{x} dx$

Hence the general solution is **$\ln(1 + y^2) = \ln x + C$**   (1)

or $\ln(1 + y^2) - \ln x = C$

from which $\ln\left(\dfrac{1 + y^2}{x}\right) = C$

and $\dfrac{1 + y^2}{x} = e^C$ (2)

Also, if in equation (1), $C = \ln A$, we have

$$\ln(1 + y^2) = \ln x + \ln A$$

$$\ln(1 + y^2) = \ln(Ax)$$

i.e.   **$1 + y^2 = Ax$**   (3)

Equations (1), (2) and (3) are all valid general solutions to the differential equation $2xy \dfrac{dy}{dx} = 1 + y^2$, none of them being any more correct than the others. Thus, by manipulation, it is possible to obtain several general solutions to a differential equation.

Problem 2. Find the particular solution of $\dfrac{dy}{dx} = 3\, e^{2x - 3y}$ given that $y = 0$ when $x = 0$.

$\dfrac{dy}{dx} = 3\, e^{2x - 3y} = 3(e^{2x})(e^{-3y})$ by the laws of indices

Separating the variables gives $\dfrac{dy}{e^{-3y}} = 3\, e^{2x} dx$

i.e. $e^{3y} dy = 3\, e^{2x} dx$

Integrating both sides gives $\int e^{3y}\, dy = 3 \int e^{2x}\, dx$

$$\frac{1}{3} e^{3y} = \frac{3}{2} e^{2x} + C$$

(This is the general solution)

When $y = 0$, $x = 0$, thus $\frac{1}{3} e^0 = \frac{3}{2} e^0 + C$

$$C = \frac{1}{3} - \frac{3}{2} = -\frac{7}{6}$$

Hence the particular solution is $\frac{1}{3} e^{3y} = \frac{3}{2} e^{2x} - \frac{7}{6}$

or $\qquad\qquad\qquad\qquad$ $2\, e^{3y} = 9\, e^{2x} - 7$

Problem 3. An electrical circuit contains inductance L and resistance R connected to a constant voltage source E. The current i is given by the differential equation $E - L\dfrac{di}{dt} = Ri$, where L and R are constants. Find the current in terms of time t given that when $t = 0$, $i = 0$.

$E - L\dfrac{di}{dt} = Ri$. Rearranging gives $\dfrac{di}{dt} = \dfrac{E - Ri}{L}$

and $\qquad\qquad\qquad$ $\dfrac{di}{E - Ri} = \dfrac{dt}{L}$

Integrating both sides gives: $\int \dfrac{di}{E - Ri} = \int \dfrac{dt}{L}$

$$-\frac{1}{R} \ln(E - Ri) = \frac{t}{L} + C$$

(This is the general solution)

$t = 0$ when $i = 0$, hence $-\dfrac{1}{R} \ln E = C$

Thus, $-\dfrac{1}{R} \ln(E - Ri) = \dfrac{t}{L} - \dfrac{1}{R} \ln E$

This particular solution must now be transposed to find i:

$$-\frac{1}{R} \ln(E - Ri) + \frac{1}{R} \ln E = \frac{t}{L}$$

$$\frac{1}{R}(\ln E - \ln(E - Ri)) = \frac{t}{L}$$

$$\ln\left(\frac{E}{E - Ri}\right) = \frac{Rt}{L}$$

$$\frac{E}{E - Ri} = e^{Rt/L}$$

$$\frac{E - Ri}{E} = e^{-Rt/L}$$

$$Ri = E - E\,e^{-Rt/L}$$

Hence current $i = \dfrac{E}{R}\left(1 - e^{-Rt/L}\right)$

This expression for current represents the natural law of growth of current in an inductive circuit.

Further problems on solving 'variable separable' types of differential equations may be found in the following Section 22.5 (Problems 30 to 46), page 337.

22.5   Further problems

Differential equations of the form $\dfrac{dy}{dx} = f(x)$

In Problems 1 to 9 solve the differential equations.

1. $\dfrac{dy}{dx} = 2x^4$ $\left[y = \dfrac{2}{5}x^5 + c\right]$

2. $\dfrac{dy}{dx} = 5x + \sin x$ $\left[y = \dfrac{5}{2}x^2 - \cos x + c\right]$

3. $x\dfrac{dy}{dx} = 4 - x^2$ $\left[y = 4\ln x - \dfrac{x^2}{2} + c\right]$

4. $\dfrac{dy}{dx} - 2x^3 = e^{3x}$ $[6y = 2e^{3x} + 3x^4 + c]$

5. $x^2\dfrac{dy}{dx} = 2 + x$ $\left[y = \ln x - \dfrac{2}{x} + c\right]$

6. $\dfrac{dy}{dx} + x = 2$ and $y = 3$ when $x = 2$ $\left[y = 2x - \dfrac{x^2}{2} + 1\right]$

7. $2\dfrac{dr}{d\theta} + \cos\theta = 0$ and $r = \dfrac{5}{2}$ when $\theta = \dfrac{\pi}{2}$ $\left[r = 3 - \dfrac{1}{2}\sin\theta\right]$

8. $x\left(x - \dfrac{dy}{dx}\right) = 3$ and $y = 1$ when $x = 1$ $[2y = x^2 - 6\ln x + 1]$

9. $\dfrac{1}{2\,e^{t}} + 4 = t - 3\dfrac{d\theta}{dt}$ and $\theta = \dfrac{1}{6}$ when $t = 0$

$\left[\theta = \dfrac{1}{3}\left(\dfrac{t^{2}}{2} + \dfrac{e^{-t}}{2} - 4t\right)\right]$

10. The acceleration of a body a is equal to its rate of change of velocity, $\dfrac{dv}{dt}$. Determine an equation for v in terms of t given that the velocity is u when $t = 0$. $[v = u + at]$

11. The velocity of a body v is equal to its rate of change of distance, $\dfrac{dx}{dt}$. Determine an equation for x in terms of t, given $v = u + at$. where u and a are constants and $x = 0$ when $t = 0$. $[x = ut + \frac{1}{2}at^{2}]$

12. The gradient of a curve is given by $\dfrac{dy}{dx} = 4x - \dfrac{x^{3}}{6}$. Determine the equation of the curve if it passes through the point $\left(2, 3\dfrac{1}{3}\right)$.

$\left[y = 2x^{2} - \dfrac{x^{4}}{24} - 4\right]$

13. The bending moment of a beam, M, and shear force F are related by the equation $\dfrac{dM}{dx} = F$, where x is the distance from one end of the beam. Determine M in terms of x when $F = -w(l - x)$ where w and l are constants and $M = \dfrac{1}{2}wl^{2}$ when $x = 0$. $\left[M = \dfrac{1}{2}w(l - x)^{2}\right]$

14. The angular velocity ω of a flywheel of moment of inertia I is given by $I\dfrac{d\omega}{dt} + N = 0$, where N is a constant. Determine ω in terms of t given that $\omega = \omega_{0}$ when $t = 0$. $\left[\omega = \omega_{0} - \dfrac{Nt}{I}\right]$

Differential equations of the form $\dfrac{dy}{dx} = f(y)$

In Problems 15 to 22 solve the differential equations.

15. $\dfrac{dy}{dx} = 3 + 2y$ $\left[\dfrac{1}{2}\ln(3 + 2y) = x + c\right]$

16. $5\dfrac{dy}{dx} = \cot 2y$ $\left[\dfrac{5}{2}\ln(\sec 2y) = x + c\right]$

17. $y\dfrac{dy}{dx} = 3 - y^2$ $\left[-\dfrac{1}{2}\ln(3 - y^2) = x + c \right]$

18. $2\dfrac{dy}{dx} + 3y = 4$ $\left[-\dfrac{2}{3}\ln(4 - 3y) = x + c \right]$

19. $\dfrac{dy}{dx} = 2\tan y.$ Find y $[y = \arcsin(e^{2x+c})]$

20. $y\dfrac{dy}{dx} = 1 - y,$ and $y = 0$ when $x = 1$ $[x + y + \ln(1 - y) = 1]$

21. $(y^2 + 1)\dfrac{dy}{dx} = 2y$ and $y = 1$ when $x = \dfrac{1}{4}$ $[y^2 + 2\ln y = 4x]$

22. $\sqrt{y}\dfrac{dy}{dx} - 1 = 0$ and $y = 4$ when $x = \dfrac{1}{3}$ $\left[\dfrac{2}{3}\sqrt{y^3} = x + 5 \right]$

23. An equation of motion may be represented by the equation
$\dfrac{dv}{dt} + kv^2 = 0$ where v is the velocity of a body travelling in a restraining medium. Show that $v = \dfrac{v_0}{1 + ktv_0}$ given that $v = v_0$ when $t = 0$.

24. The current in an electric circuit is given by $L\dfrac{di}{dt} + Ri = 0$ where L and R are constants. Solve for i given that $i = I$ when $t = 0.$ $[i = I e^{-Rt/L}]$

25. The difference in tension, T newtons, between two sides of a belt when in contact with a pulley over an angle of θ radians and when it is on the point of slipping, is given by $\dfrac{dT}{d\theta} = \mu T,$ where μ is the coefficient of friction between the material of the belt and that of the pulley at the point of slipping. When $\theta = 0$ radians, the tension is 150 N and $\mu = 0.29$ as slipping starts. Find the tension at the point of slipping when $\theta = \dfrac{2\pi}{3}$ radians. Also determine the angle of lap correct to the nearest degree to give a tension of 300 N just before slipping starts.
[275.3 N, 137°]

26. The charge Q coulombs at time t seconds for a capacitor of capacitance C farads when discharging through a resistance of R ohms is given by $R\dfrac{dQ}{dt} + \dfrac{Q}{C} = 0.$ Solve for Q given that $Q = Q_0$ when $t = 0.$ A circuit contains a resistance of 400 kilohms and a capacitance of 7.3 microfarads, and after 225 milliseconds the charge falls to 7.0 coulombs. Find the initial charge and the

charge after 2 seconds, correct to three significant figures.
$[Q = Q_0\,e^{-t\,CR}, 7.56\text{ C}, 3.81\text{ C}]$

27. In a chemical reaction in which x is the amount transformed
in time t the velocity of the reaction is given by $\dfrac{dx}{dt} = k(a - x)$
where k is a constant and a is the concentration at time $t = 0$
when $x = 0$. Find x in terms of t. $[x = a(1 - e^{-kt})]$

28. The rate of decay of a radioactive substance is given by
$\dfrac{dN}{dt} = -\lambda N$, where λ is the decay constant and λN is the
number of radioactive atoms disintegrating per second. Deter-
mine the half-life of radium in years (i.e. the time for N to
become one-half of its original value) taking the decay constant
for radium as 1.36×10^{-11} atoms per second and assuming a
'365 day' year.   [1 616 years]

29. The variation of resistance R ohms of a copper conductor with
temperature. $\theta°$C, is given by $\dfrac{dR}{d\theta} = \alpha R$, where α is the temperature coefficient of resistance of copper. If $R = R_0$ at $\theta = 0°$C,
solve the equation for R. Taking α as 39×10^{-4} per $°$C, find
the resistance of a copper conductor at $20°$C, correct to four
significant figures, when its resistance at $80°$C is 57.4 ohms.
$[R = R_0\,e^{\alpha\theta}, 45.42\text{ ohms}]$

'Variable separable' types of differential equations

In Problems 30 to 36 solve the differential equations.

30. $\dfrac{dy}{dx} = (2y)(x^2)$ $\left[\begin{array}{l} \dfrac{1}{2}\ln y = \dfrac{x^3}{3} + c \\[2mm] \text{or } \dfrac{1}{2}\ln 2y = \dfrac{x^3}{3} + k \end{array}\right]$

31. $\dfrac{dy}{dx} = y\cos x$ $[\ln y = \sin x + c]$

32. $(x + 2)\dfrac{dy}{dx} = (1 - y)$ $[\ln(x + 2)(1 - y) = c]$

33. $\dfrac{dy}{dx} = \dfrac{2x^2 - 1}{3y + 2}$ and $x = 0$ when $y = 0$ $\left[\dfrac{3y^2}{2} + 2y = \dfrac{2x^3}{3} - x\right]$

34. $\dfrac{dy}{dx} = e^{x-2y}$ and $x = 0$ when $y = 0$ $[e^{2y} = 2\,e^x - 1]$

35. $\dfrac{1}{2}\dfrac{dy}{dx} = e^{x+3y}$ and $x = 0$ when $y = 0$ $[7 - e^{-3y} = 6\,e^x]$

36. $y(1 + x) + x(1 - y)\dfrac{dy}{dx} = 0$ and $x = 1$ when $y = 1$

   $[\ln(xy) = y - x]$

37. Show that the solution of the equation $xy\dfrac{dy}{dx} = 1 + 2y^2$ may

   be of the form $y = \sqrt{\left(\dfrac{x^4 k - 1}{2}\right)}$, where k is a constant.

38. Show that the solution of the differential equation $\dfrac{y^2 + 2}{x^2 + 2} = \dfrac{y}{x}\dfrac{dy}{dx}$

   is of the form $\sqrt{\left(\dfrac{x^2 + 2}{y^2 + 2}\right)} = $ constant.

39. Prove that $y = x$ is a solution of the equation

   $x\sqrt{(y^2 - 1)} - y\sqrt{(x^2 - 1)}\dfrac{dy}{dx} = 0$ when $x = 1$ and $y = 1$.

40. Solve $xy = (1 + x^2)\dfrac{dy}{dx}$ for y. $\left[y = \dfrac{1}{k}\sqrt{(1 + x^2)}\right]$

41. Find the curve which satisfies the equation $2xy\dfrac{dy}{dx} = x^2 + 1$
   and which passes through the point $(1, 2)$.
   $[2y^2 = x^2 + 2\ln x + 7]$

42. Solve the equation $y\cos^2 x\dfrac{dy}{dx} = \tan x + 2$ given that $y = 2$

   when $x = \dfrac{\pi}{4}$. $[y^2 = \tan^2 x + 4\tan x - 1]$

43. A capacitor C is charged by applying a steady voltage E
   through a resistance R. The p.d. between the plates, V, is given

   by the differential equation $CR\dfrac{dV}{dt} + V = E$. Solve the equation

   for V given that $V = 0$ when $t = 0$ and evaluate V when
   $E = 20$ volts, $C = 25$ microfarads, $R = 300$ kilohms and $t = 2$
   seconds. $[V = E(1 - e^{-t/CR}), 4.681$ volts$]$

44. For an adiabatic expansion of a gas $C_p\dfrac{dv}{V} + C_v\dfrac{dp}{P} = 0$, where

   C_p and C_v are constants. Show that $pv^n = $ constant, where

   $n = \dfrac{C_p}{C_v}$.

45. The streamlines of a cylinder of radius a in a stream of liquid of ambient velocity v are given by the equation:

$$\frac{dr}{d\theta} = \frac{r(a^2 - r^2)}{(a^2 + r^2)} \cot \theta$$

r is the distance of the centre of the cylinder from the applied force, the line joining them being at an angle θ to the direction of the liquid flow. Solve the equation. (Hint: let

$$a^2 + r^2 = a^2 - r^2 + 2r^2) \qquad \left[\left(\frac{a^2}{r} - r \right) \sin \theta = C \right]$$

46. The equilibrium constant (K) of a chemical reaction varies with temperature (T) according to the equation: $\dfrac{d(\ln K)}{dT} = \dfrac{\Delta H}{RT^2}$. If $\Delta H = 10^4$, $R = 8.3$ and $K = 4$ when $T = 600$, solve the equation completely. $\qquad \left[\ln K = \dfrac{-\Delta H}{RT} + 3.39 \right]$

Chapter 23

Homogeneous first order differential equations

23.1 Solution of differential equations of the form $P\dfrac{dy}{dx} = Q$

Certain first order differential equations are not separable but can be made separable by a simple change of variables.

An equation of the form $P\dfrac{dy}{dx} = Q$, where P and Q are functions of both x and y of the same degree throughout, is said to be **homogeneous**. An example of a homogeneous function if $f(x, y) = x^3 + 2x^2y + y^3$, since each of the three terms are of degree 3. However, $f(x, y) = 2x^2 + xy^2 + y^2$ is not homogeneous since the second term is of degree 3 and the other two are of degree 2. Similarly, $f(x, y) = \dfrac{x + 2y}{3x - y}$ is homogeneous in x and y since each of the four terms is of degree 1.

Procedure to solve a differential equation of the form $P\dfrac{dy}{dx} = Q$

(i) Rearrange $P\dfrac{dy}{dx} = Q$ into the form $\dfrac{dy}{dx} = \dfrac{Q}{P}$.

(ii) Make the substitution $y = vx$, where v is a function of x. If $y = vx$ then, by the product rule, $\dfrac{dy}{dx} = v(1) + x\dfrac{dv}{dx}$. There is no particular reason for choosing the letter v, and in any case, it does not appear in the final solution.

(iii) Substitute for both y and $\dfrac{dy}{dx}$ in the equation $\dfrac{dy}{dx} = \dfrac{Q}{P}$. It will be found that the x terms will cancel leaving an equation in which the variables are separable.

(iv) Separate the variables and solve using the method shown in Chapter 22.

(v) The solution will be in terms of v and x. Thus, substitute $v = \dfrac{y}{x}$ to solve in terms of the original variables x and y.

Problem 1. Solve $xy\dfrac{dy}{dx} = x^2 + y^2$ given that $x = 1$ when $y = 2$.

Using the above procedure:

(i) Since $xy\dfrac{dy}{dx} = x^2 + y^2$ then $\dfrac{dy}{dx} = \dfrac{x^2 + y^2}{xy}$ which is homogeneous in x and y, since each of the three terms on the right-hand side is of the same degree (i.e. degree 2).

(ii) Let $y = vx$, then $\dfrac{dy}{dx} = v + x\dfrac{dv}{dx}$.

(iii) Substituting for y and $\dfrac{dy}{dx}$ in $\dfrac{dy}{dx} = \dfrac{x^2 + y^2}{xy}$ gives:

$$v + x\frac{dv}{dx} = \frac{x^2 + (vx)^2}{x(vx)} = \frac{x^2 + v^2x^2}{x^2v} = \frac{1 + v^2}{v}$$

Hence $v + x\dfrac{dv}{dx} = \dfrac{1 + v^2}{v}$

(iv) Separating the variables gives:

$$x\frac{dv}{dx} = \frac{1 + v^2}{v} - v = \frac{1 + v^2 - v^2}{v} = \frac{1}{v}$$

i.e. $v\,dv = \dfrac{dx}{x}$

Integrating both sides gives: $\displaystyle\int v\,dv = \int\frac{dx}{x}$

Hence $\dfrac{v^2}{2} = \ln x + c$

(v) Replacing v by $\dfrac{y}{x}$ gives: $\dfrac{y^2}{2x^2} = \ln x + c$

When $x = 1$, $y = 2$ thus $\dfrac{4}{2} = \ln 1 + c$

i.e. $c = 2$

Thus the particular solution of $xy\dfrac{dy}{dx} = x^2 + y^2$ is

$\dfrac{y^2}{2x^2} = \ln x + 2$

Problem 2. Solve the equation $(4y + 3x)\dfrac{dy}{dx} = (3x - y)$.

(i)   If $(4y + 3x)\dfrac{dy}{dx} = (3x - y)$ then $\dfrac{dy}{dx} = \dfrac{(3x - y)}{(4y + 3x)}$, which is
homogeneous in x and y since each of the four terms on the
right-hand side is of the same degree (i.e. degree 1).

(ii) Let $y = vx$ then $\dfrac{dy}{dx} = v + x\dfrac{dv}{dx}$.

(iii) Substituting for y and $\dfrac{dy}{dx}$ gives:

$$v + x\frac{dv}{dx} = \frac{3x - vx}{4vx + 3x} = \frac{3 - v}{4v + 3}$$

(iv) Separating the variables gives:

$$x\frac{dv}{dx} = \frac{3 - v}{4v + 3} - v$$

$$= \frac{(3 - v) - v(4v + 3)}{4v + 3} = \frac{3 - 4v - 4v^2}{4v + 3}$$

Hence $\dfrac{4v + 3}{(3 + 2v)(1 - 2v)}\,dv = \dfrac{dx}{x}$

Integrating both sides gives: $\displaystyle\int \frac{4v + 3}{(3 + 2v)(1 - 2v)}\,dv = \int \frac{dx}{x}$

Resolving $\dfrac{4v + 3}{(3 + 2v)(1 - 2v)}$ into partial fractions gives:

$$\frac{5}{4(1 - 2v)} - \frac{3}{4(3 + 2v)}$$

Hence $\displaystyle\int \left\{\frac{5}{4(1 - 2v)} - \frac{3}{4(3 + 2v)}\right\} dv = \int \frac{dx}{x}$

Thus $-\dfrac{5}{8}\ln(1 - 2v) - \dfrac{3}{8}\ln(3 + 2v) = \ln x + c$

(v)   Replacing v by $\dfrac{y}{x}$ gives:

$$-\frac{5}{8}\ln\left(1 - \frac{2y}{x}\right) - \frac{3}{8}\ln\left(3 + \frac{2y}{x}\right) = \ln x + c$$

or $\ln x + \dfrac{5}{8}\ln\left(\dfrac{x - 2y}{x}\right) + \dfrac{3}{8}\ln\left(\dfrac{3x + 2y}{x}\right) + c = 0$

Multiplying throughout by 8 gives:

$$8\ln x + 5\ln\left(\frac{x-2y}{x}\right) + 3\ln\left(\frac{3x+2y}{x}\right) + 8c = 0$$

By the laws of logarithms:

$$\ln x^8 + \ln\left(\frac{x-2y}{x}\right)^5 + \ln\left(\frac{3x+2y}{x}\right)^3 = K$$

$$(\text{where } K = -8c)$$

Hence $\ln\left\{\dfrac{(x^8)(x-2y)^5(3x+2y)^3}{(x^5)(x^3)}\right\} = K$

i.e. $\ln\{(x-2y)^5(3x+2y)^3\} = K$ **is the general solution**

Problem 3. Solve $(y-x)\dfrac{dy}{dx} = \dfrac{y^2}{x} - y + \dfrac{x^2}{y}$ given that $x=1$ when $y=3$.

(i) Multiplying throughout by xy gives:

$$xy(y-x)\frac{dy}{dx} = y^3 - xy^2 + x^3$$

from which $\dfrac{dy}{dx} = \dfrac{y^3 - xy^2 + x^3}{xy(y-x)} = \dfrac{y^3 - xy^2 + x^3}{xy^2 - x^2y}$,

which is homogeneous in x and y.

(ii) Let $y = vx$. then $\dfrac{dy}{dx} = v + x\dfrac{dv}{dx}$

(iii) Substituting for y and $\dfrac{dy}{dx}$ gives:

$$v + x\frac{dv}{dx} = \frac{(vx)^3 - x(vx)^2 + x^3}{x(vx)^2 - x^2(vx)} = \frac{v^3 - v^2 + 1}{v^2 - v}$$

(iv) Separating the variables gives:

$$x\frac{dv}{dx} = \frac{v^3 - v^2 + 1}{v^2 - v} - v = \frac{(v^3 - v^2 + 1) - v(v^2 - v)}{v^2 - v}$$

$$= \frac{v^3 - v^2 + 1 - v^3 + v^2}{v^2 - v} = \frac{1}{v^2 - v}$$

Hence $(v^2 - v)\,dv = \dfrac{dx}{x}$

Integrating both sides gives: $\displaystyle\int(v^2 - v)\,dv = \int\frac{dx}{x}$

$$\frac{v^3}{3} - \frac{v^2}{2} = \ln x + c$$

(v) Replacing v by $\dfrac{y}{x}$ gives: $\dfrac{y^3}{3x^3} - \dfrac{y^2}{2x^2} = \ln x + c$

(This is the general solution)

When $x = 1$, $y = 3$, thus: $\dfrac{27}{3(1)} - \dfrac{9}{2(1)} = \ln 1 + c$

i.e. $c = \dfrac{9}{2}$

Hence the particular solution is $\dfrac{y^3}{3x^3} - \dfrac{y^2}{2x^2} = \ln x + \dfrac{9}{2}$

i.e. $2y^3 - 3xy^2 = 6x^3 \ln x + 27\, x^3$

Further problems on homogeneous differential equations may be found in the following Section 23.2 (Problems 1 to 10).

23.2   Further problems

In Problems 1 to 5 find the general solution of the differential equations.

1. $x\dfrac{dy}{dx} + \dfrac{y^2}{x} = y$ $\left[\dfrac{x}{y} = \ln x + c \text{ or } x = y \ln kx \right]$

2. $(x + y)\dfrac{dy}{dx} = \dfrac{y}{x}(y - x)$

$\left[-\dfrac{1}{2}\ln\dfrac{y}{x} - \ln x = \dfrac{y}{2x} + c \text{ or } \dfrac{y}{2x} + \ln\sqrt{(xy)} = K \right]$

3. $y^2 = x^2\dfrac{dy}{dx}$ $[y = x(1 + ky)]$

4. $xy\dfrac{dy}{dx} = \dfrac{x^2 + y^2}{2}$ $[x^2 - y^2 = kx]$

5. $(x + y) = (x - y)\dfrac{dy}{dx}$ $\left[\arctan\dfrac{y}{x} = \ln\sqrt{(x^2 + y^2)} + c \right]$

6. If $x + y + x\dfrac{dy}{dx} = 0$, show that $x(x + 2y) = k$, where k is a constant.

In Problems 7 to 9 find the particular solution of the differential equations.

7. $\dfrac{dy}{dx} = \dfrac{xy}{x^2 + y^2}$ given that $x = 1$ when $y = 1$. $[y = e^{(x^2 - y^2)/2y^2}]$

8. $(x + y)\dfrac{dy}{dx} = x + \dfrac{y^2}{x}$ given that $x = 4$ when $y = 2$.

$$\left[2x \ln \left\{ \dfrac{(x - y)^2}{x} \right\} = x - 2y \right]$$

9. $\dfrac{xy^2}{(x^3 + y^3)} \dfrac{dy}{dx} = 1$ given that $x = 1$ when $y = 3$.

$$[y^3 = 3x^3(\ln x + 9)]$$

10. Given that $\dfrac{4xy}{(x^2 - y^2)} \dfrac{dy}{dx} = 1$, and $y = 0$ when $x = 1$, show that

$(\sqrt{x})(x^2 - 5y^2) = 1$.

Chapter 24

Linear first order differential equations

24.1 Solution of differential equations of the form $\dfrac{dy}{dx} + Py = Q$

An equation of the form $\dfrac{dy}{dx} + Py = Q$, where P and Q are functions of x only, is called a **linear differential equation** since y and its derivative $\left(\text{i.e. } \dfrac{dy}{dx}\right)$ are of the first degree.

The solution of $\dfrac{dy}{dx} + Py = Q$ is obtained by multiplying throughout by what is called an **integrating factor**.

Multiplying $\dfrac{dy}{dx} + Py = Q$ throughout by some function of x alone, say R, gives:

$$R\frac{dy}{dx} + RPy = RQ \tag{1}$$

Now the differential coefficient of a product Ry is given by the product rule and is:

$$\frac{d}{dx}(Ry) = R\frac{dy}{dx} + y\frac{dR}{dx}$$

$R\dfrac{dy}{dx} + y\dfrac{dR}{dx}$ is the same as the left-hand side of equation (1) when R is chosen such that RP is equal to $\dfrac{dR}{dx}$.

If $\dfrac{dR}{dx} = Rp$, separating the variables gives:

$$\frac{dR}{R} = P\,dx$$

Integrating both sides gives: $\int \dfrac{dR}{R} = \int P \, dx$

i.e. $\ln R = \int P \, dx + c$

and $R = e^{\int P \, dx + c} = e^{\int P \, dx} e^c$

i.e. $R = A \, e^{\int P \, dx}$ (where $A = e^c$, an arbitrary constant).

Substituting $R = A \, e^{\int P \, dx}$ in equation (1) gives:

$$A \, e^{\int P \, dx} \left(\frac{dy}{dx}\right) + A \, e^{\int P \, dx} \, Py = A \, e^{\int P \, dx} \, Q$$

and by cancelling the constant A:

$$e^{\int P \, dx}\left(\frac{dy}{dx}\right) + e^{\int P \, dx} \, Py = e^{\int P \, dx} \, Q \tag{2}$$

The left-hand side of equation (2) is $\dfrac{d}{dx}(y \, e^{\int P \, dx})$, which may be checked by differentiating $y \, e^{\int P \, dx}$ with respect to x, using the product rule.

Hence from equation (2):

$$\frac{d}{dx}[y \, e^{\int P \, dx}] = e^{\int P \, dx} \, Q$$

Integrating both sides gives:

$$y \, e^{\int P \, dx} = \int e^{\int P \, dx} \, Q \, dx \tag{3}$$

Procedure to solve differential equations of the form $\dfrac{dy}{dx} + Py = Q$

(i) Rearrange the differential equation, if necessary, into the form $\dfrac{dy}{dx} + Py = Q$, where P and Q are functions of x.

(ii) Find $\int P \, dx$. Since the arbitrary constant of integration disappears (as shown in equation (2) above) it is unnecessary to include the constant for this indefinite integral.

(iii) Find the integrating factor $e^{\int P \, dx}$.

(iv) Substitute $e^{\int P \, dx}$ into equation (3).

(v) Integrate the right-hand side of equation (3) to give the general solution. If boundary conditions are given the arbitrary constant of integration may be determined and the particular solution of the differential equation may be found.

Problem 1. Solve the equation $\dfrac{1}{x}\dfrac{dy}{dx} + 2y = 1$ given that when $x = 0$, $y = 1$.

Following the above procedure:

(i)   Multiplying throughout by x gives $\dfrac{dy}{dx} + 2xy = x$, which is

of the form $\dfrac{dy}{dx} + Py = Q$, where $P = 2x$ and $Q = x$.

(ii) $\int P\, dx = \int 2x\, dx = x^2$

(iii)   Integrating factor $= e^{x^2}$

(iv)   Substituting into equation (3) gives: $y\, e^{x^2} = \int e^{x^2}(x)\, dx$
The right-hand side may be integrated by using the
substitution $u = x^2$

Hence $y\, e^{x^2} = \dfrac{e^{x^2}}{2} + c$ (This is the general solution)

When $x = 0$, $y = 1$, thus (1) $e^0 = \dfrac{e^0}{2} + c$

i.e. $c = \frac{1}{2}$

Hence the particular solution is

$y\, e^{x^2} = \dfrac{e^{x^2}}{2} + \dfrac{1}{2}$ or $2y = 1 + e^{-x^2}$

Problem 2. Solve: $x\cos x \dfrac{dy}{dx} + (x\sin x + \cos x)y = 1$.

(i)   Dividing throughout by $x\cos x$ gives:

$\dfrac{dy}{dx} + \left(\dfrac{x\sin x + \cos x}{x\cos x}\right)y = \dfrac{1}{x\cos x}$

i.e. $\dfrac{dy}{dx} + \left(\tan x + \dfrac{1}{x}\right)y = \dfrac{\sec x}{x}$

which is of the form $\dfrac{dy}{dx} + Py = Q$, where $P = \left(\tan x + \dfrac{1}{x}\right)$

and $Q = \left(\dfrac{\sec x}{x}\right)$

(ii) $\int P\, dx = \int\left(\tan x + \dfrac{1}{x}\right)dx = \ln(\sec x) + \ln x = \ln(x\sec x)$

(iii)   Integrating factor $= e^{\int P\, dx} = e^{\ln(x\sec x)} = x\sec x$

(iv) Substituting in equation (3) gives:

$$y(x \sec x) = \int (x \sec x)\left(\frac{\sec x}{x}\right) dx$$

i.e. $xy \sec x = \int \sec^2 x \, dx$

(v) Integrating gives: $\qquad xy \sec x = \tan x + c$

and dividing by $\sec x$: $\qquad xy = \dfrac{\tan x}{\sec x} + \dfrac{c}{\sec x}$

i.e. the general solution is $\qquad xy = \sin x + c \cos x$

Problem 3. Find the particular solution of the differential equation $\dfrac{dy}{dx} + 2x = y$, given that $x = 0$ when $y = 2$.

(i) Rearranging gives $\dfrac{dy}{dx} - y = -2x$, which is of the form

$\dfrac{dy}{dx} + Py = Q$, where $P = -1$ and $Q = -2x$

(ii) $\int P \, dx = \int -1 \, dx = -x$

(iii) Integrating factor $= e^{\int P \, dx} = e^{-x}$

(iv) Substituting into equation (3) gives

$$y e^{-x} = \int e^{-x}(-2x) \, dx$$

i.e. $y e^{-x} = -2 \int x e^{-x} \, dx \qquad\qquad$ (1)

(v) $\int x e^{-x} \, dx$ is determined using integration by parts.

Let $u = x$ and $dv = e^{-x} \, dx$

Then $du = dx$ and $v = \int e^{-x} \, dx = -e^{-x}$

Thus $\int x e^{-x} \, dx = x(-e^{-x}) - \int (-e^{-x}) \, dx$

$= -x e^{-x} - e^{-x} + c$

Hence from equation (4):

$y e^{-x} = -2(-x e^{-x} - e^{-x} + c)$

which is the general solution.

When $x = 0, y = 2$, thus $2(1) = -2(0 - 1 + c)$

from which $\qquad\qquad\qquad c = 0$

Thus the particular solution is $y e^{-x} = 2x e^{-x} + 2 e^{-x}$

or $\qquad\qquad\qquad\qquad\qquad y = 2x + 2$

or $\qquad\qquad\qquad\qquad\qquad y = 2(x + 1)$

Problem 4. Solve: $(x + 2)\dfrac{dy}{dx} = 3 - \dfrac{2y}{x}$.

(i) Rearranging gives: $\dfrac{dy}{dx} + \dfrac{2}{x(x + 2)}y = \dfrac{3}{(x + 2)}$, which is of

the form $\dfrac{dy}{dx} + Py = Q$, where $P = \dfrac{2}{x(x + 2)}$ and $Q = \dfrac{3}{(x + 2)}$

(ii) $\int P\,dx = \int \dfrac{2}{x(x + 2)}\,dx$ which may be integrated using

partial fractions.

Let $\dfrac{2}{x(x + 2)} \equiv \dfrac{A}{x} + \dfrac{B}{(x + 2)} \equiv \dfrac{A(x + 2) + Bx}{x(x + 2)}$

Equating the numerators gives: $2 \equiv A(x + 2) + Bx$

When $x = 0$, $2 = 2A$, i.e. $A = 1$

When $x = -2$, $2 = -2B$, i.e. $B = -1$

Hence

$\int \dfrac{2}{x(x + 2)}\,dx = \int \left\{\dfrac{1}{x} - \dfrac{1}{(x + 2)}\right\}dx = \ln x - \ln(x + 2)$

$$= \ln\left(\dfrac{x}{x + 2}\right)$$

(iii) Integrating factor $= e^{\int P\,dx} = e^{\ln x/(x - 2)} = \dfrac{x}{x + 2}$

(iv) Substituting in equation (3) gives:

$$y\left(\dfrac{x}{x + 2}\right) = \int\left(\dfrac{x}{x + 2}\right)\left(\dfrac{3}{x + 2}\right)dx = \int \dfrac{3x}{(x + 2)^2}\,dx$$

(v) $\int \dfrac{3x}{(x + 2)^2}\,dx$ cannot be determined 'on sight'. It is

thus split into two fractions, the first fraction being
specifically chosen to give a logarithmic solution.

Hence $\dfrac{xy}{x + 2} = \int\left\{\dfrac{\frac{3}{2}(2x + 4)}{(x + 2)^2} - \dfrac{6}{(x + 2)^2}\right\}dx$

$$= \tfrac{3}{2}\ln(x + 2)^2 + \dfrac{6}{(x + 2)} + c$$

$\dfrac{xy}{x + 2} = \ln[(x + 2)^2]^{\frac{3}{2}} + \dfrac{6}{(x + 2)} + \ln k$ (where $\ln k = c$)

Hence the general solution is $xy = (x + 2)\ln\{k(x + 2)^3\} + 6$

---

*Further problems on solving linear differential equations may be found in the
following Section 24.2 (Problems 1 to 18).*

24.2   Further problems

In Problems 1 to 10 solve the differential equations.

1. $x\dfrac{dy}{dx} + y = 2$ $[xy = 2x + c]$

2. $y = x\left(3 - \dfrac{dy}{dx}\right)$ and $x = 1$ when $y = 2$ $[2xy = 3x^2 + 1]$

3. $x\dfrac{dy}{dx} = y + x^3 - 2x$ and $x = 1$ when $y = 3$
    $[2y = x^3 + 5x - 4x \ln x]$

4. $\dfrac{dy}{dx} + 1 = \dfrac{-y}{x}$ and $x = 2$ when $y = 1$ $[2xy + x^2 = 8]$

5. $\dfrac{dy}{dx} = x - \dfrac{2}{x}y$ and $x = 2$ when $y = 1$ $[4y = x^2]$

6. $\cos x\dfrac{dy}{dx} = (\sin x)(1 - 2y)$ $[2y = 1 + k\cos^2 x]$

7. $\tan x\dfrac{dy}{dx} + y = x^2 \tan x$
    $[(y - 2x)\sin x = (2 - x^2)\cos x + c]$

8. $\dfrac{dy}{dx} + x = y$ and $x = 0$ when $y = 2$ $[y = x + 1 + e^x]$

9. $(x + 1)\dfrac{dy}{dx} + \dfrac{y}{x} = 2$
    $[xy = 2(x + 1)\ln(x + 1) + 2 + c(x + 1)]$

10. $2(x + 1)\dfrac{dy}{dx} - (1 + 2x)y = x^2\sqrt{(1 + x)}$
    $[2y\sqrt{(x + 1)} = c\,e^x - (x^2 + 2x + 2)]$

11. In an electrical alternating current circuit containing resistance R and inductance L the current i is given by the equation
    $Ri + L\dfrac{di}{dt} = E$. If $E = E_0 \sin \omega t$ and $i = 0$ when $t = 0$ show that:
    $$i = \dfrac{E_0}{R^2 + \omega^2 L^2}(R \sin \omega t - \omega L \cos \omega t) + \dfrac{E_0 \omega L}{R^2 + \omega^2 L^2}e^{-Rt/L}$$

12. Solve the following equation for θ, given that when $t = 0$, $\theta = \frac{1}{2}$:
    $\cosh t\dfrac{d\theta}{dt} + \theta \sinh t = \sinh 2t$ $[\theta = \frac{1}{2}\cosh 2t \operatorname{sech} t]$

13. An equation of motion when a particle moves in a resisting medium is given by: $\dfrac{dv}{dt} = -(kv + bt)$, where k and b are constants. Solve the equation for v given that $v = u$ when $t = 0$.

$$\left[v = \frac{b}{k^2} - \frac{bt}{k} + \left(u - \frac{b}{k^2} \right) e^{-kt} \right]$$

14. A train of total mass m is moved from rest by the engine which exerts a time-dependent force $mk(1 - e^{-t})$ on the train. The resistance to motion is mcv, where v is the speed of the train and c is a constant. Find the subsequent speed of the train if the equation of motion is given by

$$m \frac{dv}{dt} = mk(1 - e^{-t}) - mcv$$

$$\left[v = k \left(\frac{1}{c} - \frac{e^{-t}}{(c-1)} + \frac{e^{-ct}}{c(c-1)} \right) \right]$$

15. In a unimolecular chemical reaction a substance α changes into β. If x is the number of molecules of β present at any time t, and a the initial number of molecules of α, then $\dfrac{dx}{dt} = k_1(a - x) - k_2 x$, where k_1 and k_2 are constants. Show that if $x = 0$ when $t = 0$, then

$$x = \frac{k_1 a}{k_1 + k_2} [1 - e^{-(k_1 + k_2)t}]$$

16. The instantaneous current (i) passing through a solution, in a circuit of resistance R and inductance L, whose dielectric constant is to be measured is given by $\dfrac{di}{dt} + \dfrac{R}{L} i = \dfrac{V_0}{L} \sin pt$, where t is time and V_0 and p are constants. Show that the solution of this equation is

$$i = \left\{ \frac{V_0}{R^2 + p^2 L^2} \right\} \{ R \sin pt - pL \cos pt \} + C e^{-Rt/L}$$

17. In the drain tank of an oil purifier in a ship the concentration (c) of impurities varies with time (t) according to the equation

$$M \frac{dc}{dt} = p + km - cm$$

where M, p, m and k are constants.

Solve this equation given $c = c_0$ when $t = 0$.

$$\left[c = c_0 \, e^{-mt/M} + \left(k + \frac{p}{m} \right)(1 - e^{-mt/M}) \right]$$

18. The angular velocity ω of a flywheel at time t is given by

$$I \frac{d\omega}{dt} + K\omega = A + B \sin^2 pt,$$

where I, K, A, B and p are constants. If $\omega = 0$ when $t = 0$ show that:

$$\omega = \frac{2A + B}{2K} - \frac{B(K \cos 2pt + 2pI \sin 2pt)}{2(K^2 + 4p^2I^2)}$$

$$- \frac{1}{K}\left(A + \frac{2p^2I^2B}{K^2 + 4p^2I^2} \right) e^{-Kt/I}$$

Chapter 25

The solution of linear second order differential equations of the form $a(d^2y/dx^2) + b(dy/dx) + cy = 0$

25.1 Introduction

An equation of the form:

$$a\frac{d^2y}{dx^2} + b\frac{dy}{dx} + cy = 0, \tag{1}$$

where a, b and c are constants, is called a **linear second order differential equation with constant coefficients**. It is termed 'linear' since y and its derivatives are all of the first degree, and it is of 'second order' since its highest derivative is $\frac{d^2y}{dx^2}$ (see Section 22.1)

An alternative way of stating equation (1) is:

$$(aD^2 + bD + c)y = 0, \tag{2}$$

where D indicates the operation $\frac{d}{dx}$ and D^2 symbolises $\frac{d^2}{dx^2}$. When an equation is stated as in equation (2) it is said to be in 'D **operator**' form. Equations of the form stated in equations (1) and (2) have important engineering applications, especially in electrical and mechanical work. Two examples include:

(i) $L\frac{d^2q}{dt^2} + R\frac{dq}{dt} + \frac{1}{c}q = 0$, representing an equation for charge q in an electrical circuit containing resistance R, inductance L and capacitance C in series, and

(ii) $m\frac{d^2s}{dt^2} + a\frac{ds}{dt} + ks = 0$, defining a mechanical system, where s is the distance from a fixed point after t seconds, m is a mass, a the damping factor and k the spring stiffness.

In this chapter the solution of differential equations of the type

$a \dfrac{d^2y}{dx^2} + b \dfrac{dy}{dx} + cy = 0$ is dealt with, and in Chapter 26 the more general

type $a \dfrac{d^2y}{dx^2} + b \dfrac{dy}{dx} + cy = f(x)$ is dealt with.

25.2   Types of solution of second order differential equations with constant coefficients

If in equation (1) we let $a = 0$, then $b \dfrac{dy}{dx} + cy = 0$

i.e.
$$\frac{dy}{dx} + \frac{c}{b} y = 0$$

Hence
$$\frac{dy}{dx} = -\frac{c}{b} y$$

and
$$\frac{dy}{y} = -\frac{c}{b} dx \quad \text{by separating the}$$
$$\text{variables}$$

Therefore $\displaystyle\int \frac{dy}{y} = -\frac{c}{b} \int dx$

$\ln y = -\dfrac{c}{b} x + k$, where k is a constant.

Hence, $y = e^{-(c/b)x + k} = (e^{-(c/b)x})(e^{k})$

If we let $e^k = A = $ a constant, then

$\qquad y = A\, e^{-(c/b)x}$ or $y = A\, e^{mx}$, where m is a constant.

Let us see if the solution $y = A\, e^{mx}$ also satisfies equation (1), m being a constant to be found by substituting $y = A\, e^{mx}$ into equation (1)

If $y = A\, e^{mx}$ then $\dfrac{dy}{dx} = Am\, e^{mx}$

and
$$\frac{d^2y}{dx^2} = Am^2\, e^{mx}$$

Substituting for y, $\dfrac{dy}{dx}$ and $\dfrac{d^2y}{dx^2}$ in equation (1) gives:

$$a \frac{d^2y}{dx^2} + b \frac{dy}{dx} + cy = aAm^2\, e^{mx} + bAm\, e^{mx} + cA\, e^{mx} = 0$$

i.e. $A\, e^{mx}(am^2 + bm + c) = 0$ (3)

Equation (3) is true for all values of x. Since e^{mx} is not equal to zero and A is not zero (for this would make $y = 0$ always) then:

$$am^2 + bm + c = 0 \tag{4}$$

which is a quadratic equation in m.

Equation (4) is called the **auxiliary equation** (or the characteristic equation).

From equation (4), m may be obtained by factorisation or by using the quadratic formula, $m = \dfrac{-b \pm \sqrt{[b^2 - 4ac]}}{2a}$

Since a, b and c are real the auxiliary equation may have either:

(i)   two different real roots (when $b^2 > 4ac$),
(ii) two real equal roots (when $b^2 = 4ac$), or
(iii) two complex roots (when $b^2 < 4ac$).

These three cases will now be discussed separately.

(i)   Two different real roots

Let the solutions to equation (4) be $m = \alpha$ and $m = \beta$. Then either $y = A\,e^{\alpha x}$ or $y = B\,e^{\beta x}$ is a solution of equation (1). However, it is shown below that a general property of linear equations, such as equation (4), is that the sum of the two solutions is also a solution.

Hence: $y = A\,e^{\alpha x} + B\,e^{\beta x}$ \hfill (5)

is a solution of $a\dfrac{d^2 y}{dx^2} + b\dfrac{dy}{dx} + cy = 0$, and since this expression contains two arbitrary constants, A and B, it is the **general solution**. If sufficient data are given the constants A and B may be determined and the **particular solution** found.

The general solution $y = A\,e^{\alpha x} + B\,e^{\beta x}$, stated in equation (5), may be derived directly as follows:

Let α and β be the roots of the auxiliary equation $am^2 + bm + c = 0$

i.e. $m^2 + \dfrac{b}{a}m + \dfrac{c}{a} = 0$

From the theory of quadratic equations, $\alpha + \beta = -\dfrac{b}{a}$

and $\alpha\beta = \dfrac{c}{a}$

$a\dfrac{d^2 y}{dx^2} + b\dfrac{dy}{dx} + cy = 0$ may be written as

$$\frac{d^2y}{dx^2} + \frac{b}{a}\frac{dy}{dx} + \frac{c}{a}y = 0,$$

or $$\frac{d^2y}{dx^2} - (\alpha + \beta)\frac{dy}{dx} + \alpha\beta y = 0, \text{ and by rearranging,}$$

$$\frac{d}{dx}\left(\frac{dy}{dx} - \beta y\right) = \alpha\left(\frac{dy}{dx} - \beta y\right) \tag{6}$$

Let $$\frac{dy}{dx} - \beta y = Z$$

then $$\frac{dZ}{dx} = \alpha Z \text{ from equation (6)}$$

Hence $$\int\frac{dZ}{Z} = \int\alpha \, dx, \text{ by separation of variables}$$

$$\ln Z = \alpha x + k, \text{ where } k \text{ is a constant}$$

and $$Z = e^{\alpha x + k} = (e^{\alpha x})(e^k) = C \, e^{\alpha x} \text{ where } e^k = C = \text{a constant,}$$

i.e. $$Z = \frac{dy}{dx} - \beta y = C \, e^{\alpha x} \tag{7}$$

Equation (7) is of the form of the linear first order differential equations discussed in Chapter 24 solved by introducing an integrating factor, and the solution of this equation is:

$$y \, e^{\int -\beta \, dx} = \int e^{\int -\beta \, dx}(C \, e^{\alpha x}) \, dx$$

i.e. $$y \, e^{-\beta x} = \int C \, e^{(\alpha - \beta)x} \, dx \tag{8}$$

$$y \, e^{-\beta x} = \frac{C}{\alpha - \beta}e^{(\alpha - \beta)x} + B, \text{ where } B \text{ is a constant}$$

Multiplying throughout by $e^{\beta x}$ gives:

$$y = \frac{C}{\alpha - \beta}e^{\alpha x} \, e^{-\beta x} \, e^{\beta x} + B \, e^{\beta x}$$

$$y = A \, e^{\alpha x} + B \, e^{\beta x}, \text{ where } A = \frac{C}{\alpha - \beta} = \text{a constant}$$

This is the same as equation (5) showing that the sum of the two solutions is also a solution.

Thus, given the differential equation $2\dfrac{d^2y}{dx^2} - \dfrac{dy}{dx} - 6y = 0$

then $\hspace{4cm} (2D^2 - D - 6)y = 0$

Letting $y = A\,e^{mx}$ we obtain $A\,e^{mx}(2m^2 - m - 6) = 0$, and since $A\,e^{mx} \neq 0$ then $2m^2 - m - 6 = 0$. This is the auxiliary equation. (Note that the auxiliary equation may be obtained from the original equation by replacing D by m.)

If $2m^2 - m - 6$, then $(2m + 3)(m - 2) = 0$,

i.e. $\hspace{2cm} m = -\dfrac{3}{2}$ or $m = 2$,

i.e. two different real roots.

Hence the general solution of the differential equation

$$2\frac{d^2y}{dx^2} - \frac{dx}{dy} - 6y = 0 \text{ is } y = A\,e^{-\frac{3}{2}x} + B\,e^{2x}$$

Such a solution may always be checked:

if $y = A\,e^{-\frac{3}{2}x} + B\,e^{2x}$ then $\dfrac{dx}{dx} = -\dfrac{3}{2}\,A\,e^{-\frac{3}{2}x} + 2B\,e^{2x}$,

and $\dfrac{d^2y}{dx^2} = \dfrac{9}{4}\,A\,e^{-\frac{3}{2}x} + 4B\,e^{2x}$

Substituting for y, $\dfrac{dy}{dx}$ and $\dfrac{d^2y}{dx^2}$ into $2\dfrac{d^2y}{dx^2} - \dfrac{dy}{dx} - 6y$ gives:

$$2\left(\frac{9}{4}\,A\,e^{-\frac{3}{2}x} + 4B\,e^{2x}\right) - \left(-\frac{3}{2}\,A\,e^{-\frac{3}{2}x} + 2B\,e^{2x}\right) - 6\left(A\,e^{-\frac{3}{2}x} + B\,e^{2x}\right)$$

$$= \left(\frac{9}{2} + \frac{3}{2} - 6\right)A\,e^{-\frac{3}{2}x} + (8 - 2 - 6)B\,e^{2x} = 0$$

Hence $y = A\,e^{-\frac{3}{2}x} + B\,e^{2x}$ is a solution of the differential equation

$$2\frac{d^2y}{dx^2} - \frac{dy}{dx} - 6y = 0$$

(ii) Two real equal roots

If from the auxiliary equation $\alpha = \beta$ then the general solution is given by $y = A\,e^{\alpha x} + B\,e^{\alpha x}$

i.e. $\hspace{1cm} y = (A + B)\,e^{\alpha x} = D\,e^{\alpha x}$, where $D = A + B = $ a constant

However this is no longer the general solution of equation (1) since the solution **must** contain **two** arbitrary constants. The analysis used in (i) can be used as far as equation (8), which becomes:

$$y\,e^{-\alpha x} = \int C\,dx, \text{ since } \alpha = \beta,$$

i.e. $y\,e^{-\alpha x} = Cx + B$

Multiplying throughout by $e^{\alpha x}$ gives: $y = Cx\,e^{\alpha x} + B\,e^{\alpha x}$
Replacing C by A gives:

$$y = Ax\,e^{\alpha x} + B\,e^{\alpha x}$$

i.e. $\boldsymbol{y = (Ax + B)\,e^{\alpha x}}$

This is the general solution for the case of two real equal roots.
Thus, given the differential equation $\dfrac{d^2 y}{dx^2} - 6\dfrac{dy}{dx} + 9y = 0$

i.e. $(D^2 - 6D + 9)y = 0$

The auxiliary equation is $m^2 - 6m + 9 = 0$
i.e. $(m - 3)(m - 3) = 0$

Hence $m = 3$, twice

The general solution of $\dfrac{d^2 y}{dx^2} - 6\dfrac{dy}{dx} + 9y = 0$ is $y = (Ax + B)\,e^{3x}$

This solution may be checked:

If $y = (Ax + B)\,e^{3x}$ then $\dfrac{dy}{dx} = (Ax + B)(3\,e^{3x}) + (e^{3x})(A)$

and $\dfrac{d^2 y}{dx^2} = (Ax + B)(9\,e^{3x}) + (3\,e^{3x})(A) + 3A\,e^{3x}$

Substituting for y, $\dfrac{dy}{dx}$ and $\dfrac{d^2 y}{dx^2}$ into $\dfrac{d^2 y}{dx^2} - 6\dfrac{dy}{dx} + 9y$ gives:

$$(9Ax\,e^{3x} + 9B\,e^{3x} + 3A\,e^{3x} + 3A\,e^{3x})$$
$$- 6(3Ax\,e^{3x} + 3B\,e^{3x} + A\,e^{3x}) + 9(Ax\,e^{3x} + B\,e^{3x}) = 0$$

Hence $y = (Ax + B)\,e^{3x}$ is a solution of the differential equation

$$\frac{d^2 y}{dx^2} - 6\frac{dy}{dx} + 9y = 0$$

(iii)   Two complex roots

When solving the auxiliary equation $am^2 + bm + c = 0$, then

$$m = \frac{-b \pm \sqrt{[b^2 - 4ac]}}{2a}$$

If the coefficients of the differential equation are such that $b^2 < 4ac$ then the values of m are complex.

For example, to solve $m^2 - 6m + 13 = 0$

$$m = \frac{-(-6) \pm \sqrt{[(-6)^2 - 4(1)(13)]}}{2(1)} = \frac{6 \pm \sqrt{(-16)}}{2}$$

Now $\sqrt{(-16)} = \sqrt{[(-1)(16)]} = (\sqrt{-1})(\sqrt{16})$
By definition, $\sqrt{-1} = j$
Hence $\sqrt{(-16)} = \pm j4$

Thus $m = \dfrac{6 \pm j4}{2} = 3 \pm j2$

Hence the solution of the differential equation:

$$\frac{d^2y}{dx^2} - 6\frac{dy}{dx} + 13y = 0$$

is $y = A\,e^{(3+j2)x} + B\,e^{(3-j2)x}$

i.e. $y = e^{3x}\{A\,e^{j2x} + B\,e^{-j2x}\}$

It was shown in Chapter 8 that $e^{jx} = \cos x + j \sin x$
and $e^{-jx} = \cos x - j \sin x$

Hence $y = e^{3x}\{A(\cos 2x + j \sin 2x) + B(\cos 2x - j \sin 2x)\}$

$$y = e^{3x}\{(A + B) \cos 2x + j(A - B) \sin 2x\}$$

It would appear at first sight that this is a complex solution. However, y (which represents a real voltage or current or displacement, etc.) cannot be equal to a complex expression. In practice it is always found that A and B are complex conjugate numbers. For example, let $A = 3 + j4$ and $B = 3 - j4$. Then $A + B = 6$, a real number, and $j(A - B) = j(j8) = -8$, also a real number.

Hence, given that the solution of the auxiliary equation is $m = 3 \pm j2$ then the solution may be written down directly as $y = e^{3x}\{C \cos 2x + D \sin 2x\}$, where C and D are two real unknown constants. This solution may be checked by substituting y, $\dfrac{dy}{dx}$ and $\dfrac{d^2y}{dx^2}$ into the left-hand side of the differential equation $\dfrac{d^2y}{dx^2} - 6\dfrac{dy}{dx} + 13y = 0$

Generally, if the roots of the auxiliary equation are $\alpha \pm j\beta$, then the general solution is $y = e^{\alpha x}\{C \cos \beta x + D \sin \beta x\}$

25.3 Summary of the procedure used to solve differential equations of the form $a\dfrac{d^2y}{dx^2} + b\dfrac{dy}{dx} + cy = 0$

(a) Rewrite the given differential equations $a\dfrac{d^2y}{dx^2} + b\dfrac{dy}{dx} + cy = 0$ as

$(aD^2 + bD + c)y = 0$

(b) Substitute m for D and solve the auxiliary equation $am^2 + bm + c = 0$ for m

(c) (i) If the roots of the auxiliary equation are **real and different**, say, $m = \alpha$ and $m = \beta$, then the general solution is:

$y = A\,e^{\alpha x} + B\,e^{\beta x}$

(ii) If the roots of the auxiliary equation are **real and equal**, say $m = \alpha$, twice, then the general solution is:

$y = (Ax + B)\,e^{\alpha x}$

(iii) If the roots of the auxiliary equation are **complex**, say $m = \alpha \pm j\beta$, then the general solution is:

$y = e^{\alpha x}\{C \cos \beta x + D \sin \beta x\}$

(d) If the **particular solution** of a differential equation is required then substitute the given boundary conditions to find the unknown constants (i.e. to find A and B in (c, i) and (c, ii) or C and D in (c, iii).

Worked problems on solving differential equations of the form $a\dfrac{d^2y}{dx^2} + b\dfrac{dy}{dx} + cy = 0$

Problem 1. (a) Find the general solution of

$6\dfrac{d^2y}{dx^2} + 5\dfrac{dy}{dx} - 4y = 0.$

(b) Find the particular solution of the differential equation in (a) given that when $x = 0$, $y = 11$ and $\dfrac{dy}{dx} = 0.$

(a)   In D operator form the differential equation is

$(6D^2 + 5D - 4)y = 0$ where $D \equiv \dfrac{d}{dx}$

The auxiliary equation is $6m^2 + 5m - 4 = 0$

Factorising gives: $(3m + 4)(2m - 1) = 0$

i.e. $m = -\dfrac{4}{3}$ or $m = \dfrac{1}{2}$

Since the two roots are real and different **the general solution is**

$y = A e^{-\frac{4}{3}x} + B e^{\frac{1}{2}x}$

(b)   When $x = 0$, $y = 11$, then

$11 = A + B$ (1)

Since $y = A e^{-\frac{4}{3}x} + B e^{\frac{1}{2}x}$ then

$\dfrac{dy}{dx} = -\dfrac{4}{3} A e^{-\frac{4}{3}x} + \dfrac{1}{2} B e^{\frac{1}{2}x}$

When $x = 0$, $\dfrac{dy}{dx} = 0$

Thus $0 = -\dfrac{4}{3} A + \dfrac{1}{2}B$ (2)

$2 \times$ (2) gives: $0 = -\dfrac{8}{3} A + B$ (3)

(1) $-$ (3) gives: $11 = 3\dfrac{2}{3} A$, i.e. $A = 3$

Substituting $A = 3$ in equation (1) gives $B = 8$

Hence the particular solution is $y = 3 e^{-\frac{4}{3}x} + 8 e^{\frac{1}{2}x}$

Problem 2. (a)   Determine the general solution of

$4\dfrac{d^2y}{dt^2} - 12\dfrac{dy}{dt} + 9y = 0$

(b)   Find the particular solution of the equation given in (a) which makes $y = 2$ at $t = 0$ and

$\dfrac{dy}{dt} = 4$ at $t = 0$.

(a) In D operator form the differential equation is

$(4D^2 - 12D + 9)y = 0$ where $D \equiv \dfrac{d}{dt}$

The auxiliary equation is $4m^2 - 12m + 9 = 0$

i.e. $(2m - 3)(2m - 3) = 0$

i.e. $m = \dfrac{3}{2}$ twice

Since the two roots are real and equal **the general solution is**

$y = (At + B)\, e^{\frac{3}{2}t}$

(b) When $y = 2$, $t = 0$

Thus $2 = (0 + B)\, e^0$, i.e. **$B = 2$**

$\dfrac{dy}{dt} = (At + B)\left(\dfrac{3}{2}\, e^{\frac{3}{2}t}\right) + (e^{\frac{3}{2}t})(A)$ by the product rule of
differentiation

When $\dfrac{dy}{dt} = 4$, $t = 0$

Thus $4 = (0 + B)\left(\dfrac{3}{2}\, e^0\right) + (e^0)(A)$

i.e. $4 = \dfrac{3}{2} B + A$

Since $B = 2$, **$A = 1$**
Hence the particular solution is $y = (t + 2)\, e^{\frac{3}{2}t}$

Problem 3. (a) Determine the general solution of the differential equation

$$\dfrac{d^2 y}{dx^2} + 2\dfrac{dy}{dx} + 5y = 0$$

(b) Given the boundary conditions that when $x = 0$, $y = 1$, and when $x = 0$, $\dfrac{dy}{dx} = 5$, find the particular solution of the equation given in (a).

(a) In D operator form the differential equation is

$(D^2 + 2D + 5)y = 0$ where $D \equiv \dfrac{d}{dx}$

The auxiliary equation is $m^2 + 2m + 5 = 0$

Using the quadratic formula, $m = \dfrac{-2 \pm \sqrt{[(2)^2 - 4(1)(5)]}}{2(1)}$

$$= \dfrac{-2 \pm \sqrt{(-16)}}{2}$$

i.e. $m = \dfrac{-2 \pm j4}{2} = -1 \pm j2$

Since the roots are complex **the general solution is**

$y = e^{-x}\{C \cos 2x + D \sin 2x\}$

(b)   When $x = 0$, $y = 1$
Thus $1 = e^0\{C \cos 0 + D \sin 0\}$
Hence $1 = C$

$\dfrac{dy}{dx} = e^{-x}(-2C \sin 2x + 2D \cos 2x) - e^{-x}(C \cos 2x + D \sin 2x)$,

by the product rule,

$\qquad = e^{-x}\{(2D - C) \cos 2x - (2C + D) \sin 2x\}$

When $x = 0$, $\dfrac{dy}{dx} = 5$

Hence $5 = e^0\{(2D - C) \cos 0 - (2C + D) \sin 0\}$
$\qquad 5 = 2D - C$
Since $C = 1$, $D = 3$

Hence the particular solution is $y = e^{-x}\{\cos 2x + 3 \sin 2x\}$

Since $a \cos \omega t + b \sin \omega t \equiv R \sin(\omega t + \alpha)$, when $R = \sqrt{(a^2 + b^2)}$

and $\alpha = \arctan \dfrac{a}{b}$,

$\cos 2x + 3 \sin 2x \equiv \sqrt{(1^2 + 3^2)} \sin\left(2x + \arctan \dfrac{1}{3}\right)$

$\qquad\qquad \equiv \sqrt{10} \sin(2x + 18° \, 26')$

$\qquad\qquad \equiv \sqrt{10} \sin(2x + 0.3218)$

Thus the particular solution may also be expressed as

$y = \sqrt{10}\, e^{-x} \sin(2x + 0.3218)$

Problem 4. Solve the differential equations:

(a) $\dfrac{d^2y}{dx^2} + n^2 y = 0$

(b) $\dfrac{d^2y}{dx^2} - n^2 y = 0$, where n is a constant.

(a) $\dfrac{d^2y}{dx^2} + n^2y = 0$ is a differential equation representing simple harmonic motion (S.H.M.) and has many practical applications.

In D operator form the equation is $(D^2 + n^2)y = 0$

The auxiliary equation is $m^2 + n^2 = 0$

i.e. $m^2 = -n^2$

$$m = \sqrt{(-n^2)} = \pm jn$$

Since the two roots are complex the general solution is

$y = e^0\{C \cos nx + D \sin nx\}$, i.e. $y = C \cos nx + D \sin nx$ or

$y = R \sin(nx + \alpha)$, where $R = \sqrt{(C^2 + D^2)}$ and $\alpha = \arctan \dfrac{C}{D}$

(b) In D operator form the differential equation

$\dfrac{d^2y}{dx^2} - n^2y = 0$ is $(D^2 - n^2)y = 0$

The auxiliary equation is $m^2 - n^2 = 0$

Hence $m^2 = n^2$

$$m = \pm n$$

Since the two roots are real and different **the general solution is $y = A\,e^{nx} + B\,e^{-nx}$**

Since $\sinh nx = \dfrac{1}{2}(e^{nx} - e^{-nx})$ and

$\cosh nx = \dfrac{1}{2}(e^{nx} + e^{-nx})$

then $\sinh nx + \cosh nx = e^{nx}$

and $\cosh nx - \sinh nx = e^{-nx}$

Hence the general solution may also be written as:

$\quad y = A(\sinh nx + \cosh nx) + B(\cosh nx - \sinh nx)$

i.e. $y = (A + B)\cosh nx + (A - B)\sinh nx$

i.e. $y = C \cosh nx + D \sinh nx$

Problem 5. A circuit containing resistance R, inductance L and capacitance C in series has an equation for current i given by:

$$L\dfrac{d^2i}{dt^2} + R\dfrac{di}{dt} + \dfrac{1}{C}i = 0$$

If $L = 0.25$ henry, $C = 25 \times 10^{-6}$ farads and $R = 200$ ohms, solve the equation given that when time $t = 0$, $i = 0$ and $\dfrac{di}{dt} = 50$

In D operator form the differential equation is $\left(LD^2 + RD + \dfrac{1}{C}\right)i = 0$

The auxiliary equation is $Lm^2 + Rm + \dfrac{1}{C} = 0$

Hence $m = \dfrac{-R \pm \sqrt{\left(R^2 - \dfrac{4L}{C}\right)}}{2L}$

Substituting $L = 0.25$, $C = 25 \times 10^{-6}$ and $R = 200$ gives:

$$m = \dfrac{-200 \pm \sqrt{\left[(200)^2 - \dfrac{4(0.25)}{25 \times 10^{-6}}\right]}}{2(0.25)} = \dfrac{-200 \pm \sqrt{0}}{0.5}$$

i.e. $m = \dfrac{-200}{0.5} = -400$

For a second order differential equation there must be two solutions.
Hence $m = -400$ twice. **Hence the general solution is**
$i = (At + B)\,e^{-400t}$
When $t = 0$, $i = 0$
Hence $0 = B$

$\dfrac{di}{dt} = (At + B)(-400\,e^{-400t}) + (e^{-400t})(A)$

When $t = 0$, $\dfrac{di}{dt} = 50$

Hence $50 = -400B + A$
Since $B = 0$, $A = 50$
Hence the particular solution is $i = 50t\,e^{-400t}$

Further problems on solving differential equations of the form
$a\dfrac{d^2y}{dx^2} + b\dfrac{dy}{dx} + cy = 0$ *may be found in the following Section 25.4 (Problems*
1 to 21).

25.4   Further problems

In Problems 1 to 6 find the general solution of the given differential equations.

1. (a) $\dfrac{d^2y}{dx^2} - 5\dfrac{dy}{dx} + 6y = 0$ $[y = A\,e^{3x} + B\,e^{2x}]$

    (b) $\dfrac{d^2y}{dx^2} - 2\dfrac{dy}{dx} + y = 0$ $[y = (Ax + B)\,e^x]$

2. (a) $\dfrac{d^2y}{dx^2} + 6\dfrac{dy}{dx} + 13y = 0$ $\qquad [y = e^{-3x}(C\cos 2x + D\sin 2x)]$

(b) $\dfrac{d^2x}{dt^2} + 3\dfrac{dx}{dt} + 2x = 0$ $\qquad [x = A\,e^{-t} + B\,e^{-2t}]$

3. (a) $\dfrac{d^2y}{d\theta^2} - 4\dfrac{dy}{d\theta} + 4y = 0$ $\qquad [y = (A\theta + B)\,e^{2\theta}]$

(b) $\dfrac{d^2y}{dt^2} + y = 0$ $\qquad [y = C\cos t + D\sin t]$

4. (a) $6\dfrac{d^2\theta}{dt^2} + 4\dfrac{d\theta}{dt} - 2\theta = 0$ $\qquad [\theta = A\,e^{\frac{1}{3}t} + B\,e^{-t}]$

(b) $4\dfrac{d^2y}{dx^2} + 4\dfrac{dy}{dx} + y = 0$ $\qquad [y = (Ax + B)\,e^{-\frac{1}{2}x}]$

5. (a) $(3D^2 - 2D + 5)y = 0$ where $D \equiv \dfrac{d}{dx}$

$$\left[y = e^{\frac{1}{3}x}\left(C\cos\frac{\sqrt{14}}{3}x + D\sin\frac{\sqrt{14}}{3}x \right) \right]$$

(b) $(4D^2 - 7D - 15)y = 0$ where $D \equiv \dfrac{d}{dx}$

$[y = A\,e^{-\frac{5}{4}x} + B\,e^{3x}]$

6. (a) $(16D^2 + 8D + 1)y = 0$ where $D \equiv \dfrac{d}{dx}$

$[y = (Ax + B)\,e^{-\frac{1}{4}x}]$

(b) $(D^2 + D + 1)\theta = 0$ where $D \equiv \dfrac{d}{dt}$

$$\left[\theta = e^{-\frac{1}{2}t}\left(C\cos\frac{\sqrt{3}}{2}t + D\sin\frac{\sqrt{3}}{2}t \right) \right]$$

7. Show that the solution to the differential equation

$\dfrac{d^2y}{dx^2} + (2a\cos\theta)\dfrac{dy}{dx} + a^2y = 0$, where a and θ are constants,

may be expressed as: $y = A\,e^{-ax\cos\theta}\cos(B + ax\sin\theta)$

In Problems 8 to 13, find the particular solution of the given differential equations for the stated boundary conditions.

8. $12\dfrac{d^2y}{dx^2} - 3y = 0$ $y = 3$ when $x = 0$

$\dfrac{dy}{dx} = \dfrac{1}{2}$ when $x = 0$ $\qquad [y = 2\,e^{\frac{1}{2}x} + e^{-\frac{1}{2}x}]$

9. $9\dfrac{d^2y}{dt^2} - 12\dfrac{dy}{dt} + 4y = 0$ $\begin{cases} y = 3 \text{ when } t = 0 \\ \dfrac{dy}{dt} = 4 \text{ when } t = 0 \end{cases}$ $[y = (2t + 3)\,e^{\frac{2}{3}t}]$

10. $\dfrac{d^2y}{dx^2} + 2\dfrac{dy}{dx} + 6y = 0$ $\begin{cases} y = 2 \text{ when } x = 0 \\ \dfrac{dy}{dx} = 3 \text{ when } x = 0 \end{cases}$

$[y = e^{-x}(2\cos\sqrt{5}x + \sqrt{5}\sin\sqrt{5}x)]$

11. $(35D^2 - 11D - 6)x = 0$ where $D \equiv \dfrac{d}{dt}$ $\begin{cases} x = 5 \text{ when } t = 0 \\ \dfrac{dx}{dt} = \dfrac{12}{35} \text{ when } t = 0 \end{cases}$

$[x = 3\,e^{-\frac{2}{7}t} + 2\,e^{\frac{3}{5}t}]$

12. $(2D^2 + 2D + 1)y = 0$ where $D \equiv \dfrac{d}{dx}$ $\begin{cases} y = 4 \text{ when } x = 0 \\ \dfrac{dy}{dx} = 5 \text{ when } x = 0 \end{cases}$

$\left[y = 2\,e^{-\frac{1}{2}x}\left(2\cos\dfrac{x}{2} + 7\sin\dfrac{x}{2}\right) \right]$

13. $(25D^2 - 20D + 4)y = 0$ where $D \equiv \dfrac{d}{dx}$ $\begin{cases} y = 5 \text{ when } x = 0 \\ \dfrac{dy}{dx} = 3 \text{ when } x = 0 \end{cases}$

$[y = (x + 5)\,e^{\frac{2}{5}x}]$

14. The oscillations of a heavily damped pendulum satisfy the differential equation $\dfrac{d^2x}{dt^2} + 7\dfrac{dx}{dt} + 12x = 0$, where x cm is the displacement of the bob at time t seconds. The initial displacement is equal to $+3$ cm and the initial velocity $\left(\text{i.e. } \dfrac{dx}{dt}\right)$ is 6 cm/s. Solve the equation for x. $[x = 3(6\,e^{-3t} - 5\,e^{-4t})]$

15. If $\dfrac{d^2V}{dx^2} - w^2V = 0$ where w is a constant show that:

$V = V_0 \cosh wx + \dfrac{2}{w}\sinh wx$

given that when $x = 0$, $V = V_0$ and $\dfrac{dV}{dx} = 2$.

16. The charge q on a capacitor in a certain electrical circuit satisfies the differential equation $\dfrac{d^2q}{dt^2} + 3\dfrac{dq}{dt} + 4q = 0$. Initially

(i.e. when $t = 0$), $q = Q$ and $\dfrac{dQ}{dt} = 0$. Show that the charge in the circuit may be expressed as:

$$q = \frac{4}{\sqrt{7}}\,Q\,e^{-\frac{3}{2}t}\sin\!\left(\frac{\sqrt{7}}{2}t + 0.723\right)$$

17. The differential equation $I\dfrac{d^2\theta}{dt^2} + K\dfrac{d\theta}{dt} + F\theta = 0$ represents the motion of the pointer of a galvanometer about its position of equilibrium. I is the moment of inertia of the pointer about its pivot, K is the resistance due to friction at unit angular velocity and F is the force on the spring necessary to produce unit displacement. If $I = 0.006$, $K = 0.03$ and $F = 0.187\,5$ solve the equation for θ in terms of t given that when $t = 0$, $\theta = 0.2$ and $\dfrac{d\theta}{dt} = 0$. $\qquad$ $[\theta = e^{-2.5t}(0.2\cos 5t + 0.1\sin 5t)]$

18. A body moves in a straight line so that its distance S metres from the origin after time t seconds is given by $\dfrac{d^2S}{dt^2} + n^2 S = 0$, where n is a constant. Solve the equation for S given that $S = k$ and $\dfrac{dS}{dt} = 0$ when $t = \dfrac{2\pi}{n}$. $\qquad$ $[S = k\cos nt]$

19. The equation $\dfrac{d^2i}{dt^2} + \dfrac{R}{L}\dfrac{di}{dt} + \dfrac{1}{LC}i = 0$ represents a current i flowing in an electrical circuit containing resistance R, inductance L and capacitance C connected in series. Current $i = 0$ when time $t = 0$ and $\dfrac{di}{dt} = 100$ when $t = 0$. Inductance $L = 0.20$ henrys and capacitance $C = 80 \times 10^{-6}$ farads. Solve the equation for i when (a) $R = 100$ ohms and (b) $R = 223.6$ ohms.

(a) $[i = 100t\,e^{-250t}]$ $\quad$ (b) $\left[i = \dfrac{1}{10}(e^{-59t} - e^{-1059t})\right]$

20. The equation of motion of a body oscillating on the end of a spring is: $\dfrac{d^2x}{dt^2} + 225x = 0$, where x is the displacement in metres of the body from its equilibrium position after time t seconds. Find x in terms of t given that at time $t = 0$, $x = 1$ and $\dfrac{dx}{dt} = 0$. $\qquad$ $[x = \cos 15t]$

21. The equation: $\dfrac{d^2y}{dx^2} = k\sqrt{\left\{1 + \left(\dfrac{dy}{dx}\right)^2\right\}}$ on solution will give

that for the catenary. Solve it for y given that $\dfrac{dy}{dx} = 0$ when

$x = 0.$ $\left[y = \dfrac{1}{k}\cosh(kx) + B \right]$

Chapter 26

The solution of linear second order differential equations of the form $a(d^2y/dx^2) + b(dy/dx) + cy = f(x)$

26.1 Complementary function and particular integral

The differential equation

$$a\frac{d^2y}{dx^2} + b\frac{dy}{dx} + cy = f(x) \tag{1}$$

where a, b and c are constants, is a linear, second order differential equation with constant coefficients.

In order to solve this type of equation a substitution is made as follows. Let $u + v$ be substituted for y in equation (1). The equation then becomes:

$$a\frac{d^2}{dx^2}(u + v) + b\frac{d}{dx}(u + v) + c(u + v) = f(x)$$

Hence $\quad a\dfrac{d^2u}{dx^2} + a\dfrac{d^2v}{dx^2} + b\dfrac{du}{dx} + b\dfrac{dv}{dx} + cu + cv = f(x)$

i.e. $\quad \left(a\dfrac{d^2u}{dx^2} + b\dfrac{du}{dx} + cu\right) + \left(a\dfrac{d^2v}{dx^2} + b\dfrac{dv}{dx} + cv\right) = f(x)$

Let v be any solution of the equation (1) such that:

$$a\frac{d^2v}{dx^2} + b\frac{dv}{dx} + cv = f(x) \tag{2}$$

This would mean that

$$a\frac{d^2u}{dx^2} + b\frac{du}{dx} + cu = 0 \tag{3}$$

The general solution. u, of equation (3) will contain two unknown constants as required for the general solution of equation (1). (The method of solution of this type of equation was discussed in the previous chapter.)

371

If, also, the particular solution, v, of equation (2) can be found without containing any unknown constants then $y = u + v$ will give the general solution of equation (1). Section 26.2 following discusses methods whereby v may be determined.

The function u is called the **complementary function (C.F.)**

The function v is called the **particular integral (P.I.)**

The general solution y of a linear differential equation such as equation (1) is given by the sum of the complementary function and the particular integral.

$$y = \text{C.F.} + \text{P.I.}$$

or $$y = u + v$$

26.2   Methods of finding the particular integral

The function $f(x)$ on the right-hand side of equation (1) may be any non-zero expression, the more common expressions being a constant, a polynomial (i.e. of the form $p + qx + rx^2 + \ldots$, where p, q and r are constants), an exponential function, a sine or cosine function, or a sum or product of these functions. The method of finding the particular integral differs for each expression of $f(x)$ and it is usual to assume a form of particular integral which is suggested by $f(x)$. The following examples will make this clear.

(a) $f(x)$ = a constant

(i)   Straightforward case

If $f(x)$ is a constant then first try $v = k$, where k is a constant, as the particular integral. $v = k$ is substituted into equation (2) and k may be determined.

Problem 1. Solve $\dfrac{d^2 y}{dx^2} + 4\dfrac{dy}{dx} + 3y = 6.$

Let $v = k$ then from equation (2): $(D^2 + 4D + 3)k = 6$

Since $D^2(k) = 0$ and $D(k) = 0$, then $3k = 6$, i.e. $k = 2$

Hence the particular integral $v = 2$

The complementary function u is obtained as in the previous chapter.

The auxiliary equation is $m^2 + 4m + 3 = 0$

i.e. $(m + 3)(m + 1) = 0$

i.e. $m = -3$ and $m = -1$

Hence the complementary function $u = A\, e^{-3x} + B\, e^{-x}$

The general solution is given by $y = u + v$,

i.e. $$y = A e^{-3x} + B e^{-x} + 2$$

(ii) 'Snag' case

When $f(x)$ is a constant the substitution of $y = k$ will not always work. If the complementary function is found to contain a constant term (as in Problem 2) then try $y = kx$ as the particular integral.

Problem 2. (a) Find the general solution of $\dfrac{d^2 y}{dx^2} + 4\dfrac{dy}{dx} = 6$.

(b) Find the particular solution for (a) given that when $x = 0$, $y = 0$ and $\dfrac{dy}{dx} = 0$.

(a)   In D operator form the differential equation is $(D^2 + 4D)y = 6$

The auxiliary equation is $m^2 + 4m = 0$

i.e. $m(m + 4) = 0$

Hence $m = 0$ or $m = -4$

Hence the complementary function

$$u = A e^0 + B e^{-4x} = A + B e^{-4x}$$

If we assume a particular integral of $v = k$ in this case, we obtain $(D^2 + 4D)k = 6$ from equation (2) from which $0 = 6$ which is impossible. This occurs because the particular integral assumed, namely a constant, already occurs in the complementary function (i.e. as A)

In this case let $v = kx$, then $(D^2 + 4D)kx \equiv 6$,

i.e. $D^2(kx) + 4D(kx) \equiv 6$

$$D^2(kx) = 0 \text{ and } 4D(kx) = 4k$$

Hence $4k = 6$, i.e. $k = \dfrac{3}{2}$

Hence the particular integral, $v = \dfrac{3}{2}x$

The general solution, $y = u + v = A + B e^{-4x} + \dfrac{3}{2}x$

(b)   When $x = 0$, $y = 0$ and since $y = A + B e^{-4x} + \dfrac{3}{2}x$, then

$$0 = A + B \tag{1}$$

$$\frac{dy}{dx} = 0 - 4 B e^{-4x} + \frac{3}{2}$$

$$\frac{dy}{dx} = 0 \text{ when } x = 0, \text{ thus } 0 = -4B + \frac{3}{2}$$

i.e. $B = \frac{3}{8}$

From equation (1) $A = -\frac{3}{8}$

Hence the particular solution is $y = -\frac{3}{8} + \frac{3}{8}e^{-4x} + \frac{3}{2}x$

i.e. $y = \frac{3}{8}(e^{-4x} - 1) + \frac{3}{2}x$

Typical practical examples of the type of equation solved in Problems 1 and 2 are the equations for the charge q in a series-connected electrical circuit and the force equation for a 'mass, spring, damped' mechanical system, i.e.

$$L\frac{dq^2}{dt^2} + R\frac{dq}{dt} + \frac{1}{c}q = E \text{ and } m\frac{d^2s}{dt^2} + a\frac{ds}{dt} + ks = F$$

(b) $f(x) = $ a polynomial

If $f(x) = p + qx + rx^2 + \ldots$, where p, q and r are constants, then try $v = a + bx + cx^2 + \ldots$, where a, b and c are constants, as the particular integral.

For example, if $f(x) = 4 + 3x$ then try $v = a + bx$
             if $f(x) = 4x^2 + 2x - 1$ then try $v = ax^2 + bx + c$,
             if $f(x) = 2x^3 - 3$ then try $v = ax^3 + bx^2 + cx + d$,

and so on.

After substitution into equation (2), the coefficients of similar terms are equated, as shown in Problem 3.

Problem 3. Solve the equation $2\frac{d^2y}{dx^2} - \frac{dy}{dx} - 6y = 3x + 2$.

In D operator form, the equation is $(2D^2 - D - 6)y = 3x + 2$
The auxiliary equation is $2m^2 - m - 6 = 0$
i.e. $(2m + 3)(m - 2) = 0$

i.e. $m = -\frac{3}{2} \text{ or } m = 2$

The complementary function, $u = A e^{-\frac{3}{2}x} + B e^{2x}$

Let the particular integral $v = ax + b$, then substituting into equation (2) gives:

$$(2D^2 - D - 6)(ax + b) \equiv 3x + 2$$

i.e. $2D^2[ax + b] - D[ax + b] - 6[ax + b] \equiv 3x + 2$

$$2D^2(ax + b) = 0, \; D(ax + b) = a$$

Hence $0 - a - 6ax - 6b \equiv 3x + 2$

Equating the coefficients of x gives: $-6a = 3$

i.e.

$$a = -\frac{1}{2}$$

Equating the constant terms gives: $-a - 6b = 2$

i.e.

$$-\left(-\frac{1}{2}\right) - 6b = 2$$

$$-6b = \frac{3}{2}$$

$$b = -\frac{1}{4}$$

Hence the particular integral $v = -\dfrac{1}{2}x - \dfrac{1}{4}$

The general solution $y = u + v = A e^{-\frac{3}{2}x} + B e^{2x} - \dfrac{1}{2}x - \dfrac{1}{4}$

(c) $f(x)$ = an exponential function

(i)   *Straightforward case*

If $f(x) = C e^{px}$, where C and p are constants, then try as the particular integral $v = k e^{px}$. The reason for such a substitution is that all the differential coefficients of e^{px} are multiples of e^{px}.

Problem 4. Find the general solution of the equation

$$3\frac{d^2y}{dx^2} + \frac{dy}{dx} - 4y = 2 e^{-3x}.$$

Given the boundary conditions that when $x = 0$, $y = \dfrac{3}{5}$ and $\dfrac{dy}{dx} = -6\dfrac{4}{5}$, find the particular solution.

In D operator form the equation is $(3D^2 + D - 4)y = 2e^{-3x}$

The auxiliary equation is $3m^2 + m - 4 = 0$

i.e. $\qquad\qquad\qquad (m - 1)(3m + 4) = 0$

i.e. $\qquad\qquad\qquad m = 1 \text{ or } m = -\dfrac{4}{3}$

Hence the complementary function, $u = A\,e^x + B\,e^{-\frac{4}{3}x}$

Let the particular integral $v = k\,e^{-3x}$ then

$(3D^2 + D - 4)[k\,e^{-3x}] \equiv 2\,e^{-3x}$ from equation (2)

$D(k\,e^{-3x}) = -3k\,e^{-3x}$

$D^2(k\,e^{-3x}) = 9k\,e^{-3x}$

Hence $27k\,e^{-3x} - 3k\,e^{-3x} - 4k\,e^{-3x} \equiv 2\,e^{-3x}$

i.e. $\qquad\qquad\qquad 20k\,e^{-3x} \equiv 2\,e^{-3x}$

Hence $\qquad\qquad\qquad k = \dfrac{1}{10}$

Hence the particular integral, $v = \dfrac{1}{10}e^{-3x}$

The general solution, $y = u + v = A\,e^x + B\,e^{-\frac{4}{3}x} + \dfrac{1}{10}e^{-3x}$

When $x = 0$, $y = \dfrac{3}{5}$, and substituting in the general solution gives:

$$\dfrac{3}{5} = A + B + \dfrac{1}{10}$$

i.e. $\dfrac{1}{2} = A + B$ $\qquad\qquad\qquad\qquad\qquad\qquad$ (1)

$$\dfrac{dy}{dx} = A\,e^x - \dfrac{4}{3}B\,e^{-\frac{4}{3}x} - \dfrac{3}{10}e^{-3x}$$

When $x = 0$, $\dfrac{dy}{dx} = -6\dfrac{4}{5}$. Therefore $-6\dfrac{4}{5} = A - \dfrac{4}{3}B - \dfrac{3}{10}$

i.e. $\qquad\qquad\qquad -6\dfrac{1}{2} = A - \dfrac{4}{3}B$ $\qquad\qquad\qquad$ (2)

(1) − (2) gives $\dfrac{1}{2} - -6\dfrac{1}{2} = \left(1 - -\dfrac{4}{3}\right)B$

i.e. $\qquad\qquad\qquad 7 = \dfrac{7}{3}B$

i.e. $B = 3$

From (1), when $B = 3$, $A = -2\frac{1}{2}$

Hence the particular solution is $y = 3\,e^{-\frac{4}{3}x} - \frac{5}{2}\,e^x + \frac{1}{10}\,e^{-3x}$

(ii)   'Snag' cases

If $f(x) = C\,e^{px}$, where C and p are constants, and e^{px} occurs in the complementary function the particular integral cannot be found by assuming that $v = k\,e^{px}$, for on substituting, k becomes zero. In this case try $v = kx\,e^{px}$ for the particular integral (see Problem 5). If e^{px} **and** $x\,e^{px}$ both appear in the complementary function then try $v = kx^2\,e^{px}$ as the particular integral, and so on (see Problem 6).

Problem 5. Solve the differential equation $(D^2 - D - 2)y = 6\,e^{-x}$

The auxiliary equation is $m^2 - m - 2 = 0$

i.e. $(m - 2)(m + 1) = 0$

i.e. $m = 2$ and $m = -1$

Hence the complementary function, $u = A\,e^{2x} + B\,e^{-x}$

Since e^{-x} appears in the complementary function **and** in the right-hand side of the differential equation, let the particular integral $v = kx\,e^{-x}$
Then $(D^2 - D - 2)[kx\,e^{-x}] \equiv 6\,e^{-x}$
$D(kx\,e^{-x}) = (kx)(-e^{-x}) + (e^{-x})(k) = k\,e^{-x}(1 - x)$,
                              by the product rule
$D^2(kx\,e^{-x}) = D(k\,e^{-x}(1 - x)) = (k\,e^{-x})(-1) + (1 - x)(-k\,e^{-x})$
                    by the product rule
$$= -k\,e^{-x}(2 - x)$$
Hence $(D^2 - D - 2)[kx\,e^{-x}] = -k\,e^{-x}(2 - x) - k\,e^{-x}(1 - x)$
$$- 2kx\,e^{-x} \equiv 6\,e^{-x}$$
Thus $-2k\,e^{-x} + kx\,e^{-x} - k\,e^{-x} + kx\,e^{-x} - 2kx\,e^{-x} \equiv 6\,e^{-x}$

i.e. $-3k\,e^{-x} = 6\,e^{-x}$

i.e. $k = -2$

Hence the particular integral $v = -2x\,e^{-x}$

The general solution $y = u + v = A e^{2x} + B e^{-x} - 2x e^{-x}$

Problem 6. Solve $\dfrac{d^2 y}{dx^2} - 2 \dfrac{dy}{dx} + y = 5 e^x$.

In D operator form the differential equation is
$(D^2 - 2D + 1)y = 5 e^x$
The auxiliary equation is $m^2 - 2m + 1 = 0$
i.e. $(m - 1)(m - 1) = 0$
i.e. $m = 1$ twice
Hence the complementary function $u = (Ax + B) e^x$
Since e^x **and** $x e^x$ both appear in the complementary function, let
the particular integral $v = kx^2 e^x$

$\qquad$ Thus $(D^2 - 2D + 1)[kx^2 e^x] \equiv 5 e^x$

$D(kx^2 e^x) = (kx^2)(e^x) + (e^x)(2kx) \equiv k e^x(x^2 + 2x)$

$D^2(kx^2 e^x) = D(k e^x(x^2 + 2x)) = (k e^x)(2x + 2) + (x^2 + 2x)(k e^x)$
$\qquad\qquad\qquad = k e^x(x^2 + 4x + 2)$

Hence
$(D^2 - 2D + 1)[kx^2 e^x] = k e^x(x^2 + 4x + 2) - 2k e^x(x^2 + 2x)$
$\qquad\qquad\qquad\qquad\qquad\qquad + kx^2 e^x \equiv 5 e^x$

i.e. $2k e^x = 5 e^x$

i.e. $k = \dfrac{5}{2}$

Hence the particular integral $v = \dfrac{5}{2} x^2 e^x$

The general solution $y = u + v = (Ax + B) e^x + \dfrac{5}{2} x^2 e^x$

i.e. $y = e^x\left(Ax + B + \dfrac{5}{2} x^2\right)$

(d) $f(x) =$ **a sine or cosine**

(i)   *Straightforward case*

If $f(x) = k_1 \sin px + k_2 \cos px$, where p, k_1 and k_2 are constants
and either k_1 or k_2 may be zero, then try $v = A \sin px + B \cos px$
as the particular integral. The reason for this is that all the
differential coefficients of either $\sin px$ or $\cos px$ are multiples of
either $\sin px$ or $\cos px$. (Note that in the suggested substitution for

v it is insufficient to just let $v = A \sin pt$ or $v = B \cos pt$ even if either k_1 or k_2 are zero.) Substituting for v in the differential equation and comparing coefficients of $\sin px$ and $\cos px$ on both sides produces two equations to determine A and B.

Problem 7. Solve $2 \dfrac{d^2y}{dx^2} + 5 \dfrac{dy}{dx} - 3y = 4 \sin 2x$.

In D operator form the differential equation is

$(2D^2 + 5D - 3)y = 4 \sin 2x$

The auxiliary equation is $2m^2 + 5m - 3 = 0$

i.e. $\qquad\qquad\qquad (2m - 1)(m + 3) = 0$

i.e. $\qquad\qquad\qquad m = \dfrac{1}{2} \text{ or } m = -3$

Hence the complementary function $\boldsymbol{u = A\, e^{\frac{1}{2}x} + B\, e^{-3x}}$

Since we have used A and B as the constants for the complementary function, we will use P and Q as the constants for the particular integral.

Let the particular integral be $v = P \sin 2x + Q \cos 2x$

then $(2D^2 + 5D - 3)[P \sin 2x + Q \cos 2x] \equiv 4 \sin 2x$

$D(P \sin 2x + Q \cos 2x) = 2P \cos 2x - 2Q \sin 2x$

$D^2(P \sin 2x + Q \cos 2x) = D(2P \cos 2x - 2Q \sin 2x)$
$\qquad\qquad\qquad\qquad\qquad = -4P \sin 2x - 4Q \cos 2x$

Hence

$(2D^2 + 5D - 3)[P \sin 2x + Q \cos 2x] = -8P \sin 2x - 8Q \cos 2x$
$\qquad\qquad\qquad\qquad\qquad\qquad\qquad + 10P \cos 2x - 10Q \sin 2x$
$\qquad\qquad\qquad\qquad\qquad\qquad\qquad - 3P \sin 2x - 3Q \cos 2x$
$\qquad\qquad\qquad\qquad\qquad\qquad\qquad \equiv 4 \sin 2x$

Equating coefficients of $\sin 2x$ gives: $-11P - 10Q = 4$ $\qquad$ (1)

Equating coefficients of $\cos 2x$ gives: $\quad 10P - 11Q = 0$ $\qquad$ (2)

$\qquad\qquad 10 \times (1) \text{ gives: } -110P - 100Q = 40$ $\qquad$ (3)

$\qquad\qquad 11 \times (2) \text{ gives: } \quad 110P - 121Q = 0$ $\qquad$ (4)

$\qquad\qquad (3) + (4) \text{ gives: } \qquad\quad -221Q = 40$

i.e. $\qquad\qquad\qquad\qquad\qquad\qquad Q = \dfrac{-40}{221}$

Substituting $Q = \dfrac{-40}{221}$ into equation (1) or (2) gives:

$$P = \dfrac{-44}{221}$$

Hence the particular integral $v = \dfrac{4}{221}(-11 \sin 2x - 10 \cos 2x)$

The general solution

$$y = u + v = A\,e^{\frac{1}{2}x} + B\,e^{-3x} - \dfrac{4}{221}(11 \sin 2x + 10 \cos 2x)$$

(ii) '*Snag*' *case*

If $f(x) = k_1 \sin px + k_2 \cos px$ and $\sin px$ and/or $\cos px$ occur in the complementary function then let the particular integral $v = x(A \sin px + B \cos px)$. This case is demonstrated in Problem 8.

Problem 8. Find the general solution of $\dfrac{d^2 y}{dx^2} + 9y = 12 \cos 3x$. If $y = 2$ and $\dfrac{dy}{dx} = 3$ when $x = 0$, determine the particular solution.

In D operator form the differential equation is

$(D^2 + 9)y = 12 \cos 3x$

The auxiliary equation is $m^2 + 9 = 0$
i.e. $m = \sqrt{-9} = \pm j3$

Hence the complementary function $u = e^0(C \cos 3x + D \sin 3x)$
$$= C \cos 3x + D \sin 3x$$

Since $\cos 3x$ occurs in the complementary function and in the right-hand side of the differential equation, let the particular integral $v = x(A \sin 3x + B \cos 3x)$.

Hence $(D^2 + 9)\,[x(A \sin 3x + B \cos 3x)] \equiv 12 \cos 3x$

$D[x(A \sin 3x + B \cos 3x)] = (x)(3A \cos 3x - 3B \sin 3x)$
$\qquad\qquad\qquad\qquad\qquad + (A \sin 3x + B \cos 3x)(1),$
$\qquad\qquad\qquad\qquad\qquad$ by the product rule

$D^2[x(A \sin 3x + B \cos 3x)] = (x)(-9A \sin 3x - 9B \cos 3x)$
$\qquad\qquad\qquad\qquad\qquad + (3A \cos 3x - 3B \sin 3x)(1)$
$\qquad\qquad\qquad\qquad\qquad + (3A \cos 3x - 3B \sin 3x)$

Hence
$$(D^2 + 9)[x(A \sin 3x + B \cos 3x)] = -9Ax \sin 3x - 9Bx \cos 3x$$
$$+ 6A \cos 3x - 6B \sin 3x$$
$$+ 9Ax \sin 3x + 9Bx \cos 3x$$
$$\equiv 12 \cos 3x$$

i.e. $6A \cos 3x - 6B \sin 3x \equiv 12 \cos 3x$

Equating coefficients of $\cos 3x$ gives: $\quad 6A = 12$
i.e. $\qquad\qquad\qquad\qquad\qquad\qquad A = 2$

Equating coefficients of $\sin 3x$ gives: $-6B = 0$
i.e. $\qquad\qquad\qquad\qquad\qquad\qquad B = 0$

Hence the particular integral $v = x(2 \sin 3x)$

The general solution $y = u + v = C \cos 3x + D \sin 3x + 2x \sin 3x$

When $x = 0$, $y = 2$. Hence $2 = C \cos 0 + D \sin 0 + 0$
i.e. $\qquad\qquad\qquad\qquad C = 2$

$$\frac{dy}{dx} = -3C \sin 3x + 3D \cos 3x + (2x)(3 \cos 3x) + (2 \sin 3x)$$

When $x = 0$, $\dfrac{dy}{dx} = 3$.

Hence $3 = -3C \sin 0 + 3D \cos 0 + 0 + 2 \sin 0$
i.e. $\quad 3 = 3D$
$\qquad D = 1$

Hence the particular solution is $y = 2 \cos 3x + \sin 3x + 2x \sin 3x$
i.e. $\qquad\qquad\qquad\qquad y = 2 \cos 3x + (1 + 2x) \sin 3x$

Typical practical examples of the type of equation solved in Problems 7 and 8 include:

(i)   the equation of motion of a body: $\dfrac{d^2y}{dt^2} + n^2y = k \cos \omega t$, and

(ii) the equation for variation of charge q in an alternating current circuit containing L, R and C in series:

$$L\frac{d^2q}{dt^2} + R\frac{dq}{dt} + \frac{1}{C}q = V_0 \sin \omega t$$

(e) $f(x) =$ a sum or product

In all of the straightforward cases of finding particular integrals it is noticed that a form of particular integral is assumed which is suggested by $f(x)$ but which contains undetermined coefficients.

If $f(x)$ consists of a **sum** of terms the particular integral is the sum of the particular integrals corresponding to the separate terms. Thus, for example, if $f(x) = 2x + 3 \sin 4x$ then let the particular integral $v = (ax + b) + (C \sin 2x + D \cos 2x)$

If $f(x)$ is a **product** of two terms then the particular integral assumed is that suggested by $f(x)$. Thus, for example, if $f(x) = e^{\alpha x} \cos \beta x$ then let the particular integral $v = e^{\alpha x}(A \cos \beta x + B \sin \beta x)$

Problem 9. Solve $(D^2 - 4D + 4)y = 4x + 3 \cos 2x$.

The auxiliary equation is $m^2 - 4m + 4 = 0$
i.e. $(m - 2)(m - 2) = 0$
i.e. $m = 2$ twice

The complementary function $\boldsymbol{u = (Ax + B)\, e^{2x}}$

Let the particular integral $v = ax + b + C \cos 2x + D \sin 2x$,

then $(D^2 - 4D + 4)[v] = 4x + 3 \cos 2x$

$D(v) = a - 2C \sin 2x + 2D \cos 2x$
$D^2(v) = -4C \cos 2x - 4D \sin 2x$

Hence
$(D^2 - 4D + 4)[v] = -4C \cos 2x - 4D \sin 2x - 4a + 8C \sin 2x$
$\qquad\qquad\qquad\quad - 8D \cos 2x + 4ax + 4b + 4C \cos 2x$
$\qquad\qquad\qquad\quad + 4D \sin 2x \equiv 4x + 3 \cos 2x$

Equating constant terms gives: $-4a + 4b = 0$
Equating the coefficients of x gives: $4a = 4$
i.e. $\boldsymbol{a = 1}$ and thus $\boldsymbol{b = 1}$

Equating coefficients of $\cos 2x$ gives: $-4C - 8D + 4C = 3$
i.e. $8D = 3$

$$D = \frac{-3}{8}$$

Equating coefficients of $\sin 2x$ gives: $-4D + 8C + 4D = 0$
i.e. $8C = 0$
$$C = 0$$

Hence the particular integral $\boldsymbol{v = x + 1 - \dfrac{3}{8} \sin 2x}$

The general solution $\boldsymbol{y = u + v = (Ax + B)\, e^{2x} + x + 1 - \dfrac{3}{8} \sin 2x}$

Problem 10. Solve $\dfrac{d^2 y}{dx^2} + 2\dfrac{dy}{dx} + 2y = 6\, e^x \sin 2x$.

In D operator form the differential equation is

$(D^2 + 2D + 2)y = 6 e^x \sin 2x$

The auxiliary equation is $m^2 + 2m + 2 = 0$

i.e. $m = \dfrac{-2 \pm \sqrt{[2^2 - 4(1)(2)]}}{2} = -1 \pm j1$

The complementary function $u = e^{-x}(C \cos x + D \sin x)$

Let the particular integral $v = e^x(A \cos 2x + B \sin 2x)$

then $(D^2 + 2D + 2)[v] \equiv 6 e^x \sin 2x$

$D(v) = (e^x)(-2A \sin 2x + 2B \cos 2x) + e^x(A \cos 2x + B \sin 2x)$
$[= e^x(-2A + B) \sin 2x + e^x(2B + A) \cos 2x]$

$D^2(v) = (e^x)(-4A \cos 2x - 4B \sin 2x)$
$\quad + (e^x)(-2A \sin 2x + 2B \cos 2x)$
$\quad + (e^x)(-2A \sin 2x + 2B \cos 2x)$
$\quad + (e^x)(A \cos 2x + B \sin 2x)$
$\quad = e^x[(-3A + 4B) \cos 2x + (-3B - 4A) \sin 2x]$

Hence

$(D^2 + 2D + 2)[v] = e^x(-3A + 4B) \cos 2x + e^x(-3B - 4A) \sin 2x$
$\quad + 2 e^x(-2A + B) \sin 2x + 2 e^x(2B + A) \cos 2x$
$\quad + 2 e^x A \cos 2x + 2 e^x B \sin 2x$
$\quad \equiv 6 e^x \sin 2x$

Equating coefficients of $e^x \sin 2x$ gives:

$-3B - 4A - 4A + 2B + 2B = 6$

i.e. $-8A + B = 6$ (1)

Equating coefficients of $e^x \cos 2x$ gives:

$-3A + 4B + 4B + 2A + 2A = 0$

i.e. $A + 8B = 0$ (2)

Solving equations (1) and (2) gives $A = \dfrac{-48}{65}$ and $B = \dfrac{6}{65}$

Hence the particular integral $v = e^x\left(\dfrac{-48}{65} \cos 2x + \dfrac{6}{65} \sin 2x\right)$

The general solution is

$y = u + v = e^{-x}(C \cos x + D \sin x) + \dfrac{e^x}{65}(6 \sin 2x - 48 \cos 2x)$

Further problems on solving differential equations of the type
$a\dfrac{d^2 y}{dx^2} + b\dfrac{dy}{dx} + cy = f(x)$ *may be found in Section 26.4 (Problems 1 to 26),*
page 384.

26.3   Summary of the procedure to solve differential equations of the form $a\dfrac{d^2y}{dx^2} + b\dfrac{dy}{dx} + cy = f(x)$

1. Rewrite the given differential equation as $(aD^2 + bD + c)y = f(x)$
2. Substitute m for D and solve the **auxiliary equation** $am^2 + bm + c = 0$ for m
3. Obtain the **complementary function, u.** (This is exactly the same procedure discussed in Section 25.3, although it was not referred to at that stage as the complementary function.)
4. To find the **particular integral, v,** firstly assume a particular integral which is suggested by $f(x)$ but which contains undetermined coefficients. A list of suggested substitutions is on page 385.
5. Substitute the suggested particular integral into the differential equation $a\dfrac{d^2v}{dx^2} + b\dfrac{dv}{dx} + cv = f(x)$ and equate relevant coefficients to find the constants introduced.
6. The **general solution** is given by $y =$ complementary function + particular integral, i.e. $y = u + v$
7. Given sufficient **boundary conditions** the arbitrary constants in the complementary function may be determined to give the **particular solution.**

26.4   Further problems

In Problems 1 to 12 find the general solution of the given differential equations.

1. (a) $\dfrac{d^2y}{dx^2} - 5\dfrac{dy}{dx} + 6y = 3$ $[y = A\,e^{3x} + B\,e^{2x} + \tfrac{1}{2}]$

(b) $\dfrac{d^2y}{dt^2} + 2\dfrac{dy}{dt} + 2y = 5$ $[y = e^{-t}(C\cos t + D\sin t) + \tfrac{5}{2}]$

2. (a) $(D^2 + 9D)y = 2$ where $D \equiv \dfrac{d}{dx}$ $[y = A + B\,e^{-9x} + \tfrac{2}{9}x]$

(b) $\dfrac{d^2x}{d\theta^2} + 7\dfrac{dx}{d\theta} = 7$ $[x = A + B\,e^{-7\theta} + \theta]$

3. (a) $4\dfrac{d^2y}{dx^2} - 12\dfrac{dy}{dx} + 9y = x$

$\left[y = (A + Bx)\,e^{\frac{3}{2}x} + \dfrac{1}{27}(3x + 4) \right]$

Type	Straightforward cases Try as particular integral:	'Snag' cases Try as particular integral:
(a) $f(x) = $ a constant	$v = k$	$v = kx$ (used when C.F. contains a constant)
(b) $f(x) = $ polynomial $(f(x) = L + Mx + Nx^2$ $+ \ldots$ where any of the coefficients may be zero)	$v = a + bx + cx^2 + \ldots$	
(c) $f(x) = $ an exponential function $(f(x) = A\,e^{ax})$	$v = k\,e^{ax}$	$v = kx\,e^{ax}$ (used when e^{ax} appears in the C.F.) $v = kx^2\,e^{ax}$ (used when e^{ax} **and** $x\,e^{ax}$ both appear in the C.F.), and so on
(d) $f(x) = $ a sine or cosine $(f(x) = a \sin px +$ $b \cos px,$ where a or b may be zero)	$v = A \sin px +$ $B \cos px$	$v = x(A \sin px +$ $B \cos px)$ (used when $\sin px$ and/or $\cos px$ appears in the C.F.)

(e) $f(x) = $ a sum
e.g.
  (i) $f(x) = 2x^2 + 5 \cos 3x$ (i) $v = ax^2 + bx + c$
                                            $+ d \cos 3x$
                                            $+ e \sin 3x$
  (ii) $f(x) = x + 1 - e^{-x}$ (ii) $v = ax + b + ce^{-x}$
  $f(x) = $ a product $v = e^{2x}(A \cos 4x +$
  e.g. $f(x) = 3\,e^{2x} \sin 4x$ $B \sin 4x)$

     (b) $(D^2 + 3D - 4)y = 2x + 3$
     $[y = A\,e^x + B\,e^{-4x} - \frac{1}{8}(4x + 9)]$

  4. (a) $2\dfrac{d^2y}{dx^2} + 3\dfrac{dy}{dx} + 2y = x^2 + 2x + 1$

$$\left[y = e^{-\frac{3}{4}x}\left(C \cos \frac{\sqrt{7}}{4}x + D \sin \frac{\sqrt{7}}{4}x\right) + \frac{1}{2}x^2 - \frac{1}{2}x + \frac{1}{4}\right]$$

     (b) $(2D^2 + 5D - 3)y = 2x^3 - x + 7$

$$\left[y = A\,e^{\frac{1}{2}x} + B\,e^{-3x} - \frac{1}{27}(18x^3 + 90x^2 + 363x + 788)\right]$$

5. (a) $\dfrac{d^2y}{dx^2} + 3\dfrac{dy}{dx} - 10y = 2\,e^{3x}$

$[\,y = A\,e^{2x} + B\,e^{-5x} + \tfrac{1}{4}e^{3x}\,]$

(b) $9\dfrac{d^2x}{dt^2} - 6\dfrac{dx}{dt} + x = 5\,e^{2t}$ $[\,x = (A + Bt)\,e^{\frac{1}{3}t} + \tfrac{1}{5}e^{2t}\,]$

6. (a) $\dfrac{d^2y}{dx^2} + \dfrac{dy}{dx} - 6y = e^{-3x}$

$[\,y = A\,e^{2x} + B\,e^{-3x} - \tfrac{1}{5}x\,e^{-3x}\,]$

(b) $(2D^2 - D - 3)y = 5\,e^{-x}$ $[\,y = A\,e^{-x} + B\,e^{\frac{3}{2}x} - x\,e^{-x}\,]$

7. (a) $\dfrac{d^2y}{dx^2} + 4\dfrac{dy}{dx} + 4y = 4\,e^{-2x}$

$[\,y = (A + Bx)\,e^{-2x} + 2x^2\,e^{-2x}\,]$

(b) $(4D^2 + 20D + 25)y = 8\,e^{-\frac{5}{2}x}$

$[\,y = (A + Bx)\,e^{-\frac{5}{2}x} + x^2\,e^{-\frac{5}{2}x}\,]$

8. (a) $2\dfrac{d^2x}{dt^2} + 5\dfrac{dx}{dt} - 3x = 2\sin 3t$

$\left[\,x = A\,e^{\frac{1}{2}t} + B\,e^{-3t} - \dfrac{1}{111}(7\sin 3t + 5\cos 3t)\,\right]$

(b) $(9D^2 - 6D + 1)y = 40\cos 2x$

$[\,y = (A + Bx)\,e^{\frac{1}{3}x} - (0.350\,6\sin 2x + 1.022\,6\cos 2x)\,]$

9. (a) $\dfrac{d^2y}{dx^2} + 4y = 2\sin 2x$

$[\,y = C\cos 2x + D\sin 2x - \tfrac{1}{2}x\cos 2x\,]$

(b) $(D^2 + 6D + 13)y = \cos 2x$

$\left[\,y = e^{-3x}(C\cos 2x + D\sin 2x) + \dfrac{x}{26}(2\sin 2x + 3\cos 2x)\,\right]$

10. (a) $15\dfrac{d^2y}{dx^2} - 2\dfrac{dy}{dx} - y = 3x + 65\sin x$

$[\,y = A\,e^{-\frac{1}{5}x} + B\,e^{\frac{1}{3}x} - 3x + 6 - 4\sin x + \tfrac{1}{2}\cos x\,]$

(b) $(4D^2 + 12D + 9)y = e^x\cos x$

$\left[\,y = (A + Bx)\,e^{-\frac{3}{2}x} + \dfrac{e^x}{841}(20\sin x + 21\cos x)\,\right]$

11. (a) $\dfrac{d^2x}{dt^2} - 2\dfrac{dx}{dt} + 2x = e^t\sin t$

$[\,x = e^t(C\cos t + D\sin t) - \tfrac{1}{2}e^t t\cos t\,]$

(b) $(2D^2 - 5D - 7)y = 3\sin x + 4\cos x$

$\left[\,y = A\,e^{\frac{7}{2}x} + B\,e^{-x} - \dfrac{1}{106}(47\sin x + 21\cos x)\,\right]$

12. (a) $\dfrac{d^2y}{dx^2} - 3\dfrac{dy}{dx} + 2y = x^2 + e^x$

$$\left[y = A\,e^x + B\,e^{2x} + \frac{1}{2}x^2 + \frac{3}{2}x + \frac{7}{4} - x\,e^x \right]$$

(b) $(D^2 - 1)y = \sinh 2x$ $\left[y = A\,e^x + B\,e^{-x} + \dfrac{1}{3}\sinh 2x \right]$

In Problems 13 to 19 find the particular solution of the given differential equations for the stated boundary conditions.

13. $2\dfrac{d^2y}{dx^2} + 5\dfrac{dy}{dx} - 3y = 9$: when $x = 0$, $y = 0$ and $\dfrac{dy}{dx} = 2$

$$\left[y = \frac{1}{7}(22\,e^{\frac{1}{2}x} - e^{-3x}) - \cdot \right]$$

14. $(D^2 + 1)y = 4$; when $x = 0$, $y = 0$ and $D(y) = 3$ $\left(D \equiv \dfrac{d}{dx} \right)$

$[y = 3\sin x - 4\cos x + 4]$

15. $3\dfrac{d^2x}{dt^2} + \dfrac{dx}{dt} - 4x = 2\,e^t$; when $t = 0$, $x = 0$ and $\dfrac{dx}{dt} = -5$

$[x = 3\,e^{-\frac{4}{3}t} - 2\,e^t - e^{-t}]$

16. $(D^2 - 2D + 1)y = 5\,e^x$; when $x = 0$, $y = 2$ and $D(y) = 3$

$$\left[y = (2 + x)\,e^x + \frac{5}{2}x^2\,e^x \right]$$

17. $(D^2 - 5D + 6)y = -3\sin x$; when $x = 0$, $y = 0$ and $D(y) = \dfrac{2}{5}$

$$\left[y = \frac{1}{10}\{e^{3x} + 2\,e^{2x} - 3(\sin x + \cos x)\} \right]$$

18. $\dfrac{d^2\theta}{dt^2} - 6\dfrac{d\theta}{dt} + 10\theta = 20 - e^{2t}$; when $t = 0$, $\theta = 4$ and $\dfrac{d\theta}{dt} = \dfrac{25}{2}$

$$\left[\theta = e^{3t}\!\left(\frac{5}{2}\cos t + 6\sin t \right) + 2 - \frac{1}{2}e^{2t} \right]$$

19. $2\dfrac{d^2x}{dt^2} - \dfrac{dx}{dt} - 6x = 12\,e^t\cos t$; when $t = 0$, $x = \dfrac{-13}{29}$ and

$\dfrac{dx}{dt} = \dfrac{-169}{29}$ $\left[x = 2\,e^{-\frac{3}{2}t} - e^{2t} + \dfrac{e^t}{29}(18\sin t - 42\cos t) \right]$

20. The charge q in an electrical circuit of time t satisfies the differential equation $L\dfrac{d^2q}{dt^2} + R\dfrac{dq}{dt} + \dfrac{1}{C}q = E$, where L, R, C and E are constants. $L = 0.9$, $C = 40 \times 10^{-6}$ and $E = 100$.

Solve the equation for q when (a) $R = 300$ and (b) R is negligible. Assume that when $t = 0$, $q = 0$ and $\dfrac{dq}{dt} = 0$.

(a) $\left[q = \dfrac{1}{250} - \left(\dfrac{1}{250} + \dfrac{2}{3} t \right) e^{-\frac{500}{3} t} \right]$

(b) $\left[q = \dfrac{1}{250} \left(1 - \cos \dfrac{500}{3} t \right) \right]$

21. Solve the following differential equation representing the motion of a body:

$$\dfrac{d^2 y}{dt^2} + n^2 y = a \sin pt$$

where $n \neq 0$ and $p^2 \neq n^2$, given that when $t = 0$, $y = 0$ and $\dfrac{dy}{dt} = 0$. $\left[y = \dfrac{a}{n^2 - p^2} \left(\sin pt - \dfrac{p}{n} \sin nt \right) \right]$

22. In a galvanometer the deflection θ satisfies the differential equation:

$$\dfrac{d^2 \theta}{dt^2} + 2 \dfrac{d\theta}{dt} + \theta = 4$$

Solve the equation for θ given that when $t = 0$, $\theta = 0$ and $\dfrac{d\theta}{dt} = 0$. $[\theta = 4 - 4 e^{-t}(1 + t)]$

23. Solve the equation $\dfrac{d^2 x}{dt^2} + n^2 x = k \cos \omega t$ representing the equation of motion of a body, given that when $t = 0$, $x = 0$ and $\dfrac{dx}{dt} = 0$. $\left[x = \dfrac{k}{n^2 - \omega^2} (\cos \omega t - \cos nt) \right]$

24. The differential equation describing the variation of capacitor charge in an alternating current circuit containing inductance L, resistance R and capacitance C in series is:

$$L \dfrac{d^2 q}{dt^2} + R \dfrac{dq}{dt} + \dfrac{1}{C} q = V_0 \sin \omega t$$

Find an expression for the charge q of the capacitor in the circuit at any time t seconds, given that when $t = 0$, $q = 0$ and $\dfrac{dq}{dt} = -44.29$, in the following case: $R = 20$ ohms, $L = 10 \times 10^{-3}$ henrys, $C = 100 \times 10^{-6}$ farads, $V_0 = 500$ volts and $\omega = 250$

$[q = (0.022\,15 - 32.53t) e^{-1\,000t} + 0.0415\,2 \sin 250t$
$\qquad\qquad\qquad\qquad\qquad\qquad - 0.022\,15 \cos 250t]$

25. The equation of motion of a particle moving in a straight line under damping is:

$$2\frac{d^2x}{dt^2} + 2\frac{dx}{dt} + x = k \sin t$$

where k is a constant. Show that if when $t = 0$, $x = n$ and $\frac{dx}{dt} = 0$, x is given by:

$$x = e^{-\frac{1}{2}t}\left\{\left(n + \frac{2}{5}k\right)\cos\frac{t}{2} + \left(n + \frac{4}{5}k\right)\sin\frac{t}{2}\right\}$$
$$- \frac{k}{5}\sin t - \frac{2k}{5}\cos t$$

26. The motion of a mass vibrating in a given way is shown by the equation:

$$\frac{d^2y}{dt^2} + 10\frac{dy}{dt} + 81y = 4 \sin 3t$$

Solve the equation for y.

$$\left[y = e^{-5}(C \cos \sqrt{56}\,t + D \sin \sqrt{56}\,t) + \frac{2}{507}(12 \sin 3t - 5 \cos 3t)\right]$$

Chapter 27

The Fourier series for periodic functions
of period 2π

27.1   Introduction

There are many phenomena that are studied in engineering and science which
are periodic in nature. Some examples include the current and voltage in an
alternating current circuit, the displacement, velocity and acceleration of a
slider-crank mechanism of a reciprocating engine, acoustic waves and
vibrating systems in general.

The best known of mathematical periodic functions are the sine and cosine
functions and it thus seems logical that these functions be used to model
such practical situations as mentioned above. Fourier series is a mode of
analysing a periodic function into its constituent components (called its
fundamental and harmonics – see Section 27.2) and is a technique developed
and named after the French mathematician and physicist Jean Baptiste
Fourier (1768–1830). Fourier used the series initially to solve problems in
the theory of heat conduction. However, Fourier series have found application in many other fields, such as analysis work in electrical and mechanical
vibrations, bending of beams, and in radio waves. It is an important function
of many electronic devices to send signals without distortion. The result of
amplification of a waveform can be studied by Fourier analysis. The original
wave may be analysed into its harmonics and the amplification of each
harmonic calculated from the characteristics of the amplifying device. It is
possible to determine from a Fourier analysis how much of each harmonic
is present in the amplified wave and hence the degree of distortion produced.
The range of freedom from such distortion is termed the bandwidth of
the device. It is important to know how much the amplitude and phase of
the harmonics can be changed (many harmonics can be discarded) before the
quality of reproduction suffers noticeably. The effects of filters can also
be studied by first analysing the input wave into its constituent
components.

Fourier series can be applied to many more functions than can Taylor's
and Maclaurin's series (see Chapter 6), and while the theory of the analysis
is complicated the application of these series is quite simple. To appreciate
Fourier series it is first essential to

(a)   understand what a periodic function is (see Section 27.2, following),

(b)   be able to evaluate definite integrals, such as

$$\int_{-\pi}^{\pi} \sin mx \cos nx \, dx$$

(see Section 27.3, following), and sometimes to
(c)   use integration by parts (see Chapter 21)

This chapter, together with Chapters 28 to 31, attempts to present the basic theory of Fourier series, the understanding of which leads to the application of the analysis to the behaviour of the physical systems mentioned above.

27.2   Periodic functions

A function $f(x)$ is said to be periodic if $f(x + l) = f(x)$, for all values of x, where l is some positive number. l is the interval between two successive repetitions and is called the **period** of the function $f(x)$. Figure 27.1 shows a periodic function $f(x)$ having a period l.

Familiar examples of periodic functions are the sine and cosine functions and graphs of $y = \sin x$ and $y = \cos x$, as shown in Figs 27.2(a) and (b). Each of the graphs repeats itself at regular intervals of 2π, i.e. $\sin x = \sin(x + 2\pi) = \sin(x + 4\pi)$, etc. and $\cos(x + 2\pi) = \cos(x + 4\pi)$, etc.

It was shown in *Technician mathematics 3*, Chapter 14, that generally, if $y = \sin \omega t$ or $y = \cos \omega t$ then the period of the waveform is $\dfrac{2\pi}{\omega}$. A graph of $y = \sin 2x$ is shown in Fig. 27.3(a) and is of period $\dfrac{2\pi}{2}$, i.e. π. A graph of $y = 2 \cos 3x$ is shown in Fig. 27.3(b) and is of period $\dfrac{2\pi}{3}$.

If a function is a constant, such as $y = 3$, then this is also considered as a periodic function, the period being infinite.

Fig. 27.1

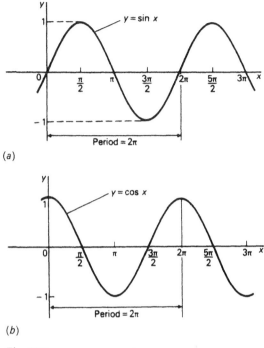

(a)

(b)

Fig. 27.2

Continuity

If a graph of a function has no sudden jumps or breaks it is called a **continuous function**, examples being the graphs shown in Figs 27.2 and 27.3 of sine and cosine functions. It is possible for a function to have what is called a **finite discontinuity**, which means that the function makes a finite jump at some point, or points in the interval. A typical square wave function is shown in Fig. 27.4 where finite discontinuities are found at $x = \pi$, 2π, 3π and so on. The function has a period of 2π and is defined by $f(x) = \begin{cases} +k, & 0 < x < \pi \\ -k, & \pi < x < 2\pi \end{cases}$.

Trigonometric series

If, in the equation $y = x$, for example, we let x be any given value then there is only one corresponding value of y. The function y is then called a **one-valued function**. In contrast, in the equation $x = \sin y$, i.e. $y = \arcsin x$, for example, to each value of x between -1 and $+1$ there corresponds an infinite number of values of y. Thus the function y is then called a **many-valued function**.

    If a finite, one-valued function $f(x)$ recurs periodically over successive

Fig. 27.3

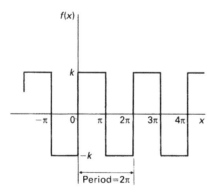

Fig. 27.4

intervals of 2π it is possible to represent it as a series of the form:

$$f(x) = a_0 + a_1 \cos x + a_2 \cos 2x + a_3 \cos 3x + \cdots$$
$$+ b_1 \sin x + b_2 \sin 2x + b_3 \sin 3x + \cdots \qquad (1)$$

where $a_0, a_1, a_2, \ldots b_1, b_2 \ldots$ are real constants.

Such a series is called a **trigonometric series** and a_n and b_n are called the **coefficients** of the series. If the coefficient can be determined, the series is then called the **Fourier series** corresponding to $f(x)$ and its coefficients are called the **Fourier coefficients**.

Alternatively equation (1) may be expressed as:

$$f(x) = a_0 + (a_1 \cos x + b_1 \sin x) + (a_2 \cos 2x + b_2 \sin 2x) + \cdots$$

i.e. $$f(x) = a_0 + c_1 \sin(x + \alpha_1) + c_2 \sin(2x + \alpha_2) + \cdots$$

$$+ c_n \sin(nx + \alpha_n) \quad (2)$$

where $c_1 = \sqrt{(a_1^2 + b_1^2)}$, $c_n = \sqrt{(a_n^2 + b_n^2)}$ and $\alpha_n = \arctan \dfrac{a_n}{b_n}$

[Note that if $R \sin(\omega t + \alpha) = b \sin \omega t + a \cos \omega t$ then $b = R \cos \alpha$, $a = R \sin \alpha$, $R = \sqrt{(a^2 + b^2)}$ and $\alpha = \arctan \dfrac{a}{b}$. See *Technician mathematics 3*, Chapter 16.]

Periodic functions that occur in practice are often rather complicated and it is therefore desirable to represent these functions in terms of simple periodic functions. Almost any periodic function $f(x)$ of period 2π that appears in applications can be represented by a trigonometric series such as that of equation (1) or equation (2).

A great advantage of a Fourier series is that it can be applied to functions which are discontinuous, that is, functions that have a finite discontinuity.

Harmonics
In the trigonometric series of equations (1) and (2), $(a_1 \cos x + b_1 \sin x)$ or $c_1 \sin(x + \alpha_1)$ is called the **first harmonic** or the **fundamental**, $(a_2 \cos 2x + b_2 \sin 2x)$ or $c_2 \sin(2x + \alpha_2)$ is called the **second harmonic**, and so on. $c_1, c_2, \ldots$ are the **amplitudes** of the various harmonics.

27.3   Evaluation of definite integrals involving trigonometric functions

The following definite integrals, discussed in Chapters 17 and 18, are used when calculating the coefficients $a_0, a_1, a_2, \ldots b_1, b_2, \ldots$ of the trigonometric series of the form stated in equation (1). If m and n are any integers (except zero) then:

(a) $$\int_{-\pi}^{\pi} \cos nx \, dx = 0 \qquad (3)$$

(b)
$$\int_{-\pi}^{\pi} \sin nx \, dx = 0 \tag{4}$$

(c)
$$\int_{-\pi}^{\pi} \cos^2 nx \, dx = \int_{-\pi}^{\pi} \tfrac{1}{2}(1 + \cos 2 \, nx) \, dx$$

$$= \frac{1}{2}\left[x + \frac{\sin 2 \, nx}{2n} \right]_{-\pi}^{\pi}$$

$$= \pi \quad \text{if } n \neq 0 \tag{5}$$

(d)
$$\int_{-\pi}^{\pi} \sin^2 nx \, dx = \int_{-\pi}^{\pi} \tfrac{1}{2}(1 - \cos 2 \, nx) \, dx$$

$$= \frac{1}{2}\left[x - \frac{\sin 2 \, nx}{2n} \right]_{-\pi}^{\pi}$$

$$= \pi \quad \text{if } n \neq 0 \tag{6}$$

(e)
$$\int_{-\pi}^{\pi} \cos mx \cos nx \, dx = \int_{-\pi}^{\pi} \tfrac{1}{2}[\cos(m + n)x + \cos(m - n)x] \, dx$$

$$= \frac{1}{2}\left[\frac{\sin(m + n)x}{(m + n)} + \frac{\sin(m - n)x}{(m - n)} \right]_{-\pi}^{\pi}$$

$$\left. \begin{aligned} &= 0 \quad \text{if } m \neq n \\ &= \pi \quad \text{if } m = n \end{aligned} \right\} \qquad \begin{aligned} &\text{(7(a))} \\ &\text{(7(b))} \end{aligned}$$

[Note that if $m = n$, the initial integral becomes $\displaystyle\int_{-\pi}^{\pi} \cos^2 nx \, dx = \pi$ if $n \neq 0$
(see (c) above)]

(f)
$$\int_{-\pi}^{\pi} \sin mx \sin nx \, dx = \int_{-\pi}^{\pi} -\tfrac{1}{2}[\cos(m + n)x - \cos(m - n)x] \, dx$$

$$= -\frac{1}{2}\left[\frac{\sin(m + n)x}{(m + n)} - \frac{\sin(m - n)x}{(m - n)} \right]_{-\pi}^{\pi}$$

$$\left. \begin{aligned} &= 0 \quad \text{if } m \neq n \\ &= \pi \quad \text{if } m = n \end{aligned} \right\} \qquad \begin{aligned} &\text{(8(a))} \\ &\text{(8(b))} \end{aligned}$$

[Note that if $m = n$, the initial integral becomes $\displaystyle\int_{-\pi}^{\pi} \sin^2 nx \, dx = \pi$ if $n \neq 0$
(see (d) above)]

(g) $\displaystyle\int_{-\pi}^{\pi} \sin mx \cos nx \, \mathrm{d}x = \int_{-\pi}^{\pi} \tfrac{1}{2}[\sin(m+n)x + \sin(m-n)x] \, \mathrm{d}x$

$$= -\frac{1}{2}\left[\frac{\cos(m+n)x}{(m+n)} + \frac{\cos(m-n)x}{(m-n)}\right]_{-\pi}^{\pi}$$

$$= 0 \tag{9}$$

27.4   Fourier series

Let $f(x)$ be a periodic function with period 2π such that:

$$f(x) = a_0 + a_1 \cos x + a_2 \cos 2x + a_3 \cos 3x + \cdots$$
$$+ \, b_1 \sin x + b_2 \sin 2x + b_2 \sin 3x + \cdots$$

i.e. $\displaystyle f(x) = a_0 = \sum_{n=1}^{\infty} (a_n \cos nx + b_n \sin nx) \tag{10}$

To find the coefficient a_0
Integrating both sides of equation (10) from $-\pi$ to π gives:

$$\int_{-\pi}^{\pi} f(x) \, \mathrm{d}x = \int_{-\pi}^{\pi} a_0 \, \mathrm{d}x, \quad \begin{array}{l}\text{all the other terms becoming zero}\\ \text{(from equations (3) and (4))}\end{array}$$

$\therefore$ $\displaystyle\int_{-\pi}^{\pi} f(x) \, \mathrm{d}x = a_0[x]_{-\pi}^{\pi} = 2\pi a_0 \tag{11}$

Hence $\displaystyle a_0 = \frac{1}{2\pi}\int_{-\pi}^{\pi} f(x) \, \mathrm{d}x$ (i.e. $a_0 =$ the area under the curve $f(x)$ from $-\pi$ to π, divided by 2π)

To find the coefficient a_n $(n \neq 0)$
Multiplying both sides of equation (10) by $\cos mx$ and integrating from $-\pi$ to π gives:

$$\int_{-\pi}^{\pi} f(x) \cos mx \, \mathrm{d}x = \int_{-\pi}^{\pi} a_0 \cos mx \, \mathrm{d}x$$

$$+ \sum_{n=1}^{\infty} \left\{\int_{-\pi}^{\pi} (a_n \cos nx \cos mx + b_n \sin nx \cos mx) \, \mathrm{d}x\right\}$$

All the terms on the right-hand side become zero (from equations (3), (7(a)) and (9)) with the exception of $\displaystyle\int_{-\pi}^{\pi} a_n \cos nx \cos mx \, \mathrm{d}x$, when $m = n$. This equals π from equation 7(b).

Hence $\displaystyle\int_{-\pi}^{\pi} f(x) \cos mx \, dx = a_m(\pi)$

i.e. $\displaystyle a_m = \frac{1}{\pi}\int_{-\pi}^{\pi} f(x) \cos mx \, dx$

i.e. $\displaystyle a_n = \frac{1}{\pi}\left\{\int_{-\pi}^{\pi} f(x) \cos nx \, dx, \text{ where } n = 1, 2, 3, \dots\right\}$ (12)

To find the coefficient b_n
Multiplying both sides of equation (10) by $\sin mx$ and integrating from $-\pi$ to π gives:

$$\int_{-\pi}^{\pi} f(x) \sin mx \, dx = \int_{-\pi}^{\pi} a_0 \sin mx \, dx$$

$$+ \sum_{n=1}^{\infty}\left\{\int_{-\pi}^{\pi} (a_n \cos nx \sin mx + b_n \sin nx \sin mx) \, dx\right\}$$

All the terms on the right-hand side become zero (from equations (4), (8(a)) and (9)) with the exception of $\displaystyle\int_{-\pi}^{\pi} b_n \sin nx \sin mx \, dx$ when $m = n$. This equals π from equation (8(b)).

Hence $\displaystyle\int_{-\pi}^{\pi} f(x) \sin mx \, dx = b_m(\pi)$

i.e. $\displaystyle b_m = \frac{1}{\pi}\int_{-\pi}^{\pi} f(x) \sin mx \, dx$

or $\displaystyle b_n = \frac{1}{\pi}\int_{-\pi}^{\pi} f(x) \sin nx \, dx, \text{ where } n = 1, 2, 3, \dots$ (13)

Hence, given a periodic function $f(x)$ with period 2π, a_n and b_n may be determined and the trigonometric series of equation (10) may be found. This series is then called the **Fourier series** corresponding to $f(x)$ and its coefficients (from equations (11), (12) and (13)) are called the **Fourier coefficients**. Since $f(x)$ is periodic the same results as in equations (11), (12) and (13) would be achieved if the range is taken between say 0 and 2π instead of $-\pi$ and π, or any other interval of length 2π.

Summary
If $f(x)$ is a periodic function of period 2π then its Fourier series is given by:

$$f(x) = a_0 + \sum_{n=1}^{\infty} (a_n \cos nx + b_n \sin nx)$$

where, for range $-\pi$ to $+\pi$

$$a_0 = \frac{1}{2\pi} \int_{-\pi}^{\pi} f(x) \, dx$$

$$a_n = \frac{1}{\pi} \int_{-\pi}^{\pi} f(x) \cos nx \, dx \quad (n = 1, 2, 3, \ldots)$$

$$b_n = \frac{1}{\pi} \int_{-\pi}^{\pi} f(x) \sin nx \, dx \quad (n = 1, 2, 3, \ldots)$$

Worked problems on Fourier series of periodic functions of period 2π

Problem 1. Obtain a Fourier series for the periodic function $f(x)$ defined as follows:

$$f(x) = \begin{cases} -1 & \text{when } -\pi < x < 0 \\ 1 & \text{when } \ \ \ 0 < x < \pi \end{cases}$$

The function is periodic outside of that range with period 2π (i.e. $f(x) = f(x + 2\pi)$).

The square wave function defined above is shown in Fig. 27.5 and such functions may occur on external forces acting on mechanical systems, or as electromotive forces in electrical circuits.

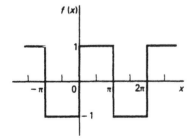

Fig. 27.5

Since $f(x)$ is given by two different expressions in the two halves of the range the integration is divided into two parts, one from $-\pi$ to 0 and the other from 0 to π.

$$a_0 = \frac{1}{2\pi} \int_{-\pi}^{\pi} f(x) \, dx = \frac{1}{2\pi} \left\{ \int_{-\pi}^{0} -1 \, dx + \int_{0}^{\pi} 1 \, dx \right\}$$

$$= \frac{1}{2\pi} \left\{ \left[-x \right]_{-\pi}^{0} + \left[x \right]_{0}^{\pi} \right\} = 0$$

(This step was really unnecessary since from the graph in Fig. 27.5 it is seen that the area under the curve $f(x)$ between $-\pi$ and π is zero and hence $a_0 = 0$.)

$$a_n = \frac{1}{\pi} \int_{-\pi}^{\pi} f(x) \cos nx \, dx$$

$$= \frac{1}{\pi} \left\{ \int_{-\pi}^{0} (-1) \cos nx \, dx + \int_{0}^{\pi} (1) \cos nx \, dx \right\}$$

$$= \frac{1}{\pi} \left\{ \left[\frac{-\sin nx}{n} \right]_{-\pi}^{0} + \left[\frac{\sin nx}{n} \right]_{0}^{\pi} \right\} = 0$$

Hence $a_1, a_2, a_3, \ldots$ are all zero and no cosine terms will appear in the Fourier series.

$$b_n = \frac{1}{\pi} \int_{-\pi}^{\pi} f(x) \sin nx \, dx$$

$$= \frac{1}{\pi} \left\{ \int_{-\pi}^{0} (-1) \sin nx \, dx + \int_{0}^{\pi} (1) \sin nx \, dx \right\}$$

$$= \frac{1}{\pi} \left\{ \left[\frac{\cos nx}{n} \right]_{-\pi}^{0} + \left[\frac{-\cos nx}{n} \right]_{0}^{\pi} \right\}$$

When n is odd:

$$b_n = \frac{1}{\pi} \left\{ \left[\frac{1}{n} - \left(-\frac{1}{n} \right) \right] + \left[-\left(-\frac{1}{n} \right) - \left(-\frac{1}{n} \right) \right] \right\} = \frac{1}{\pi} \left(\frac{2}{n} + \frac{2}{n} \right)$$

i.e. $b_n = \dfrac{4}{n\pi}$

Hence $b_1 = \dfrac{4}{\pi} \sin x$, $b_3 = \dfrac{4}{3\pi} \sin 3x$, $b_5 = \dfrac{4}{5\pi} \sin 5x, \ldots$

When n is even: $b_n = \dfrac{1}{\pi} \left\{ \left(\dfrac{1}{n} - \dfrac{1}{n} \right) + \left(-\dfrac{1}{n} - -\dfrac{1}{n} \right) \right\} = 0$

Hence the Fourier series is given by:

$$f(x) = \frac{4}{\pi} \left(\sin x + \frac{1}{3} \sin 3x + \frac{1}{5} \sin 5x + \cdots \right)$$

$\dfrac{4}{\pi} \sin x$ is termed the first **partial sum** of the Fourier series of $f(x)$,

$\dfrac{4}{\pi} \left(\sin x + \dfrac{1}{3} \sin 3x \right)$ is termed the second partial sum of the

Fourier series, $\dfrac{4}{\pi}\left(\sin x + \dfrac{1}{3}\sin 3x + \dfrac{1}{5}\sin 5x\right)$ is termed the third partial sum of the Fourier series, and so on.

Problem 2. (i) Show, by plotting the first three partial sums of the Fourier series for Problem 1, that, as the Fourier series is added together term by term, the result approximates more and more closely to the function it represents.

(ii)   Deduce from the Fourier series of Problem 1 that

$$\frac{\pi}{4} = 1 - \frac{1}{3} + \frac{1}{5} - \frac{1}{7} + \cdots$$

(i)   Let $P_1 = \dfrac{4}{\pi}\sin x$, $P_2 = \dfrac{4}{\pi}\left(\sin x + \dfrac{1}{3}\sin 3x\right)$,

$P_3 = \dfrac{4}{\pi}\left(\sin x + \dfrac{1}{3}\sin 3x + \dfrac{1}{5}\sin 5x\right)$, and so on.

Graphs of P_1, P_2 and P_3, obtained by drawing up tables of values, are shown in Figs 27.6(a), (b) and (c) and they indicate that the series is convergent, that is, continually approximating towards a definite limit, as more and more partial sums are taken and in the limit will have the sum $f(x)$. [From the graphs of Fig. 27.6 it is seen that at the points of discontinuity of $f(x)$, i.e. at $-\pi$, 0, π, ... all the partial sums have the value zero. This is, in fact, the **arithmetic mean of the two limiting values** of $f(x)$ as x approaches the point of discontinuity from the two sides $\left(\text{i.e. }\dfrac{1 + -1}{2} = 0\right)$. This is true for any point of discontinuity of a periodic function. Figure 27.7 shows two functions, one a square wave and the other a saw-tooth waveform. For Fig. 27.7(a) the sum of the Fourier series at the point of discontinuity (i.e. at 0, π, 2π, ...) is given by $\dfrac{15 + 0}{2}$, i.e. $7\dfrac{1}{2}$. For Fig. 27.7(b) the sum of the Fourier series at the point of discontinuity is given by $\dfrac{8 + 0}{2}$, i.e. 4].

(ii) $f(x) = \dfrac{4}{\pi}\left(\sin x + \dfrac{1}{3}\sin 3x + \dfrac{1}{5}\sin 5x + \dfrac{1}{7}\sin 7x + \cdots\right)$

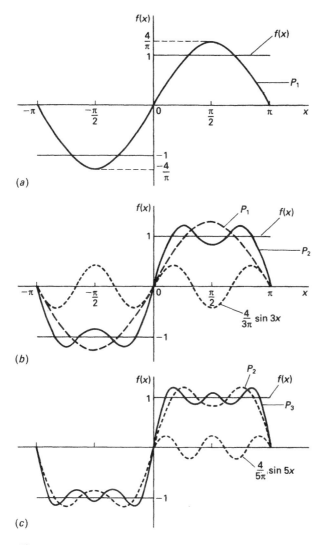

Fig. 27.6

From Fig. 27.5, when $x = \dfrac{\pi}{2}$, $f(x) = 1$ and $\sin \dfrac{\pi}{2} = 1$,

$\sin \dfrac{3\pi}{2} = -1$, $\sin \dfrac{5\pi}{2} = 1$, $\sin \dfrac{7\pi}{2} = -1$, and so on.

Hence $1 = \dfrac{4}{\pi} \left\{ 1 + \dfrac{1}{3}(-1) + \dfrac{1}{5}(1) + \dfrac{1}{7}(-1) + \cdots \right\}$

(a)

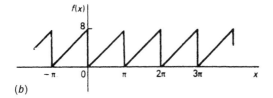

(b)

Fig. 27.7

Hence $\dfrac{\pi}{4} = 1 - \dfrac{1}{3} + \dfrac{1}{5} - \dfrac{1}{7} + \cdots$

Thus the values of various series with constant terms can be obtained by evaluating Fourier series at specific points.

Problem 3. Find the Fourier series for the rectified sine wave $v = 10 \sin \dfrac{\theta}{2}$, shown in Fig. 27.8.

v is a periodic function of period 2π.

Thus $v = f(\theta) = a_0 + \displaystyle\sum_{n=1}^{\infty} (a_n \cos n\theta + b_n \sin n\theta)$

In this case it is better to take the range from 0 to 2π instead of $-\pi$ to $+\pi$ since the waveform is continuous between 0 and 2π.

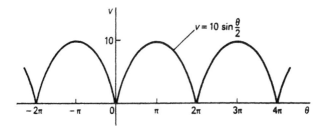

Fig. 27.8

Thus

$$a_0 = \frac{1}{2\pi} \int_{-\pi}^{\pi} f(x)\, dx = \frac{1}{2\pi} \int_0^{2\pi} f(x)\, dx = \frac{1}{2\pi} \int_0^{2\pi} 10 \sin \frac{\theta}{2}\, d\theta$$

$$= \frac{10}{2\pi} \left[-2 \cos \frac{\theta}{2} \right]_0^{2\pi} = \frac{-10}{\pi} \left[\cos \frac{\theta}{2} \right]_0^{2\pi} = \frac{-10}{\pi} (\cos \pi - \cos 0)$$

$$= \frac{-10}{\pi} (-1 - 1) = \frac{20}{\pi}$$

$$a_n = \frac{1}{\pi} \int_0^{2\pi} 10 \sin \frac{\theta}{2} \cos n\theta\, d\theta$$

$$= \frac{10}{\pi} \int_0^{2\pi} \frac{1}{2} \left[\sin\left(\frac{\theta}{2} + n\theta\right) + \sin\left(\frac{\theta}{2} - n\theta\right) \right] d\theta$$

$$= \frac{10}{2\pi} \left[\frac{-\cos[\theta(\frac{1}{2} + n)]}{(\frac{1}{2} + n)} - \frac{\cos[\theta(\frac{1}{2} - n)]}{(\frac{1}{2} - n)} \right]_0^{2\pi}$$

$$= \frac{5}{\pi} \left[\left\{ \frac{-\cos[2\pi(\frac{1}{2} + n)]}{(\frac{1}{2} + n)} - \frac{\cos[2\pi(\frac{1}{2} - n)]}{(\frac{1}{2} - n)} \right\} \right.$$
$$\left. - \left\{ \frac{-\cos 0}{(\frac{1}{2} + n)} - \frac{\cos 0}{(\frac{1}{2} - n)} \right\} \right]$$

When n is both odd and even,

$$a_n = \frac{5}{\pi} \left[\left\{ \frac{1}{(\frac{1}{2} + n)} + \frac{1}{(\frac{1}{2} - n)} \right\} - \left\{ \frac{-1}{(\frac{1}{2} + n)} - \frac{1}{(\frac{1}{2} - n)} \right\} \right]$$

$$= \frac{5}{\pi} \left[\frac{2}{(\frac{1}{2} + n)} + \frac{2}{(\frac{1}{2} - n)} \right] = \frac{10}{\pi} \left[\frac{1}{(\frac{1}{2} + n)} + \frac{1}{(\frac{1}{2} - n)} \right]$$

Hence,

$$a_1 = \frac{10}{\pi} \left(\frac{1}{\frac{3}{2}} + \frac{1}{-\frac{1}{2}} \right) = \frac{10}{\pi} \left(\frac{2}{3} - \frac{2}{1} \right) = \frac{-40}{3\pi}$$

$$a_3 = \frac{10}{\pi} \left(\frac{1}{2\frac{1}{2}} + \frac{1}{-1\frac{1}{2}} \right) = \frac{10}{\pi} \left(\frac{2}{5} - \frac{2}{3} \right) = \frac{-40}{(3)(5)\pi}$$

$$a_3 = \frac{10}{\pi} \left(\frac{1}{3\frac{1}{2}} + \frac{1}{-2\frac{1}{2}} \right) = \frac{10}{\pi} \left(\frac{2}{7} - \frac{2}{5} \right) = \frac{-40}{(5)(7)\pi} \quad \text{and so on.}$$

$$b_n = \frac{1}{\pi} \int_0^{2\pi} 10 \sin \frac{\theta}{2} \sin n\theta\, d\theta$$

$$= \frac{10}{\pi} \int_0^{2\pi} -\left[\frac{1}{2} \cos \theta \left(\frac{1}{2} + n \right) - \cos \theta \left(\frac{1}{2} - n \right) \right] d\theta$$

$$= \frac{10}{2\pi} \left[\frac{\sin[\theta(\frac{1}{2} - n)]}{(\frac{1}{2} - n)} - \frac{\sin[\theta(\frac{1}{2} + n)]}{(\frac{1}{2} + n)} \right]_0^{2\pi}$$

$$= \frac{5}{\pi}\left[\left\{\frac{\sin 2\pi(\frac{1}{2} - n)}{(\frac{1}{2} - n)} - \frac{\sin 2\pi(\frac{1}{2} + n)}{(\frac{1}{2} + n)}\right\} - \left\{\frac{\sin 0}{(\frac{1}{2} - n)} - \frac{\sin 0}{(\frac{1}{2} + n)}\right\}\right]$$

When n is both odd and even, $b_n = 0$ since $\sin 0$, $\sin \pi$, $\sin 3\pi$, ... are all zero.

Hence the Fourier series for the rectified sine wave $v = 10 \sin \dfrac{\theta}{2}$ is given by:

$$v = f(\theta) = \frac{20}{\pi} - \frac{40}{3\pi}\cos\theta - \frac{40}{(3)(5)\pi}\cos 2\theta - \frac{40}{(5)(7)\pi}\cos 3\theta - \cdots$$

i.e.

$$v = \frac{20}{\pi}\left\{1 - \frac{2}{3}\cos\theta - \frac{2}{(3)(5)}\cos 2\theta - \frac{2}{(5)(7)}\cos 3\theta - \cdots\right\}$$

Further problems on Fourier series of periodic functions of period 2π may be found in the following Section 27.5 (Problems 1 to 6).

27.5   Further problems

In Problems 1 to 3 find the Fourier series for the given periodic functions of period 2π.

1. $f(x) = \begin{cases} 0 \text{ when} & -\pi < x < 0 \\ 1 \text{ when} & 0 < x < \pi \end{cases}$

$$\left[f(x) = \frac{1}{2} + \frac{2}{\pi}\left(\sin x + \frac{1}{3}\sin 3x + \frac{1}{5}\sin 5x + \cdots\right)\right]$$

2. $f(x) = \begin{cases} 0 \text{ when } -\pi < x < 0 \\ \sin x \text{ when } 0 < x < \pi \end{cases}$

$$\left[f(x) = \frac{1}{\pi} - \frac{2}{\pi}\left(\frac{\cos 2x}{3} + \frac{\cos 4x}{(3)(5)} + \frac{\cos 6x}{(5)(7)} + \cdots\right)\right]$$

3. $f(t) = \begin{cases} 0 \text{ when } -\pi < t < -\dfrac{\pi}{2} \\ 1 \text{ when } -\dfrac{\pi}{2} < t < \dfrac{\pi}{2} \\ 0 \text{ when } \dfrac{\pi}{2} < t < \pi \end{cases}$

$$\left[f(t) = \frac{1}{2} + \frac{2}{\pi}\left(\cos t - \frac{1}{3}\cos 3t + \frac{1}{5}\cos 5t - \cdots\right)\right]$$

4. For the waveform shown in Fig. 27.9 show that:

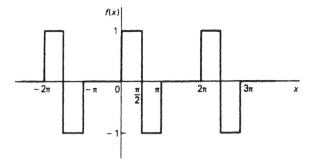

Fig. 27.9

$$f(x) = \frac{2}{\pi}\left(\cos x - \frac{1}{3}\cos 3x + \frac{1}{5}\cos 5x \cdots + \sin 2x + \frac{1}{3}\sin 6x \right.$$
$$\left. + \frac{1}{5}\sin 10x \ldots \right)$$

5.   Obtain the Fourier series for the periodic function defined by:

$$f(t) = \begin{cases} -2, & -\pi < t < -\dfrac{\pi}{2} \\[2mm] 2, & -\dfrac{\pi}{2} < t < \dfrac{\pi}{2} \\[2mm] -2, & \dfrac{\pi}{2} < t < \pi \end{cases}$$

The function has a period of 2π.

$$\left[f(t) = \frac{8}{\pi}\left(\cos t - \frac{1}{3}\cos 3t + \frac{1}{5}\cos 5t - \cdots \right) \right]$$

6.   (*a*) For Problems 1, 3, and 5 draw graphs of the first three partial sums of the Fourier series and show that as the series is added together term by term the result approximates more and more closely to the function it represents.

   (*b*) For Problem 3, find the sum of the Fourier series at the point of discontinuity at $\dfrac{\pi}{2}$. $\left[\dfrac{1}{2} \right]$.

The Fourier series for a non-periodic function over range 2π

28.1 Expansion of non-periodic functions

If a function $f(x)$ is not periodic then it cannot be expanded in a Fourier series for **all** values of x. However, it is possible to find a Fourier series to represent the function over any range of width 2π, such as from $-\pi$ to $+\pi$ or from 0 to 2π. Thus a new function, say $g(x)$, is constructed by taking the values of $f(x)$ in the given range and then repeating them outside of the given range at intervals of 2π. Since, by construction, the function $g(x)$ is periodic with period 2π it may then be expanded in a Fourier series for all values of x. Since $g(x) = f(x)$ in the given range, the sum of the series is equal to $f(x)$ at all points in the given range, but it is **not** equal to $f(x)$ at points outside of the given range. We use exactly the same formulae for the Fourier series and its coefficients as used in the previous chapter. The following worked problems show how Fourier series for non-periodic functions over a range of 2π may be determined.

Worked problems on Fourier series of non-periodic functions over a range of 2π

Problem 1. (a) Obtain a Fourier series to represent $f(x) = x$ in the range $-\pi$ to $+\pi$.

(b) In the Fourier series obtained in (a) let $f(x) = \dfrac{\pi}{2}$ and deduce a series for $\dfrac{\pi}{4}$.

(a) The function $f(x) = x$ is not periodic. The function is shown in the range $-\pi$ to $+\pi$ in Fig. 28.1 and is then purposely constructed outside of that range so that it is periodic with period 2π. This is known as a saw-tooth waveform. Points of discontinuity occur at $-\pi, \pi, 3\pi, \ldots$ and so on.

$$f(x) = a_0 + \sum_{n=1}^{\infty} (a_n \cos nx + b_n \sin nx)$$

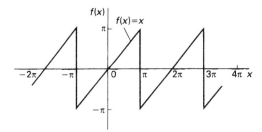

Fig. 28.1

$$a_0 = \frac{1}{2\pi} \int_{-\pi}^{\pi} f(x)\, dx = \frac{1}{2\pi} \int_{-\pi}^{\pi} x\, dx = \frac{1}{2\pi} \left[\frac{x^2}{2} \right]_{-\pi}^{\pi} = 0$$

$$a_n = \frac{1}{\pi} \int_{-\pi}^{\pi} f(x) \cos nx\, dx = \frac{1}{\pi} \int_{-\pi}^{\pi} x \cos nx\, dx$$

$$= \frac{1}{\pi} \left[\frac{x \sin nx}{n} - \int \frac{\sin nx}{n}\, dx \right]_{-\pi}^{\pi} \quad \text{by parts}$$

$$= \frac{1}{\pi} \left[\frac{x \sin nx}{n} + \frac{\cos nx}{n^2} \right]_{-\pi}^{\pi}$$

$$= \frac{1}{\pi} \left[\left(0 + \frac{\cos n\pi}{n^2} \right) - \left(0 + \frac{\cos n(-\pi)}{n^2} \right) \right]$$

$$= 0$$

$$b_n = \frac{1}{\pi} \int_{-\pi}^{\pi} f(x) \sin nx\, dx$$

$$= \frac{1}{\pi} \int_{-\pi}^{\pi} x \sin nx\, dx$$

$$= \frac{1}{\pi} \left[\frac{-x \cos nx}{n} - \int \left(\frac{-\cos nx}{n} \right) dx \right]_{-\pi}^{\pi} \quad \text{by parts}$$

$$= \frac{1}{\pi} \left[\frac{-x \cos nx}{n} + \frac{\sin nx}{n^2} \right]_{-\pi}^{\pi}$$

$$= \frac{1}{\pi} \left[\left(\frac{-\pi \cos n\pi}{n} + \frac{\sin n\pi}{n^2} \right) \right.$$

$$\left. - \left(\frac{-(-\pi) \cos n(-\pi)}{n} + \frac{\sin n(-\pi)}{n^2} \right) \right]$$

$$= \frac{1}{\pi} \left[\frac{-\pi \cos n\pi}{n} - \frac{\pi \cos n\pi}{n} \right]$$

$$= -\frac{2}{n} \cos n\pi$$

When n is odd, $b_n = \dfrac{2}{n}$. Thus $b_1 = 2, b_3 = \dfrac{2}{3}, b_5 = \dfrac{2}{5}$, and so on.

When n is even, $b_n = -\dfrac{2}{n}$. Thus $b_2 = -1, b_4 = -\dfrac{1}{2}, b_6 = -\dfrac{1}{3}$, and so on. Thus

$$f(x) = x = 2 \sin x - \sin 2x + \frac{2}{3} \sin 3x - \frac{1}{2} \sin 4x + \frac{2}{5} \sin 5x$$
$$- \frac{1}{3} \sin 6x + \cdots$$

i.e. $x = 2\left(\sin x - \frac{1}{2} \sin 2x + \frac{1}{3} \sin 3x - \frac{1}{4} \sin 4x + \frac{1}{5} \sin 5x \right.$
$$\left. - \frac{1}{6} \sin 6x + \cdots \right)$$

for values of x between $-\pi$ and $+\pi$. For values of x outside of the range $-\pi$ to $+\pi$ the sum of the series is not equal to x.

(b) If $x = \dfrac{\pi}{2}$ then $f(x) = \dfrac{\pi}{2}$.

Thus $\dfrac{\pi}{2} = 2\left(\sin \dfrac{\pi}{2} - \dfrac{1}{2} \sin \pi + \dfrac{1}{3} \sin \dfrac{3\pi}{2} - \dfrac{1}{4} \sin 2\pi \right.$
$$\left. + \frac{1}{5} \sin \frac{5\pi}{2} - \frac{1}{6} \sin 3\pi + \cdots \right)$$

i.e. $\dfrac{\pi}{4} = 1 - \dfrac{1}{3} + \dfrac{1}{5} - \dfrac{1}{7} + \cdots$

Problem 2. Expand $f(x) = x^2$ in the range $0 < x < 2\pi$ in a Fourier series. By letting $x = \pi$ in the series obtained, find a series for $\dfrac{\pi^2}{12}$.

$f(x) = x^2$ is not a periodic function but is constructed in the range 0 to 2π and periodic with period 2π outside that range, as shown in Fig. 28.2.

$$f(x) = a_0 + \sum_{n=1}^{\infty} (a_n \cos nx + b_n \sin nx)$$

$$a_0 = \frac{1}{2\pi} \int_0^{2\pi} f(x)\, dx = \frac{1}{2\pi} \int_0^{2\pi} x^2\, dx = \frac{1}{2\pi} \left[\frac{x^3}{3} \right]_0^{2\pi} = \frac{1}{2\pi} \left(\frac{8\pi^3}{3} \right) = \frac{4\pi^2}{3}$$

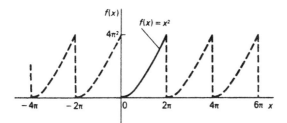

Fig. 28.2

$$a_n = \frac{1}{\pi} \int_0^{2\pi} f(x) \cos nx \, dx = \frac{1}{\pi} \int_0^{2\pi} x^2 \cos nx \, dx$$

$$= \frac{1}{\pi} \left[\frac{x^2 \sin nx}{n} + \frac{2x \cos nx}{n^2} - \frac{2 \sin nx}{n^3} \right]_0^{2\pi} \text{ by parts}$$

$$= \frac{1}{\pi} \left[\left(0 + \frac{4\pi \cos 2\pi n}{n^2} - 0 \right) - (0 + 0 - 0) \right]$$

$$= \frac{4}{n^2} \cos 2\pi n = \frac{4}{n^2} \text{ where } n = 1, 2, 3, \ldots$$

Hence $a_1 = \frac{4}{1^2}, a_2 = \frac{4}{2^2}, a_3 = \frac{4}{3^2}, \ldots$ and so on.

$$b_n = \frac{1}{\pi} \int_0^{2\pi} f(x) \sin nx \, dx = \frac{1}{\pi} \int_0^{2\pi} x^2 \sin nx \, dx$$

$$= \frac{1}{\pi} \left[\frac{-x^2 \cos nx}{n} + \frac{2x \sin nx}{n^2} + \frac{2 \cos nx}{n^3} \right]_0^{2\pi} \text{ by parts}$$

$$= \frac{1}{\pi} \left[\left(\frac{-4\pi^2 \cos 2\pi n}{n} + 0 + \frac{2 \cos 2\pi n}{n^3} \right) - \left(0 + 0 - \frac{2 \cos 0}{n^3} \right) \right]$$

$$= \frac{1}{\pi} \left[\frac{-4\pi^2}{n} + \frac{2}{n^3} - \frac{2}{n^3} \right] = \frac{-4\pi}{n}$$

Hence $b_1 = \frac{-4\pi}{1}, b_2 = \frac{-4\pi}{2}, b_3 = \frac{-4\pi}{3}, \ldots$ and so on.

Thus $f(x) = x^2 = \frac{4\pi^2}{3} + \sum_{n=1}^{\infty} \left(\frac{4}{n^2} \cos nx - \frac{4\pi}{n} \sin nx \right)$

i.e. $x^2 = \frac{4\pi^2}{3} + 4 \left(\cos x + \frac{1}{2^2} \cos 2x + \frac{1}{3^2} \cos 3x + \cdots \right)$

$$- 4\pi \left(\sin x + \frac{1}{2} \sin 2x + \frac{1}{3} \sin 3x + \cdots \right)$$

for values of x between 0 and 2π.

When $x = \pi$, $f(x) = \pi^2$. Hence

$$\pi^2 = \frac{4\pi^2}{3} + 4\left(\cos \pi + \frac{1}{4}\cos 2\pi + \frac{1}{9}\cos 3\pi + \frac{1}{16}\cos 4\pi + \cdots\right)$$

$$- 4\pi\left(\sin \pi + \frac{1}{2}\sin 2\pi + \frac{1}{3}\sin 3\pi + \cdots\right)$$

i.e. $\pi^2 - \dfrac{4\pi^2}{3} = 4\left(-1 + \dfrac{1}{4} - \dfrac{1}{9} + \dfrac{1}{16} - \cdots\right) - 4\pi(0)$

$\dfrac{-\pi^2}{3} = 4\left(-1 + \dfrac{1}{4} - \dfrac{1}{9} + \dfrac{1}{16} - \cdots\right)$

$\dfrac{\pi^2}{3} = 4\left(1 - \dfrac{1}{4} + \dfrac{1}{9} - \dfrac{1}{16} + \cdots\right)$

Thus $\dfrac{\pi^2}{12} = 1 - \dfrac{1}{4} + \dfrac{1}{9} - \dfrac{1}{16} + \cdots$

Further problems on Fourier series of non-periodic functions over a range of 2π may be found in the following Section 28.2 (Problems 1 to 9).

28.2   Further problems

1. If $f(x) = x + \pi$ when $-\pi < x < \pi$ find the Fourier series for $f(x)$ in that range. Sketch the graph of the function within and outside of the given range.

$$\left[f(x) = \pi + 2\left(\sin x - \frac{1}{2}\sin 2x + \frac{1}{3}\sin 3x - \cdots\right)\right]$$

2. (a) Find the Fourier series for the function $f(x) = x^2$ when $-\pi < x < \pi$ and sketch the function within and outside of this range.

    (b) In the series obtained in (a) let $x = \pi$ and deduce the series for $\displaystyle\sum_{n=1}^{\infty} \frac{1}{n^2}$.

    (a) $\left[f(x) = \dfrac{\pi^2}{3} - 4\left(\cos x - \dfrac{1}{2^2}\cos 2x + \dfrac{1}{3^2}\cos 3x - \cdots\right)\right]$

    (b) $\left[1 + \dfrac{1}{4} + \dfrac{1}{9} + \dfrac{1}{16} + \dfrac{1}{25} + \cdots = \dfrac{\pi^2}{6}\right]$

3. (a) Obtain a Fourier series for the function defined by

$$f(x) = \begin{cases} x, & 0 < x < \pi \\ 0, & \pi < x < 2\pi \end{cases}$$

(b) What is the sum of the series in (a) at the point of discontinuity (i.e. at $x = \pi$)?

(c) Using the answer to (b), deduce a series for $\dfrac{\pi^2}{8}$.

(a) $$\left[f(x) = \frac{\pi}{4} - \frac{2}{\pi} \left(\cos x + \frac{1}{3^2} \cos 3x + \frac{1}{5^2} \cos 5x + \cdots \right) \right.$$
$$\left. + \left(\sin x - \frac{1}{2} \sin 2x + \frac{1}{3} \sin 3x - \cdots \right) \right]$$

(b) $$\left[\frac{\pi}{2} \right]$$

(c) $$\left[\frac{\pi^2}{8} = 1 + \frac{1}{9} + \frac{1}{25} + \frac{1}{49} + \cdots \right]$$

4. Find the Fourier series expansion to represent

$$f(x) = \begin{cases} x & \text{when } 0 < x < \pi \\ 2\pi - x & \text{when } \pi < x < 2\pi \end{cases}$$

and sketch the graph of the function within and outside of the given range, assuming the period is 2π.

$$\left[f(x) = \frac{\pi}{2} - \frac{4}{\pi} \left(\cos x + \frac{\cos 3x}{3^2} + \frac{\cos 5x}{5^2} + \cdots \right) \right]$$

5. (a) Obtain a Fourier series for the triangular wave function shown in Fig. 28.3

Note that $$f(x) = \begin{cases} 1 + \dfrac{2x}{\pi}, & -\pi < x < 0 \\ 1 - \dfrac{2x}{\pi}, & 0 < x < \pi \end{cases}$$

(b) Deduce from the result in (a) the series for $\dfrac{\pi^2}{8}$.

(a) $$\left[f(x) = \frac{8}{\pi^2} \left(\cos x + \frac{1}{3^2} \cos 3x + \frac{1}{5^2} \cos 5x \right. \right.$$
$$\left. \left. + \frac{1}{7^2} \cos 7x + \cdots \right) \right]$$

(b) $$\left[\frac{\pi^2}{8} = 1 + \frac{1}{3^2} + \frac{1}{5^2} + \frac{1}{7^2} + \cdots \right]$$

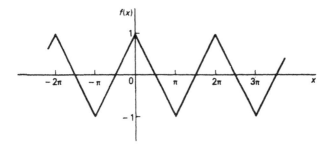

Fig. 28.3

In Problems 6 and 7 obtain the Fourier series for the given functions.

6. $f(x) = \begin{cases} 1 - x, & -\pi < x < 0 \\ 1 + x, & 0 < x < \pi \end{cases}$

$$\left[f(x) = \frac{\pi}{2} + 1 - \frac{4}{\pi}\left(\cos x + \frac{\cos 3x}{3^2} + \frac{\cos 5x}{5^2} + \cdots \right) \right]$$

7. $f(x) = x,\ 0 < x < 2\pi$

$$\left[f(x) = \pi - 2\left(\sin x + \frac{1}{2}\sin 2x + \frac{1}{3}\sin 3x + \frac{1}{4}\sin 4x + \cdots \right) \right]$$

8. Show that the first four terms of the Fourier series for $f(x) = x^4$ in the range $-\pi$ to $+\pi$ is given by:

$$f(x) = \frac{\pi^4}{5} + 8(6 - \pi^2)\cos x + (2\pi^2 - 3)\cos 3x$$

$$+ \frac{8}{27}(2 - 3\pi^2)\cos 3x + \cdots$$

9. Show that the Fourier series for $f(x) = e^x$ in the range $-\pi$ to $+\pi$ is given by:

$$f(x) = \frac{e^\pi - e^{-\pi}}{\pi}\left[\frac{1}{2} + \sum_{n=1}^{\infty} \frac{(-1)^n}{1 + n^2}(\cos nx - n\sin nx) \right]$$

What are the terms comprising the second harmonic?

$$\left[\frac{1}{5}(\cos 2x - 2\sin 2x) \right]$$

Chapter 29

The Fourier series for even and odd functions and half range series

29.1   Even and odd functions

In some of the problems in the preceding two chapters either the Fourier coefficient a_n or the Fourier coefficient b_n became zero after integration. Determination of a zero coefficient is time consuming and could have been avoided, for, with a knowledge of even and odd functions, a zero coefficient may be predicted without performing the often arduous integration.

Even functions
A function $y = f(x)$ is said to be **even** if:

$$f(-x) = f(x), \text{ for all values of } x$$

Examples of even functions include x^2, x^4, $\cos x$, $e^x + e^{-x}$ and so on. Four even functions are shown in Fig. 29.1 and it is found that a graph of such a function is always **symmetrical about the y-axis** (i.e. a mirror image).

Odd functions
A function $y = f(x)$ is said to be **odd** if:

$$f(-x) = -f(x), \text{ for all values of } x$$

Examples of odd functions include x, x^3, $\sin x$, $\tan 3x$ and so on. Four odd functions are shown in Fig. 29.2 and it is found that a graph of such a function is always **symmetrical about the origin**.

Functions that are neither even nor odd
Many functions are neither even nor odd, two such examples being shown in Fig. 29.3. In these cases it cannot be predicted whether sine or cosine terms will be absent in the Fourier series.

Worked problems on even and odd functions

Problem 1. Determine by sketching graphs, whether the following functions are even, odd or neither even nor odd:

Fig. 29.1

Fig. 29.2

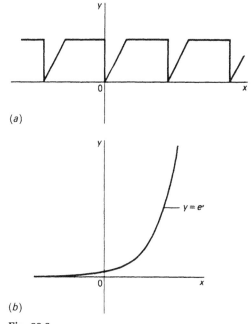

(a)

(b)

Fig. 29.3

(a) $\tan x$ (b) $\theta \sin \theta$ (c) e^{2x}

(d) $f(x) = \begin{cases} 1 - x & \text{when } -\pi < x < 0 \\ 1 + x & \text{when } \quad 0 < x < \pi \end{cases}$

and is periodic, of period 2π.

A sketch of each of the functions is shown in Fig. 29.4

(a)   The graph of $y = \tan x$, shown in Fig. 29.4(a), is symmetrical about the origin and thus is an **odd function**. Alternatively, $\tan(-x) = -\tan x$.

(b)   The graph of $y = \theta \sin \theta$, shown in Fig. 29.4(b), is symmetrical about the y-axis and thus is an **even function**. Alternatively, $-\theta \sin(-\theta) = \theta \sin \theta$.

(c)   The graph of $y = e^{2x}$ (Fig. 29.4(c)) is neither symmetrical about the y-axis nor symmetrical about the origin. Hence e^{2x} is **neither even nor odd**.

(d)   The graph of $f(x)$ (Fig. 29.4(d)) is symmetrical with respect to the $f(x)$ axis hence the function is **even**. Alternatively, $f(-x) = f(x)$.

Further problems on even and odd functions may be found in Section 29.4 (Problems 1 to 3), page 426.

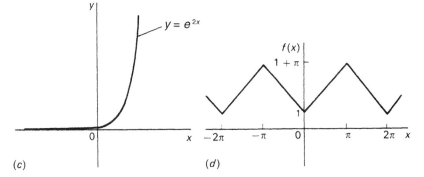

(a)

(b)

(c)

(d)

Fig. 29.4

29.2   Fourier cosine and Fourier sine series

If $f(x)$ is an **even** function then $\displaystyle\int_{-\pi}^{\pi} f(x)\,dx = 2\int_{0}^{\pi} f(x)\,dx$ (1)

If $f(x)$ is an **odd** function then $\displaystyle\int_{-\pi}^{\pi} f(x)\,dx = 0$ (2)

Equations (1) and (2) are obvious from the graphs of Figs 29.1 and 29.2.
The product $f(x)\,g(x)$ of an even function $f(x)$ and an odd function $g(x)$
is odd since $f(-x)g(-x) = f(x)[-g(x)] = -f(x)g(x)$.
Thus if $f(x)$ is even then $f(x)\sin nx$ is odd and since

$$b_n = \int_{-\pi}^{\pi} f(x)\sin nx\,dx \text{ then } b_n = 0 \qquad (3)$$

Similarly, if $f(x)$ is odd then $f(x) \cos nx$ is odd and since

$$a_n = \int_{-\pi}^{\pi} f(x) \cos nx \, dx \text{ then } a_n = 0 \tag{4}$$

The following series therefore follow from the results of equations (1) to (4):

Fourier cosine series
The Fourier series of an even periodic function $f(x)$ having period 2π is a 'Fourier cosine series' (that is, it contains no sine terms).

Hence $f(x) = a_0 + \sum\limits_{n=1}^{\infty} a_n \cos nx$

where $a_0 = \dfrac{1}{2\pi} \int_{-\pi}^{\pi} f(x) \, dx = \dfrac{1}{\pi} \int_{0}^{\pi} f(x) \, dx$ $\tag{5}$

and $a_n = \dfrac{1}{\pi} \int_{-\pi}^{\pi} f(x) \cos nx \, dx = \dfrac{2}{\pi} \int_{0}^{\pi} f(x) \cos nx \, dx$ $\tag{6}$

Fourier sine series
The Fourier series of an odd periodic function $f(x)$ having period 2π is a 'Fourier sine series' (that is, it contains no constant term and no cosine terms).

Hence $f(x) = \sum\limits_{n=1}^{\infty} b_n \sin nx$

where $b_n = \dfrac{1}{\pi} \int_{-\pi}^{\pi} f(x) \sin nx \, dx = \dfrac{2}{\pi} \int_{0}^{\pi} f(x) \sin nx \, dx$ $\tag{7}$

Worked problems on Fourier cosine and Fourier sine series

Problem 1. Obtain the Fourier series for each of the square wave functions shown in Fig. 29.5. From the series of the waveform shown in Fig. 29.5(b) deduce a series for $\dfrac{\pi}{4}$.

(a)   The square wave function shown in Fig. 29.5(a) is an **even** function since it is symmetrical about the y-axis. Hence the Fourier series is given by $f(x) = a_0 + \sum\limits_{n=1}^{\infty} a_n \cos nx$ (i.e. the Fourier series will contain no sine terms).

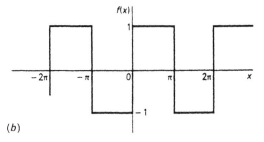

(a)

(b)

Fig. 29.5

The function $f(x)$ is defined by

$$f(x) = \begin{cases} -1, & -\pi < x < -\dfrac{\pi}{2} \\[2mm] 1, & -\dfrac{\pi}{2} < x < \dfrac{\pi}{2} \\[2mm] -1, & \dfrac{\pi}{2} < x < \pi \end{cases}$$

in the range $-\pi$ to $+\pi$. From equation (5),

$$a_0 = \frac{1}{\pi} \int_0^\pi f(x) \, dx = \frac{1}{\pi} \left\{ \int_0^{\pi/2} 1 \, dx + \int_{\pi/2}^\pi -1 \, dx \right\}$$

$$= \frac{1}{\pi} \left\{ \left[x \right]_0^{\pi/2} + \left[-x \right]_{\pi/2}^\pi \right\}$$

$$= \frac{1}{\pi} \left\{ \frac{\pi}{2} + \left[(-\pi) - \left(-\frac{\pi}{2} \right) \right] \right\} = 0$$

From equation (6),

$$a_n = \frac{2}{\pi} \int_0^\pi f(x) \cos nx \, dx$$

$$= \frac{2}{\pi} \left\{ \int_0^{\pi/2} 1 \cos nx \, dx + \int_{\pi/2}^{\pi} -1 \cos nx \, dx \right\}$$

$$= \frac{2}{\pi} \left\{ \left[\frac{\sin nx}{n} \right]_0^{\pi/2} + \left[\frac{-\sin nx}{n} \right]_{\pi/2}^{\pi} \right\}$$

$$= \frac{2}{\pi} \left\{ \left(\frac{\sin(\pi/2)n}{n} - 0 \right) + \left(0 - \frac{-\sin(\pi/2)n}{n} \right) \right\}$$

$$= \frac{2}{\pi} \left\{ \frac{2\sin(\pi/2)n}{n} \right\} = \frac{4}{\pi n} \left(\sin \frac{n\pi}{2} \right)$$

When n is even, $a_n = 0$

When n is odd, $a_n = \dfrac{4}{\pi n}$ for $n = 1, 5, 9, \ldots$

and $\qquad\qquad a_n = \dfrac{-4}{\pi n}$ for $n = 3, 7, 11, \ldots$

Hence $\qquad\qquad a_1 = \dfrac{4}{\pi}, a_3 = \dfrac{-4}{3\pi}, a_5 = \dfrac{4}{5\pi}, \ldots$

Hence the Fourier series for the waveform shown in Fig. 29.5(a) is given by:

$$f(x) = \frac{4}{\pi} \left(\cos x - \frac{1}{3} \cos 3x + \frac{1}{5} \cos 5x - \frac{1}{7} \cos 7x + \cdots \right)$$

(b)   The square wave function shown in Fig. 29.5(b) is an **odd** function since it is symmetrical about the origin. Hence the Fourier series is given by

$$f(x) = \sum_{n=1}^{\infty} b_n \sin nx$$

The function $f(x)$ is defined by:

$$f(x) = \begin{cases} -1, & -\pi < x < 0 \\ +1, & 0 < x < \pi \end{cases}$$

From equation (7),

$$b_n = \frac{2}{\pi} \int_0^{\pi} f(x) \sin nx \, dx = \frac{2}{\pi} \int_0^{\pi} 1 \sin nx \, dx$$

$$= \frac{2}{\pi} \left[\frac{-\cos nx}{n} \right]_0^{\pi}$$

$$= \frac{2}{\pi} \left[\frac{-\cos n\pi}{n} - \frac{1}{n} \right]$$

$$= \frac{2}{\pi n} (1 - \cos n\pi)$$

When n is even, $b_n = 0$

When n is odd, $b_n = \dfrac{2}{\pi n}(1 - -1) = \dfrac{4}{\pi n}$

Hence $b_1 = \dfrac{4}{\pi}$, $b_3 = \dfrac{4}{3\pi}$, $b_5 = \dfrac{4}{5\pi}$, ...

Hence the Fourier series for the waveform shown in Fig. 29.5(*b*) is given by:

$$f(x) = \frac{4}{\pi}\left(\sin x + \frac{1}{3}\sin 3x + \frac{1}{5}\sin 5x + \frac{1}{7}\sin 7x + \cdots\right)$$

When $x = \dfrac{\pi}{2}$, $f(x) = 1$

Thus $1 = \dfrac{4}{\pi}\left(\sin \dfrac{\pi}{2} + \dfrac{1}{3}\sin \dfrac{3\pi}{2} + \dfrac{1}{5}\sin \dfrac{5\pi}{2} + \dfrac{1}{7}\sin \dfrac{7\pi}{2} + \cdots\right)$

$1 = \dfrac{4}{\pi}\left(1 - \dfrac{1}{3} + \dfrac{1}{5} - \dfrac{1}{7} + \cdots\right)$

i.e. $\dfrac{\pi}{4} = 1 - \dfrac{1}{3} + \dfrac{1}{5} - \dfrac{1}{7} + \cdots$

Problem 2. Expand the function $f(x) = x^2$ as a Fourier series in the range $-\pi < x < \pi$ given that $f(x + 2\pi) = f(x)$. Sketch the function within and outside the given range. In the series obtained let $x = \pi$ and show that

$$\sum_{n=1}^{\infty} \frac{1}{n^2} = \frac{\pi^2}{6}$$

A graph of $f(x) = x^2$ in the range $x = -\pi$ to $x = +\pi$ with period 2π is shown in Fig. 29.6 and it is seen that the function is an **even** one since it is symmetrical about the y-axis. Thus a

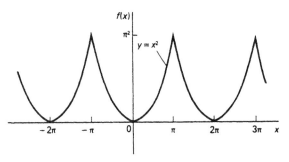

Fig. 29.6

Fourier cosine series results and is of the form:

$$f(x) = a_0 + \sum_{n=1}^{\infty} a_n \cos nx \quad \text{(i.e. } b_n = 0)$$

From equation (5),

$$a_0 = \frac{1}{\pi} \int_0^{\pi} f(x)\,dx = \frac{1}{\pi} \int_0^{\pi} x^2\,dx = \frac{1}{\pi}\left[\frac{x^3}{3}\right]_0^{\pi} = \frac{\pi^2}{3}$$

From equation (6),

$$a_n = \frac{2}{\pi} \int_0^{\pi} f(x) \cos nx\,dx = \frac{2}{\pi} \int_0^{\pi} x^2 \cos nx\,dx$$

$$= \frac{2}{\pi}\left[\frac{x^2 \sin nx}{n} + \frac{2x \cos nx}{n^2} - \frac{2 \sin nx}{n^3}\right]_0^{\pi} \text{ by parts}$$

$$= \frac{2}{\pi}\left[\left(0 + \frac{2\pi \cos n\pi}{n^2} - 0\right) - (0 + 0 - 0)\right]$$

$$= \frac{4}{n^2} \cos n\pi$$

When n is odd, $a_n = -\dfrac{4}{n^2}$. Hence $a_1 = \dfrac{-4}{1^2}, a_3 = \dfrac{-4}{3^2}, a_5 = \dfrac{-4}{5^2}, \ldots$

When n is even. $a_n = \dfrac{4}{n^2}$. Hence $a_2 = \dfrac{4}{2^2}, a_4 = \dfrac{4}{4^2}, a_6 = \dfrac{4}{6^2}, \ldots$

Hence $f(x) = x^2$

$$= \frac{\pi^2}{3} - 4\left(\cos x - \frac{1}{4}\cos 2x + \frac{1}{9}\cos 3x - \frac{1}{16}\cos 4x + \frac{1}{25}\cos 5x - \cdots\right)$$

When $x = \pi$, $f(x) = \pi^2$. Hence

$$\pi^2 = \frac{\pi^2}{3} - 4\left(-1 - \frac{1}{4} - \frac{1}{9} - \frac{1}{16} - \frac{1}{25} - \cdots\right)$$

i.e. $\quad \pi^2 - \dfrac{\pi^2}{3} = 4\left(1 + \dfrac{1}{4} + \dfrac{1}{9} + \dfrac{1}{16} + \dfrac{1}{25} + \cdots\right)$

$$\frac{2\pi^2}{3(4)} = 1 + \frac{1}{4} + \frac{1}{9} + \frac{1}{16} + \frac{1}{25} + \cdots$$

$$\frac{\pi^2}{6} = 1 + \frac{1}{2^2} + \frac{1}{3^2} + \frac{1}{4^2} + \frac{1}{5^2} + \cdots$$

Hence $\displaystyle\sum_{n=1}^{\infty} \frac{1}{n^2} = 1 + \frac{1}{2^2} + \frac{1}{3^2} + \frac{1}{4^2} + \frac{1}{5^2} + \cdots = \frac{\pi^2}{6}$

Further problems on Fourier cosine and Fourier sine series may be found in Section 29.4 (Problems 4 to 8), page 426.

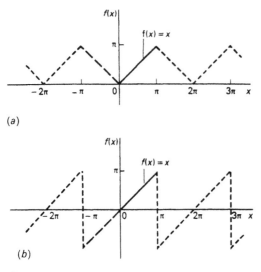

Fig. 29.7

29.3   Half range Fourier series

If a function is defined over the range say 0 to π, instead of from 0 to 2π, or from $-\pi$ to $+\pi$, it may be expanded in a series of sine terms only or of cosine terms only. This is particularly useful in the solution of some partial differential equations where the boundary conditions may restrict us to a series which contains sine terms only or cosine terms only. The series produced is then called a **half range Fourier series**. For example, if a half range cosine series is required for the function $f(x) = x$ in the range 0 to π, then an even periodic function is required. In Fig. 29.7(a), $f(x) = x$ is sketched from $x = 0$ to $x = \pi$. An even function means that it must be symmetrical about the $f(x)$ axis and this is shown by the broken line between $x = -\pi$ and $x = 0$. It is then assumed that the 'triangular waveform' produced is periodic of period 2π outside of this range as shown by the dotted lines.

When a '**half range**' **cosine series** is required then the same formula as used for the Fourier cosine series (see Section 29.2) is applied, that is:

$$f(x) = a_0 + \sum_{n=1}^{\infty} a_n \cos nx, \text{ where } a_0 = \frac{1}{\pi} \int_0^\pi f(x)\,\mathrm{d}x$$

$$\text{and } a_n = \frac{2}{\pi} \int_0^\pi f(x) \cos nx\,\mathrm{d}x$$

If a '**half range**' **sine series** is required for the function $f(x) = x$ in the range 0 to π, then an odd periodic function is required. In the sketch shown in Fig. 29.7(b), $f(x) = x$ is sketched from $x = 0$ to $x = \pi$ as before. An odd function means that it must be symmetrical about the origin and this is shown by the broken line between $x = -\pi$ and $x = 0$. It is then assumed that the 'saw-tooth waveform' produced is periodic, of period 2π outside of this range as shown by the dotted lines.

When a half range sine series is required then the same formula as used for the Fourier sine series is applied, that is:

$$f(x) = \sum_{n=1}^{\infty} b_n \sin nx, \text{ where } b_n = \frac{2}{\pi} \int_0^{\pi} f(x) \sin nx \, dx$$

Worked problems on half range Fourier series

Problem 1. Expand the function $f(x) = x$ in the range $0 < x < \pi$ into (a) a half range cosine series, and (b) a half range sine series.

(a)   The function $f(x) = x$ is shown in Fig. 29.7(a) as an even function, and, for a half range cosine series,

$$f(x) = a_0 + \sum_{n=1}^{\infty} a_n \cos nx$$

$$a_0 = \frac{1}{\pi} \int_0^{\pi} f(x) \, dx = \frac{1}{\pi} \int_0^{\pi} x \, dx = \frac{1}{\pi} \left[\frac{x^2}{2} \right]_0^{\pi} = \frac{\pi}{2}$$

$$a_n = \frac{2}{\pi} \int_0^{\pi} f(x) \cos nx \, dx = \frac{2}{\pi} \int_0^{\pi} x \cos nx \, dx$$

$$= \frac{2}{\pi} \left[\frac{x \sin nx}{n} + \frac{\cos nx}{n^2} \right]_0^{\pi} \text{ by parts}$$

$$= \frac{2}{\pi} \left[\left(\frac{\pi \sin n\pi}{n} + \frac{\cos n\pi}{n^2} \right) - \left(0 + \frac{\cos 0}{n^2} \right) \right]$$

$$= \frac{2}{\pi} \left(0 + \frac{\cos n\pi}{n^2} - \frac{\cos 0}{n^2} \right) = \frac{2}{\pi n^2} (\cos n\pi - 1)$$

When n is odd, $a_n = \frac{2}{\pi n^2} (-1 - 1) = \frac{-4}{\pi n^2}$. Hence $a_1 = \frac{-4}{\pi}$,

$$a_3 = \frac{-4}{\pi 3^2}, \dots$$

When n is even, $a_n = \frac{2}{\pi} (1 - 1) = 0$.

Hence the Fourier half range cosine series is

$$f(x) = x = \frac{\pi}{2} - \frac{4}{\pi} \left(\cos x + \frac{1}{3^2} \cos 3x + \frac{1}{5^2} \cos 5x + \cdots \right)$$

(b) The function $f(x) = x$ is shown in Fig. 29.7(b) as an odd function, and, for a half range sine series, $f(x) = \sum\limits_{n=1}^{\infty} b_n \sin nx$

$$b_n = \frac{2}{\pi} \int_0^{\pi} f(x) \sin nx \, dx = \frac{2}{\pi} \int_0^{\pi} x \sin nx \, dx$$

$$= \frac{2}{\pi} \left[\frac{-x \cos nx}{n} + \frac{\sin nx}{n^2} \right]_0^{\pi} \text{ by parts}$$

$$= \frac{2}{\pi} \left[\left(\frac{-\pi \cos n\pi}{n} + \frac{\sin n\pi}{n^2} \right) - (0 + 0) \right]$$

$$= -\frac{2}{n} \cos n\pi$$

When n is odd, $b_n = \frac{2}{n}$. Hence $b_1 = 2$, $b_3 = \frac{2}{3}$, $b_5 = \frac{2}{5}, \ldots$

When n is even, $b_n = -\frac{2}{n}$. Hence $b_2 = -\frac{2}{2}$, $b_4 = -\frac{2}{4}$,

$b_6 = -\frac{2}{6}, \ldots$

Hence the Fourier half range sine series is

$$f(x) = x = 2 \left(\sin x - \frac{1}{2} \sin 2x + \frac{1}{3} \sin 3x - \frac{1}{4} \sin 4x + \cdots \right)$$

Problem 2. Expand $f(x) = \sin^2 x$ as a half range Fourier sine series in the range $0 < x < \pi$ and sketch the function within and outside the given range.

When a half range sine series is required then an odd function is implied, that is a function symmetrical about the origin. The graph of $f(x) = \sin^2 x$ is shown in Fig. 29.8 in the range $x = 0$ to

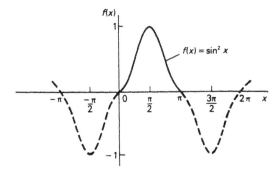

Fig. 29.8

$x = \pi$. For $\sin^2 x$ to be symmetrical about the origin the function is as shown by the broken line in Fig. 29.8 outside the given range.

For a half range Fourier sine series, $f(x) = \sum\limits_{n=1}^{\infty} b_n \sin nx \, dx$

$$b_n = \frac{2}{\pi} \int_0^\pi f(x) \sin nx \, dx = \frac{2}{\pi} \int_0^\pi \sin^2 x \sin nx \, dx$$

Since $\cos 2x = 1 - 2\sin^2 x$ then $\sin^2 x = \dfrac{1 - \cos 2x}{2}$

Hence

$$b_n = \frac{2}{\pi} \int_0^\pi \left(\frac{1 - \cos 2x}{2} \right) \sin nx \, dx = \frac{1}{\pi} \int_0^\pi (\sin nx - \sin nx \cos 2x) \, dx$$

$$= \frac{1}{\pi} \int_0^\pi \sin nx - \frac{1}{2} \Big(\sin(n+2)x + \sin(n-2)x \Big) \, dx$$

$$= \frac{1}{\pi} \left[\frac{-\cos nx}{n} + \frac{\cos(n+2)x}{2(n+2)} + \frac{\cos(n-2)x}{2(n-2)} \right]_0^\pi \text{ if } n \neq 2$$

$$= \frac{1}{\pi} \left[\left(\frac{-\cos n\pi}{n} + \frac{\cos(n+2)\pi}{2(n+2)} + \frac{\cos(n-2)\pi}{2(n-2)} \right) \right.$$
$$\left. - \left(\frac{-\cos 0}{n} + \frac{\cos 0}{2(n+2)} + \frac{\cos 0}{2(n-2)} \right) \right]$$

If n is odd,

$$b_n = \frac{1}{\pi} \left[\frac{2}{n} - \frac{2}{2(n+2)} - \frac{2}{2(n-2)} \right]$$

$$= \frac{2(n^2 - 4) - n(n-2) - n(n+2)}{\pi n(n+2)(n-2)}$$

$$= \frac{-8}{\pi n(n+2)(n-2)}$$

Hence,

$$b_1 = \frac{-8}{-3\pi} = \frac{8}{3\pi}, \quad b_3 = \frac{-8}{\pi(3)(5)}, \quad b_5 = \frac{-8}{\pi(5)(7)(3)}, \ldots$$

If n is even, $b_n = 0$ if $n \neq 2$

$$b_2 = \frac{1}{\pi} \int_0^\pi \left(\sin 2x - \frac{1}{2} \sin 4x \right) dx = 0$$

(since $\sin 4x = 2 \sin 2x \cos 2x$)

Hence the half range Fourier sine series is given by:

$$f(x) = \sin^2 x = \frac{8}{\pi} \left\{ \frac{\sin x}{(1)(3)} - \frac{\sin 3x}{(1)(3)(5)} - \frac{\sin 5x}{(3)(5)(7)} - \frac{\sin 7x}{(5)(7)(9)} - \cdots \right\}$$

Further problems on half range Fourier series may be found in the following Section 29.4 (Problems 9 to 16), page 427.

29.4   Further problems

Even and odd functions
In Problems 1 and 2 determine whether the given functions are even, odd or neither even nor odd.

1.   (a) e^x (b) $\sin nx$ (c) $\ln x$ (d) $\cos 2x$ (e) $3x^3$
     (a) [neither]   (b) [odd]   (c) [neither]   (d) [even]   (e) [odd]

2.   (a) e^{x^2} (b) $2x \sin x$ (c) $\dfrac{\cos x}{x}$ (d) $\sin x^2$ (e) $\sin^2 x$
     (a) [even]   (b) [even]   (c) [odd]   (d) [even]   (e) [even]

3.   Are the following functions which are assumed to be periodic, of period 2π, even, odd or neither even nor odd?

   (a) $f(x) = \begin{cases} x, & -\pi < x < 0 \\ -x, & 0 < x < \pi \end{cases}$

   (b) $f(x) = \begin{cases} e^x, & -\pi < x < 0 \\ e^{-x}, & 0 < x < \pi \end{cases}$

   (c) $f(x) = \begin{cases} x, & -\dfrac{\pi}{2} < x < \dfrac{\pi}{2} \\ \\ 0, & \dfrac{\pi}{2} < x < \dfrac{3\pi}{2} \end{cases}$

   (d) $f(t) = \begin{cases} t - 1, & -\pi < t < 0 \\ t + 1, & 0 < t < \pi \end{cases}$

   (a) [even]   (b) [even]   (c) [odd]   (d) [odd]

Fourier cosine and Fourier sine series

4.   Sketch the function defined by $f(x) = \begin{cases} x + \pi, & -\pi < x < 0 \\ x - \pi, & 0 < x < \pi \end{cases}$
     which is periodic of period 2π, and obtain the Fourier series representing the given function.

$$\left[f(x) = -2\left(\sin x + \frac{1}{2}\sin 2x + \frac{1}{3}\sin 3x + \frac{1}{4}\sin 4x + \cdots \right) \right]$$

5.   Find a Fourier sine series for the function

   $f(x) = \begin{cases} x - 1, & -\pi < x < 0 \\ x + 1, & 0 < x < \pi \end{cases}$

which is periodic of period 2π.

$$\left[f(x) = \frac{2}{\pi} \left((2 + \pi) \sin x - \frac{\pi}{2} \sin 2x + \frac{1}{3} (2 + \pi) \sin 3x \right. \right.$$

$$\left. \left. - \frac{\pi}{4} \sin 4x + \cdots \right) \right]$$

6. Find the Fourier cosine series for the function

$$f(x) = \begin{cases} -3, & -\pi < x < -\dfrac{\pi}{2} \\[2mm] 3, & -\dfrac{\pi}{2} < x < \dfrac{\pi}{2} \\[2mm] -3, & \dfrac{\pi}{2} < x < \pi \end{cases}$$

which is periodic outside of this range, of period 2π.

$$\left[f(x) = \frac{12}{\pi} \left(\cos x - \frac{1}{3} \cos 3x + \frac{1}{5} \cos 5x - \frac{1}{7} \cos 7x + \cdots \right) \right]$$

7. A function of t, which is periodic, of period 2π is defined by:

$$f(t) = \begin{cases} 1 - t, & -\pi < t < 0 \\ 1 + t, & 0 < t < \pi \end{cases}$$

Obtain a Fourier series for $f(t)$.

$$\left[f(t) = \frac{\pi}{2} + 1 - \frac{4}{\pi} \left(\cos t + \frac{1}{9} \cos 3t + \frac{1}{25} \cos 5t + \cdots \right) \right]$$

8. Find a Fourier sine series for the function $f(x) = \dfrac{1}{4}$ in the range $x = 0$ to $x = \pi$. By plotting graphs of the partial sums of the series show that the result approximates more and more closely to the function it represents. Deduce a series for $\dfrac{\pi}{4}$ from the Fourier expansion.

$$\left[\begin{array}{l} f(x) = \dfrac{1}{\pi} \left(\sin x + \dfrac{\sin 3x}{3} + \dfrac{\sin 5x}{5} + \dfrac{\sin 7x}{7} + \cdots \right) \\[3mm] \dfrac{\pi}{4} = 1 - \dfrac{1}{3} + \dfrac{1}{5} - \dfrac{1}{7} + \dfrac{1}{9} - \cdots \end{array} \right]$$

Half range Fourier series

9. Find (a) the half range sine series and (b) the half range cosine series for the function defined by:

$$f(x) = \begin{cases} 0, & 0 < x < \dfrac{\pi}{2} \\[2mm] 1, & \dfrac{\pi}{2} < x < \pi \end{cases}$$

(a) $\left[f(x) = \dfrac{2}{\pi} \left(\sin x - \sin 2x + \dfrac{\sin 3x}{3} + \dfrac{\sin 5x}{5} \right.\right.$

$$\left.\left. - \dfrac{\sin 6x}{3} + \cdots \right) \right]$$

(b) $\left[f(x) = \dfrac{1}{2} - \dfrac{2}{\pi} \left(\cos x - \dfrac{\cos 3x}{3} + \dfrac{\cos 5x}{5} - \cdots \right) \right]$

10. Expand $f(x) = \sin^2 x$ between $x = 0$ and $x = \pi$ into a Fourier half range cosine series.

$$\left[f(x) = \dfrac{1}{2}(1 - \cos 2x) \right]$$

11. Show that the first five terms of the half range Fourier sine series for $f(x) = x^2$, when $0 < x < \pi$. is given by:

$$f(x) = \left(2\pi - \dfrac{8}{\pi} \right) \sin x - \pi \sin 2x + \left(\dfrac{2\pi}{3} - \dfrac{8}{3^3 \pi} \right) \sin 3x$$

$$- \dfrac{\pi}{2} \sin 4x + \left(\dfrac{2\pi}{5} - \dfrac{8}{5^3 \pi} \right) \sin 5x - \cdots$$

12. Show that the half range sine series for:

$$f(x) = \begin{cases} x, & 0 < x < \dfrac{\pi}{2} \\[2mm] 0, & \dfrac{\pi}{2} < x < \pi \end{cases}$$

is given by:

$$f(x) = \dfrac{2}{\pi} \left\{ \sin x + \dfrac{\pi}{4} \sin 2x - \dfrac{\sin 3x}{9} - \dfrac{\pi}{8} \sin 4x \ldots \right\}$$

13. Determine the half range Fourier cosine series for $f(x) = \sin x$ in the range 0 to π. Illustrate graphically the function represented by the series within and outside the defined range.

$$\left[f(x) = \dfrac{4}{\pi} \left\{ \dfrac{1}{2} - \dfrac{\cos 2x}{(1)(3)} - \dfrac{\cos 4x}{(3)(5)} - \dfrac{\cos 6x}{(5)(7)} \cdots \right\} \right]$$

14. Find the half range Fourier sine series for $f(x) = \cos x$ in the

range 0 to π.

$$\left[f(x) = \frac{4}{\pi} \left\{ \frac{2 \sin 2x}{3} + \frac{4 \sin 4x}{(3)(5)} + \frac{6 \sin 6x}{(5)(7)} + \cdots \right\} \right]$$

15. (a) Obtain the half range Fourier cosine series in the range $x = 0$ to $x = \pi$ for:

$$f(x) = \begin{cases} x, & 0 < x < \dfrac{\pi}{2} \\ (\pi - x), & \dfrac{\pi}{2} < x < \pi \end{cases}$$

(b) From the results in (a) obtain a series for $\dfrac{\pi^2}{8}$ by letting $x = \dfrac{\pi}{2}$.

(a) $\left[f(x) = \dfrac{\pi}{4} - \dfrac{2}{\pi} \left(\cos 2x + \dfrac{\cos 6x}{9} + \dfrac{\cos 10x}{25} + \cdots \right) \right]$

(b) $\left[\dfrac{\pi^2}{8} = 1 + \dfrac{1}{3^2} + \dfrac{1}{5^2} + \dfrac{1}{7^2} + \cdots \right]$

16. Show that the Fourier sine series expansion of e^x in the range 0 to π is given by:

$$\frac{2}{\pi} \sum_{n=1}^{\infty} \frac{n}{1 + n^2} \left\{ 1 + (-1)^{n-1} e^{\pi} \right\} \sin nx$$

Chapter 30

Fourier series over any range

30.1 Expansion of a periodic function of period l

A periodic function $f(x)$ of period l repeats itself when x increases by l, that is, $f(x + l) = f(x)$. The transition from functions having period 2π to functions having any period l is straightforward since it may be achieved by a change of scale.

To find a Fourier series for a function $f(x)$ in the range $-\dfrac{l}{2} \leqslant x \leqslant \dfrac{l}{2}$ a new variable, u, is introduced such that $f(x)$, as a function of u, has period 2π. Let $u = \dfrac{2\pi x}{l}$. Then, when $x = -\dfrac{l}{2}$, $u = -\pi$ and when $x = \dfrac{l}{2}$, $u = +\pi$.

Also if $f(x) = f\left(\dfrac{lu}{2\pi}\right) = F(u)$, then the Fourier series for $F(u)$ is given by:

$$F(u) = a_0 + \sum_{n=1}^{\infty} (a_n \cos nu + b_n \sin nu)$$

where, from Section 27.4:

$$a_0 = \frac{1}{2\pi} \int_{-\pi}^{\pi} F(u) \, du$$

$$a_n = \frac{1}{\pi} \int_{-\pi}^{\pi} F(u) \cos nu \, du$$

and $\quad b_n = \dfrac{1}{\pi} \displaystyle\int_{-\pi}^{\pi} F(u) \sin nu \, du$

It is usually easier to change the above formulae to terms of x. Since $u = \dfrac{2\pi x}{l}$ then $du = \dfrac{2\pi}{l} \, dx$ and the limits of integration are $-\dfrac{l}{2}$ to $+\dfrac{l}{2}$ instead of $-\pi$ to $+\pi$.

Hence the Fourier series expressed in terms of x is given by:

$$f(x) = a_0 + \sum_{n=1}^{\infty} \left\{ a_n \cos\left(\frac{2\pi nx}{l}\right) + b_n \sin\left(\frac{2\pi nx}{l}\right) \right\}$$

where $a_0 = \dfrac{1}{2\pi} \displaystyle\int_{-\pi}^{\pi} F(u)\,du = \dfrac{1}{2\pi} \displaystyle\int_{-1/2}^{1/2} f\left(\dfrac{lu}{2\pi}\right)\left(\dfrac{2\pi}{l}\,dx\right)$

i.e. $a_0 = \dfrac{1}{l} \displaystyle\int_{-1/2}^{1/2} f(x)\,dx$ (1)

$a_n = \dfrac{1}{\pi} \displaystyle\int_{-\pi}^{\pi} F(u)\cos nu\,du = \dfrac{1}{\pi} \displaystyle\int_{-1/2}^{1/2} f\left(\dfrac{lu}{2\pi}\right)\cos\left(\dfrac{2\pi nx}{l}\right)\left(\dfrac{2\pi}{l}\,dx\right)$

i.e. $a_n = \dfrac{2}{l} \displaystyle\int_{-1/2}^{1/2} f(x)\cos\left(\dfrac{2\pi nx}{l}\right)dx$ (2)

in the range $-\dfrac{l}{2}$ to $+\dfrac{1}{2}$

and $b_n = \dfrac{1}{\pi} \displaystyle\int_{-\pi}^{\pi} F(u)\sin nu\,du = \dfrac{1}{\pi} \displaystyle\int_{-1/2}^{1/2} f\left(\dfrac{lu}{2\pi}\right)\sin\left(\dfrac{2\pi nx}{l}\right)\left(\dfrac{2\pi}{l}\,dx\right)$

i.e. $b_n = \dfrac{2}{l} \displaystyle\int_{-1/2}^{1/2} f(x)\sin\left(\dfrac{2\pi nx}{l}\right)dx$ (3)

in the range $-\dfrac{l}{2}$ to $+\dfrac{l}{2}$.

The limits of integration in equations (1) to (3) may be replaced by any interval of length l; for example, the interval 0 to l.

Worked problems on Fourier series over any range

Problem 1. Find the Fourier series defined by:

$$f(x) = \begin{cases} 0 & \text{when } -2 < x < -1 \\ 3 & \text{when } -1 < x < 1 \\ 0 & \text{when } 1 < x < 2 \end{cases}$$

and which is periodic outside of this range, of period 4.

The function $f(x)$ is shown in Fig. 30.1 where $l = 4$.

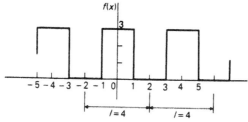

Fig. 30.1

Since the function is symmetrical about the $f(x)$ axis it is an even function and the Fourier series contains no sine terms (i.e. $b_n = 0$).

Thus $f(x) = a_0 + \sum\limits_{n=1}^{\infty} a_n \cos\left(\dfrac{2\pi n x}{l}\right)$

From equation (1),

$$a_0 = \frac{1}{l}\int_{-l/2}^{l/2} f(x)\,dx = \frac{1}{4}\int_{-2}^{2} f(x)\,dx$$

$$= \frac{1}{4}\left[\int_{-1}^{1} 3\,dx + \int_{-2}^{-1} 0\,dx + \int_{1}^{2} 0\,dx\right]$$

$$= \frac{1}{4}[3x]_{-1}^{1} = \frac{1}{4}[(3)-(-3)]$$

$$= \frac{6}{4} = \frac{3}{2}$$

From equation (2),

$$a_n = \frac{2}{l}\int_{-l/2}^{l/2} f(x)\cos\left(\frac{2\pi n x}{l}\right)dx$$

$$= \frac{2}{4}\int_{-2}^{2} f(x)\cos\left(\frac{2\pi n x}{4}\right)dx$$

$$= \frac{1}{2}\left[\int_{-2}^{-1} 0\cos\left(\frac{\pi n x}{2}\right)dx + \int_{-1}^{1} 3\cos\left(\frac{\pi n x}{2}\right)dx\right.$$

$$\left. + \int_{0}^{2} 0\cos\left(\frac{\pi n x}{2}\right)dx\right]$$

$$= \frac{3}{2}\left[\frac{\sin\left(\dfrac{\pi n x}{2}\right)}{\dfrac{\pi n}{2}}\right]_{-1}^{1} = \frac{3}{\pi n}\left[\sin\left(\frac{\pi n}{2}\right) - \sin\left(\frac{-\pi n}{2}\right)\right]$$

When n is even, $a_n = 0$

$$a_1 = \frac{3}{\pi}(1--1) = \frac{6}{\pi}, \quad a_3 = \frac{3}{3\pi}(-1-1) = \frac{-2}{\pi},$$

$$a_5 = \frac{3}{5\pi}(1--1) = \frac{6}{5\pi},$$

and so on.

Hence

$$f(x) = \frac{3}{2} + \frac{6}{\pi}\cos\left(\frac{\pi x}{2}\right) - \frac{2}{\pi}\cos\left(\frac{3\pi x}{2}\right) + \frac{6}{5\pi}\cos\left(\frac{5\pi x}{2}\right)$$

$$- \frac{6}{7\pi}\cos\left(\frac{7\pi x}{2}\right) + \cdots$$

$$f(x) = \frac{3}{2} + \frac{6}{\pi}\left\{\cos\left(\frac{\pi x}{2}\right) - \frac{1}{3}\cos\left(\frac{3\pi x}{2}\right) + \frac{1}{5}\cos\left(\frac{5\pi x}{2}\right)\right.$$

$$\left. - \frac{1}{7}\cos\left(\frac{7\pi x}{2}\right) + \cdots\right\}$$

Problem 2. Find the Fourier series expansion of the function $f(x) = x$ in the range $x = 0$ to $x = 1$.

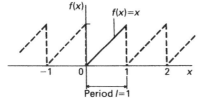

Fig. 30.2

$f(x) = x$ in the interval 0 to 1 is as shown in Fig. 30.2. Although the function is not periodic it may be constructed so, of period 1, outside of the given range as shown by the broken lines.

The Fourier series is given by:

$$f(x) = a_0 + \sum_{n=1}^{\infty}\left\{a_n\cos\left(\frac{2\pi nx}{l}\right) + b_n\sin\left(\frac{2\pi nx}{l}\right)\right\}$$

From equation (1),

$$a_0 = \frac{1}{l}\int_{-l/2}^{l/2} f(x)\,dx = \frac{1}{l}\int_0^l f(x)\,dx = \frac{1}{1}\int_0^1 x\,dx = \left[\frac{x^2}{2}\right]_2^1 = \frac{1}{2}$$

From equation (2),

$$a_n = \frac{2}{l}\int_{-l/2}^{l/2} f(x)\cos\left(\frac{2\pi nx}{l}\right)dx = \frac{2}{l}\int_0^l x\cos\left(\frac{2\pi nx}{l}\right)dx$$

$$= \frac{2}{1}\int_0^1 x\cos(2\pi nx)\,dx$$

$$= 2\left[\frac{x\sin(2\pi nx)}{2\pi n} + \frac{\cos(2\pi nx)}{(2\pi n)^2}\right]_0^1 \text{ by parts,}$$

$$= 2\left[\left(\frac{\sin(2\pi n)}{2\pi n} + \frac{\cos(2\pi n)}{(2\pi n)^2}\right) - \left(0 + \frac{\cos 0}{(2\pi n)^2}\right)\right]$$

i.e. $a_n = 0$

From equation (3),

$$b_n = \frac{2}{l} \int_{-1/2}^{1/2} f(x) \sin\left(\frac{2\pi n x}{l}\right) dx = \frac{2}{l} \int_0^l x \sin\left(\frac{2\pi n x}{l}\right) dx$$

$$= \frac{2}{l} \int_0^1 x \sin(2\pi n x)\, dx$$

$$= 2\left[\frac{-x\cos(2\pi n x)}{2\pi n} + \frac{\sin(2\pi n x)}{(2\pi n)^2}\right]_0^1$$

$$= 2\left[\left(\frac{-\cos(2\pi n)}{2\pi n} + \frac{\sin 2\pi n}{(2\pi n)^2}\right) - \left(0 + \frac{\sin 0}{(2\pi n)^2}\right)\right]$$

$$= \frac{-\cos 2\pi n}{\pi n} = \frac{-1}{\pi n}$$

Hence, $b_1 = \dfrac{-1}{\pi}$, $b_2 = \dfrac{-1}{2\pi}$, $b_3 = \dfrac{-1}{3\pi}, \cdots$

Thus $f(x) = \dfrac{1}{2} - \dfrac{1}{\pi}\left(\sin 2\pi x + \dfrac{1}{2}\sin 4\pi x + \dfrac{1}{6}\sin 6\pi x + \cdots\right)$ **in the range 0 to 1.**

Further problems on Fourier series over any range may be found in Section 30.3 (Problems 1 to 4), page 438.

30.2 Half range Fourier series for functions defined over range *l*

By making the substitution $u = \dfrac{\pi x}{l}$ we can make the range $x = 0$ to $x = l$ correspond to the range $u = 0$ to $u = \pi$ and hence expand the function in a series of sines or cosines alone of multiples of u. These are called the half range series. Thus a **half range cosine series** in the range 0 to l can be expanded as:

$$f(x) = a_0 + \sum_{n=1}^{\infty} a_n \cos\left(\frac{n\pi x}{l}\right)$$

By similar substitutions as used in section (1) it may be shown that the Fourier coefficients are given by:

$$a_0 = \frac{1}{l} \int_0^l f(x)\, dx \tag{4}$$

and $\quad a_n = \dfrac{2}{l} \displaystyle\int_0^l f(x) \cos\left(\dfrac{n\pi x}{l}\right) dx \tag{5}$

The **half range sine series** in the range 0 to *l* can be expanded as:

$$f(x) = \sum_{n=1}^{\infty} b_n \sin\left(\frac{n\pi x}{l}\right)$$

where $\quad b_n = \dfrac{2}{l}\displaystyle\int_0^l f(x) \sin\left(\frac{n\pi x}{l}\right) \mathrm{d}x$ (6)

Summary

A function of *x* can be expanded in the range $-\dfrac{l}{2}$ to $\dfrac{l}{2}$ or from 0 to *l*, or any range of width *l*, in any of three forms. These are:

(i)   as a series of sines and cosines of multiples of $\dfrac{2\pi x}{l}$, using equations (1)–(3) to determine the Fourier coefficients,

(ii)   as a series of cosines of multiples of $\dfrac{\pi x}{l}$, using equations (4) and (5) to determine the Fourier coefficients, and

(iii)   as a series of sines of multiples of $\dfrac{\pi x}{l}$, using equation (6) to determine the Fourier coefficients.

Worked problem on half range Fourier series for functions defined over range *l*

Problem 1. Expand to the first 4 terms the function $f(x) = x$, when $0 < x < 4$ in a half range Fourier (*a*) sine series and (*b*) cosine series.

(*a*)   A half range sine series indicates an odd function. Thus the graph of $f(x) = x$ in the range 0 to 4 is shown in Fig. 30.3 and is extended outside of this range so as to be symmetrical about the origin as shown by the broken lines.

For a half range sine series: $f(x) = \displaystyle\sum_{n=1}^{\infty} b_n \sin\left(\frac{n\pi x}{l}\right)$

From equation (6),

$$b_n = \frac{2}{l}\int_0^l f(x) \sin\left(\frac{n\pi x}{l}\right) \mathrm{d}x = \frac{2}{4}\int_0^4 x \sin\left(\frac{n\pi x}{4}\right) \mathrm{d}x, \text{ since } l = 4$$

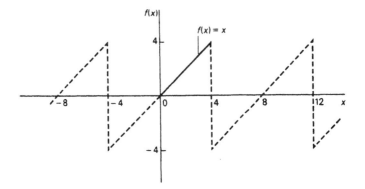

f(x) = x

Fig. 30.3

$$= \frac{1}{2}\left[\frac{-x\cos\left(\dfrac{n\pi x}{4}\right)}{\dfrac{n\pi}{4}} - \frac{\sin\left(\dfrac{n\pi x}{4}\right)}{\left(\dfrac{n\pi}{4}\right)^2}\right]_0^4$$

$$= \frac{1}{2}\left[\left(\frac{-4\cos n\pi}{\dfrac{n\pi}{4}} - \frac{\sin n\pi}{\left(\dfrac{n\pi}{4}\right)^2}\right) - \left(0 - \frac{\sin 0}{\left(\dfrac{n\pi}{4}\right)^2}\right)\right]$$

$$= \frac{1}{2}\left(\frac{-4\cos n\pi}{\dfrac{n\pi}{4}}\right)$$

$$= \frac{-8}{n\pi}\cos n\pi$$

When $n = 1$, $b_1 = \dfrac{-8}{\pi}(-1) = \dfrac{8}{\pi}$

When $n = 2$, $b_2 = \dfrac{-8}{2\pi}(1) = \dfrac{-4}{\pi}$

When $n = 3$, $b_3 = \dfrac{-8}{3\pi}(-1) = \dfrac{8}{3\pi}$

When $n = 4$, $b_4 = \dfrac{-8}{4\pi}(1) = \dfrac{-2}{\pi}$

Hence

$$f(x) = \frac{8}{\pi}\left(\sin\left(\frac{\pi x}{4}\right) - \frac{1}{2}\sin\left(\frac{\pi x}{2}\right) + \frac{1}{3}\sin\left(\frac{3\pi x}{4}\right) - \frac{1}{4}\sin(\pi x) + \cdots\right)$$

in the range 0 to 4.

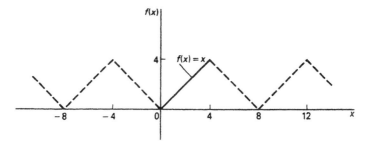

Fig. 30.4

(b)   A half range cosine series indicates an even function. Thus the graph of $f(x) = x$ in the range 0 to 4 is shown in Fig. 30.4 and is extended outside of this range so as to be symmetrical with the $f(x)$ axis as shown by the broken lines.

For a half range cosine series, $f(x) = a_0 + \sum\limits_{n=1}^{\infty} a_n \cos\left(\dfrac{n\pi x}{l}\right)$

From equation (4),

$$a_0 = \frac{1}{l}\int_0^l f(x)\,\mathrm{d}x = \frac{1}{4}\int_0^4 x\,\mathrm{d}x = \frac{1}{4}\left[\frac{x^2}{2}\right]_0^4 = 2$$

From equation (5),

$$a_n = \frac{2}{l}\int_0^l f(x)\cos\left(\frac{n\pi x}{l}\right)\mathrm{d}x = \frac{2}{4}\int_0^4 x\cos\left(\frac{n\pi x}{4}\right)\mathrm{d}x$$

$$= \frac{1}{2}\left[\frac{x\sin\left(\dfrac{n\pi x}{4}\right)}{\dfrac{n\pi}{4}} - \frac{\cos\left(\dfrac{n\pi x}{4}\right)}{\left(\dfrac{n\pi}{4}\right)^2}\right]_0^4$$

$$= \frac{1}{2}\left[\left(\frac{4\sin n\pi}{\dfrac{n\pi}{4}} - \frac{\cos n\pi}{\left(\dfrac{n\pi}{4}\right)^2}\right) - \left(0 - \frac{\cos 0}{\left(\dfrac{n\pi}{4}\right)^2}\right)\right]$$

$$= \frac{1}{2}\left[\frac{-\cos n\pi}{\left(\dfrac{n\pi}{4}\right)^2} + \frac{1}{\left(\dfrac{n\pi}{4}\right)^2}\right]$$

$$= \frac{1}{2}\left(\frac{4}{n\pi}\right)^2[1 - \cos n\pi]$$

$$= \frac{8}{n^2\pi^2}(1 - \cos n\pi)$$

When n is even, $a_n = 0$.

When $n = 1$, $a_1 = \dfrac{8}{\pi^2}(2) = \dfrac{16}{\pi^2}$

When $n = 3$, $a_3 = \dfrac{8}{9\pi^2}(2) = \dfrac{16}{9\pi^2}$

When $n = 5$, $a_5 = \dfrac{8}{25\pi^2}(2) = \dfrac{16}{25\pi^2}$

When $n = 7$, $a_7 = \dfrac{8}{49\pi^2}(2) = \dfrac{16}{49\pi^2}$

Hence

$$f(x) = 2 + \frac{16}{\pi^2}\left\{ \cos\left(\frac{\pi x}{4}\right) + \frac{1}{9}\cos\left(\frac{3\pi x}{4}\right) + \frac{1}{25}\cos\left(\frac{5\pi x}{4}\right) \right.$$
$$\left. + \frac{1}{49}\cos\left(\frac{7\pi x}{4}\right) + \cdots \right\} \text{ in the range 0 to 4}$$

Further problems on half range Fourier series over range l may be found in the following Section 30.3 (Problems 5 to 7).

30.3   Further problems

Fourier series over any range

1. The voltage from a square wave generator is of the form:
$$V(t) = \begin{cases} 0, & -5 < t < 0 \\ 3, & 0 < t < 5 \end{cases}$$
    Find the Fourier series for this periodic function having period 10.
$$\left[V(t) = \frac{3}{2} + \frac{6}{\pi}\left\{ \sin\left(\frac{\pi t}{5}\right) + \frac{1}{3}\sin\left(\frac{3\pi t}{5}\right) + \frac{1}{5}\sin\left(\frac{5\pi t}{5}\right) + \cdots \right\} \right]$$

2. Figure 30.5(a) represents a half wave rectified sinusoidal voltage $E\sin\omega t$. Develop the Fourier series for the function.
$$\left[f(t) = \frac{E}{\pi} + \frac{E}{2}\sin\omega t - \frac{2E}{\pi}\left(\frac{\cos 2\omega t}{(1)(3)} + \frac{\cos 4\omega t}{(3)(5)} + \cdots \right) \right]$$

3. Obtain the Fourier series for $f(x)$ given that
$$f(x) = \begin{cases} -1, & -2 < x < 0 \\ +1, & 0 < x < 2 \end{cases}$$
$$\left[f(x) = \frac{4}{\pi}\left\{ \sin\left(\frac{\pi x}{2}\right) + \frac{1}{3}\sin\left(\frac{3\pi x}{2}\right) + \frac{1}{5}\sin\left(\frac{5\pi x}{2}\right) + \cdots \right\} \right]$$

(a)

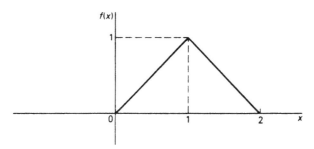

(b)

Fig. 30.5

4. Find the Fourier series for $f(x) = x$ in the range $x = -1$ to $x = +1$

$$\left[f(x) = \frac{2}{\pi}\left\{ \sin \pi x - \frac{1}{2}\sin 2\pi x + \frac{1}{3}\sin 3\pi x - \cdots \right\} \right]$$

Half range Fourier series over range *l*

5. Find the half range cosine series for the function $f(x) = x^2$ in the range 0 to 4. Sketch the function within and outside of the given range.

$$\left[f(x) = \frac{16}{3} - \frac{64}{\pi^2}\left\{ \cos\left(\frac{\pi x}{4}\right) - \frac{1}{4}\cos\left(\frac{\pi x}{2}\right) + \frac{1}{9}\cos\left(\frac{3\pi x}{4}\right) - \cdots \right\} \right]$$

6. Expand $f(x) = x$ in the range $0 \leqslant x \leqslant 2$ in a half range (a) sine series, (b) cosine series. Sketch the function in each case.

(a) $$\left[f(x) = \frac{4}{\pi}\left\{ \sin\left(\frac{\pi x}{2}\right) - \frac{1}{2}\sin(\pi x) + \frac{1}{3}\sin\left(\frac{3\pi x}{2}\right) - \cdots \right\} \right]$$

(b) $$\left[f(x) = 1 - \frac{8}{\pi^2}\left\{ \cos\left(\frac{\pi x}{2}\right) + \frac{1}{3^2}\cos\left(\frac{3\pi x}{2}\right) \right.\right.$$
$$\left.\left. + \frac{1}{5^2}\cos\left(\frac{5\pi x}{2}\right) + \cdots \right\} \right]$$

7. Show that the half range sine series for the function shown in Fig. 30.5(*b*) is given by:

$$f(x) = \frac{8}{\pi^2}\left\{\sin\left(\frac{\pi x}{2}\right) - \frac{1}{3^2}\sin\left(\frac{3\pi x}{2}\right) + \frac{1}{5^2}\sin\left(\frac{5\pi x}{2}\right) - \cdots\right\}$$

Chapter 31

A numerical method of harmonic analysis

31.1   Introduction

Many of the waveforms met in practice can be represented by simple mathematical expressions and the magnitudes of their harmonic components can be conveniently found by Fourier series. Numerical methods are used to analyse waveforms for which simple mathematical expressions cannot be obtained. **Harmonic analysis** is the process of resolving a periodic, non-sinusoidal quantity into a series of sinusoidal components of ascending order of frequency. This chapter deals with a straightforward numerical method of analysing irregular waveforms.

31.2   Harmonic analysis on data given in tabular or graphical form

The general Fourier series is given by:

$$f(x) = a_0 + a_1 \cos x + a_2 \cos 2x + a_3 \cos 3x + \cdots$$
$$+ b_1 \sin x + b_2 \sin 2x + b_3 \sin 3x + \cdots$$

i.e. $$f(x) = a_0 + \sum_{n=1}^{\infty} (a_n \cos nx + b_n \sin nx) \tag{1}$$

and the Fourier coefficients are given by:

$$a_0 = \frac{1}{2\pi} \int_{-\pi}^{\pi} f(x)\, dx = \frac{1}{2\pi} \int_{0}^{2\pi} f(x)\, dx = \text{the mean value of } f(x) \text{ in}$$
$$\text{the range } -\pi \text{ to } \pi \text{ or } 0 \text{ to } 2\pi$$

$$a_n = \frac{1}{\pi} \int_{-\pi}^{\pi} f(x) \cos nx\, dx = \frac{1}{\pi} \int_{0}^{2\pi} f(x) \cos nx\, dx = \text{twice the mean}$$
$$\text{value of } f(x) \cos nx \text{ in the range } 0 \text{ to } 2\pi$$

and $$b_n = \frac{1}{\pi} \int_{-\pi}^{\pi} f(x) \sin nx\, dx = \frac{1}{\pi} \int_{0}^{2\pi} f(x) \sin nx\, dx = \text{twice the mean}$$
$$\text{value of } f(x) \sin nx \text{ in the range } 0 \text{ to } 2\pi$$

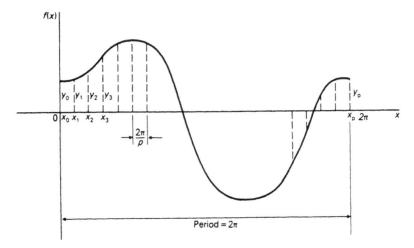

Fig. 31.1

However if an irregular graph is given we cannot evaluate the above integrals by calculus since $f(x)$ is not given by a mathematical expression. In these cases approximate methods, such as the **trapezoidal rule**, are used to evaluate the Fourier coefficients a_0, a_n and b_n. Most of the graphs to be analysed in practice are periodic. Let us assume the period of a waveform is 2π. (If it is not of period 2π then, by a simple change of variable the period may be reduced to 2π.) Let the range 0 to 2π be divided into p equal parts, the points of division being $x_0, x_1, x_2, \ldots, x_p$, as shown in Fig. 31.1. The width of each interval is then $\dfrac{2\pi}{p}$. Let the ordinates at these points be labelled $y_0, y_1, y_2, \ldots, y_p$. (Note that $y_0 = y_p$ since the waveform is assumed periodic.)

The trapezoidal rule states:

$$\text{Area} \approx (\text{width of interval}) \left[\frac{1}{2}(\text{first} + \text{last ordinate}) + \text{sum of remaining ordinates}\right]$$

$$\approx \left(\frac{2\pi}{p}\right)\left[\frac{1}{2}(y_0 + y_p) + y_1 + y_2 + y_3 + \cdots\right]$$

Since $y_0 = y_p$ then $\dfrac{1}{2}(y_0 + y_p) = y_0 = y_p$.

Hence Area $\approx \dfrac{2\pi}{p} \displaystyle\sum_{K=1}^{p} y_K$ $\left(\text{or Area} \approx \dfrac{2\pi}{p} \displaystyle\sum_{K=0}^{p-1} y_K\right)$

$$\text{Mean value} = \frac{\text{area}}{\text{length of base}} \approx \frac{1}{2\pi}\left(\frac{2\pi}{p}\sum_{K=1}^{p} y_K\right)$$

$$\approx \frac{1}{p}\sum_{K=1}^{p} y_K$$

However, a_0 = the mean value of $f(x)$ in the range 0 to 2π.

Hence $a_0 \approx \dfrac{1}{p}\sum_{K=1}^{p} y_K$ (2)

Similarly, since a_n = twice the mean value of $f(x)\cos nx$ in the range 0 to 2π then

$$a_n \approx \frac{2}{p}\sum_{K=1}^{p} y_K \cos nx_K \qquad (3)$$

and since b_n = twice the mean value of $f(x)\sin nx$ in the range 0 to 2π then

$$b_n \approx \frac{2}{p}\sum_{K=1}^{p} y_K \sin nx_K \qquad (4)$$

Equations (2) to (4) are used to determine the Fourier coefficients for irregular waveforms and this is shown in the following worked problem.

Problem 1. The values of the current i amperes at different moments in a cycle are given by:

θ (degrees)	30	60	90	120	150	180	210	240	270	300	330	360
i (amperes)	-8	10	25	31	32	25	17	10	-8	-19	-17	-13

Draw the graph of current i against angle θ and analyse the current into its first three constituent harmonics, each correct to two decimal places.

The graph of current i against angle θ is shown in Fig. 31.2.
The range 0 to 2π is divided into 12 equal parts (i.e. an interval width of $\dfrac{2\pi}{12}$, i.e. $\dfrac{\pi}{6}$ or 30°) and the values of the ordinates $y_1, y_2, y_3, \ldots$ are $-8, 10, 25 \ldots$ from the given table of values. (Note that either y_0 or y_{12} is taken, but not both.)
It is usually easiest to tabulate the data as shown in Table 31.1. From equation (2),

$$a_0 \approx \frac{1}{p}\sum_{K=1}^{p} y_K = \frac{1}{12}(85) = 7.08, \text{ since } p = 12$$

From equation (3), $a_n \approx \dfrac{2}{p}\sum_{K=1}^{p} y_K \cos nx_K$

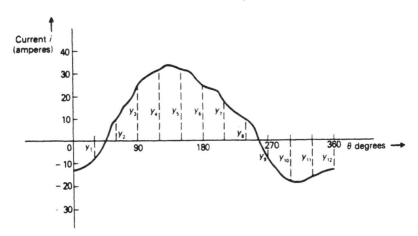

Fig. 31.2   Graph of current against angle θ (Problem 1)

Hence $a_1 \approx \dfrac{2}{12}(-127.08) = -21.18$

$a_2 \approx \dfrac{2}{12}(-9) = -1.50$

$a_3 \approx \dfrac{2}{12}(12) = 2.00$

From equation (4), $b_n \approx \dfrac{2}{p}\sum\limits_{K=1}^{p} y_K \sin nx_K$

Hence $b_1 \approx \dfrac{2}{12}(88.3) = 14.72$

$b_2 \approx \dfrac{2}{12}(1.72) = 0.29$

$b_3 \approx \dfrac{2}{12}(-9) = -1.50$

Substituting values of the Fourier coefficients into equation (1) gives:

$i = 7.08 - 21.18 \cos\theta - 1.50 \cos 2\theta + 2.00 \cos 3\theta + \cdots$
$\quad\quad + 14.72 \sin\theta + 0.29 \sin 2\theta - 1.50 \sin 3\theta \cdots$

From Problem 1 it can be appreciated that the arithmetic involved is not too difficult and a proforma of the type shown in Table 31.1 could be readily produced to tabulate results and calculations. Although 12 ordinates are used in Problem 1, the method is, in fact, independent of the number of ordinates. The more ordinates used the greater will be the accuracy of the

Table 31.1

Ordinates	$\theta°$	i	$\cos\theta$	$i\cos\theta$	$\sin\theta$	$i\sin\theta$	$\cos2\theta$	$i\cos2\theta$	$\sin2\theta$	$i\sin2\theta$	$\cos3\theta$	$i\cos3\theta$	$\sin3\theta$	$i\sin3\theta$
y_1	30	−8	0.866	−6.93	0.5	−4	0.5	−4	0.866	−6.93	0	0	1	−8
y_2	60	10	0.5	5	0.866	8.66	−0.5	−5	0.866	8.66	−1	−10	0	0
y_3	90	25	0	0	1	25	−1	−25	0	0	0	0	−1	−25
y_4	120	31	−0.5	−15.5	0.866	26.85	−0.5	−15.5	−0.866	−26.85	1	31	0	0
y_5	150	32	−0.866	−27.71	0.5	16	0.5	16	−0.866	−27.71	0	0	1	32
y_6	180	25	−1	−25	0	0	1	25	0	0	−1	−25	0	0
y_7	210	17	−0.866	−14.72	−0.5	−8.5	0.5	8.5	0.866	14.72	0	0	−1	−17
y_8	240	10	−0.5	−5	−0.886	−8.66	−0.5	−5	0.866	8.66	1	10	0	0
y_9	270	−8	0	0	−1	8	−1	8	0	0	0	0	1	−8
y_{10}	300	−19	0.5	−9.5	−0.866	16.45	−0.5	9.5	−0.866	16.45	−1	19	0	0
y_{11}	330	−17	0.866	−14.72	−0.5	8.5	0.5	−8.5	−0.866	14.72	0	0	−1	17
y_{12}	360	−13	1	−13	0	0	1	−13	0	0	1	−13	0	0
		$\sum\limits_{K=1}^{12} y_K = 85$		$\sum\limits_{K=1}^{12} y_K\cos\theta_K$ $= -127.08$		$\sum\limits_{K=1}^{12} y_K\sin\theta_K$ $= 88.3$		$\sum\limits_{K=1}^{12} y_K\cos2\theta_K$ $= -9$		$\sum\limits_{K=1}^{12} y_K\sin2\theta_K$ $= 1.72$		$\sum\limits_{K=1}^{12} y_K\cos3\theta_K$ $= 12$		$\sum\limits_{K=1}^{12} y_K\sin3\theta_K$ $= -9$

results. Only the first three harmonics were required in Problem 1. The method can be continued for further harmonics with the arithmetic generally becoming easier the higher the harmonic.

31.3   Complex waveform considerations

Examination of a complex waveform often can reveal much about the harmonic content and the following five points should be considered before analysis is attempted.

(i) $a_0 = \dfrac{1}{2\pi} \displaystyle\int_0^{2\pi} f(x)\,dx$. This is the mean value of $f(x)$ in the range 0 to 2π. Thus if the average value of $f(x)$ over one cycle is zero (i.e. the area above the x-axis is equal to the area below it) then $a_0 = 0$. Hence for a waveform of the form shown in Fig. 31.3(a), $a_0 = 0$.

(ii)   An **even** function is defined by $f(-x) = f(x)$ and it is symmetrical about the $f(x)$-axis. **An even function contains no sine terms**. A typical even waveform is shown in Fig. 31.3(b).

(iii)   An **odd** function is defined by $f(-x) = -f(x)$ and it is symmetrical about the origin. **An odd function contains no cosine terms**. A typical odd function is shown in Fig. 31.3(c).

(iv) $f(x) = f(x+\pi)$ represents a wave which repeats after half a cycle. Only even harmonics can be present in such a waveform, i.e. $a_1, a_3, a_5, \ldots,$ $b_1, b_3, b_5, \ldots$ are all zero. A typical wave of this form is shown in Fig. 31.3(d).

(v) $f(x) = -f(x+\pi)$ represents a wave for which the positive and negative half cycles are identical in shape. Only odd harmonics can be present in such a waveform, i.e. $a_0, a_2, a_4, \ldots b_2, b_4, \ldots,$ are all zero. A typical wave of this form is shown in Fig. 31.3(e).

Problem 2. An alternating emf E volts is shown in Fig. 31.4. Analyse the waveform into its constituent harmonics as far as, and including, the fifth harmonic by taking 30° intervals.

(i)   The mean value of a cycle is seen to be zero, since the area above the θ-axis is equal to the area under it. Thus the constant term, or d.c. component, $a_0 = 0$.

(ii)   Since the waveform is symmetrical about the E-axis the function E is even, which means that there are no sine terms present in the Fourier series.

(iii)   The waveform is of the form $f(\theta) = -f(\theta + \pi)$ which means that only odd harmonics are present.

These considerations can save a lot of time doing unnecessary calculations, for in this case we find that only odd cosine terms

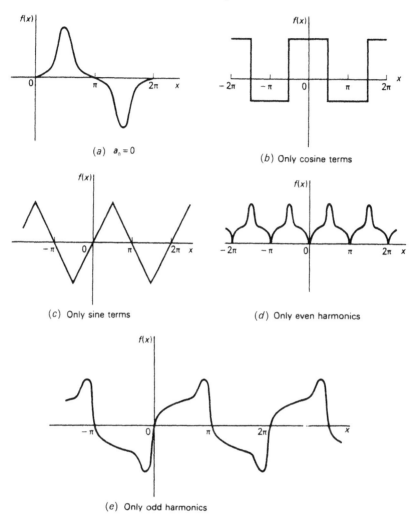

(a) $a_n = 0$

(b) Only cosine terms

(c) Only sine terms

(d) Only even harmonics

(e) Only odd harmonics

Fig. 31.3

will appear in the Fourier series, that is,

$$E = a_1 \cos \theta + a_3 \cos 3\theta + a_5 \cos 5\theta + \cdots$$

A proforma, similar to Table 31.1, but without the 'sine term' columns and without the 'even cosine term' columns is shown in Table 31.2 up to and including the fifth harmonic, from which the Fourier coefficients a_1, a_3 and a_5 can be determined. Twelve ordinates are chosen and labelled $y_1, y_2, \ldots, y_{12}$ in Fig. 31.4.

From equation (3), $a_n \approx \dfrac{2}{p} \displaystyle\sum_{K=1}^{p} E_K \cos n\theta_K$, where $p = 12$

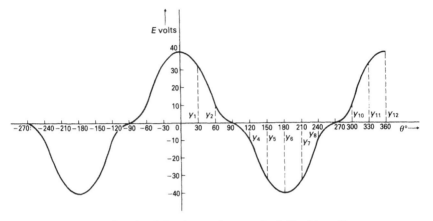

Fig. 31.4 Graph of E volts against angle θ (Problem 2)

Table 31.2

Ordinate	$\theta°$	E	$\cos\theta$	$E\cos\theta$	$\cos 3\theta$	$E\cos 3\theta$	$\cos 5\theta$	$E\cos 5\theta$
y_1	30	33	0.866	28.58	0	0	−0.866	−28.58
y_2	60	10	0.5	5	−1	−10	0.5	5
y_3	90	0	0	0	0	0	0	0
y_4	120	−10	−0.5	5	1	−10	−0.5	5
y_5	150	−33	−0.866	28.58	0	0	0.866	−28.58
y_6	180	−40	−1	40	−1	40	−1	40
y_7	210	−33	−0.866	28.58	0	0	0.866	−28.58
y_8	240	−10	−0.5	5	1	−10	−0.5	5
y_9	270	0	0	0	0	0	0	0
y_{10}	300	10	0.5	5	−1	−10	0.5	5
y_{11}	330	33	0.866	28.58	0	0	−0.866	−28.58
y_{12}	360	40	1	40	1	40	1	40

$$\sum_{K=1}^{12} E_K \cos\theta_K \qquad \sum_{K=1}^{12} E_K \cos 3\theta_K \qquad \sum_{K=1}^{12} E_K \cos 5\theta_K$$

$$= 214.32 \qquad\qquad = 40 \qquad\qquad = -14.32$$

Hence $a_1 \approx \dfrac{2}{12}(214.32) = 35.72$

$a_3 \approx \dfrac{2}{12}(40) = 6.67$

$a_5 \approx \dfrac{2}{12}(-14.32) = -2.39$

Thus $E = 35.72\cos\theta + 6.67\cos 3\theta - 2.39\cos 5\theta + \cdots$

Further problems on harmonic analysis on data given in tabular and graphical form may be found in the following Section 31.4 (Problems 1 to 10).

31.4   Further problems

1. Without performing calculations, state which harmonics will
    be present in the waveforms shown in Fig. 31.5.

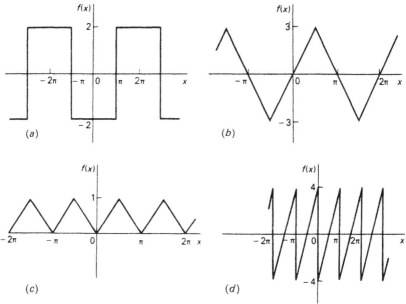

Fig. 31.5

(a) [Only odd cosine terms present]
(b) [Only odd sine terms present]
(c) [Only even cosine harmonics present]
(d) [Only even sine terms present]

Find the Fourier series to represent the periodic functions
given by the tables of values in Problems 2 to 4 up to and
including the third harmonic and each coefficient correct to two
decimal places. Use 12 ordinates in each case.

2. | θ | 0 | 30 | 60 | 90 | 120 | 150 | 180 |
    |---|---|---|---|---|---|---|---|
    | $f(\theta)$ | -1.7 | 2.9 | 4.4 | 4.7 | 5.3 | 6.1 | 4.9 |

θ	210	240	270	300	330
$f(\theta)$	1.5	-2.9	-7.0	-8.5	-5.4

    $[f(\theta) = 0.36 - 3.10 \cos \theta + 6.07 \sin\theta + 1.48 \cos 2\theta$
    $\qquad\qquad + 1.21 \sin 2\theta - 0.02 \cos 3\theta + 0.20 \sin 3\theta]$

3. | θ | 30 | 60 | 90 | 120 | 150 | 180 | 210 |
    |---|---|---|---|---|---|---|---|
    | y | 14.6 | 17.2 | 21.6 | 28.8 | 42.3 | 51.6 | 62.1. |

θ	240	270	300	330	360
y	49.3	31.6	21.4	9.1	7.4

    $[y = 29.75 - 22.31 \cos \theta - 6.42 \sin \theta + 1.92 \cos 2\theta$
    $+ 6.01 \sin 2\theta - 0.78 \cos 3\theta - 0.72 \sin 3\theta]$

4. | θ | 0 | 30 | 60 | 90 | 120 | 150 | 180 |
    |---|---|---|---|---|---|---|---|
    | I | 11 | 2 | -21 | -21 | -12 | 10 | 29 |

θ	210	240	270	300	330
I	40	43	40	18	13

    $[I = 12.67 - 10.89 \cos \theta - 27.15 \sin \theta$
    $+ 6.58 \cos 2\theta + 5.05 \sin 2\theta + 2.67 \cos 3\theta + 3.33 \sin 3\theta]$

5. Analyse the periodic waveform of y against angle θ in Fig. 31.6(a) into its constituent harmonics as far as and including the third harmonic by taking 30° intervals.
    $[y = 6.75 + 25.81 \cos \theta + 44.51 \sin \theta - 10.67 \cos 2\theta$
    $- 4.62 \sin 2\theta + 4.17 \cos 3\theta - 8.33 \sin 3\theta]$

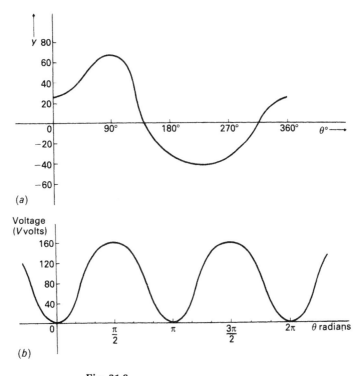

Fig. 31.6

6. For the waveform shown in Fig. 31.6(b) state why only a d.c. component and even cosine terms will be present in the Fourier series and determine the first three terms of the series using 30° intervals.

   $[V = 86.67 - 86.67 \cos 2\theta - 6.67 \cos 4\theta]$

7. Find a Fourier series as far as the third harmonic to represent the periodic function y given by the waveform in Fig. 31.7(a). Use 30° intervals.

   $[y = -25.21 \cos \theta + 73.42 \sin \theta + 10.00 \cos 3\theta + 3.33 \sin 3\theta]$

8. For the waveform shown in Fig. 31.7(b) state why only odd cosine harmonics appear in the Fourier series. By taking 12

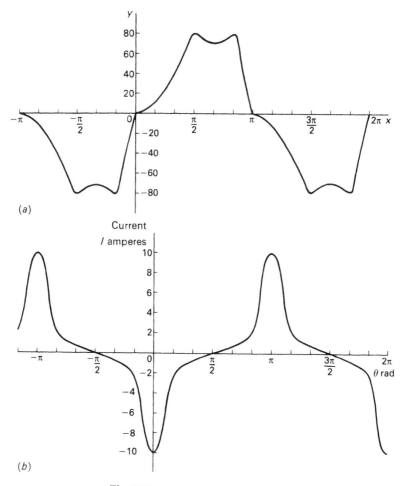

(a)

(b)

Fig. 31.7

intervals analyse the waveform and find the first three terms of the Fourier series.

$[I = -4.75 \cos \theta - 2.80 \cos 3\theta - 2.45 \cos 5\theta]$

9. Displacement y on a point on a pulley turned through an angle of θ degrees is given by:

θ	30	60	90	120	150	180
y	7.976	8.026	7.204	5.676	3.674	1.764

θ	210	240	270	300	330	360
y	0.552	0.262	0.904	2.492	4.736	6.824

Construct a Fourier series for the first three harmonics.

$[y = 4.17 + 2.45 \cos \theta + 0.12 \cos 2\theta + 0.08 \cos 3\theta$
$+ 3.16 \sin \theta + 0.03 \sin 2\theta + 0.01 \sin 3\theta]$

10. The anode current pulse in amperes in a class C power amplifier is given by:

$i = -1 + 2 \cos \omega t$, when i is positive. At all other times $i = 0$. Perform a Fourier analysis on the waveform up to and including the third harmonic.

$[i = 10^{-3}(215 + 388 \cos \omega t + 279 \cos 2\omega t + 144 \cos 3\omega t]$

Chapter 32

Introduction to Laplace transforms

32.1 Introduction

The solution of most electrical circuit problems reduces ultimately to the solution of differential equations. Methods of solving certain first and second order differential equations are discussed in Chapters 22 to 26. The operation, called the Laplace transformation, provides a powerful and much used alternative method of solving linear differential equations. The method has many practical applications, including aspects of control engineering and coupled electrical circuits, beam problems, various electrical systems, mechanical vibration systems and in servomechanisms, and was developed by the French mathematician Pierre Simon de Laplace (1749–1827).

Before using Laplace transforms to solve differential equations it is necessary to define the Laplace transform, to derive the transforms of elementary functions, to appreciate certain of their properties and to define and derive the inverse Laplace transform, including the use of partial fractions.

In this chapter the Laplace transform is defined and the use of a table of standard transforms to evaluate the transforms of elementary functions is discussed.

Some simple properties of Laplace transforms are discussed in Chapter 33.

In Chapter 34 the inverse Laplace transform is defined and determined for elementary functions using partial fractions and a table of standard transforms. Also included in Chapter 34 is an introduction to solving linear second order differential equations with constant coefficients.

32.2 Definition of the Laplace transform

Let $f(t)$ be a given function of time which is defined for all positive values of t. Let $f(t)$ be multiplied by e^{-st} (where s is a parameter introduced by us and assumed to be a real number) and then the product integrated with respect to t from zero to infinity. Then, if the resulting integral exists, it is a function of s, say $F(s)$.

That is, $F(s) = \displaystyle\int_0^\infty e^{-st} f(t)\, dt$

The function $F(s)$ is called the **Laplace transform** of the original function $f(t)$. There are various commonly used notations for the Laplace transform of $f(t)$ and these include:

(i) $\mathscr{L}(f)$ or Lf
(ii) $\mathscr{L}\{f(t)\}$ or $L\{f(t)\}$
(iii) $\bar{f}(s)$ or $\tilde{f}(s)$

Also, in some texts, the letter p is used instead of s as the parameter.

The notation adopted in this text will be to use small letters for the original function (i.e. $f(t)$) and $\mathscr{L}\{f(t)\}$ for its Laplace transform.

Hence, $F(s) = \mathscr{L}\{f(t)\} = \displaystyle\int_0^\infty e^{-st}f(t)\,dt$ (1)

where $\mathscr{L}\{\ \}$ represents symbolically the operation of taking the Laplace transform of whatever function occurs inside the bracket. The operation on the original function is called the **Laplace transformation**.

32.3   Linearity property of the Laplace transform

From equation (1), $\mathscr{L}\{kf(t)\} = \displaystyle\int_0^\infty e^{-st}\,kf(t)\,dt = k\int_0^\infty e^{-st}\,f(t)\,dt$

i.e. $\mathscr{L}\{kf(t)\} = k\mathscr{L}\{f(t)\}$ (2)

where k is any constant.
Similarly,

$$\mathscr{L}\{af(t) + b\,g(t)\} = \int_0^\infty e^{-st}[af(t) + b\,g(t)]\,dt$$

$$= a\int_0^\infty e^{-st}\,f(t)\,dt + b\int_0^\infty e^{-st}g(t)\,dt$$

i.e. $\mathscr{L}\{af(t) + b\,g(t)\} = a\mathscr{L}\{f(t)\} + b\mathscr{L}\{g(t)\}$ (3)

where a and b are any constants.

Because of the properties shown in equations (2) and (3) the Laplace transform is termed a **linear operator**.

32.4   Laplace transforms of elementary functions

Using the definition of the Laplace transform given in equation (1) a number of elementary functions may be transformed.

(i) $f(t) = 1$

From equation (1)

$$\mathscr{L}\{1\} = \int_0^\infty e^{-st}(1)\,dt = \left[\frac{e^{-st}}{-s}\right]_0^\infty$$

$$= -\frac{1}{s}[e^{-s(\infty)} - e^0]$$

$$= -\frac{1}{s}[0 - 1]$$

$$= \frac{1}{s}$$

(provided $s > 0$, so that $e^{-s(\infty)} = 0$)

Hence the Laplace transform of 1 is $\dfrac{1}{s}$.

(ii) $f(t) = k$ (where k is a constant)

From equation (1), $\mathscr{L}\{k\} = \int_0^\infty e^{-st} k\,dt$

$$k\mathscr{L}\{1\} = k\int_0^\infty e^{-st}\,dt, \text{ from equation (2)}$$

$$= k\left(\frac{1}{s}\right). \text{ from (i) above.}$$

Hence the Laplace transform of k is $\dfrac{k}{s}$ (provided $s > 0$).

(iii) $f(t) = e^{at}$ (where a is a real constant $\neq 0$)

From equation (1),

$$\mathscr{L}\{e^{at}\} = \int_0^\infty e^{-st}(e^{at})\,dt$$

$$= \int_0^\infty e^{-(s-a)t}\,dt$$

$$= \left[\frac{e^{-(s-a)t}}{-(s-a)}\right]_0^\infty$$

$$= \frac{1}{-(s-a)}[0 - 1]$$

$$= \frac{1}{s-a} \text{ (provided } s - a > 0, \text{ i.e. } s > a)$$

Hence the Laplace transform of e^{at} is $\dfrac{1}{s-a}$.

(iv) $f(t) = \sin at$ (where a is a real constant)

From equation (1),

$$\mathscr{L}\{\sin at\} = \int_0^\infty e^{-st} \sin at \; dt$$

$$= \left[\frac{e^{-st}}{s^2 + a^2} \left(-s \sin at - a \cos at \right) \right]_0^\infty$$

This is obtained by integrating by parts twice (see Chapter 21, Problem 8, page 322).

Hence $\mathscr{L}\{\sin at\} = \dfrac{1}{s^2 + a^2} \left[e^{-s(\infty)} \left(-s \sin a(\infty) - a \cos a(\infty) \right) \right.$

$$\left. - e^0 \left(-s \sin 0 - a \cos 0 \right) \right]$$

$$= \frac{1}{s^2 + a^2} [0 - 1(-a)] = \frac{a}{s^2 + a^2} \quad \text{(provided } s > 0\text{)}$$

Hence the Laplace transform of $\sin at$ is $\dfrac{a}{s^2 + a^2}$.

(v) $f(t) = \cos at$ (where a is a real constant)

From equation (1),

$$\mathscr{L}\{\cos at\} = \int_0^\infty e^{-st} \cos at \; dt$$

$$= \left[\frac{e^{-st}}{s^2 + a^2} \left(a \sin at - s \cos at \right) \right]_0^\infty$$

by integrating by parts twice

$$= \frac{s}{s^2 + a^2} \quad \text{(provided } s > 0\text{)}$$

Hence the Laplace transform of $\cos at$ is $\dfrac{s}{s^2 + a^2}$.

(vi) $f(t) = t$

From equation (1),

$$\mathcal{L}\{t\} = \int_0^\infty e^{-st} t \, dt$$

$$= \left[\frac{t e^{-st}}{-s} \right]_0^\infty - \int_0^\infty \frac{e^{-st}}{-s} \, dt, \text{ by integration by parts}$$

$$= \left[\frac{\infty \, e^{-s(\infty)}}{-s} - 0 \right] - \left[\frac{e^{-st}}{s^2} \right]_0^\infty$$

$$= [0 - 0] - \frac{1}{s^2} [e^{-s(\infty)} - e^0], \text{ since } \infty \times 0 = 0,$$

$$= -\frac{1}{s^2} (0 - 1) = \frac{1}{s^2} \text{ (provided } s > 0)$$

Hence the Laplace transform of t is $\dfrac{1}{s^2}$.

(vii) $f(t) = t^2$

From equation (1),

$$\mathcal{L}\{t^2\} = \int_0^\infty e^{-st} t^2 \, dt$$

$$= \left[\frac{t^2 e^{-st}}{-s} - \frac{2t e^{-st}}{s^2} - \frac{2 e^{-st}}{s^3} \right]_0^\infty \text{ by integration by parts twice}$$

$$= (0 - 0 - 0) - \left(0 - 0 - \frac{2}{s^3} \right)$$

$$= \frac{2}{s^3} \text{ (provided } s > 0)$$

Hence the Laplace transform of t^2 is $\dfrac{2}{s^3}$.

(viii) $f(t) = t^3$

From equation (1),

$$\mathcal{L}\{t^3\} = \int_0^\infty e^{-st} t^3 \, dt$$

$$= \left[\frac{t^3 e^{-st}}{-s} - \frac{3t^2 e^{-st}}{s^2} - \frac{(3)(2)t e^{-st}}{s^3} - \frac{(3)(2) e^{-st}}{s^4} \right]_0^\infty$$

$$= \frac{(3)(2)}{s^4} = \frac{3!}{s^4} \text{ (provided } s > 0)$$

Hence the Laplace transform of t^3 is $\dfrac{3!}{s^4}$.

(ix) $f(t) = t^n$ (where $n = 0, 1, 2, 3, \ldots$)

The results of (vi), (vii) and (viii) can be extended to n being any positive integer, giving:

$$\mathcal{L}\{t^n\} = \frac{n!}{s^{n+1}} \quad \text{(provided } s > 0)$$

Thus, for example,

$$\mathcal{L}\{t^4\} = \frac{4!}{s^{4+1}} = \frac{(4)(3)(2)(1)}{s^5} = \frac{24}{s^5}$$

and $\mathcal{L}\{t^5\} = \dfrac{5!}{s^{5+1}} = \dfrac{(5)(4)(3)(2)(1)}{s^6} = \dfrac{120}{s^6}$

(x) $f(t) = \cosh at$

$$\cosh at = \frac{1}{2}(e^{at} + e^{-at})$$

Hence $\mathcal{L}\{\cosh at\} = \mathcal{L}\left\{\dfrac{1}{2}(e^{at} + e^{-at})\right\}$

$$= \frac{1}{2}\mathcal{L}\{e^{at}\} + \frac{1}{2}\mathcal{L}\{e^{-at}\}, \text{ from equations (2) and (3)}$$

$$= \frac{1}{2}\left(\frac{1}{s-a}\right) + \frac{1}{2}\left(\frac{1}{s--a}\right), \text{ from (iii) above}$$

$$= \frac{1}{2}\left[\frac{1}{s-a} + \frac{1}{s+a}\right] = \frac{1}{2}\left[\frac{(s+a)+(s-a)}{(s-a)(s+a)}\right]$$

$$= \frac{s}{s^2 - a^2} \quad \text{(provided } s > a)$$

Hence the Laplace transform of $\cosh at = \dfrac{s}{s^2 - a^2}$.

(xi) $f(t) = \sinh at$

$$\sinh at = \frac{1}{2}(e^{at} - e^{-at})$$

Hence $\mathscr{L}\{\sinh at\} = \dfrac{1}{2}\mathscr{L}\{e^{at}\} - \dfrac{1}{2}\mathscr{L}\{e^{-at}\}$, from equations (2) and (3)

$$= \frac{1}{2}\left(\frac{1}{s-a}\right) - \frac{1}{2}\left(\frac{1}{s+a}\right) = \frac{1}{2}\left(\frac{1}{s-a} - \frac{1}{s+a}\right)$$

$$= \frac{a}{s^2 - a^2} \quad \text{(provided } s > a)$$

Hence the Laplace transform of sinh $at = \dfrac{a}{s^2 - a^2}$.

For ease of reference, the above Laplace transforms are summarised in Table 32.1.

With a working knowledge of the transforms listed in Table 32.1 nearly all of the transforms needed in practice can be obtained through the use of simple general properties, one of these being considered in Chapter 33.

Table 32.1 Elementary standard Laplace transforms

Function $f(t)$	Laplace transforms $\mathscr{L}\{f(t)\} = \displaystyle\int_0^{\infty} e^{-st} f(t)\,\mathrm{d}t$
(i) 1	$\dfrac{1}{s}$
(ii) k	$\dfrac{k}{s}$
(iii) e^{at}	$\dfrac{1}{s-a}$
(iv) $\sin at$	$\dfrac{a}{s^2 + a^2}$
(v) $\cos at$	$\dfrac{s}{s^2 + a^2}$
(vi) t	$\dfrac{1}{s^2}$
(vii) t^2	$\dfrac{2!}{s^3}$
(viii) t^3	$\dfrac{3!}{s^4}$
(ix) t^n (n = positive integer)	$\dfrac{n!}{s^{n+1}}$
(x) $\cosh at$	$\dfrac{s}{s^2 - a^2}$
(xi) $\sinh at$	$\dfrac{a}{s^2 - a^2}$

Worked problems on standard Laplace transforms

Problem 1. Find the Laplace transforms of: (a) $1 + 3t - 2t^2 + \dfrac{t^6}{12}$, (b) $3e^{2t} - 4e^{-5t}$, (c) $2\sin 4t - 3\cos 2t$

(a) $\quad \mathcal{L}\left\{1 + 3t - 2t^2 + \dfrac{t^6}{12}\right\} = \mathcal{L}\{1\} + 3\mathcal{L}\{t\}$

$$- 2\mathcal{L}\{t^2\} + \frac{1}{12}\mathcal{L}\{t^6\}$$

from equations (2) and (3)

$$= \frac{1}{s} + \frac{3}{s^2} - \frac{2 \cdot 2}{s^3} + \frac{6 \cdot 5 \cdot 4 \cdot 3 \cdot 2 \cdot 1}{12 s^7}$$

from (1), (vi), (vii) and (ix) of Table 32.1

$$= \frac{1}{s} + \frac{3}{s^2} - \frac{4}{s^3} + \frac{60}{s^7}$$

(b) $\quad \mathcal{L}\{3e^{2t} - 4e^{-5t}\} = 3\mathcal{L}\{e^{2t}\} - 4\mathcal{L}\{e^{-5t}\}$

from equations (2) and (3)

$$= \frac{3}{(s-2)} - \frac{4}{(s--5)}, \text{ from (iii) of Table 32.1}$$

$$= \frac{3(s+5) - 4(s-2)}{(s-2)(s+5)} = \frac{23 - s}{(s-2)(s+5)}$$

(c) $\quad \mathcal{L}\{2\sin 4t - 3\cos 2t\} = 2\mathcal{L}\{\sin 4t\} - 3\mathcal{L}\{\cos 2t\}$,

from equations (2) and (3)

$$= 2\left(\frac{4}{s^2 + 4^2}\right) - 3\left(\frac{s}{s^2 + 2^2}\right),$$

from (iv) and (v) of Table 32.1

$$= \frac{8}{(s_2 + 16)} - \frac{3s}{(s^2 + 4)}$$

Problem 2. Find the Laplace transforms of:

(a) $x^5 - 2x^4$, (b) $\cosh\theta - \sinh 2\theta$, (c) $\cos^2 t$

(a) $\mathscr{L}\{x^5 - 2x^4\} = \mathscr{L}\{x^5\} - 2\mathscr{L}\{x^4\}$

$$= \frac{5!}{s^6} - 2\frac{4!}{s^5} \quad \text{from (ix) of Table 32.1}$$

$$= \frac{(5 \cdot 4 \cdot 3 \cdot 2 \cdot 1) - 2s(4 \cdot 3 \cdot 2 \cdot 1)}{s^6} = \frac{120 - 48s}{s^6}$$

$$= \frac{24(5 - 2s)}{s^6}$$

(b) $\mathscr{L}\{\cosh\theta - \sinh 2\theta\} = \mathscr{L}\{\cosh\theta\} - \mathscr{L}\{\sinh 2\theta\}$

$$= \frac{s}{s^2 - 1^2} - \frac{2}{s^2 - 2^2}$$

from (x) and (xi) of Table 32.1

$$= \frac{s(s^2 - 4) - 2(s^2 - 1)}{(s^2 - 1)(s^2 - 4)}$$

$$= \frac{s^3 - 2s^2 - 4s + 2}{(s^2 - 1)(s^2 - 4)}$$

(c) $\mathscr{L}\{\cos^2 t\} = \mathscr{L}\left\{\frac{1}{2}(1 + \cos 2t)\right\}$

$$= \frac{1}{2}\mathscr{L}\{1\} + \frac{1}{2}\mathscr{L}\{\cos 2t\}$$

$$= \frac{1}{2}\left(\frac{1}{s}\right) + \frac{1}{2}\left(\frac{s}{s^2 + 4}\right)$$

from (i) and (v) of Table 32.1

$$= \frac{(s^2 + 4) + (s^2)}{2s(s^2 + 4)} = \frac{2(s^2 + 2)}{2s(s^2 + 4)} = \frac{(s^2 + 2)}{s(s^2 + 4)}$$

Problem 3. Find the Laplace transforms of: (a) $\sinh^2 2t$, (b) $2\sin(\omega t + \alpha)$, where ω and α are constants.

(a) $\mathscr{L}\{\sinh^2 2t\} = \mathscr{L}\left\{\frac{1}{2}(\cosh 4t - 1)\right\}$

since $\cosh 2t = 1 + 2\sinh^2 t$ (see Section 9.5)

$$= \frac{1}{2}\mathscr{L}\{\cosh 4t\} - \frac{1}{2}\mathscr{L}\{1\}$$

$$= \frac{1}{2}\left(\frac{s}{s^2 - 16}\right) - \frac{1}{2}\left(\frac{1}{s}\right)$$

from (x) and (i) of Table 32.1

$$= \frac{(s^2) - (s^2 - 16)}{2s(s^2 - 16)} = \frac{16}{2s(s^2 - 16)} = \frac{8}{s(s^2 - 16)}$$

(b) $\mathcal{L}\{2 \sin(\omega t + \alpha)\} = \mathcal{L}\{2(\sin \omega t \cos \alpha + \cos \omega t \sin \alpha)\}$

$$= (2 \cos \alpha)\mathcal{L}\{\sin \omega t\} + (2 \sin \alpha)\mathcal{L}\{\cos \omega t\}$$

$$= (2 \cos \alpha)\left(\frac{\omega}{s^2 + \omega^2}\right) + (2 \sin \alpha)\left(\frac{s}{s^2 + \omega^2}\right)$$

from (iv) and (v) of Table 32.1

$$= \frac{2}{s^2 + \omega^2}(\omega \cos \alpha + s \sin \alpha)$$

Further problems on standard Laplace transforms may be found in the following Section 32.5 (Problems 1 to 10).

32.5   Further problems

In Problems 1 to 9 find the Laplace transforms.

1. (a) $3t - 5$ (b) $4t^2 + 2t - 1$

    (a) $\left[\dfrac{3}{s^2} - \dfrac{5}{s}\right]$ (b) $\left[\dfrac{8}{s^3} + \dfrac{2}{s^2} - \dfrac{1}{s}\right]$

2. (a) $3t^3 - 4t + 7$ (b) $t^6 - 3t^4 + t^2$

    (a) $\left[\dfrac{18}{s^4} - \dfrac{4}{s^2} + \dfrac{7}{s}\right]$ (b) $\left[\dfrac{720}{s^7} - \dfrac{72}{s^5} + \dfrac{2}{s^3}\right]$

3. (a) $2 e^{4t}$ (b) $3 e^{-2t}$

    (a) $\left[\dfrac{2}{s - 4}\right]$ (b) $\left[\dfrac{3}{s + 2}\right]$

4. (a) $5 \sin 2t$ (b) $4 \cos 3t$

    (a) $\left[\dfrac{10}{s^2 + 4}\right]$ (b) $\left[\dfrac{4s}{s^2 + 9}\right]$

5. (a) 2 cosh 2t   (b) 3 sinh t

   (a) $\left[\dfrac{2s}{s^2 - 4}\right]$ (b) $\left[\dfrac{3}{s^2 - 1}\right]$

6. (a) $\sin^2 t$ (b) $\sin t \cos t$

   (a) $\left[\dfrac{2}{s(s^2 + 4)}\right]$ (b) $\left[\dfrac{1}{s^2 + 4}\right]$

7. (a) $\cosh^2 \theta$ (b) $\sinh^2 3\theta$

   (a) $\left[\dfrac{s^2 - 2}{s(s^2 - 4)}\right]$ (b) $\left[\dfrac{18}{s(s^2 - 36)}\right]$

8. (a) $3 \sin(at + b)$, where a and b are constants,
    (b) $4 \cos(\omega t - \alpha)$, where ω and α are constants.

   (a) $\left[\dfrac{3}{s^2 + a^2}(a \cos b + s \sin b)\right]$

   (b) $\left[\dfrac{4}{s^2 + \omega^2}(s \cos \alpha + \omega \sin \alpha)\right]$

9. (a) $\sin\left(t + \dfrac{\pi}{4}\right)$ (b) $2 \cos(2t + \beta)$, where β is a constant.

   (a) $\left[\dfrac{s + 1}{\sqrt{2(s^2 + 1)}}\right]$ (b) $\left[\dfrac{2}{s^2 + 4}(s \cos \beta - 2 \sin \beta)\right]$

10. Show that (a) $\mathcal{L}\left\{\dfrac{1}{w} \sin wt - \dfrac{1}{\beta} \sin \beta t\right\} = \dfrac{\beta^2 - w^2}{(s^2 + w^2)(s^2 + \beta^2)}$

    (b) $\mathcal{L}\{\cos^2 2t - \sin^2 2t\} = \dfrac{s}{(s^2 + 16)}$

Chapter 33

Properties of Laplace transforms

33.1 The Laplace transform of $e^{at} f(t)$

In Chapter 32 the Laplace transforms of elementary functions were considered. Further transforms can be obtained by the following important property of the Laplace transformation.

If
$$F(s) = \mathcal{L}\{f(t)\} = \int_0^\infty e^{-st} f(t)\, dt \tag{1}$$

then
$$\mathcal{L}\{e^{at} f(t)\} = \int_0^\infty e^{-st} (e^{at} f(t))\, dt$$

$$= \int_0^\infty e^{-(s-a)t} f(t)\, dt$$

i.e.
$$\mathcal{L}\{e^{at} f(t)\} = F(s - a) \tag{2}$$

where a is a real constant.

Hence the substitution of $s - a$ for s in the transform shown in equation (1) corresponds to the multiplication of the original functon by e^{at}. This is often known as a **shift theorem**.

If a is negative it follows that:
$$\mathcal{L}\{e^{-at} f(t)\} = F(s + a) \tag{3}$$

33.2 Laplace transforms of functions of the type $e^{at} f(t)$

Laplace transforms of functions of the type $e^{at} f(t)$, where a is a positive real constant, may be obtained by using Table 32.1 and equations (2) and (3) above.

(i) $\mathcal{L}\{e^{at} t^n\}$

Since
$$\mathcal{L}\{t^n\} = \frac{n!}{s^{n+1}},\ \text{from (ix) of Table 32.1}$$

then
$$\mathcal{L}(e^{at} t^n) = \frac{n!}{(s - a)^{n+1}},\ \text{from equation (2) (provided } s > a)$$

and $\quad \mathscr{L}\{e^{-at} t^n\} = \dfrac{n!}{(s+a)^{n+1}}$, from equation (3) (provided $s > -a$)

(ii) $\mathscr{L}\{e^{at} \sin \omega t\}$

Since $\quad \mathscr{L}\{\sin \omega t\} = \dfrac{\omega}{s^2 + \omega^2}$, from (iv) of Table 32.1

then $\quad \mathscr{L}\{e^{at} \sin \omega t\} = \dfrac{\omega}{(s-a)^2 + \omega^2}$, from equation (2) (provided $s > a$)

and $\quad \mathscr{L}\{e^{-at} \sin \omega t\} = \dfrac{\omega}{(s+a)^2 + \omega^2}$, from equation (3) (provided $s > -a$)

(iii) $\mathscr{L}\{e^{at} \cos \omega t\}$

Since $\quad \mathscr{L}\{\cos \omega t\} = \dfrac{s}{s^2 + \omega^2}$, from (v) of Table 32.1

then $\quad \mathscr{L}\{e^{at} \cos \omega t\} = \dfrac{s-a}{(s-a)^2 + \omega^2}$, from equation (2) (provided $s > a$)

and $\quad \mathscr{L}\{e^{-at} \cos \omega t\} = \dfrac{s+a}{(s+a)^2 + \omega^2}$,

from equation (3) (provided $s > -a$)

(iv) $\mathscr{L}\{e^{at} \cosh \omega t\}$

Since $\quad \mathscr{L}\{\cosh \omega t\} = \dfrac{s}{s^2 - \omega^2}$, from (x) of Table 32.1

then $\quad \mathscr{L}\{e^{at} \cosh \omega t\} = \dfrac{s-a}{(s-a)^2 - \omega^2}$, from equation (2) (provided $s > a$)

and $\quad \mathscr{L}\{e^{-at} \cosh \omega t\} = \dfrac{s+a}{(s+a)^2 - \omega^2}$, from equation (3) (provided $s > -a$)

(v) $\mathscr{L}\{e^{at} \sinh \omega t\}$

Since $\quad \mathscr{L}\{\sinh \omega t\} = \dfrac{\omega}{s^2 - \omega^2}$, from (xi) of Table 32.1

then $\mathscr{L}\{e^{at}\sinh\omega t\} = \dfrac{\omega}{(s-a)^2-\omega^2}$, from equation (2) (provided $s > a$)

and $\mathscr{L}\{e^{-at}\sinh\omega t\} = \dfrac{\omega}{(s+a)^2-\omega^2}$, from equation (3) (provided $s > -a$)

For ease of reference the above Laplace transforms are summarised in Table 33.1.

Table 33.1

Function $e^{at}\,f(t)$ (a is a positive real constant)	Laplace transform $\mathscr{L}\{e^{at}\,f(t)\}$
(i) $e^{at}\,t^n$	$\dfrac{n!}{(s-a)^{n+1}}$
$e^{-at}\,t^n$	$\dfrac{n!}{(s+a)^{n+1}}$
(ii) $e^{at}\sin\omega t$	$\dfrac{\omega}{(s-a)^2+\omega^2}$
$e^{-at}\sin\omega t$	$\dfrac{\omega}{(s+a)^2+\omega^2}$
(iii) $e^{at}\cos\omega t$	$\dfrac{s-a}{(s-a)^2+\omega^2}$
$e^{-at}\cos\omega t$	$\dfrac{s+a}{(s+a)^2+\omega^2}$
(iv) $e^{at}\cosh\omega t$	$\dfrac{s-a}{(s-a)^2-\omega^2}$
$e^{-at}\cosh\omega t$	$\dfrac{s+a}{(s+a)^2-\omega^2}$
(v) $e^{at}\sinh\omega t$	$\dfrac{\omega}{(s-a)^2-\omega^2}$
$e^{-at}\sinh\omega t$	$\dfrac{\omega}{(s+a)^2-\omega^2}$

Worked problems on Laplace transforms of functions of the type $e^{at}\,f(t)$

Problem 1. Find the Laplace transforms of: (a) $t^3\,e^{2t}$, (b) $e^{3t}\sin 4t$, (c) $2\,e^{-2t}\cos 2t$.

(a) $\mathcal{L}\{t^3\} = \dfrac{3!}{s^4} = \dfrac{6}{s^4}$, from (viii) of Table 32.1

Hence $\mathcal{L}\{t^3 e^{2t}\} = \dfrac{6}{(s-2)^4}$, from (i) of Table 33.1

(b) $\mathcal{L}\{\sin 4t\} = \dfrac{4}{s^2 + 4^2} = \dfrac{4}{s^2 + 16}$, from (iv) of Table 32.1

Hence $\mathcal{L}\{e^{3t} \sin 4t\} = \dfrac{4}{(s-3)^2 + 16}$, from (ii) of Table 33.1

$$= \dfrac{4}{s^2 - 6s + 25}$$

(c) $\mathcal{L}\{2 \cos 2t\} = \dfrac{2s}{s^2 + 4}$, from (v) of Table 32.1

Hence $\mathcal{L}\{2e^{-2t} \cos 2t\} = \dfrac{2(s+2)}{(s+2)^2 + 4}$, from (iii) of Table 33.1

$$= \dfrac{2(s+2)}{s^2 + 4s + 8}$$

Problem 2. Find the Laplace transforms of: (a) $2 e^{4t} \sin^2 t$, (b) $e^{-3t} \cosh 2t$, (c) $e^{4t} (2 \cos 3t - 3 \sin 3t)$.

(a) $\sin^2 t = \dfrac{1}{2}(1 - \cos 2t)$ since $\cos 2t = 1 - 2 \sin^2 t$

Hence $\mathcal{L}\{2 e^{4t} \sin^2 t\} = \mathcal{L}\left\{2 e^{4t} \dfrac{1}{2}(1 - \cos 2t)\right\}$

$$= \mathcal{L}\{e^{4t}\} - \mathcal{L}\{e^{4t} \cos 2t\}$$

$$= \dfrac{1}{s-4} - \dfrac{s-4}{(s-4)^2 + 4} \text{ from (iii) of Table}$$

32.1 and (iii) of Table 33.1

$$= \dfrac{1}{s-4} - \dfrac{s-4}{s^2 - 8s + 20}$$

(b) $\mathcal{L}\{e^{-3t} \cosh 2t\} = \dfrac{s+3}{(s+3)^2 - 2^2}$ from (iv) of Table 33.1

$$= \dfrac{s+3}{s^2 + 6s + 5}$$

(c) $\mathscr{L}\{e^{4t}(2\cos 3t - 3\sin 3t)\} = 2\mathscr{L}\{e^{4t}\cos 3t\} - 3\mathscr{L}\{e^{4t}\sin 3t\}$

$$= \frac{2(s-4)}{(s-4)^2 + 3^2} - \frac{3(3)}{(s-4)^2 + 3^2}$$

from (iii) and (ii) of Table 33.1

$$= \frac{2s - 8 - 9}{s^2 - 8s + 25} = \frac{2s - 17}{s^2 - 8s + 25}$$

Further problems on Laplace transforms of functions of the type $e^{at} f(t)$ may be found in Section 33.5 (Problems 1 to 10), page 474.

33.3   The Laplace transforms of derivatives

First derivative

Let the first derivative of $f(t)$ be $f'(t)$.

Since $\quad \mathscr{L}\{f(t)\} = \displaystyle\int_0^\infty e^{-st} f(t)\,dt$ by definition,

then $\quad \mathscr{L}\{f'(t)\} = \displaystyle\int_0^\infty e^{-st} f'(t)\,dt$

When integrating by parts, $\displaystyle\int u \frac{dv}{dt}\,dt = uv - \int v \frac{du}{dt}$ (see Chapter 21).

When evaluating $\displaystyle\int_0^\infty e^{-st} f'(t)\,dt$, let $u = e^{-st}$ and $\dfrac{dv}{dt} = f'(t)$, from which

$\dfrac{du}{dt} = -s\,e^{-st}$ and $v = \displaystyle\int f'(t)\,dt = f(t)$

Hence $\displaystyle\int_0^\infty e^{-st} f'(t)\,dt = [e^{-st} f(t)]_0^\infty - \int_0^\infty f(t)(-s\,e^{-st})\,dt$

$$= [0 - f(0)] + s \int_0^\infty e^{-st} f(t)\,dt$$

$$= -f(0) + s\mathscr{L}\{f(t)\}$$

assuming that $e^{-st} f(t)$ approaches zero as t approaches zero, and $f0)$ is the value of $f(t)$ at $t = 0$.

Hence $$\mathcal{L}\{f'(t)\} = s\mathcal{L}\{f(t)\} - f(0)\Big\}$$

of alternatively, $\mathcal{L}\left\{\dfrac{dy}{dx}\right\} = s\mathcal{L}\{y\} - y(0)$ (4)

where $y(0)$ is the value of y at $x = 0$.

Second derivative

Let the second derivative of $f(t)$ be $f''(t)$.

By definition, $\mathcal{L}\{f''(t)\} = \displaystyle\int_{0}^{\infty} e^{-st} f''(t)\, dt$

Integration by parts gives:

$$\int_{0}^{\infty} e^{-st} f''(t)\, dt = [e^{-st} f'(t)]_{0}^{\infty} + s\int_{0}^{\infty} e^{-st} f'(t)\, dt$$

$$= (0 - f'(0)) + s\mathcal{L}\{f'(t)\}, \text{ assuming } e^{-st} f'(t)$$
approaches zero as t approaches zero, and
$f'(0)$ is the value of $f'(t)$ at $t = 0$,

$$= -f'(0) + s[s\mathcal{L}\{f(t)\} - f(0)], \text{ from equation (4)}$$

Hence $$\mathcal{L}\{f''(t)\} = s^2\mathcal{L}\{f(t)\} - sf(0) - f'(0)\Big\}$$

or alternatively, $\mathcal{L}\left\{\dfrac{d^2y}{dx^2}\right\} = s^2\mathcal{L}\{y\} - sy(0) - y'(0)$ (5)

where $y'(0)$ is the value of $\dfrac{dy}{dx}$ at $x = 0$.

Higher derivatives

Laplace transforms of higher derivatives may be determined in the same way as above, and in general it is found:

$$\mathcal{L}\{f^n(t)\} = s^n\mathcal{L}\{f(t)\} - s^{n-1}f(0) - s^{n-2}f'(0)\cdots - f^{n-1}(0)\Big\}$$

or $\mathcal{L}\left\{\dfrac{d^ny}{dx^n}\right\} = s^n\mathcal{L}\{y\} - s^{n-1}y(0) - s^{n-2}y'(0)\cdots y^{n-1}(0)$ (6)

Equations (4), (5) and (6) are very important and are used in the solution of differential equations (see Section 34.4).

Worked problems on the Laplace transforms of derivatives

Problem 1. Use the Laplace transform of the first derivative (i.e. equation (4)) to derive the following:

(a) $\mathscr{L}\{1\} = \dfrac{1}{s}$ (b) $\mathscr{L}\{t\} = \dfrac{1}{s^2}$ (c) $\mathscr{L}\{e^{at}\} = \dfrac{1}{s-a}$

(a) From equation (4), $\mathscr{L}\{f'(t)\} = s\mathscr{L}\{f(t)\} - f(0)$.
Let $f(t) = 1$, then $f'(t) = 0$ and $f(0) = 1$.
Therefore, substituting in equation (4) gives:

$$\mathscr{L}\{0\} = s\mathscr{L}\{1\} - 1$$
$$0 = s\mathscr{L}\{1\} - 1$$
$$1 = s\mathscr{L}\{1\}$$

Hence $\mathscr{L}\{1\} = \dfrac{1}{s}$

(b) Let $f(t) = t$, then $f'(t) = 1$ and $f(0) = 0$.
Therefore, substituting in equation (4) gives:

$$\mathscr{L}\{1\} = s\mathscr{L}\{t\} - 0$$

i.e. $\dfrac{1}{s} = s\mathscr{L}\{t\}$

Hence $\mathscr{L}\{t\} = \dfrac{1}{s^2}$

(c) Let $f(t) = e^{at}$, then $f'(t) = a\, e^{at}$ and $f(0) = 1$.
Therefore, substituting in equation (4) gives:

$$\mathscr{L}\{a\, e^{at}\} = s\mathscr{L}\{e^{at}\} - 1$$
$$a\mathscr{L}\{e^{at}\} = s\mathscr{L}\{e^{at}\} - 1$$
$$1 = (s - a)\mathscr{L}\{e^{at}\}$$

Hence $\mathscr{L}\{e^{at}\} = \dfrac{1}{s-a}$.

Problem 2. Use the Laplace transform of the second derivative (i.e. equation (5)) to derive: $\mathscr{L}\{\sin at\} = \dfrac{a}{s^2 + a^2}$.

Let $f(t) = \sin at$, then $f'(t) = a \cos at$ and $f''(t) = -a^2 \sin at$, $f(0) = 0$ and $f'(0) = a$.

From equation (5), $\mathscr{L}\{f''(t)\} = s^2\mathscr{L}\{f(t)\} - sf(0) - f'(0)$

Hence $\mathscr{L}\{-a^2 \sin at\} = s^2\mathscr{L}\{\sin at\} - s(0) - a$

$-a^2\mathscr{L}\{\sin at\} = s^2\mathscr{L}\{\sin at\} - a$

$a = (s^2 + a^2)\mathscr{L}\{\sin at\}$

Hence $\mathscr{L}\{\sin at\} = \dfrac{a}{s^2 + a^2}$

Further problems on the Laplace transforms of derivatives may be found in Section 33.5 (Problems 11 to 13), page 475.

33.4   The initial and final value theorems

The initial value and final value theorems are just two of many Laplace transform theorems used to simplify and interpret the solution of certain problems.

The initial value theorem states:

$$\text{limit}_{t \to 0} [f(t)] = \text{limit}_{s \to \infty} [sF(s)], \text{ where } F(s) = \mathscr{L}\{f(t)\}$$

This is proved as follows:

From the previous section, equation (4):

$$\mathscr{L}\{f'(t)\} = \int_0^\infty e^{-st} f'(t)\, dt = sF(s) - f(0)$$

Now as s approaches infinity, e^{-st} approaches zero and $\displaystyle\int_0^\infty e^{-st} f'(t)\, dt$ will approach zero.

Hence $0 = \text{limit}_{s \to \infty} [sF(s) - f(0)]$

$f(0)$ is independent of s, thus

$$\text{limit}_{s \to \infty} [sF(s)] = f(0) = \text{limit}_{t \to 0} [f(t)]$$

For example, if $f(t) = 4 e^{-2t}$

then $F(s) = \mathscr{L}\{f(t)\} = \mathscr{L}\{4 e^{-2t}\} = \dfrac{4}{s + 2}$ from (iii) of Table 32.1.

By the initial value theorem:

$$\underset{t \to 0}{\text{limit}} \,[4\,e^{-2t}] = \underset{s \to \infty}{\text{limit}} \left[\frac{4s}{s+2}\right]$$

i.e. $$4\,e^0 = \frac{4\infty}{\infty + 2}$$

i.e. $4 = 4$, which illustrates the theorem.

The final value theorem states:

$$\underset{t \to \infty}{\text{limit}} \,[f(t)] = \underset{s \to 0}{\text{limit}} \,[sF(s)]$$

This is proved as follows:
From the previous section, equation (4):

$$\mathcal{L}\{f'(t)\} = \int_0^\infty e^{-st} \, f'(t) \, dt = sF(s) - f(0) \qquad (7)$$

Now as s approaches zero the left-hand side of equation (7),

i.e. $\displaystyle\int_0^\infty e^{-st} f'(t) \, dt$, will approach $\displaystyle\int_0^\infty f'(t) \, dt$,

and $\displaystyle\int_0^\infty f'(t) \, dt = \underset{t \to \infty}{\text{limit}} \int_0^t f'(t) \, dt$

$$= \underset{t \to \infty}{\text{limit}} \,[f(t) - f(0)] = \underset{t \to \infty}{\text{limit}} \,[f(t)] - f(0)$$

Now as s approaches zero the right-hand side of equation (7),

i.e. $sF(s) - f(0)$, approaches $\underset{s \to 0}{\text{limit}} \,[sF(s)] - f(0)$

Equating these results gives:

$$\underset{t \to \infty}{\text{limit}} \,[f(t)] - f(0) = \underset{s \to 0}{\text{limit}} \,[sF(s)] - f(0)$$

i.e. $$\underset{t \to \infty}{\text{limit}} \,[f(t)] = \underset{s \to 0}{\text{limit}} \,[sF(s)]$$

For example, if $f(t) = 4\,e^{-2t}$

then $$\underset{t \to \infty}{\text{limit}} \,[4\,e^{-2t}] = \underset{s \to 0}{\text{limit}} \left[\frac{4s}{s+2}\right]$$

i.e. $$4(0) = \frac{0}{2}$$

i.e. $0 = 0$, which illustrates the theorem.

The initial and final value theorems are used in particular in pulse circuit applications where the response of a circuit for very small intervals of time, or the behaviour immediately after the switch is closed are of interest. The initial value theorem shows that there is a direct connection between the behaviour of a transform as s approaches infinity and the behaviour of the corresponding time function as t approaches zero. The final value theorem is particularly useful in investigating the stability of systems and is concerned with the steady state response for large values of t, i.e. after all transient effects have died away. The behaviour of $sF(s)$ as s approaches zero is seen to be the same as that of $f(t)$ as t approaches infinity. A particular practical application of the final value theorem is in automatic aircraft-landing systems.

Worked problems on the initial and final value theorems

Problem 1. Verify the initial value theorem for the following:
(a) $4 - 2\cos t$,   (b) $(3t + 4)^2$

(a) $f(t) = 4 - 2\cos t$

$$F(s) = \mathcal{L}\{f(t)\} = \mathcal{L}\{4 - 2\cos t\} = \frac{4}{s} - \frac{2s}{s^2 + 1}, \text{ from (ii) and (v)}$$
$$\text{of Table 32.1}$$

By the initial value theorem, $\displaystyle\lim_{t \to 0}\ [f(t)] = \lim_{s \to \infty}\ [sF(s)]$

i.e. $\displaystyle\lim_{t \to 0}\ [4 - 2\cos t] = \lim_{s \to \infty}\left[\frac{4s}{s} - \frac{2s^2}{s^2 + 1}\right]$

$$= 4 - \lim_{s \to \infty}\left[\frac{2s^2}{s^2 + 1}\right]$$

As $s \to \infty$, $s^2 + 1 \to s^2$

Thus $4 - 2\cos 0 = 4 - \displaystyle\lim_{s \to \infty}\left[\frac{2s^2}{s^2}\right]$

i.e. $4 - 2 = 4 - 2$

i.e. $\mathbf{2 = 2}$, which verifies the theorem in this case.

(b) $f(t) = (3t + 4)^2 = 9t^2 + 24t + 16$

$$F(s) = \mathcal{L}\{f(t)\} = \mathcal{L}\{9t^2 + 24t + 16\} = 9\left(\frac{2}{s^3}\right) + \frac{24}{s^2} + \frac{16}{s}\text{ from (vii),}$$
$$\text{(vi) and (ii) of Table 32.1}$$

By the initial value theorem,

$$\underset{t\to 0}{\text{limit}}\,[(3t + 4)^2] = \underset{s\to\infty}{\text{limit}}\left[\frac{18s}{s^3} + \frac{24s}{s^2} + \frac{16s}{s}\right]$$

i.e. $(0 + 4)^2 = (0 + 0 + 16)$

i.e.   **16 = 16**, which verifies the theorem in this case.

Problem 2. Verify the final value theorem for the function $3 + e^{-t}(\sin t + \cos t)$

$f(t) = 3 + e^{-t}(\sin t + \cos t)$

$F(s) = \mathscr{L}\{f(t)\} = \mathscr{L}\{3 + e^{-t}\sin t + e^{-t}\cos t\}$

$$= \frac{3}{s} + \frac{1}{(s + 1)^2 + 1} + \frac{s + 1}{(s + 1)^2 + 1}$$

from (ii) of Table 32.1 and (ii) and (iii) of Table 33.1

By the final value theorem,

$$\underset{t\to\infty}{\text{limit}}\,[f(t)] = \underset{s\to 0}{\text{limit}}\,[sF(s)]$$

i.e. $\underset{t\to\infty}{\text{limit}}\,[3 + e^{-t}(\sin t + \cos t)]$

$$= \underset{s\to 0}{\text{limit}}\left[3 + \frac{s}{(s + 1)^2 + 1} + \frac{s(s + 1)}{(s + 1)^2 + 1}\right]$$

i.e. $(3 + 0) = (3 + 0 + 0)$

i.e. **3 = 3**, which verifies the theorem.

Further problems on the initial and final value theorems may be found in the following Section 33.5 (Problems 14 to 17), page 476.

33.5   Further problems

Laplace transforms of functions of the type $e^{at}\,f(t)$

In Problems 1 to 8, find the Laplace transforms.

1. (a) $t\,e^t$ (b) $t^2\,e^{3t}$ (a) $\left[\dfrac{1}{(s - 1)^2}\right]$ (b) $\left[\dfrac{2}{(s - 3)^3}\right]$

2. (a) $2t^3\,e^{-2t}$ (b) $e^t\sin t$

   (a) $\left[\dfrac{12}{(s + 2)^4}\right]$ (b) $\left[\dfrac{1}{s^2 - 2s + 2}\right]$

3. (a) $3e^{2t}\cos 3t$ (b) $e^{-2t}\sin 2t$

   (a) $\left[\dfrac{3(s - 2)}{s^2 - 4s + 13}\right]$ (b) $\left[\dfrac{2}{s^2 + 4s + 8}\right]$

4. (a) $4 e^{-3t} \cos 2t$ (b) $2 e^t \cos^2 t$

(a) $\left[\dfrac{4(s+3)}{s^2 + 6s + 13}\right]$ (b) $\left[\dfrac{1}{s-1} + \dfrac{s-1}{s^2 - 2s + 5}\right]$

5. (a) $e^t \sin^2 t$ (b) $2 e^{2t} \cosh t$

(a) $\left[\dfrac{1}{2}\left\{\dfrac{1}{s-1} - \dfrac{s-1}{s^2 - 2s + 5}\right\}\right]$ (b) $\left[\dfrac{2(s-2)}{s^2 - 4s + 3}\right]$

6. (a) $e^t \sinh 2t$ (b) $3 e^{-2t} \cosh 3t$

(a) $\left[\dfrac{2}{s^2 - 2s - 3}\right]$ (b) $\left[\dfrac{3(s+2)}{s^2 + 4s - 5}\right]$

7. (a) $4 e^{-3t} \sinh 4t$ (b) $e^t(2 \cos 3t - 4 \sin 3t)$

(a) $\left[\dfrac{16}{s^2 + 6s - 7}\right]$ (b) $\left[\dfrac{2s - 14}{s^2 - 2s + 10}\right]$

8. (a) $e^{-2t}(\cosh t - \sinh t)$ (b) $3 e^{3t}(\sinh t - \sin t)$

(a) $\left[\dfrac{s+1}{s^2 + 4s + 3}\right]$ (b) $\left[\dfrac{6}{(s^2 - 6s + 10)(s^2 - 6s + 8)}\right]$

9. Show that $\mathscr{L}\{(\sin t - \cos t)^2\} = \dfrac{s^2 - 2s + 4}{s(s^2 + 4)}$.

10. Show that $\mathscr{L}\{\cosh^2 4t\} = \dfrac{s^2 - 32}{s(s^2 - 64)}$.

Laplace transforms of derivatives

11. Use the Laplace transform of the first derivative to derive the following transforms: (a) $\mathscr{L}\{t^2\} = \dfrac{2}{s^3}$,   (b) $\mathscr{L}\{e^{-at}\} = \dfrac{1}{s+a}$,

(c) $\mathscr{L}\{3t^3\} = \dfrac{18}{s^4}$.

12. Prove that $\mathscr{L}\left\{\dfrac{d^3 y}{dt^3}\right\} = s^3 \mathscr{L}\{y\} - s^2 y(0) - s y'(0) - y''(0)$, where

$y(0)$, $y'(0)$ and $y''(0)$ are the values of y, $\dfrac{dy}{dt}$ and $\dfrac{d^2 y}{dt^2}$ respectively

when $t = 0$.

13. Use the Laplace transform of the second derivative to derive the following transforms: (a) $\mathscr{L}\{\cos at\} = \dfrac{s}{s^2 + a^2}$,

(b) $\mathscr{L}\{\cosh at\} = \dfrac{s}{s^2 - a^2}$,   (c) $\mathscr{L}\{\sinh at\} = \dfrac{a}{s^2 - a^2}$.

Initial and final value theorems

14. Verify the initial value theorem for the functions:
    (a) $3 - 2 \sin t$ (b) $(t + 3)^2$

15. Verify the initial value theorem for the functions:
    (a) $t + \sin 2t$ (b) $1 + 3 \cos t$

16. Verify the final value theorem for the function $t^2 e^{-2t}$.

17. Verify the final value theorem for the function $2 + e^{-2t} \sin 2t$.

Chapter 34

Inverse Laplace transforms and the use of Laplace transforms to solve differential equations

34.1 Definition of the inverse Laplace transform

If the Laplace transform of a function $f(t)$ is $F(s)$, i.e. $\mathcal{L}\{f(t)\} = F(s)$, then $f(t)$ is called **the inverse Laplace transform** of $F(s)$ and is written as $f(t) = \mathcal{L}^{-1}\{F(s)\}$, where $\mathcal{L}^{-1}\{\ \}$ is called the inverse Laplace transformation operator. The inverse Laplace transform is used in the solution of differential equations (see Section 34.4).

For example, since $\mathcal{L}\{1\} = \dfrac{1}{s}$

then $\qquad \mathcal{L}^{-1}\left\{\dfrac{1}{s}\right\} = 1$

Similarly, since $\quad \mathcal{L}\{e^{at}\} = \dfrac{1}{s-a}$

then $\qquad \mathcal{L}^{-1}\left\{\dfrac{1}{s-a}\right\} = e^{at}$

34.2 Inverse Laplace transforms of simple functions

Inverse Laplace transforms may be most conveniently found from tables of standard transforms such as Table 32.1 and Table 33.1. For ease of reference Table 34.1 lists some inverse Laplace transforms.

Worked problems on finding inverse Laplace transforms

Problem 1. Find the following inverse Laplace transforms:

(a) $\mathcal{L}^{-1}\left\{\dfrac{1}{s^2+4}\right\}$ (b) $\mathcal{L}^{-1}\left\{\dfrac{3}{4s-2}\right\}$ (c) $\mathcal{L}^{-1}\left\{\dfrac{1}{s^3}\right\}$ (d) $\mathcal{L}^{-1}\left\{\dfrac{2}{s^4}\right\}$

Table 34.1 Inverse Laplace transforms

	$F(s) = \mathcal{L}\{f(t)\}$	$\mathcal{L}^{-1}\{F(s)\} = f(t)$
1.	$\dfrac{1}{s}$	1
2.	$\dfrac{k}{s}$	k
3.	$\dfrac{1}{s-a}$	e^{at}
4.	$\dfrac{a}{s^2+a^2}$	$\sin at$
5.	$\dfrac{s}{s^2+a^2}$	$\cos at$
6.	$\dfrac{1}{s^2}$	t
7.	$\dfrac{2!}{s^3}$	t^2
8.	$\dfrac{3!}{s^4}$	t^3
9.	$\dfrac{n!}{s^{n+1}}$	t^n (n = positive integer)
10.	$\dfrac{s}{s^2-a^2}$	$\cosh at$
11.	$\dfrac{a}{s^2-a^2}$	$\sinh at$
12.	$\dfrac{n!}{(s-a)^{n+1}}$	$e^{at} t^n$
13.	$\dfrac{\omega}{(s-a)^2+\omega^2}$	$e^{at} \sin \omega t$
14.	$\dfrac{s-a}{(s-a)^2+\omega^2}$	$e^{at} \cos \omega t$
15.	$\dfrac{s-a}{(s-a)^2-\omega^2}$	$e^{at} \cosh \omega t$
16.	$\dfrac{\omega}{(s-a)^2-\omega^2}$	$e^{at} \sinh \omega t$

(a) Since, from 4 of Table 34.1, $\mathcal{L}^{-1}\left\{\dfrac{a}{s^2+a^2}\right\} = \sin at$,

$$\mathcal{L}^{-1}\left\{\dfrac{1}{s^2+a^2}\right\} = \dfrac{1}{a}\sin at$$

Thus $\mathscr{L}^{-1}\left\{\dfrac{1}{s^2 + 4}\right\} = \mathscr{L}^{-1}\left\{\dfrac{1}{s^2 + 2^2}\right\} = \dfrac{1}{2}\sin 2t$.

(b) $\mathscr{L}^{-1}\left\{\dfrac{3}{4s - 2}\right\} = \mathscr{L}^{-1}\left\{\dfrac{3}{4(s - \frac{1}{2})}\right\} = \dfrac{3}{4}\mathscr{L}^{-1}\left\{\dfrac{1}{s - \frac{1}{2}}\right\} = \dfrac{3}{4}e^{\frac{1}{2}t}$,

from 3 of Table 34.1

(c) Since, from 7 of Table 34.1, $\mathscr{L}^{-1}\left\{\dfrac{2}{s^3}\right\} = t^2$ then

$\mathscr{L}^{-1}\left\{\dfrac{1}{s^3}\right\} = \dfrac{1}{2}t^2$

(d) Since, from 8 of Table 34.1, $\mathscr{L}^{-1}\left\{\dfrac{3!}{s^4}\right\} = t^3$ then

$\mathscr{L}^{-1}\left\{\dfrac{1}{s^4}\right\} = \dfrac{t^3}{3!}$ and $\mathscr{L}^{-1}\left\{\dfrac{2}{s^4}\right\} = \dfrac{2t^3}{3!} = \dfrac{1}{3}t^3$

Problem 2. Find the following inverse Laplace transforms:

(a) $\mathscr{L}^{-1}\left\{\dfrac{2s}{s^2 + 9}\right\}$ (b) $\mathscr{L}^{-1}\left\{\dfrac{3s}{s^2 - 25}\right\}$ (c) $\mathscr{L}^{-1}\left\{\dfrac{4}{s^2 - 5}\right\}$

(d) $\mathscr{L}^{-1}\left\{\dfrac{1}{(s - 2)^4}\right\}$

(a) $\mathscr{L}^{-1}\left\{\dfrac{2s}{s^2 + 9}\right\} = 2\mathscr{L}^{-1}\left\{\dfrac{s}{s^2 + 3^2}\right\} = \mathbf{2\cos 3t}$, from 5 of Table 34.1

(b) $\mathscr{L}^{-1}\left\{\dfrac{3s}{s^2 - 25}\right\} = 3\mathscr{L}^{-1}\left\{\dfrac{s}{s^2 - 5^2}\right\} = \mathbf{3\cosh 5t}$, from 10 of

Table 34.1

(c) Since, from 11 of Table 34.1, $\mathscr{L}^{-1}\left\{\dfrac{a}{s^2 - a^2}\right\} = \sinh at$ then

$\mathscr{L}^{-1}\left\{\dfrac{1}{s^2 - a^2}\right\} = \dfrac{1}{a}\sinh at$

Thus $\mathscr{L}^{-1}\left\{\dfrac{4}{s^2 - 5}\right\} = 4\mathscr{L}^{-1}\left\{\dfrac{1}{s^2 - (\sqrt{5})^2}\right\} = \dfrac{4}{\sqrt{5}}\sinh\sqrt{5}t$

(d) Since, from 12 of Table 34.1, $\mathscr{L}^{-1}\left\{\dfrac{n!}{(s - a)^{n+1}}\right\} = e^{at}t^n$ then

$\mathscr{L}^{-1}\left\{\dfrac{1}{(s - a)^{n+1}}\right\} = \dfrac{1}{n!}e^{at}t^n$

Thus, given $\mathscr{L}^{-1}\left\{\dfrac{1}{(s-2)^4}\right\}$ means that $n=3$ and $a=2$

Hence $\mathscr{L}^{-1}\left\{\dfrac{1}{(s-2)^4}\right\} = \dfrac{1}{3!}e^{2t}t^3 = \dfrac{1}{6}e^{2t}t^3$

Problem 3. Find the following inverse Laplace transforms:

(a) $\mathscr{L}^{-1}\left\{\dfrac{2}{(s^2-6s+13)}\right\}$ (b) $\mathscr{L}^{-1}\left\{\dfrac{s+2}{(s^2+4s+20)}\right\}$

(c) $\mathscr{L}^{-1}\left\{\dfrac{3s-1}{(s^2-2s-24)}\right\}$ (d) $\mathscr{L}^{-1}\left\{\dfrac{2}{(s^2+4s+3)}\right\}$

(a) $\mathscr{L}^{-1}\left\{\dfrac{2}{(s^2-6s+13)}\right\} = \mathscr{L}^{-1}\left\{\dfrac{2}{(s-3)^2+2^2}\right\} = e^{3t}\sin 2t,$

from 13 of Table 34.1

(b) $\mathscr{L}^{-1}\left\{\dfrac{s+2}{(s^2+4s+20)}\right\} = \mathscr{L}^{-1}\left\{\dfrac{s+2}{(s+2)^2+4^2}\right\} = e^{-2t}\cos 4t,$

from 14 of Table 34.1

(c) $\mathscr{L}^{-1}\left\{\dfrac{3s-1}{(s^2-2s-24)}\right\} = \mathscr{L}^{-1}\left\{\dfrac{3s-1}{(s-1)^2-5^2}\right\}$

$= \mathscr{L}^{-1}\left\{\dfrac{3(s-1)+2}{(s-1)^2-5^2}\right\}$

$= \mathscr{L}^{-1}\left\{\dfrac{3(s-1)}{(s-1)^2-5^2}\right\}$

$+ \mathscr{L}^{-1}\left\{\dfrac{2}{(s-1)^2-5^2}\right\}$

since $\mathscr{L}^{-1}\{\ \}$ is a linear operator.

$= 3\,e^t\cosh 5t + \mathscr{L}^{-1}\left\{\dfrac{\dfrac{2}{5}(5)}{(s-1)^2-5^2}\right\},$

from 15 of Table 34.1

$= 3\,e^t\cosh 5t + \dfrac{2}{5}e^t\sinh 5t,$ from 16 of

Table 34.1

(d) $\mathscr{L}^{-1}\left\{\dfrac{2}{(s^2+4s+3)}\right\} = \mathscr{L}^{-1}\left\{\dfrac{2}{(s+2)^2-1^2}\right\} = 2\,e^{-2t}\sinh t,$

from 16 of Table 34.1

Further problems on finding inverse Laplace transforms may be found in Section 34.5 (Problems 1 to 13), page 490.

34.3   Inverse Laplace transforms using partial fractions

From Section 34.1, if the Laplace transform of a function $f(t)$ is $F(s)$, then the inverse Laplace transform of $F(s)$ is given by:

$$f(t) = \mathcal{L}^{-1}\{F(s)\}$$

When $F(s)$ is a rational function of s but is not immediately recognisable as a standard type (from Table 34.1), it may often be expressed, using partial fractions, as the sum of a number of terms which may be inverted at sight.

The process of resolving an algebraic expression into partial fractions is discussed in Chapter 3.

Summarising:

Provided that the numerator $f(s)$ is of less degree than the relevant denominator, the following identities are typical examples of the form of partial fraction used:

$$\frac{f(s)}{(s+a)(s+b)(s+c)} \equiv \frac{A}{(s+a)} + \frac{B}{(s+b)} + \frac{C}{(s+c)}$$

$$\frac{f(s)}{(s-a)^3(s+b)} \equiv \frac{A}{(s-a)} + \frac{B}{(s-a)^2} + \frac{C}{(s-a)^3} + \frac{D}{(s+b)}$$

$$\frac{f(s)}{(as^2+bs+c)(s-d)} \equiv \frac{As+B}{(as^2+bs+c)} + \frac{D}{(s-d)}$$

The technique of finding inverse Laplace transforms using partial fractions is demonstrated in the following worked problems.

Worked problems on finding inverse Laplace transforms using partial fractions

Problem 1. Find $\mathcal{L}^{-1}\left\{\dfrac{3s+1}{s^2+2s-3}\right\}$.

$$\frac{3s+1}{s^2+2s-3} \equiv \frac{3s+1}{(s-1)(s+3)} \equiv \frac{A}{(s-1)} + \frac{B}{(s+3)}$$

Hence $3s + 1 \equiv A(s + 3) + B(s - 1)$,

from which $A = 1$ and $B = 2$ (for method, see Chapter 3)

Hence $\mathscr{L}^{-1}\left\{\dfrac{3s+1}{s^2+2s-3}\right\} \equiv \mathscr{L}^{-1}\left\{\dfrac{1}{s-1}+\dfrac{2}{s+3}\right\}$

$$= \mathscr{L}^{-1}\left\{\dfrac{1}{s-1}\right\}+\mathscr{L}^{-1}\left\{\dfrac{2}{s+3}\right\}$$

$$= e^t + 2\,e^{-3t}, \text{ from 3 of Table 34.1}$$

Problem 2. Find $\mathscr{L}^{-1}\left\{\dfrac{5s^2-2s-4}{s(s+1)(s-2)}\right\}$.

$$\dfrac{5s^2-2s-4}{s(s+1)(s-2)} \equiv \dfrac{A}{s}+\dfrac{B}{(s+1)}+\dfrac{C}{(s-2)}$$

Hence $5s^2-2s-4 \equiv A(s+1)(s-2) + B(s)(s-2) + C(s)(s+1)$

from which $A = 2$, $B = 1$ and $C = 2$

Hence $\mathscr{L}^{-1}\left\{\dfrac{5s^2-2s-4}{s(s+1)(s-2)}\right\} \equiv \mathscr{L}^{-1}\left\{\dfrac{2}{s}+\dfrac{1}{(s+1)}+\dfrac{2}{(s-2)}\right\}$

$$= 2 + e^{-t} + 2\,e^{2t}, \text{ from 2 and 3 of}$$
$$\text{Table 34.1}$$

Problem 3. Find $\mathscr{L}^{-1}\left\{\dfrac{4s^3-14s^2+3s+13}{(s+2)(s-1)^3}\right\}$.

$$\dfrac{4s^3-14s^2+3s+13}{(s+2)(s-1)^3} \equiv \dfrac{A}{(s+2)}+\dfrac{B}{(s-1)}+\dfrac{C}{(s-1)^2}+\dfrac{D}{(s-1)^3}$$

Hence
$$4s^3 - 14s^2 + 3s + 13 \equiv A(s-1)^3 + B(s+2)(s-1)^2$$
$$+ C(s+2)(s-1) + D(s+2)$$
$$\text{from which } A = 3, B = 1, C = -5 \text{ and } D = 2$$

Hence $\mathscr{L}^{-1}\left\{\dfrac{4s^3-14s^2+3s+13}{(s+2)(s-1)^3}\right\} \equiv \mathscr{L}^{-1}\left\{\dfrac{3}{(s+2)}+\dfrac{1}{(s-1)}\right.$

$$\left.-\dfrac{5}{(s-1)^2}+\dfrac{2}{(s-1)^3}\right\}$$

$$= 3\,e^{-2t} + e^t - 5\,e^t t + e^t t^2$$
$$\text{from 3 and 12 of Table 34.1}$$

Problem 4. Find $\mathscr{L}^{-1}\left\{\dfrac{5s^2+7s-1}{(s+4)(s^2+1)}\right\}$.

$$\frac{5s^2 + 7s - 1}{(s + 4)(s^2 + 1)} \equiv \frac{A}{(s + 4)} + \frac{Bs + C}{(s^2 + 1)}$$

Hence $5s^2 + 7s - 1 \equiv A(s^2 + 1) + (Bs + C)(s + 4)$

from which $A = 3$, $B = 2$ and $C = -1$

Hence $\mathscr{L}^{-1}\left\{\dfrac{5s^2 + 7s - 1}{(s + 4)(s^2 + 1)}\right\} \equiv \mathscr{L}^{-1}\left\{\dfrac{3}{s + 4} + \dfrac{2s - 1}{(s^2 + 1)}\right\}$

$$= \mathscr{L}^{-1}\left\{\frac{3}{s + 4}\right\} + \mathscr{L}^{-1}\left\{\frac{2s}{s^2 + 1}\right\}$$

$$- \mathscr{L}^{-1}\left\{\frac{1}{s^2 + 1}\right\}$$

$$= 3\,e^{-4t} + 2\cos t - \sin t, \text{ from 3, 5}$$
and 4 of Table 34.1

Problem 5. Find $\mathscr{L}^{-1}\left\{\dfrac{3s + 10}{s(s^2 + 2s + 10)}\right\}$.

$$\frac{3s + 10}{s(s^2 + 2s + 10)} \equiv \frac{A}{s} + \frac{Bs + C}{(s^2 + 2s + 10)}$$

Hence $3s + 10 \equiv A(s^2 + 2s + 10) + (Bs + C)(s)$

from which $A = 1$, $B = -1$ and $C = 1$

Hence $\mathscr{L}^{-1}\left\{\dfrac{3s + 10}{s(s^2 + 2s + 10)}\right\} \equiv \mathscr{L}^{-1}\left\{\dfrac{1}{s} + \dfrac{1 - s}{(s^2 + 2s + 10)}\right\}$

$$= \mathscr{L}^{-1}\left\{\frac{1}{s}\right\} + \mathscr{L}^{-1}\left\{\frac{1 - s}{(s + 1)^2 + 3^2}\right\}$$

$$= \mathscr{L}^{-1}\left\{\frac{1}{s}\right\} + \mathscr{L}^{-1}\left\{\frac{-(s + 1) + 2}{(s + 1)^2 + 3^2}\right\}$$

$$= \mathscr{L}^{-1}\left\{\frac{1}{s}\right\} - \mathscr{L}^{-1}\left\{\frac{(s + 1)}{(s + 1)^2 + 3^2}\right\}$$

$$+ \mathscr{L}^{-1}\left\{\frac{\dfrac{2}{3}(3)}{(s + 1)^2 + 3^2}\right\}$$

$$= 1 - e^{-t}\cos 3t + \frac{2}{3}e^{-t}\sin 3t,$$
from 1, 14 and 13 of Table 34.1

Further problems on finding inverse Laplace transforms using partial fractions may be found in Section 35.5 (Problems 14 to 23), page 491.

34.4   Use of Laplace transforms to solve second order differential equations with constant coefficients

Consider the second order differential equation

$$\frac{d^2 y}{dx^2} - 4\frac{dy}{dx} + 3y = 0$$

with given boundary conditions that when $x = 0$, $y = 2$ and $\dfrac{dy}{dx} = 1$.

Taking the Laplace transform of both sides gives:

$$\mathscr{L}\left\{\frac{d^2 y}{dx^2}\right\} - 4\mathscr{L}\left\{\frac{dy}{dx}\right\} + 3\mathscr{L}\{y\} = \mathscr{L}\{0\}$$

From Section 33.3, equations (4) and (5) we have:

$$\mathscr{L}\left\{\frac{dy}{dx}\right\} = s\mathscr{L}\{y\} - y(0), \text{ where } y(0) \text{ is the value of } y \text{ at}$$

$x = 0$ and $\mathscr{L}\left\{\dfrac{d^2 y}{dx^2}\right\} = s^2\mathscr{L}\{y\} - sy(0) - y'(0)$, where $y'(0)$ is the value of $\dfrac{dy}{dx}$ at $x = 0$.

Hence, substituting these equations gives:

$$[s^2\mathscr{L}\{y\} - sy(0) - y'(0)] - 4[s\mathscr{L}\{y\} - y(0)] + 3[\mathscr{L}\{y\}] = 0$$

However, from the given boundary conditions, $y(0) = 2$ and $y'(0) = 1$.

Hence $s^2\mathscr{L}\{y\} - 2s - 1 - 4s\mathscr{L}\{y\} + 8 + 3\mathscr{L}\{y\} = 0$.

The given differential equation has now been converted into an algebraic equation in $\mathscr{L}\{y\}$.

Rearranging gives: $(s^2 - 4s + 3)\mathscr{L}\{y\} = 2s - 7$

i.e.

$$\mathscr{L}\{y\} = \frac{2s - 7}{(s^2 - 4s + 3)}$$

Thus

$$y = \mathscr{L}^{-1}\left\{\frac{2s - 7}{s^2 - 4s + 3}\right\}$$

$\dfrac{2s - 7}{s^2 - 4s + 3}$ may be resolved into partial fractions.

Thus $\dfrac{2s-7}{s^2-4s+s} \equiv \dfrac{2s-7}{(s-3)(s-1)} \equiv \dfrac{A}{(s-3)} + \dfrac{B}{(s-1)}$

Hence $2s-7 \equiv A(s-1) + B(s-3)$,

from which $A = -\dfrac{1}{2}$ and $B = \dfrac{5}{2}$.

Hence $\mathscr{L}^{-1}\left\{\dfrac{2s-7}{s^2-4s+3}\right\} \equiv \mathscr{L}^{-1}\left\{\dfrac{-\dfrac{1}{2}}{s-3}\right\} + \mathscr{L}^{-1}\left\{\dfrac{\dfrac{5}{2}}{s-1}\right\}$

Therefore $y = -\dfrac{1}{2}e^{3x} + \dfrac{5}{2}e^{x}$, which is the solution of the differential equation $\dfrac{d^2y}{dx^2} - 4\dfrac{dy}{dx} + 3y = 0$.

This demonstrates how Laplace transforms are used to solve differential equations, thus providing an alternative method of solution to those discussed in Chapters 25 and 26. One of the main advantages of Laplace transforms in the solution of differential equations is that the initial conditions are automatically included, thus making the use of the general solution with arbitrary constants unnecessary.

Procedure to solve differential equations by using Laplace transforms
From the above example, the following procedure for solving differential equations of any order, with constant coefficients and given initial conditions may be used:

1. Take the Laplace transform of both sides of the differential equation by

    (a) applying the formulae for the Laplace transforms of derivatives (i.e. by equations (4) and (5) of Section 33.3), and
    (b) using a list of standard transforms (such as Tables 32.1 and 33.1)

2. Insert the given initial conditions, i.e. $y(0)$, $y'(0)$, etc. (At this stage the given differential equation has been converted into an algebraic equation in $\mathscr{L}\{y\}$ and is often called the **subsidiary equation** of the given differential equation.)
3. Rearrange the equation to make $\mathscr{L}\{y\}$ the subject.
4. Find y by

    (a) using partial fractions to rearrange the expression to give standard forms (as in Tables 32.1 and 33.1) and then
    (b) taking the inverse of each term by using Table 34.1

This procedure is demonstrated in the following worked problems.

Worked problems on solving differential equations using Laplace transforms

Problem 1. Use Laplace transforms to solve

$$4\frac{d^2y}{dt^2} - 12\frac{dy}{dt} + 9y = 0$$

given that when $t = 0$, $y = 2$ and $\dfrac{dy}{dt} = 4$.

This is the same problem as Problem 2 of Chapter 25, page 362, and a comparison of methods can be made.

Following the above procedure:

1. $4\mathscr{L}\left\{\dfrac{d^2y}{dt^2}\right\} - 12\mathscr{L}\left\{\dfrac{dy}{dt}\right\} + 9\mathscr{L}\{y\} = 0$

$4[s^2\mathscr{L}\{y\} - sy(0) - y'(0)] - 12[s\mathscr{L}\{y\} - y(0)] + 9\mathscr{L}\{y\} = 0$

2. $y(0) = 2$ and $y'(0) = 4$

Thus $[4s^2\mathscr{L}\{y\} - 8s - 16] - [12s\mathscr{L}\{y\} - 24] + 9\mathscr{L}\{y\} = 0$

3. $(4s^2 - 12s + 9)\mathscr{L}\{y\} = 8s + 16 - 24$

    i.e. $\qquad\qquad \mathscr{L}\{y\} = \dfrac{8s - 8}{4s^2 - 12s + 9}$

4. $\qquad\qquad y = \mathscr{L}^{-1}\left\{\dfrac{8s - 8}{4s^2 - 12s + 9}\right\}$

$\dfrac{8s - 8}{4s^2 - 12s + 9} \equiv \dfrac{8s - 8}{(2s - 3)^2} \equiv \dfrac{A}{(2s - 3)} + \dfrac{B}{(2s - 3)^2}$

Hence $\qquad\qquad 8s - 8 \equiv A(2s - 3) + B$

from which $\qquad\qquad A = B = 4$

Hence $y = \mathscr{L}^{-1}\left\{\dfrac{8s - 8}{4s^2 - 12s + 9}\right\} = \mathscr{L}^{-1}\left\{\dfrac{4}{(2s - 3)} + \dfrac{4}{(2s - 3)^2}\right\}$

$= \mathscr{L}^{-1}\left\{\dfrac{4}{2\left(s - \dfrac{3}{2}\right)}\right\} + \mathscr{L}^{-1}\left\{\dfrac{4}{\left[2\left(s - \dfrac{3}{2}\right)\right]^2}\right\}$

$= \mathscr{L}^{-1}\left\{\dfrac{2}{\left(s - \dfrac{3}{2}\right)}\right\} + \mathscr{L}^{-1}\left\{\dfrac{1}{\left(s - \dfrac{3}{2}\right)^2}\right\}$

i.e. $y = 2\,e^{\frac{3}{2}t} + e^{\frac{3}{2}t}\,t$, from 3 and 12 of Table 34.1

i.e. $y = (t + 2)\,e^{\frac{3}{2}t}$

Problem 2. Use Laplace transforms to solve $\dfrac{d^2y}{dx^2} + 4\dfrac{dy}{dx} = 6$, given that when $x = 0$, $y = 0$ and $\dfrac{dy}{dx} = 0$.

This is the same problem as Problem 2 of Chapter 26, page 373

1. $\mathscr{L}\left\{\dfrac{d^2y}{dx^2}\right\} + 4\mathscr{L}\left\{\dfrac{dy}{dx}\right\} = \mathscr{L}\{6\}$

$$[s^2\mathscr{L}\{y\} - sy(0) - y'(0)] + 4[s\mathscr{L}\{y\} - y(0)] = \dfrac{6}{s}$$

2. $y(0) = 0$ and $y'(0) = 0$

Hence $s^2\mathscr{L}\{y\} + 4s\mathscr{L}\{y\} = \dfrac{6}{s}$

3. $(s^2 + 4s)\mathscr{L}\{y\} = \dfrac{6}{s}$

i.e. $\mathscr{L}\{y\} = \dfrac{6}{s(s^2 + 4s)} = \dfrac{6}{s^2(s + 4)}$

4. $y = \mathscr{L}^{-1}\left\{\dfrac{6}{s^2(s + 4)}\right\}$

$\dfrac{6}{s^2(s + 4)} \equiv \dfrac{A}{s} + \dfrac{B}{s^2} + \dfrac{C}{s + 4}$

Hence $6 \equiv As(s + 4) + B(s + 4) + Cs^2$

from which $A = -\dfrac{3}{8}$, $B = \dfrac{3}{2}$ and $C = \dfrac{3}{8}$

Hence $y = \mathscr{L}^{-1}\left\{\dfrac{6}{s^2(s + 4)}\right\} = \mathscr{L}^{-1}\left\{\dfrac{-\dfrac{3}{8}}{s} + \dfrac{\dfrac{3}{2}}{s^2} + \dfrac{\dfrac{3}{8}}{(s + 4)}\right\}$

i.e. $y = -\dfrac{3}{8} + \dfrac{3}{2}x + \dfrac{3}{8}e^{-4x}$, from 2, 6 and 3 of Table 34.1

i.e. $y = \tfrac{3}{8}(e^{-4x} - 1) + \dfrac{3}{2}x$

Problem 3. Solve, using Laplace transforms,

$\dfrac{d^2\theta}{dt^2} - 6\dfrac{d\theta}{dt} + 10\theta = 20 - e^{2t}$ given that when $t = 0$, $\theta = 4$ and

$\dfrac{d\theta}{dt} = \dfrac{25}{2}$.

1. $\mathscr{L}\left\{\dfrac{d^2\theta}{dt^2}\right\} - 6\mathscr{L}\left\{\dfrac{d\theta}{dt}\right\} + 10\mathscr{L}\{\theta\} = \mathscr{L}\{20\} - \mathscr{L}\{e^{2t}\}$

$[s^2\mathscr{L}\{\theta\} - s\theta(0) - \theta'(0)] - 6[s\mathscr{L}\{\theta\} - \theta(0)] + 10\mathscr{L}\{\theta\}$

$$= \dfrac{20}{s} - \dfrac{1}{(s-2)}$$

2. $\left[s^2\mathscr{L}\{\theta\} - 4s - \dfrac{25}{2}\right] - [6s\mathscr{L}\{\theta\} - 24] + 10\mathscr{L}\{\theta\} = \dfrac{20}{s} - \dfrac{1}{(s-2)}$

since $\theta(0) = 4$ and $\theta'(0) = \dfrac{25}{2}$

3. $(s^2 - 6s + 10)\mathscr{L}\{\theta\} = \dfrac{20}{s} - \dfrac{1}{(s-2)} + 4s + \dfrac{25}{2} - 24$

Hence $\mathscr{L}\{\theta\} = \left\{\dfrac{\dfrac{20}{s} - \dfrac{1}{(s-2)} + 4s - \dfrac{23}{2}}{(s^2 - 6s + 10)}\right\}$

$$= \dfrac{20(s-2) - s + 4s(s)(s-2) - \dfrac{23}{2}(s)(s-2)}{s(s-2)(s^2 - 6s + 10)}$$

4. $\theta = \mathscr{L}^{-1}\left\{\dfrac{4s^3 - \dfrac{39}{2}s^2 + 42s - 40}{s(s-2)(s^2 - 6s + 10)}\right\}$

$\dfrac{4s^3 - \dfrac{39}{2}s^2 + 42s - 40}{s(s-2)(s^2 - 6s + 10)} \equiv \dfrac{A}{s} + \dfrac{B}{(s-2)} + \dfrac{Cs + D}{(s^2 - 6s + 10)}$

Hence

$4s^3 - \dfrac{39}{2}s^2 + 42s - 40 \equiv A(s-2)(s^2 - 6s + 10)$

$$+ B(s)(s^2 - 6s + 10) + (Cs + D)(s)(s-2)$$

from which $A = 2$, $B = -\dfrac{1}{2}$, $C = \dfrac{5}{2}$ and $D = -\dfrac{3}{2}$

Hence $\theta = \mathcal{L}^{-1}\left\{ \dfrac{2}{s} - \dfrac{\dfrac{1}{2}}{(s-2)} + \dfrac{\dfrac{5}{2}s - \dfrac{3}{2}}{(s^2 - 6s + 10)} \right\}$

$= \mathcal{L}^{-1}\left\{ \dfrac{2}{s} - \dfrac{\dfrac{1}{2}}{(s-2)} + \dfrac{\dfrac{5}{2}(s-3) + 6}{(s-3)^2 + 1^2} \right\}$

$= \mathcal{L}^{-1}\left\{ \dfrac{2}{s} - \dfrac{\dfrac{1}{2}}{(s-2)} + \dfrac{\dfrac{5}{2}(s-3)}{(s-3)^2 + 1^2} + \dfrac{6}{(s-3)^2 + 1^2} \right\}$

i.e. $\theta = 2 - \dfrac{1}{2}e^{2t} + \dfrac{5}{2}e^{3t}\cos t + 6\,e^{3t}\sin t,$

from 2, 3, 14 and 13 of Table 34.1

Problem 4. The current i flowing in the circuit shown in Fig. 34.1 is given by the differential equation $L\dfrac{di}{dt} + Ri = E$, where E is the constant applied voltage, L the inductance and R the resistance. Using Laplace transforms solve the equation for i given that when $t = 0$, $i = 0$.

Fig. 34.1

This is the same problem as Problem 3 of Chapter 22, page 333, and a comparison of methods can be made.

1. $\mathcal{L}\left\{ L\dfrac{di}{dt} \right\} + \mathcal{L}\{Ri\} = \mathcal{L}\{E\}$

   $L[s\mathcal{L}\{i\} - i(0)] + R\mathcal{L}\{i\} = \dfrac{E}{s}$

2. $i(0) = 0$

   Hence $Ls\mathcal{L}\{i\} + R\mathcal{L}\{i\} = \dfrac{E}{s}$

3. $\mathscr{L}\{i\}[Ls + R] = \dfrac{E}{s}$

i.e. $\mathscr{L}\{i\} = \dfrac{E}{s(Ls + R)}$

4. $i = \mathscr{L}^{-1}\left\{\dfrac{E}{s(Ls + R)}\right\}$

$= \mathscr{L}^{-1}\left\{\dfrac{E}{Rs} - \dfrac{LE}{R(Ls + R)}\right\}$, by partial fractions

$= \mathscr{L}^{-1}\left\{\dfrac{E}{R}\left(\dfrac{1}{s}\right) - \dfrac{E}{R}\left(\dfrac{1}{s + \dfrac{R}{L}}\right)\right\}$

i.e. $i = \dfrac{E}{R} - \dfrac{E}{R}e^{-(R/L)t}$ from 1 and 3 of Table 34.1

Hence $i = \dfrac{E}{R}\left(1 - e^{-(R/L)t}\right)$

Further problems on solving differential equations using Laplace transforms may be found in the following Section 34.5 (Problems 24 to 33), page 492.

34.5   Further problems

Inverse Laplace transforms

Find the inverse Laplace transforms of the functions in Problems 1 to 13.

1. (a) $\dfrac{1}{s^2 + 9}$ (b) $\dfrac{2}{s^2 + 16}$ (a) $\left[\dfrac{1}{3}\sin 3t\right]$ (b) $\left[\dfrac{1}{2}\sin 4t\right]$

2. (a) $\dfrac{3}{s}$ (b) $\dfrac{4}{s - 3}$ (a) [3]   (b) $[4\,e^{3t}]$

3. (a) $\dfrac{5}{2s + 1}$ (b) $\dfrac{3s}{s^2 + 4}$ (a) $\left[\dfrac{5}{2}e^{-\frac{1}{2}t}\right]$ (b) $[3\cos 2t]$

4. (a) $\dfrac{2s}{2s^2 + 8}$ (b) $\dfrac{3}{s^2}$ (a) $[\cos 2t]$ (b) $[3t]$

5. (a) $\dfrac{4}{s^3}$ (b) $\dfrac{5}{s^4}$ (a) $[2t^2]$ (b) $\left[\dfrac{5}{6}t^3\right]$

6. (a) $\dfrac{6}{s^5}$ (b) $\dfrac{3s}{s^2 - 16}$ (a) $\left[\dfrac{1}{4}t^4\right]$ (b) $[3\cosh 4t]$

7. (a) $\dfrac{4s}{3s^2 - 27}$ (b) $\dfrac{2}{s^2 - 4}$ (a) $\left[\dfrac{4}{3}\cosh 3t\right]$ (b) $[\sinh 2t]$

8. (a) $\dfrac{3}{2s^2 - 6}$ (b) $\dfrac{2}{(s-2)^3}$ (a) $\left[\dfrac{\sqrt{3}}{2}\sinh\sqrt{3t}\right]$ (b) $[e^{2t}\,t^2]$

9. (a) $\dfrac{1}{(s-3)^4}$ (b) $\dfrac{3}{(s+2)^5}$ (a) $\left[\dfrac{1}{6}e^{3t}\,t^3\right]$ (b) $\left[\dfrac{1}{8}e^{-2t}\,t^4\right]$

10. (a) $\dfrac{4}{s^2 - 4s + 12}$ (b) $\dfrac{3}{2s^2 - 8s + 10}$

    (a) $[\sqrt{2}\,e^{2t}\sin 2\sqrt{2t}]$ (b) $\left[\dfrac{3}{2}e^{2t}\sin t\right]$

11. (a) $\dfrac{s+1}{s^2 + 2s + 5}$ (b) $\dfrac{5}{s^2 + 4s - 5}$

    (a) $[e^{-t}\cos 2t]$ (b) $\left[\dfrac{5}{3}e^{-2t}\sinh 3t\right]$

12. (a) $\dfrac{s-4}{s^2 - 8s + 20}$ (b) $\dfrac{4}{s^2 - 2s - 3}$

    (a) $[e^{4t}\cos 2t]$ (b) $[2\,e^{t}\sinh 2t]$

13. (a) $\dfrac{2s+3}{s^2 + 6s + 8}$ (b) $\dfrac{2s-5}{s^2 - 6s + 13}$

    (a) $[2\,e^{-3t}\cosh t - 3\,e^{-3t}\sinh t]$

    (b) $\left[2\,e^{3t}\cos 2t + \dfrac{1}{2}e^{3t}\sin 2t\right]$

Inverse Laplace transforms using partial fractions
Find the inverse Laplace transforms of the functions in Problems 14 to 23.

14. $\dfrac{5s - 1}{s^2 - s - 2}$ $[2\,e^{-t} + 3\,e^{2t}]$

15. $\dfrac{14 - 3s}{s^2 - s - 6}$ $[e^{3t} - 4\,e^{-2t}]$

16. $\dfrac{2s^2 + 11s - 9}{s(s-1)(s+3)}$ $[3 + e^{t} - 2\,e^{-3t}]$

17. $\dfrac{3s^2 + 11s + 16}{(s-2)(s+3)^2}$ $[2\,e^{2t} + e^{-3t} - 2t\,e^{-3t}]$

18. $\dfrac{3s^3 - 10s^2 + 6s + 3}{(s+1)(s-1)^3}$ $\left[2\,e^{-t} + e^{t} - 3t\,e^{t} + \dfrac{1}{2}t^2\,e^{t}\right]$

19. $\dfrac{5s^2 + 9s + 8}{(s+3)(s^2 + 4)}$ $[2\,e^{-3t} + 3\cos 2t]$

20. $\dfrac{3s^2 - 5s + 3}{(s - 2)(s^2 + 1)}$ $[e^{2t} + 2\cos t - \sin t]$

21. $\dfrac{13 - s^2}{s(s^2 + 4s + 13)}$ $[1 - 2\,e^{-2t}\cos 3t]$

22. $\dfrac{7s^2 - 16s + 17}{(s - 3)(s^2 - 2s + 5)}$ $[4\,e^{3t} + 3\,e^t\cos 2t + 2\,e^t\sin 2t]$

23. $\dfrac{5s^3 + 5s^2 + 33s + 20}{s(s + 1)(s^2 - 2s + 10)}$ $[2 + e^{-t} + 2\,e^t\cos 3t + 3\,e^t\sin 3t]$

Use of Laplace transforms to solve differential equations

In Problems 24 to 30 use Laplace transforms to solve the given differential equations.

24. $\dfrac{d^2y}{dx^2} + 25y = 0$, given $y(0) = -2$

and $y'(0) = 20$ $[y = 4\sin 5x - 2\cos 5x]$

25. $\dfrac{d^2y}{dt^2} + 4\dfrac{dy}{dt} + 5y = 0$, given $y(0) = 1$

and $y'(0) = -4$

$[y = e^{-2t}(\cos t - 2\sin t)]$

26. $\dfrac{d^2y}{dx^2} + 2\dfrac{dy}{dx} + 5y = 0$, given $y(0) = 1$

and $y'(0) = 5$

$[y = e^{-x}(\cos 2x + 3\sin 2x)]$

27. $\dfrac{d^2y}{dt^2} + 4y = 12t$, given $y(0) = 0$

and $y'(0) = 7$ $[y = 3t + 2\sin 2t]$

28. $\dfrac{d^2\theta}{dt^2} + 6\dfrac{d\theta}{dt} + 9\theta = 3\,e^{2t}$, given $\theta(0) = 0$

and $\theta'(0) = 0$

$\left[\theta = \dfrac{3}{25}\{e^{2t} - e^{-3t}(1 + 5t)\}\right]$

29. $(D^2 + D - 2)y = 3\cos 3x - 11\sin 3x$, given $y(0) = 0$
and $y'(0) = 6$
$[y = e^x - e^{-2x} + \sin 3x]$

30. $\dfrac{d^2y}{dx^2} - 2\dfrac{dy}{dx} + y = e^x$, given $y(0) = -2$

and $y'(0) = -3$.

$\left[y = e^x\left(\dfrac{1}{2}x^2 - x - 2\right)\right]$

31. The equation of motion of an atom is given by:

$$m \frac{d^2 x}{dt^2} + kx = F_0 \sin wt.$$

Given that $t = 0$ when $x = 0$ and $\dfrac{dx}{dt} = 0$, show, using

Laplace transforms that

$$x = \frac{wF_0}{k - w^2 m} \left\{ \sin wt - w \sqrt{\left(\frac{m}{k}\right)} \sin\left(\sqrt{\frac{k}{m}}\, t\right) \right\}$$

32. Solve, using Laplace transforms, Further Problems 8 to 21 of Chapter 25, page 367.

33. Solve, using Laplace transforms, Further Problems 13 to 26 of Chapter 26, page 387.

Index

Printed and bound by CPI Group (UK) Ltd, Croydon, CR0 4YY

01/11/2024

01782615-0017